THE HAMSTER
Reproduction and Behavior

THE HAMSTER
Reproduction and Behavior

Edited by
Harold I. Siegel

Institute of Animal Behavior and Department of Psychology
University College
Rutgers University
Newark, New Jersey

PLENUM PRESS • NEW YORK AND LONDON

Library of Congress Cataloging in Publication Data

Main entry under title:

The Hamster: reproduction and behavior.

Includes bibliographies and index.
1. Hamsters—Reproduction. 2. Hamsters—Behavior. 3. Hamsters—Physiology. 4. Mammals—Reproduction. 5. Mammals—Behavior. 6. Mammals—Physiology. I. Siegel, Harold I. [DNLM: 1. Behavior, Animal. 2. Hamsters. 3. Reproduction. QL 737.R638 H232]
QL737.R638H34   1985   599.32′33   84-24860
ISBN 0-306-41791-X

©1985 Plenum Press, New York
A Division of Plenum Publishing Corporation
233 Spring Street, New York, N.Y. 10013

Printed in the United States of America

Contributors

ANDRZEJ BARTKE   Department of Obstetrics and Gynecology, The University of Texas Health Science Center at San Antonio, San Antonio, Texas 78284. *Present Address:* Department of Physiology, Southern Illinois University, School of Medicine, Carbondale, Illinois 62901

CLAIRE A. BERIAN   Department of Psychology, Cornell University, Ithaca, New York 14853

KATARINA T. BORER   Department of Kinesiology, University of Michigan, Ann Arbor, Michigan 48109

C. SUE CARTER   Departments of Psychology and Ecology, Ethology and Evolution, University of Illinois, Champaign, Illinois 61820

LYNWOOD G. CLEMENS   Department of Zoology, Michigan State University, East Lansing, Michigan 48823

BARBARA L. FINLAY   Department of Psychology, Cornell University, Ithaca, New York 14853

GILBERT S. GREENWALD   Department of Physiology, University of Kansas Medical Center, Kansas City, Kansas 66103

ROBERT E. JOHNSTON   Department of Psychology, Cornell University, Ithaca, New York, 14853

CHRISTIANA M. LEONARD   Department of Neuroscience, University of Florida College of Medicine, Gainesville, Florida 32610

ROBERT D. LISK   Department of Biology, Princeton University, Princeton, New Jersey 08544

CHARLES W. MALSBURY   Department of Psychology, Memorial University of Newfoundland, St. John's, Newfoundland, Canada A1B 3X9

MARIO O. MICELI   Department of Psychology, Memorial University of Newfoundland, St. John's, Newfoundland, Canada A1B 3X9

LAWRENCE P. MORIN   Department of Psychiatry, State University of New York, Stony Brook, New York 11794

MICHAEL R. MURPHY   System Research Laboratories, Box 35313, Brooks Air Force Base, San Antonio, Texas 78235

RUSSEL J. REITER   Department of Cellular and Structural Biology, The University of Texas Health Science Center at San Antonio, San Antonio, Texas 78284

THOMAS A. SCHOENFELD   Worcester Foundation for Experimental Biology, Shrewsbury, Massachusetts 01545

CHARLES W. SCOUTEN   Department of Psychology, Memorial University of Newfoundland, St. John's, Newfoundland, Canada A1B 3X9

HAROLD I. SIEGEL   Institute of Animal Behavior and Department of Psychology, University College, Rutgers University, Newark, New Jersey 07102

JEFFREY A. WITCHER   Department of Zoology, Michigan State University, East Lansing, Michigan 48823

Preface

My major objective in putting together this volume on the reproductive endocrinology and psychobiology of the hamster was to fill a growing gap in the scientific literature. The only previous book devoted entirely to the hamster (*The Golden Hamster: Its Biology and Use in Medical Research,* edited by Hoffman, Robinson, and Magalhaes, and published by the Iowa State University Press) appeared in 1968 and dealt with topics such as anatomy, genetics, reproduction, hematology, virology, parasitology, and neoplasms. It is clear that a great deal of research on the reproduction and behavior of the hamster has been performed during the past two decades. In fact, a recent computer search of *Psychological Abstracts* since 1968 showed a near linear increase in the number of hamster citations per year.

The contributors to this volume are familiar with the reproductive endocrinology and reproductive and nonreproductive behavior of the hamster and share a high degree of respect for and a certain loyalty to the species. Particular scientific and methodological reasons for choosing the hamster for research can be found in many of the chapters. The topics chosen for discussion reflect areas of experimentation in which the hamster has been extensively studied. This book is potentially useful to those already investigating the hamster, to those studying similar topics in other species, and to those considering alternative research areas or species.

The book is divided into four sections. The first section consists of a historical account by Michael Murphy of the origins of the hamster as a laboratory species from its initial capture to its subsequent domestication. The second section covers reproductive endocrinology and includes chapters on the estrous cycle by Robert Lisk, pregnancy by Gilbert Greenwald, the endocrinology of the male by Andrzej Bartke, and the pineal by Russel Reiter, who is the only author to have contributed to both the 1968 volume and the present volume.

The third division discusses social behavior and begins with a chapter on communication by Robert Johnston. This is followed by a chapter on sexual differentiation by Lynwood Clemens and Jeffrey Witcher and then by separate chapters on female and male sexual behavior by Sue Carter and myself, respectively. The remaining topics include parental and aggressive behavior written by me and the neural basis of reproductive behavior by Charles Malsbury, Mario Miceli, and Charles Scouten.

The final section consists of a chapter on development by Thomas Schoenfeld and Christiana Leonard and three chapters devoted to functions of individuals. These include biological rhythms by Lawrence Morin, energy balance by Katarina Borer, and the visual and somatosensory systems by Barbara Finlay and Claire Berian.

I am extremely delighted and proud to have played a role in this work and I wish to thank each contributor, Plenum Press (especially Kirk Jensen), Jay Rosenblatt, and the Rutgers University Research Council. In addition, I extend my thanks and my love to my family—Marilyn, Matt, Paul, Scott, and my mother and father, who taught me to appreciate animals.

Harold I. Siegel

Newark, New Jersey

Contents

I. Origin of the Hamster

Chapter 1

History of the Capture and Domestication of the Syrian Golden Hamster (Mesocricetus auratus Waterhouse)

MICHAEL R. MURPHY

II. Reproductive Endocrinology

Chapter 2

The Estrous Cycle

ROBERT D. LISK

Chapter 3

Endocrinology of the Pregnant Hamster

GILBERT S. GREENWALD

Chapter 4

Male Hamster Reproductive Endocrinology

ANDRZEJ BARTKE

Chapter 5

Pineal—Reproductive Interactions

RUSSEL J. REITER

III. Social Behaviors

Chapter 6

Communication

ROBERT E. JOHNSTON

Chapter 7

Sexual Differentiation and Development

LYNWOOD G. CLEMENS and JEFFREY A. WITCHER

Chapter 8

Female Sexual Behavior

C. SUE CARTER

Chapter 9

Male Sexual Behavior

HAROLD I. SIEGEL

Chapter 10

Parental Behavior

HAROLD I. SIEGEL

Chapter 11

Neural Basis of Reproductive Behavior

CHARLES W. MALSBURY, MARIO O. MICELI, and CHARLES W. SCOUTEN

Chapter 12

Aggressive Behavior

HAROLD I. SIEGEL

IV. Development and Individual Function

Chapter 13

Behavioral Development in the Syrian Golden Hamster

THOMAS A. SCHOENFELD and CHRISTIANA M. LEONARD

Chapter 14

Biological Rhythms

LAWRENCE P. MORIN

Chapter 15

Regulation of Energy Balance in the Golden Hamster

KATARINA T. BORER

Chapter 16

Visual and Somatosensory Processes

BARBARA L. FINLAY AND CLAIRE A. BERIAN

THE HAMSTER
Reproduction and Behavior

I

Origin of the Hamster

1

History of the Capture and Domestication of the Syrian Golden Hamster (Mesocricetus auratus Waterhouse)

MICHAEL R. MURPHY

I. INTRODUCTION

In most research involving Syrian golden hamsters, the species is used as an animal model or "subject" and as such is totally unidimensional. There is no consideration of its evolution, of its domestication history, or of its ecology and ethology; nor is there any obvious need for these factors to be considered. We know that the hamster comes from the breeder, or comes from a cage, and that is enough. This situation was indeed the way in which I first encountered the hamster, but I soon found the relationship to be incomplete. My search for the real hamster began after my first talk at a scientific convention when I was "informed" by a member of the audience that the hamster was a cross between a rat and a guinea pig, and although there were many doubts expressed, neither I nor any other of those present were confident enough to challenge this absurd hybrid notion.

The search for the real hamster sent me off on a quest for information that eventually took me to England, Rumania, Israel, and Syria, as well as to numerous libraries in the United States. The purpose of the current chapter is to summarize some of the information I have obtained on the capture and domestication history of the Syrian hamster, and perhaps to provide you with a new perspective and that will give you something to muse on during seemingly endless hours of weighing hamsters adrenals, slicing hamster brains, or counting hamster ejaculations.

MICHAEL R. MURPHY • System Research Laboratories, Box 35313, Brooks Air Force Base, San Antonio, Texas 78235.

2. THE EARLY YEARS: DISCOVERY AND DESCRIPTION

The earliest known description of the Syrian golden hamster was published in 1797 in the second edition of *The Natural History of Aleppo* by Alexander Russell, with additions by his younger brother, Patrick Russell. The elder Russell was a practicing physician in Aleppo, Syria, for ten years (1740–1750), where he became an expert on the plague (Fothergill, 1770). During his residence in Aleppo, Russell kept records on the climate, flora, fauna, and culture of the region and published the first edition of *The Natural History of Aleppo* in 1756. Patrick Russell lived in Aleppo from 1750 to 1781 and published the second edition well after his brother's death. Since the hamster was not mentioned in the first edition, it might be suspected that it was Patrick who captured hamsters. This idea is only conjecture, since previously unpublished notes of Alexander's might have been used instead.

The Natural History of Aleppo is a massive work and was highly considered in its own time (Fothergill, 1770); it is still greatly prized by the residents of Aleppo and scholars of Middle Eastern natural history. The passage on the hamster, on page 181 of the English edition (Russell and Russell, 1797) and page 60 of the German translation (Russell and Russell, 1797), is as follows:

> The short tailed field mouse is the animal most pernicious to the fields; the dormouse, the greater dormouse, and the hamster are chiefly hurtful to gardens. The water rat is common about the garden houses near the river.
>
> The Hamster is less common than the field mouse. I once found upon dissecting one of them, the pouch on each side stuffed with young french beans, arranged lengthways so exactly, and close to each other, that it appeared strange by what mechanism it had been effected; for the membrane which forms the pouch, though muscular, is thin, and the most expert fingers could not have packed the beans in more regular order. When they were laid loosely on the table, they formed a heap 3 times the bulk of the animal's body.

Footnote 68 references the hamster as follows:

> 68 *Mus Cricetus* Linn. S.N. p82: Hamster, Buffon (H.N. XIII. p117).

By referencing Linnaeus and Buffon, Russell mistakenly accepted the Syrian hamster as the same species as the much larger European hamster (the "common hamster," which is not native to the Middle East) and did not claim to have discovered a new species. Therefore, the Syrian hamster was named neither by nor after Russell. Indeed, except for the comments of a resident of Aleppo I met, whose only knowledge of Syrian hamsters came from having caught them as a boy and having read Russell's description, I have never encountered another mention of this earliest reference to Syrian hamsters.

Rather than Russell, the Syrian hamster was named by and after George Robert Waterhouse, who at the age of 29 presented it as a new species at a meeting of the London Zoological Society. Waterhouse (Fig. 1) led a long and distinguished career, and, perhaps surprisingly, is not chiefly known because of his description of the Syrian hamster. The son of a solicitor's clerk, who was also an avid lepidopterist, Waterhouse had an unusual early life. Gunther (1980) relates that "as a boy he was given to straying away from home. On the first occasion, at the age of two, his parents had to advertise for his return. As a cure a friend suggested that the boy be sent to school in Brussels,

Figure 1. George Robert Waterhouse: Curator of the Zoological Society of London. Formally described and named the Syrian golden hamster as a new species from Aleppo. (This photo is a copy of a lithograph by Maguire from the General Library of the British Museum, received through the kindness of Miss D. Norman, Assistant Archivist, British Museum.)

which he described as like going to prison. This taught him fluent French which later put him on intimate terms with French naturalists."

When he was 21, Waterhouse was invited by Charles Darwin to join him on the expedition of H.M.S. Beagle (1831–1835), but Waterhouse declined and instead became apprenticed to an architect. He was not left completely out of the expedition, since Darwin later asked Waterhouse to catalogue and describe the mammals and coleopterous insects (beetles) collected on the voyage.

While Darwin was still out sailing, the turning point in Waterhouse's career came when he gave up architecture and entered academic life in 1834 as Curator of the Museum of the Liverpool Royal Institute and a year later as Curator of the London Zoological Society. In this later position, he catalogued the Society's mammals (Anonymous, 1888).

In 1843, Waterhouse joined the British Museum as Keeper of the Department of Mineralogy and Geology (which then also included paleontology), a position that he held until 1880. Interestingly, none of his more than 120 published scientific papers concerned any of the three disciplines represented in his department, but instead were on natural history, chiefly entomology (Waterhouse, 1889). He was a founding member of the London Entomological Society.

On April 9, 1839, while Waterhouse was still Curator of the London Zoological Society, he presented a new species of hamster at the Society's meeting based on a single, elderly female specimen received from Aleppo, Syria (Fig. 2). His description was later published in the Society's proceedings, as follows (Waterhouse, 1840):

> This species is less than the common Hamster (*Cricetus vulgaris*), and is remarkable for its deep golden yellow colouring. The fur is moderately long and very soft, and has a silk-like gloss; the deep golden yellow colouring extends over the upper parts and sides of the head and body, and also over the outer side of the limbs: on the back the hairs are brownish at the tip, hence in this part the fur assumes a deeper hue than on the sides of the body: the sides of the muzzle, throat, and upper parts of the body are white, but faintly tinted with yellow: on the back, and sides of the body, all the hairs are of a deep grey or lead colour at the base. The feet and tail are white. The ears are of moderate size, furnished externally with deep golden-coloured hairs, and internally with whitish hairs. The moustaches consist of black and white hairs intermixed.

A description of the skull of the animal was also provided.

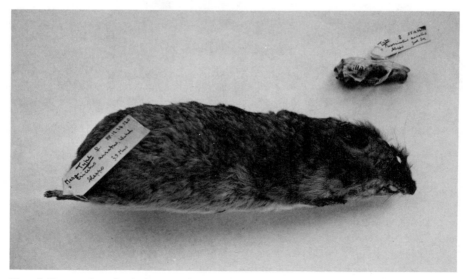

Figure 2. Type specimen for the Syrian golden hamster. *Mesocricetus auratus* Waterhouse, described by G. R. Waterhouse in 1839. Collector unknown. [Located in the British Museum (Natural History), Cromwell Road, London, item number BM(NH) 1855.12.24.120. Photo taken by the author.]

The collector and donor of the specimen was either unknown or unacknowledged, but following Waterhouse's description, this hamster became the type specimen for a new species named *Cricetus auratus,* Waterhouse (the genus name *Mesocricetus* was a later modification). You may pay your respects to this grande dame of hamsters, as I did, at the British Museum on Cromwell Road, London, this institution having purchased the collection of the Zoological Society in 1853 (Gunther, 1980).

3. ON THE ROAD TO JERUSALEM: CAPTURE AND BREEDING

After Waterhouse, interest in the Syrian hamster does not appear again for 91 years, when it is picked up in the laboratory of parasitologist Saul Adler (Fig. 3) at the Hebrew University of Jerusalem. Adler had been conducting research on the prevalent disease, leishmaniasis (Kala Azar, Black Fever, Jericho Rose) (Scott, 1945), for which the Chinese hamster (*Cricetulus griseus*) had been shown to be an excellent animal model. However, Adler was unable to breed the Chinese hamsters in captivity and found it unacceptable to rely on shipments from China. Therefore, he wanted to obtain a species of hamster that was endemic to the Middle East. According to Adler's son, Jonathan Adler, his father knew of the existence of the Syrian hamster from reading the zoological literature derived from Waterhouse's description of 1839 (Fraser, 1849; Giebel, 1855; Tristram, 1884). However, it is not clear from Adler's own writing (Adler, 1948) that

Figure 3. Saul Adler, Professor of Parasitology. Hebrew University of Jerusalem, inspired the first live capture of Syrian golden hamsters and was the first to use them for research. He started hamster colonies in Europe, Asia, and America. [Photo by W. Braun and received through the kindness of J. Adler. Published in Anonymous (1981).]

this was indeed the case. Yerganian (1972) has reviewed the available history and proposes that Adler was seeking a different hamster [*Cricetulus phaeus* (same as current *C. migratorius*)], an animal closely related to the Chinese hamster. Whether Adler actually knew of the existence of the Waterhouse hamster before they were brought to him may never be known. (Colleagues of Adler warned me about depending too much on Adler's own writing because he was known as an extremely modest man who often shared the credit to such an extent that he reserved none for himself.) In either case, Adler was frustrated with using the Chinese hamster, and asked a colleague from the Zoology Department to try to capture endemic hamsters for research.

The zoologist was Israel Aharoni (Figs. 4,5), perhaps the most colorful individual in hamster history. He was born in the town of Widzi on the Russian/Polish border and was educated in Prague. Aharoni is known as the first Hebrew zoologist because he rediscovered or assigned Hebrew names for the animals of the Holy Land. This endeavor had both zoological and religious significance. Hebrew had been an unspoken language for over 2,000 years and many animal names had been lost or misinterpreted. Aharoni was a masterful linguist, speaking and reading Arabic, Latin, Greek, Aramaic, and many European languages, as well as Hebrew (Aharoni, 1975). He identified the Hebrew names of many species by comparing the name in other languages spoken in the region and considering zoological groupings that were connected with various dietary laws in the Bible and Talmud. The results of Aharoni's analyses are currently being used

Figure 4. Israel Aharoni (formal portrait): Head of the Department of Zoology, Hebrew University. Directed the capture of the first wild-caught Syrian golden hamsters. With his wife, raised these infant hamsters to maturity and gave them into the care of H. Ben-Menachen and S. Adler. [Copied by the author from the original Hebrew edition of *Memoirs of a Hebrew Zoologist* (Aharoni, 1942).]

Figure 5. Israel Aharoni: Snapshot taken by a student during a classroom lecture. Aharoni exhibits specimens he has collected, while on the blackboard are written their names in several languages. (Photo copied by the author from the original hanging in the Zoological collection at the Russian Compound, Hebrew University.)

to re-establish in nature the fauna of Biblical times that have since vanished from Israel, but are still available from other parts of the world.

At the time of Aharoni's early life in Jerusalem, the whole region was still under Turkish rule. Aharoni, a Jew in a Moslem-dominated world, was able to travel freely under the protection of the Turkish sultan, which he received because he obtained specimens for the sultan's butterfly collection. On extensive zoological expeditions, often lasting many months, Aharoni collected every kind of animal available (Aharoni, 1930). Assisted by local guides, the initial preparation of the specimens was done in the field, and then they were sent to Berlin for final taxidermy. Many of his specimens are still available for study at the Russian Compound of the Hebrew University.

A colleague of Aharoni's, George Hass, who remembered Aharoni well, told me that he and others always wondered how Aharoni could go out in the field, he was "such a terrible coward." It was only out of "love of nature and excessive curiosity that Aharoni was able to force himself to make the trip." On one expedition, Aharoni was gone so long without word that he was assumed dead. He returned just in time to read his own obituary in the Jerusalem papers.

It was on such an expedition in 1930 that Aharoni decided to heed Saul Adler's request and look for hamsters. An account of this effort was published 12 years after the fact in Aharoni's memoirs (1942, 1972). I obtained an original edition of this work from the Harvard University Library and my friend and collegue, M. Devor, translated the hamster chapter from the original Hebrew into English. I recommend the whole chapter to you highly, but will attempt to present the gist and flavor of it with selected quotes.

First, it should be noted that although there may be some question of whether Adler knew of the existence of the Syrian golden hamster before he received them, there is no doubt that Aharoni knew about and was looking for this particular animal. On arriving in the region of Aleppo, he instructed his Syrian guide, Georgius Khalil Tah'an (Fig. 6), to go to a certain farm and entreat the local sheik (sheik El-Beled) to provide information of the location of golden hamsters. On April 12, 1930 (Adler and Theodor, 1931) the sheik called a meeting. Aharoni writes:

> At the meeting, it was decided to hunt out this creature in one of the best fields, a field that the hamster had chosen to colonize. The sheik hired a few laborers and they dug in many places, destroying a good part of the wheat field. After several hours of hard work, they succeeded in raising from a depth of 8 feet, a complete nest, nicely uphol- stered, with a mother and her 11 young! Thinking that the mother would care for her

Figure 6. Georgius Khalil Tah'an: Syr- ian guide and hunter. At the request of I. Aharoni, located and supervised the excavation of the first wild caught live Syrian hamsters. [Copied by the author from the original Hebrew edition of Memoirs of a Hebrew Zoologist (Aharoni, 1942).]

infants and feed them, Georgeus put the whole family into a colony box. But his hopes were not fulfilled. . . . I saw the mother hamster (a creature whose evolutionary level is not high) harden her heart and sever with ugly cruelty the head of the pup that approached her most closely (each of the young measured at the time just under 2½ cm). Natural mother-love led her to kill her dear child: "It is better that my infant die than that it be the object of an experiment performed on it by a member of the accursed human race."

When Georgeus saw this act of savagery, he quickly removed the murdering mother hamster (for she would surely kill them all!) and put her in a bottle of cyanide to kill her.

So, because the first live captured wild hamster was an antivivisectionist, she was killed, and Aharoni had the ten remaining pups to raise by hand. We do not know the exact age of the pups, but their eyes were still closed when they were captured. However, Aharoni and his wife did succeed as foster parents, and the litter survived. After a hair-raising escape and recapture of all but one, the remaining nine animals were given over to Heim Ben-Menachen (Fig. 7), the founder and head of the Hebrew University animal facilities on Mt. Scopas.

Ben-Menachen placed the hamsters in a cage with a wooden floor. Aharoni (1942, 1972) described what happened the next day when Ben-Menachem entered the animal colony:

When he found out about the great catastrophe (the escape of five of the hamsters who chewed their way through the bottom of the cage and got out), he was aghast. Anyone who did not see the shock on that man's face has never seen a man smitten, shaken

Figure 7. Haim Ben-Menachem: Director of the original animal facilities at the Hebrew University, Mt. Scopas, Jerusalem. Breeder of the first "domesticated" Syrian hamsters. (Photo from Mrs. H. Ben-Menachem, through the kindness of J. Adler.)

to the depths . . . I pitied him. His dismay increased as I described how difficult it was to
get the creatures out of the depths of the earth, the great value of the discovery of this
beautiful animal, that in the whole wide world, the only suitable habitat it could find was
a lone region between Aleppo and Homs; of all the bundles of dried grass, all the hay, all
the sheves of wheat . . .

It is interesting that Ben-Menachen would react so to the loss of a few rodents,
since he was no mere mild-mannered animal caretaker. According to his son (Y. Ben-
Menachem, personal communication), his father was also an officer in the Haganah, an
underground secret resistance movement, and during World War II was sent as an
agent to Europe. After the war, he procured arms for the Haganah and dispatched
"illegal" immigrants to Palestine. Soon after the foundation of the State of Israel, he
was appointed Director General of the Ministry of Posts and Telecommunications.
Later, he served for many years as Director of the Academy of the Hebrew Language at
the Hebrew University.

None of the five escapees were recovered alive. According to Aharoni, the remaining four consisted of three males and one female. [This statement is contradicted in
papers by B. Aharoni (1932) and by S. Adler (1948) in which it is indicated that one
male and three females survived the escape attempt, and one female was later killed by
the male.] Aharoni was skeptical that the remaining animals would breed. He wrote:

> . . . I knew very well from reviewing the literature and from hearsay that despite all
> the efforts of professionals in the field, not a single one had succeeded in breeding any
> variety of female hamsters in captivity; they do not become pregnant and they do not give
> birth. But then again, every rule has an exception.

The exception was due to special breeding techniques developed by H. Ben-
Menachen (1934), which were described as follows to me by his son (letter, January 19,
1982):

> . . . he filled a large wire-mesh cage with tightly packed hay, leaving only a 5 cm,
> brightly illuminated space on top. In this space he placed his female. Seeking darkness,
> the female began to burrow into the hay.
> A day or two later the male was placed in the cage. It proceeded to chase the
> female—who was far more familiar with the environment than her assigned mate—and
> finally (how long it was I don't know) caught up with her. By then both were tired and the
> male was presumably quite aroused. I assume that their position in the burrow was more
> favorable to mating than to slaughter, and they mated.

Aharoni concludes the story:

> Out of love for science and for the broadening of mankind, the Allpowerful nudged a
> single wheel of the uncountable wheels of nature—and a miracle happened!
> Only someone who has tasted true happiness, heavenly joy, can appreciate our elation
> over the fact that our great effort did not prove to be in vain. Our goal was achieved. From
> now on there will be a species of hamster that will be fruitful and multiply even in
> captivity, and will be convenient for endless laboratory experimentation.
> The supervisor of the animal colony devoted himself to raising the infants with love
> and admirable selflessness. Instead of the water she had received as a pup, the mother fed
> the infants milk, known since ancient times to be more beneficial than water. The sons
> sired and the daughters gave birth to "countless" new sons and daughters. And with the

aid of God (not just by luck) the hamster that was brought from Aleppo proved to be
incredibly prolific, and all from one mother!
How marvelous are thy works, O Lord!

The first hamster colony was indeed prolific and numbered 150 specimens within
a year (Aharoni, 1932). The first laboratory-bred hamsters were given to Adler, who
published the report of the first research using Syrian hamsters a short time later (Adler
and Theodore, 1931).

Adler led a distinguished scientific career. He was born in Karelitz, Russia, in
1895 and immigrated with his family to England in 1900. He was educated in Leeds
and received his medical degree in 1919. In 1924 he joined the faculty of the Hebrew
University, where he remained until his death in 1966. His colleagues remember him
as a gentle, well-liked man who loved poetry and had a reputation for absent-minded-
ness. The dominant theme of his work was the *Leishmania* organism and the disease
leshmaniasis, but among his more than 200 published papers are articles on evolution,
Darwin's mysterious disease, and a Hebrew translation of the *Origin of Species,* which
won him the Israeli literary prize, the Tchernichovsky. He was elected a Fellow of the
Royal Society in 1957. His biographer thought it worthwhile to mention that Adler
took particular pleasure in his role in domesticating the Syrian hamster (Shortt, 1967).

Realizing the fragility of a single colony, and with characteristic generosity, Adler
distributed breeding stock of Syrian hamsters to other laboratories. Hamsters that went
to England in 1931 were literally smuggled into the country in Adler's coat pockets and
given to E. Hindle of the London Zoological Society (D. Adler, 1973; J. Adler,
personal communication).

Although he did not mention it in his memoirs, Aharoni apparently captured
more Syrian hamsters than just the mother and her ten pups, since three old female
specimens, captured April 27 and 29, 1930, are attributed to him in the collection of
the Berlin Zoological Museum (Aharoni, 1932).

4. THE INVASION OF AMERICA: DISSEMINATION AND PROLIFERATION

There is general agreement that hamsters were first imported into the United
States in the summer of 1938; however, the exact nature of the entry is somewhat
confused. There is a claim that the hamsters were sent to the U.S. from British stock
(Bond, 1945), but this report is in error, since an earlier paper by Doull and Megrail
(1939) of the Department of Hygiene and Bacteriology, Western Reserve School of
Medicine, Cleveland states "Our original colony of 13 animals was secured from stock
at the Hebrew University on July 26, 1938 through the kindness of Dr. Adler." There
is even some internal contradiction in the Bond (1945) article in that she writes, ". . .
in a letter to Dr. Moore in 1942, Dr. G. H. Faget of the United States Public Health
Service, Carville, Louisiana states that in 1938 a shipment of hamsters was received at
their laboratory from Dr. S. Adler of Palestine and that in the same shipment animals
were sent to Western Reserve University." These animals were used at the USPHS

facility in Carville for research on leprosy and Dr. Black (1939) of that laboratory wrote "We started the colony in July, 1938, with 12 animals. . . ."

Yet a third group of hamsters went to the Rockefeller Foundation as noted by Poiley (1950), who was then head of the NIH small animal colony, "The hamster first made its appearance in the US in 1938 through consignments to the Rockefeller Institute in New York and the Western Reserve University in Cleveland." In 1940 and 1941 several papers reporting the use of hamsters for research on influenza virus were published from the Rockefeller Foundation and from an associated laboratory in Berkeley, California (Horsfall and Hahn, 1940; Pearson and Eaton, 1940; Taylor, 1940; Eaton and Beck, 1941; Lennette, 1941). Hamsters derived from the Rockefeller group were used to form the NIH colony (Poiley, 1950) and Tumblebrook Farms, West Brookfield, Massachusetts. (V. Schwentker, personal communication).

The year of entry of Syrian hamsters into America is without doubt, but the exact person responsible for the shipment is in question. Although, two of the three labs receiving hamsters attributed them directly to Saul Adler. Dr. Adler (1948) later wrote, "Shortly before the Second World War, I sent animals to India for kala azar research, and the late Prof. I. J. Kligler sent a batch to America." Since Adler mentions no other shipments to the USA (except one for Army research during the war) and since World War II started in 1939, it was probably Kligler who sent hamsters from Adler's colony to the United States. There is a further connection since Kligler was at the Rockefeller Foundation before joining the Hebrew University in Jerusalem, and might therefore have known of the possible use for hamsters in research in the United States. Kligler founded the Department of Microbiology at the Hebrew University and headed it until his death.

An early figure, who was more responsible than any other American for hamster promulgation, was Albert Marsh (Fig. 8) of Mobile, Alabama. Marsh was an unemployed highway engineer when he won his first hamster on a bet. Intrigued by the animal's cute appearance and docility, Marsh obtained some breeding stock (possibly from the leprosy labs at Carville, Louisiana) and began promoting and selling hamsters for pet and laboratory use.

For a while, Marsh made a big business out of little hamsters, founding the Gulf Hamstery and Marsh Enterprises. Through advertisements in popular magazines, comic books, and commercial animal journals (Fig. 9), he sold hamsters to would-be breeders and pet owners. He also functioned as a clearinghouse, taking orders from laboratories, buying hamsters from remote breeders, and having these suppliers ship the animals directly from the breeding site to the customer without ever passing through Marsh's operation.

Marsh's success was in large part due to his professional and intellectual approach to hamsters. His self-published book, *The Hamster Manual,* is 87 pages of the best written, most sound advice on hamsters that I have read. Eighty thousand copies had been printed by the sixth edition in 1951. The extent of Marsh's hamster business may be judged from statements from *The Hamster Manual* (1951): "The Gulf Hamstery maintains the world's largest herd of breeding hamsters" (p. 43). Quoted from Purina Feeds Promotional Department, St. Louis, "Mobile seems to be the 'capital' of the

Figure 8. Albert Marsh: Animal breeder, entrepreneur. Started the first large commercial hamstery and promoted the use of hamsters as pets and laboratory animals. [Photo copied from page 32 of *The Hamster Manual* (Marsh, 1951).]

hamster industry in America. Gulf Hamstery is credited with being the principal fountainhead of breeding stock to hamster raisers in this country" (p. 69). "We regularly ship to Canada, South America and Europe" (p. 73).

Californians might be interested to know that until February 10, 1948, hamsters were forbidden entry into their state, except with permit for immediate laboratory use. Through personal effort over two years, and with assistance from the Governor of Alabama, among others, Marsh succeeded in getting the California State Department of Agriculture to accord "laboratory reared Golden Hamsters" the status of a "normally domesticated mammal" and thereby exempt it from the California law (quoted from Marsh, 1951; also see Marsh, 1948). Wild hamsters were still restricted from California, and may still be so.

According to E. Engle (personal communication), founder of Engle Laboratories, it was probably Marsh's success that was his downfall. The large, quality-controlled, hamstery business peaked in 1948 to 1951 and in the early 1950s began to crash. Hamsteries (most started by Marsh) had sprung up everywhere. Both Engle Labs and Lakeview got their first hamsters from Marsh in 1949. Furthermore, small local breeding operations, operating from basements or garages, were supplying much of the pet

HAMSTERS
PROFITABLE FUR AND LABORATORY ANIMALS

Hamsters as Pets
By Dr. G. O. Passmore

The little Golden Hamster which originated in Syria and brought to this country in 1938 for laboratory work, is also gaining fast recognization as a pet for children as well as grown-ups.

They are naturally tame and in a few weeks training time pets can be made of them. The male hamster makes the nicest pet, they climb very carefully and will not try to climb to any object out of reach, when holding in the hands.

Hamsters are very clean and have no offensive odor. When full grown they weight about five ounces and are about five to six inches in length.

A CORRECTION
HAMSTERS IN CALIF.

Just before going to press we received a letter from Albert F. Marsh, Prop. Gulf Hamstery, Mobile, Ala. asking that we correct the statement published in last month's (April) issue to the effect that breeders of Hamsters in California required a permit to raise these animals. Mr. Marsh feels that the statement is harmful to those who receive orders from California since the Dept. of Agriculture of the state of California removed the restrictions on February 10, 1948, to support the claim a letter from the California Agricultrure Department accompanied the request and we quote from it as follows."

"Dear Mr. Marsh,

Yesterday we sent you a copy of a modified regulation governing hamsters. You will note that this exempts laboratory reared Golden Hamsters from the necessity of a permit requirement in California."

The letter was signed by Mr. W. C. Jacobson, Chief, Division of Plant Industry. A P.S. was added to the letter stating the following. Be sure that your shipping tags carry the words "Laboratory Reared Golden Hamsters".

4-H CLUB NEWS
By C. A. Henry, Chairman of the 4-H Committee of the A.R. & C.B.A.

What I would like to point out to the entire 4-H Committee is the opposition that we have been getting from some county agents and a few state agents. They say that these rabbit clubs are kid stuff, well, who are the 4-H members but kids and also our future breeders. Taking project for project in going over the records submitted to the town committee, I've found that dollar for dollar rabbits more than take care of themselves. A boy or girl with a rabbit project, a good many times has a bigger investment than quite a few poultry and some hog and sheep 4-H children and one county agent tells me as much as the original investment on some of the baby beef projects. Another thing is that we are producers of good meat and fur products so let's let them know about it.

Last year, there were quite a few boys and girls from our state, Massachusetts, who were sent to the 4-H Youth Congress, still there were quite a few scholarships that were not used. I was wondering if some of the grain companies with their advertising expense sheets could not set up a trip to the youth congress from each state on a rabbit project, if not, how about a few scholarships.

Now, to the 4-H youth. There were a few of both trips and scholarships

Figure 9. Advertisements placed in the "American Small Stock Farmer" in June 1948 by Albert Marsh, who started the commercial hamster industry in the United States.

Figure 10. Michael R. Murphy (1971): Then a graduate student, Department of Psychology, Massachusetts Institute of Technology. Made the second capture of live Syrian hamsters. Researched hamster history. (Photo by J. Murphy, taken near Aleppo, Syria.)

trade. Overhead and shipping costs were so low for the small operations that they could sell at much lower rates than big companies such as Gulf Hamstery. Gil Slater (personal communication) told me that about the same time as the pet market was dwindling, laboratory use also declined due to the advent of the notorious wet-tail disease.

The pinch was too much for Marsh and the Gulf Hamstery closed in the 1950s. He was reputed to be in difficult straits for a while, at one time operating a motel in Mobile (Steve Slater, personal communication), but later operated a quail breedery in California until his recent death.

5. NEW GENES ON THE LINE: A RECONNAISSANCE EXPEDITION TO ALEPPO

When I started working with Syrian hamsters in 1967, there had been until then only three documented captures of wild hamsters in history—one before 1781, one around 1839 and one in 1930. Also, all domesticated golden hamsters were descendants of the three sibling survivors of Aharoni's ten orphans. So, to satisfy my own curiosity and obtain wild hamsters for comparative research (Murphy, 1977), my wife, Janet, and I made a reconnaissance expedition to Aleppo, Syria, in May/June, 1971. Figure 10 shows the first wild hamster captured since 1930. In all, 13 wild hamsters were captured and 12 of them (four males and eight females) were brought back to the

United States (Murphy, 1971). The full description of this trip may have to await the writing of my own memoirs, but an observation derived from the experience is worth noting now. This comment concerns why the Syrian hamster became such a successful laboratory animal. I believe it did so because it adapts so well to laboratory existence. After only three days of handling, the wild hamsters I captured in Syria were tame and gentle. All of the animals mated within four weeks of being captured, and all eight females produced litters. The average littersize was 11, and every pup was raised to weaning, in spite of the fact that they were often disturbed for observations and weighing. In breeding wild Syrian hamsters and their descendants, I have never had any problems with their fertility or with disease, yet, in contrast, I have had considerable difficulty with Turkish and Rumanian hamsters and eventually lost my colonies of both of these species. Therefore, I am inclined to agree with Aharoni that the Syrian hamster is a truly miraculous laboratory animal, and I believe it has changed very little from the wildtype, especially with respect to its behavior, because the wildtype was originally so well suited for the laboratory.

I also note that from an evolutionary point of view, the rare and obscure Syrian hamster made a grand adaptive radiation when it became a laboratory animal. Certainly, as long as medical and behavioral research continues, it will not become extinct in the laboratory, even if it does cease to exist in the wild.

Descendants of the animals I captured in Syria are being maintained by Andrew Lewis, National Institutes of Health, Bethesda, Maryland. In 1978, Bill Duncan, Southwestern Medical School, Dallas, Texas, made the historically third capture of live wild hamsters, returning two females to the United States.

6. CONCLUSION

Thus concludes my first attempt at a complete history of hamster domestication. Hopefully, it will continue to be revised and enlarged as new information becomes available. As a scientist doing historical research, I occasionally have encountered some "real" historians and it has struck me that there is a fundamental difference between the scientific and the historical approach, especially between behavioral science and history. The historian expects to discover the truth, and deserves to do so, since there is truth to be found in history. "Significant at the 0.05 level" simply does not satisfy an historian, as it must satisfy a behavioral scientist. One of the disappointments in my inquiries to date has been my inability to determine who captured and donated the hamster that Waterhouse described in 1839. One hypothesis that I have held is that the specimen was donated by one of the Russell brothers; this is not completely out of the question since Alexander Russell is known to be an early benefactor of the British Museum (Gunther, 1980), proving that the Russells were connected with the Museum circle. Still, there is no evidence, and neither scientist nor historian is supposed to indulge in pure speculation. Therefore, I will forsake either title and switch to that of "romantic" in order to conclude.

The history of hamster domestication is fraught with critical points in time when circumstances might have been different, very different, and unlike the answer to the question that goes something like "if Edison had died young, would we still be without the light bulb", the domestication of the rare Syrian golden hamster was not a historical inevitability. Granted, even without Waterhouse, someone would have eventually described the Syrian hamster in the zoological literature, but, as Yerganian (1972) has pointed out, if Adler had been clever enough to breed Chinese hamsters in captivity, he would never have looked for a substitute hamster for his research. If the resourceful Aharoni had let his natural cowardice prevail instead of his love of nature, he would not have strayed from Jerusalem to tramp the wheat fields of North Syria. Then if Araroni and his wife had not persevered in hand-raising preweanling hamster pups, a very difficult task, the first live capture would have been dead within a few days.

If Ben-Menachen had been captured in his underground political/military activities, he would never have developed his creative and highly successful breeding techniques for the Syrian hamster, and this species might have suffered the same fate as *Cricetulus phaeus* and been considered unsuitable for research use (Adler, 1948). Adler made certain that valuable breeding stock was distributed to other laboratories in the world. If the hamster had remained only in Jerusalem, it would probably have been lost, as was most of the Mount Scopus animal colony, during the military conflicts associated with the formation of the current state of Israel.

And Albert Marsh! If he had not excited young men such as Gil Slater and Everett Engle, would the Syrian hamster have struck it big in pet stores and laboratories, or remained relatively obscure?

Was the hamster a historical inevitability (like the light bulb)? I don't think so. The domestication of the Syrian hamster is the story of observant people dedicated to the description of nature, of creative medical researchers seeking cures for disease, and of those who simply have fallen in love with the little creature. In retrospect, the hamster story certainly is a tale of serendipity and one wonders how many other stories similar to it there might have been, or are yet to be, about other animal species.

ACKNOWLEDGMENTS. Collecting information on the history of hamsters has been a labor of over 15 years and in that time I have had the assistance of many generous persons. I thank the following in particular: At the British Museum, G. B. Corbet, D. Hills, and D. Norman; In Israel, J. Adler, J. Aharoni, R. Ashkell, Mrs. H. Ben-Menachen, M. Devor, A. Gunders, G. Hass, Prof. Tchernov, and Y. Werner; In the United States, Y. Ben-Menachen, E. Engle, H. Magalhaes, J. Murphy, C. W. Nixon, V. Schwenker, G. Slater, and S. Slater.

REFERENCES

Adler, D., 1973, Letter to the Times, *The Times,* Jan. 27, London.
Adler, S., 1948, Origin of the golden hamster *Cricetus auratus* as a laboratory animal, *Nature* 162:256.
Adler, S., and Theodor, O., 1931, Investigations on Mediterranean Kala Azar, *Proc. R. Soc. (London)* 108:459.
Aharoni, B., 1932, *Die Muriden von Palästina und Syrien,* Reinhold Berger, Leipzig.

Aharoni, I., 1930, Die Säugetiere Palästinas, Z. Säugetierkunde 5:327–343.

Aharoni, I., 1942. Memoirs of a Hebrew Zoologist (in Hebrew), Am Oved Limited, Tel Aviv.

Aharoni, I., 1972, Excerpt from Memoirs of a Hebrew Zoologist, in: Pathology of the Syrian Hamster, Volume 16: Progress in Experimental Tumor Research, (F. Homburger, ed.; M. Devor, trans.), S. Karger, Basel, pp. 35–41.

Aharoni, J., 1975, Israel Aharoni (biographical note), in: The Sound of Hebrew, Rehovot, Israel.

Anonymous, 1888, Obituary of George Robert Waterhouse, Ent. Month. Mag. 24: 233–234.

Anonymous, 1981, A golden tradition: Desert rodents for research, Quest Jerusalem: Research at the Hebrew University 4:2–5.

Ben-Menachem, H., 1934, Notes sur l'élevage du hamster de Syrie, Cricetus auratus, Arch. Inst. Pasteur Algér. 12:403–407.

Black, S. H., 1939, Breeding hamsters, Int. J. Lepr. 7:412–414.

Bond, C. R., 1945, The golden hamster (Cricetus auratus). Care, breeding, breeding, and growth, Physiol. Zool. 18:52–59.

Doull, J. A., and Megrail, E., 1939, Inoculation of human leprosy into the Syrian hamster, Int. J. Lepr. 7:509–512.

Eaton, M. D., and Beck, M. D., 1941, A new strain of virus of influenza B isolated during an epidemic in California, Proc. Soc. Exp. Biol. Med. 48:177–180.

Fothergill, J., 1770, An Essay on the Character of the Late Alexander Russell, M.D., F.R.S., Society of Physicians, London.

Fraser, L., 1849, Zoologia Typica, Zoological Society of London, p. 59.

Giebel, C. G., 1855, Säugetiere, Verlag Ambrosius Abel, Leipzig, p. 577.

Gunther, A. E., 1980. The Founders of Science at the British Museum 1753–1900, The Halesworth Press, Suffolk, England.

Horsfall, F. L., Jr., and Hahn, R. G., 1940, A latent virus in normal mice capable of producing pneumonia in its natural host, J. Exp. Med. 71:391–408.

Lennette, E. H., 1941, Susceptibility of Syrian hamster (Cricetus auratus) to viruses of St. Louis and Japanese B encephalitis, Proc. Soc. Exp. Biol. Med. 47:178–181.

Marsh, A. F., 1948, California no longer requires import permits on hamsters, Am. Small Stock Farmer 32:16.

Marsh, A. F., 1951, The Hamster Manual, The Gulf Hamstery, Mobile, Alabama.

Murphy, M. R., 1971, Natural history of the Syrian golden hamster—A reconnaissance expedition, Am. Zool. 11:632.

Murphy, M. R., 1977, Intraspecific sexual preferences of female hamsters. J. Comp. Physiol. Psychol. 91:1337–1346.

Pearson, H. E., and Eaton, M. D., 1940, A virus pneumonia of Syrian hamsters, Proc. Soc. Exp. Biol. Med. 45:677–679.

Poiley, S. M., 1950, Breeding and care of the Syrian hamster, Cricetus auratus, in: The Care and Breeding of Laboratory Animals. (E. J. Farris, ed.), John Wiley and Sons, New York. pp. 118–152.

Russell, A., 1756, The Natural History of Aleppo, 1st Edition, A. Millar, London.

Russell, A., and Russell, P., 1797, The Natural History of Aleppo, 2nd Edition, A. Millar, London.

Russell, A., and Russell, P., 1797. Naturgeschichte von Aleppo, 2nd Edition. (J. F. Gmelin, trans.), Johann Georg Kosenbusch, Gòttingen.

Scott, H., 1945, A History of Tropical Medicine, Edward Arnold and Co., London.

Shortt, H. E., 1967. Saul Adler, Biographical Memoirs of Fellows of the Royal Society 13:1–34.

Taylor, R. M., 1940, Detection of human influenza virus in throat washings in immunity response in Syrian hamster (Cricetus auratus), Proc. Soc. Exp. Biol. Med. 43:541–542.

Tristram, H. B., 1884, Fauna and Flora of Palestine, Committee of the Palestine Exploration Fund, London, pp. 10–14.

Waterhouse, C. O., 1889, Memoir of George Robert Waterhouse by his son, Trans. Ent. Soc. Lond. 5:70–76.

Waterhouse, G. R., 1840, Proceedings of learned societies, Zoological Society, April 9, 1839, Ann. Mag. Nat. Hist. 4:445–446.

Yerganian, G., 1972, History and cytogenetics of hamsters, in: Pathology of the Syrian Hamster, Volume 16: Progress in Experimental Tumor Research (F. Homburger, ed.), S. Karger, Basel, pp. 2–34.

II

Reproductive Endocrinology

2

The Estrous Cycle

ROBERT D. LISK

1. INTRODUCTION

The golden hamster *Mesocricetus auratus* was described as a new species and originally named *Cricetus auratus* by Waterhouse (1839). Laboratory colonies of the golden hamster both in Europe and America originate from a single capture from the wild by Mr. I. Aharoni of the Department of Zoology, Hebrew University, Jerusalem, who dug up a litter of eight animals near Aleppo, Syria in 1930. One male and two females survived and were bred (Adler, 1948). The animals were utilized by Dr. C. Adler working on parasitological problems in Jerusalem with support from the Royal Society. Offspring from these three original litter mates were shipped by Dr. Adler to the College de France, Paris, Medical Research Council and Welcome Bureau of Scientific Research, England, and eventually to America during World War II.

Several unique characteristics of the hamster resulted in adoption of this animal for many experimental studies in medical research. Its possession of reversible cheek pouches that formed easily accessible implant sites for tumors and embryonic organs, plus its tolerance to homologous and heterologous tissue transplant as well as to human tumors, viruses, and bacteria made the hamster the species of choice for a wide variety of studies. As researchers began to study the basic biology of the golden hamster in order to make more animals available for use in medical research, it was recognized that the species was very interesting in its own right for studies of a variety of problems in biology.

A summary of the published studies utilizing the golden hamster with a bibliography that is complete through 1965, can be found in *The Golden Hamster, Its Biology and Use in Medical Research* (Hoffman *et al.*, 1968).

The hamster is uniquely suited to studies in reproductive biology due to its invariant four-day estrous cycle. The small variance in activity rhythms among animals in a population has resulted in use of the hamster for studies in chronobiology. In

---

ROBERT D. LISK • Department of Biology, Princeton University, Princeton, New Jersey 08544.

addition, the hamster has the shortest gestation length (16 days) of any eutherian mammal, making it an excellent subject for developmental studies since early access to the young can be achieved without the problems of cesarean surgery.

2. SEXUAL MATURATION—THE FIRST ESTROUS CYCLE

Sexual maturity occurs following maturation of all other organ systems and after the rapid phase of body growth has been completed. A number of hypotheses have been advanced to account for the onset of sexual maturity. These have been extensively reviewed by Ramaley (1979). There is considerable evidence that all the components of the system, namely, gonads, pituitary, and brain, are capable of response to those signals that regulate reproduction well in advance of the occurrence of the first ovulatory cycle. It is not clear which factor(s) is responsible for failure of the ovulatory cycle to occur prior to an age that is characteristic for each species. The general series of events culminating in sexual maturity in female mammals starts with a rise in serum follicle-stimulating hormone (FSH) often to levels higher than those that are found following sexual maturation. The high FSH levels are probably essential for initiation of follicular development and steroid production by the ovaries. This is followed by daily pulses of luteinizing hormone (LH) release and further steroid secretion. At this time the gonadotropin release mechanism may come under the control of a clock mechanism that is entrained by the light-dark period each 24 hr. Finally, a change in the feedback sensitivity to estrogen must occur such that estrogen secretion from mature follicles can initiate an ovulatory surge of gonadotropin to complete the first ovulatory cycle.

The hamster is born after the shortest gestation of any eutherian mammal (16 days) and reaches sexual maturity at one month of age. The earliest reported mating resulting in a successful pregnancy with birth of a litter occurred at 28 days of age (Selle, 1945). There is general agreement that the first ovulation occurs between 26 and 30 days of age (Bond, 1945; Oritiz, 1947; Nakano, 1960; Greenwald and Peppler, 1968; Bex and Goldman, 1977; Smith et al., 1980). Vaginal patency occurs very early (between 8 to 14 days of age) (Bond, 1945; Oritiz, 1947) and can be advanced to as early as six days of age by estrogen treatment (Oritiz, 1947). Ovulation can be advanced in the immature female by six to ten days as a result of pregnant mare's serum (PMS) treatment followed by LH (Bodemer et al., 1959). Thus, ovulation can be induced by 20 days of age. However, PMS treatment by itself (Greenwald and Peppler, 1968) will not result in ovulation occurring prior to day 26. Although the ratio of light to dark each 24 hr has a profound effect on estrous cyclicity and the timing of ovulation (see Section 3.1), variation of lighting conditions including total darkness has no effect on the age at sexual maturation (Darrow et al., 1980).

In the hamster the first set of follicles matures just prior to the first ovulation. Antral follicles are first observed on day 26 and injection of PMS, a preparation high in follicle-stimulating potency, is not effective in facilitating ovulation prior to day 26. By day 30 the ovaries become highly sensitive to PMS with up to 55 eggs ovulated (Greenwald and Peppler, 1968). Examination of serum levels of gonadotropins indicates that during the second and third weeks after birth, FSH levels are higher even than those found during the preovulatory surge in the adult animal. Luteinizing hormone levels are also above baseline for the adult (Bex and Goldman, 1977; Vomachka and Greenwald, 1979). Following ovariectomy on day 18, there is a significant elevation in serum levels of both FSH and LH suggesting that the feedback system between ovary and pituitary is already functional. In the intact animal, increased serum estrogen is found between days 13 to 16 suggesting that the increasing FSH output during this period is able to stimulate steroid secretion by the ovary (Vomachka and Greenwald, 1979).

Starting by day 16, a surge of FSH and LH occur at the same clock time daily (Smith et al., 1980, 1982). It is not known whether this is the result of turning on the daily clock (see 3.3) as a result of maturation of neurons or the completion of neural connections between the clock and other parts of the system. One structure that affects reproduction under short photoperiods, the pineal, does show the mature pattern of innervation by day 15 (VanVeen et al., 1978). However, in immature hamsters the daily cyclic release of gonadotropins occurs independent of photoperiod (Smith et al., 1982). When the pituitary's response to gonadotropin-releasing hormone from the brain is tested, no response is found in two-day-old animals, but rapid increases in response occur between days six to ten (Smith et al., 1982).

These findings in aggregate suggest that all parts of the sytem are capable of response well in advance of completion of the first ovulatory cycle. As the system begins to function, rapid changes in sensitivity to the hormonal stimuli that regulate the ovulatory cycle appear to be occurring. This is suggested by the fact that at the first ovulatory cycle, only 50% of the normal adult number of ova are released (Bex and Goldman, 1977). Between day 26 and 30, a tremendous change in sensitivity to PMS for facilitation of ovulation occurs (Greenwald and Peppler, 1968). Although first ovulation occurs about day 29, consistent, successful pregnancies are not seen until after day 35 suggesting a period of reduced fertility (adolescent sterility) that has been reported for a number of species (Diamond and Yanagimachi, 1970).

The major change in pattern of gonadotropin release that occurs at the onset of estrous cyclicity is suppression of the daily surge of LH and FSH release so that it occurs once each four days, creating the estrous cycle. Regression to a daily surge in LH and FSH release occurs many times during the life history of the female, namely during each lactation and during the anestrus induced by short photoperiods (Smith et al., 1980). Thus, understanding how the daily surge of gonadotropin release is suppressed and a surge initiated every four days that results in ovulation may unlock the key to understanding how puberty is brought about in this species.

In those species having short estrous cycles, examination of the cellular and mucous material sloughed off in the vagina can be used as reliable external markers of the stage of follicle development and the occurrence of ovulation. Two conventions have been employed to designate the days of the cycle. In the rat, which serves as the model for most laboratory work with rodents, the stage of follicle development and ovulation is identified by daily examination of the cell types found (Long and Evans, 1922) when a small amount of water is flushed into the vagina and aspirated with the aid of a medicine dropper. Examination of a drop of this material at 100× under the microscope reveals the following. During the first half of the cycle, when ovarian secretion of estrogen is low, the fluid aspirated from the vagina contains predominantly white blood cells (polymorphic leukocytes); these days are labeled diestrus day 1 and diestrus day 2. On day 3 of the cycle when the maturing follicles are producing increasing levels of estrogen, fluid aspirated from the vagina contains mainly large round nucleated epithelial cells; this is labeled proestrus. On the final day of the cycle (day 4) following the peak in estrogen production and ovulation, fluid aspirated from the vagina contain mainly sheets of keratinized epithelium with few or no nuclei visible; this is labeled estrus.

In the hamster, because of the copious discharge of cellular and mucous material, it is not very convenient to determine cell composition under the microscope. However, on the day of ovulation, there is a postovulatory discharge from the vagina that is readily recognizable because of its copious nature, tacky consistency, and strong, pungent odor. Since this discharge occurs only when ovulation has taken place and since the cycle is invariant in length at four days, it has become common practice to date cycle events from the postovulatory discharge that is labeled day 1 of the cycle (Orsini, 1961). This is likely to prove confusing to those who are accustomed to the rat or mouse system in which the day of ovulation (estrus) would be considered day 4. On Fig. 1 where I have plotted serum gonadotropin and steroid levels throughout the four-day estrous cycle, I have given both conventions for labeling the days of the cycle. However, the convention employed throughout this chapter will be that typically used in studies of the hamster. Thus, the cycle will be labeled days 1 to 4 with day 1 being the day on which ovulation occurs, and on which the postovulatory discharge is found on examination of the vagina.

2.1. The Ovary

Histological analysis of ovaries during development and in the adult hamster in conjunction with a variety of experimental manipulations provides a picture of follicle growth and atresia and the factors that regulate the number of follicles ovulating each estrous cycle. The hamster ovary is encapsulated by the fallopian tube. Primary oogonia are about 15 μ in diameter increasing to a mature size of 70 μ that is reached prior to the initiation of antral development in the follicle (Knigge and Leatham, 1956) that occurs when follicles have reached a diameter of about 350 μ. At ovulation follicle diameter is about 700 μ. In a detailed study of follicle development, Greenwald (1960, 1961, 1962a,b) showed that on day 2 of the cycle each ovary contains 10 to 12 large follicles that grow rapidly in size between day 3 and early on day 4 mainly due to expansion of the antral cavity. Between late on day 4 and early on day 1, about 50% of the large follicles become atretic while the other 50% ovulate. If one ovary is removed on day 2 or 3 or early on day 4, none of the large follicles in the remaining ovary undergo atresia with the result that the normal number of ova are shed. If exogenous gonadotropin (FSH) is supplied early in the cycle, a further 25 follicles per ovary enlarge and can be ovulated (reserve follicles). Thus, the number of follicles maturing each cycle

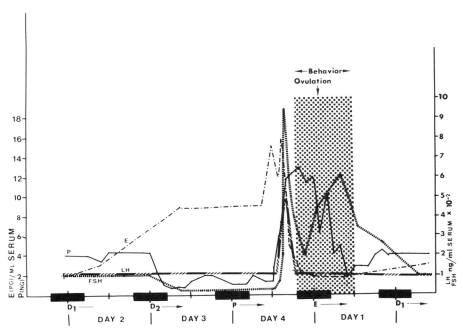

Figure 1. Serum profiles of the gonadotropins and the steroids are indicated for the estrous cycle of the hamster along with the potential period of sexual receptivity and the time of ovulation. The standard designation for each day of the rodent cycle (D1 = diestrus day 1, D2 = diestrus day 2, P = proestrus, E = estrus) is indicated and beneath this the terminology for the days of the estrous cycle found in most references on the hamster has been indicated. Serum profiles of gonadotropins are based on Bast and Greenwald (1974), estrogen on Saidapur and Greenwald (1978), and progesterone on Ridley and Greenwald (1975), with the receptivity and ovulation time based on Reuter *et al.* (1970). Heavy black bar, 10-hr dark period. Black line, 14-hr light period.

and the number ovulating are regulated by the availability of gonadotropins (particularly FSH) during the cycle.

Study of the ovulatory process with the aid of scanning electron microscopy indicates that stigma formation is first observable about 4 hr before the expected time of ovulation (Pandergrass and Reber, 1980). The stigma remains small suggesting that at ovulation the oocyte and surrounding cumulus must be ejected rather than oozing out of the ruptured follicle.

Estrogen levels in ovarian blood and the peripheral circulation correlate with follicular development during the cycle (Baranczuk and Greenwald, 1973; Saidapur and Greenwald, 1978). Estrogen levels are lowest when antral follicles are absent (day 1 and day 2). Starting on day 2 and correlated with follicle antral growth, estrogen levels increase and plateau over day 3 to early day 4. By noon of day 4, estrogen levels peak for the cycle and then decline to baseline levels within 4 hr (Fig. 1). The abrupt decline in estrogen synthesis is likely the result of the increased progesterone synthesis that results from the LH surge that occurs early on the afternoon of day 4. It has been shown that the follicular synthesis of estrogen on day 4 is very sensitive to the amount of progesterone present (Saidapur and Greenwald, 1979).

The source of progesterone during the ovulatory cycle has been disputed. Following ovulation, the corpus luteum develops rapidly and regresses equally rapidly with the result that only a single generation of corpora lutea are recognizable (Leavitt et al., 1973). The interstitial tissue (Norman and Greenwald, 1971) is suggested to be the major source of preovulatory progesterone; however, in another study the follicles and interstitial tissue were found to contribute approximately equally to the preovulatory progesterone pool (Leavitt et al., 1971). In a study of in vitro steroidogenesis in corpora lutea and nonluteal ovarian tissue (Terranova et al., 1978) no synthesis of progesterone was detected either by corpora lutea or nonluteal tissue on day 1, a time of high serum levels of progesterone. On day 2 both tissue compartments produced progesterone. No progesterone production was found for luteal tissue on days 3 and 4, whereas the nonluteal tissue continued to have a similar capability to day 2 for production of progesterone. These findings indicate that nonluteal tissue is a major source of progesterone throughout the cycle. The failure to identify a source of progesterone production on day 1 is probably the failure to supply an incubation medium containing the factors (FSH, LH) necessary for progesterone production. It is unlikely that an extra ovarian source is responsible for progesterone production at this time, because following ovariectomy progesterone rapidly declines to baseline levels (Vomachka and Greenwald, 1978). Plasma progesterone levels have been assayed at 2-hr intervals across the entire estrous cycle (Ridley and Greenwald, 1975; Bast, 1979). Progesterone is close to peak values for the cycle (12 ng/ml) at the start of day 1 and falls steadily to near baseline at 2 ng/ml by noon of that day. A gradual increase to 4 ng/ml occurs by the evening of day 1 with serum progesterone plateauing at this value across day 2. Early on day 3, progesterone falls to baseline values for the cycle (1 ng/ml) and fluctuates between 1 and 2 ng/ml across day 2 and until noon on day 4. Then within 4 hr, progesterone increases to peak values of 12 to 13 ng/ml serum where it remains (Fig. 1).

Antibodies to the gonadotropins have been used to study the necessity of both FSH and LH for follicle development and ovulation. On the afternoon of day 4, a surge release of LH and FSH occurs. Injection of specific antisera to FSH or LH at the beginning of the surge followed by FSH or LH injection at the end of the surge to neutralize any remaining antibody results in failure of follicle development for the next cycle (Rani and Moudgal, 1977). Thus, both FSH and LH are necessary for the normal process of follicle development. The need for LH could not be substituted by estrogen. Pregnant mare's serum, which serves as a follicle-stimulating hormone, is effective in increasing the number of ovulating follicles (by causing development of the reserve follicles) when injected on days 1 to 3 of the cycle, indicating that FSH levels determine the number of follicles ovulating. Injection of antibody to PMS on day 3 delays ovulation by one day (Greenwald, 1973). Thus, a continuous source of FSH is essential to maintain follicle development.

Addition of exogenous FSH early in the cycle results in an increased number of ova being shed (Bex and Goldman, 1975). Following unilateral ovariectomy, a transient increase in serum FSH is observed that is probably responsible for the increased number of follicles ovulating from the remaining ovary. At the next cycle there is a prolonged elevation of FSH during days 1 and 2 (Bast and Greenwald, 1977).

Recent studies indicate that the ovaries produce a nonsteroidal substance named inhibin (Schwartz and Channing, 1977) that can decrease the release of FSH from the pituitary. Inhibin does not affect LH release. Extracts of hamster ovary on day 4 of the cycle caused a significant decrease in serum FSH, whereas similar extracts made on other days of the cycle had no effect on FSH release (Chappel, 1979). When repeated injection of inhibin were made on day 4 and day 1 of the cycle, serum FSH was decreased on day 1 and the number of follicles ovulating on the next cycle was reduced. Employing inhibin from bovine follicular fluid, it was possible to produce a dose-dependent decrease in FSH secretion (Chappel and Selker, 1979). When FSH release was blocked on day 1, nine of ten animals failed to ovulate four days later. Therefore, the periovulatory release of FSH is necessary to initiate development of the follicles due to ovulate four days later, and one function of inhibin may be to limit the amount of FSH released on day 1 of the cycle, thereby determining the number of follicles that will develop during that cycle.

To determine specific sites of FSH action in the ovary, [^{125}I]-FSH was employed in an autoradiographic study to localize binding sites at various stages of the estrous cycle (Oxberry and Greenwald, 1982). On day 1, FSH binding was found in the granulosa of all viable preantral follicles with the larger follicles appearing to have more label. Atretic follicles were not labeled. Corpora lutea and oocytes also showed binding. By day 2, FSH binding to the corpora lutea was greatly reduced, whereas the viable follicles continued to show binding that appeared to be greatest in the granulosa and oocytes of the newly formed antral follicles. On day 3, the granulosa and oocytes of the viable antral follicles showed intense binding. By the morning of day 4, those follicles destined to ovulate showed heavy FSH binding over the oocyte and cumulus and moderate binding over the granulosa. In summary, FSH binding remains high in healthy follicles and is absent in atretic follicles. However, it is not clear if atresia is produced by a disappearance of FSH receptors. Follicle-stimulating hormone binding to oocytes and cumulus suggests that FSH may play a role in oocyte growth and maturation. In addition, the binding of FSH to corpora lutea during the period of progesterone production in the cycle may signify a role for FSH in corpus luteum function.

2.2. The Pituitary

The studies reviewed in the previous section indicate that early in the cycle high levels of FSH should be present to initiate development of the set of antral follicles due to ovulate four days later. With the development of radioimmunoassay, it became possible to measure the amount of gonadotropins in the systemic circulation throughout the estrous cycle. Several laboratories have examined LH, FSH, and prolactin levels (Goldman and Porter, 1970; Bast and Greenwald, 1974; Varavudhi and Meites, 1974; Bex and Goldman, 1975) as frequently as every 4 hr for the entire estrous cycle. Values for LH and FSH based on Bast and Greenwald (1974) are plotted in Fig. 1. Elevated FSH secretion reaching 12 times baseline occurs on day 1 with a gradual fall-off that plateaus across day 2 at about 2 times the baseline level of secretion. The baseline level of FSH secretion occurs by midmorning of day 3 and continues until shortly after noon on day 4 when within an hour secretion increases to about 20 times baseline and then falls rapidly within 4 hr to about 4 times baseline early on the evening of day 4.

Luteinizing hormone remains constant at baseline level across the entire cycle except for shortly after noon on day 4 when a rapid 5 times increase in systemic hormone level is found. By early evening, LH is back to baseline level (Fig. 1).

Prolactin levels show slight fluctuations: 1 times around baseline across the cycle except for the early afternoon of day 4 when a rapid increase to 3 times baseline occurs followed by a return to baseline within 5 hr (Bast and Greenwald, 1974).

A synchronous peak in serum levels of all three gonadotropins occurs early on the afternoon of day 4. This is the surge output required for ovulation. Otherwise, the gonadotropins remain at baseline levels except for FSH that has a second peak on day 1 and gradually decreases to baseline levels early on day 3. Thus, maintenance of cycle events and follicle maturation appears to require fluctuating levels of FSH in the presence of constant baseline levels of LH and prolactin.

To determine whether the surge in FSH plus LH are necessary for ovulation, hamsters were hypophysectomized early on day 4 and purified FSH or LH was injected to induce ovulation (Greenwald and Papkoff, 1980). Follicle-stimulating hormone proved more potent than LH for the induction of ovulation.

Gonadotropin release from the pituitary is stimulated via gonadotropin-releasing hormone (GnRH) from the brain. Therefore, one wants to know whether changes in serum levels of gonadotropins are the result of changed sensitivity of the pituitary to GnRH or does the amount of GnRH reaching the pituitary change? In a test of pituitary responsiveness to GnRH (Arimura et al., 1972), the largest response was found on day 4, with the smallest response on day 1 and day 2 and an intermediate response on day 3. This suggests a changing sensitivity of the pituitary to GnRH across the estrous cycle. A similar conclusion was reached when GnRH was infused into animals in which spontaneous release of gonadotropin was inhibited by phenobarbital injection (Shander and Goldman, 1979). Infusion on the afternoon of day 4 resulted in a normal preovulatory surge of gonadotropin, whereas similar infusion on day 3 resulted in much less secretion of FSH and LH. When GnRH was infused into ovariectomized hamsters, LH release was only 25% of that seen for intact animals. Following treatment with estradiol in silastic capsules, the response was equal to that which occurs on proestrus. This suggests that the increasing amount of estrogen secretion that occurs on day 3 and day 4 results in an increased sensitivity of the pituitary to GnRH (Shander and Goldman, 1978). The binding characteristics of the pituitary GnRH receptor sites have been examined at different times in the estrous cycle to determine whether changes in affinity or concentration might account for the difference in effectiveness of GnRH at different times in the cycle (Adams and Spies, 1981a). No changes in affinity were found. Marked changes in concentration of receptors were noted with a 70% decrease in receptor concentration during the 24 hours following the ovulatory surge of gonadotropin on day 4. However, during the remainder of the cycle, receptor concentration remained at 80% to 90% of the day 4 level. Thus, the decreased effectiveness of GnRH in causing release of FSH and LH from the pituitary early in the estrous cycle cannot be accounted for by change in affinity or concentration of receptors for GnRH (Adams and Spies, 1981a).

When phenobarbital was injected to block the surge of gonadotropin that is necessary for ovulation, those animals failing to ovulate also failed to show a decline in GnRH receptor concentration (Adams and Spies, 1981b). One might interpret this as evidence for phenobarbital action at the brain, with its mode of action being to inhibit the release of GnRH from axon terminals in the median eminence.

There are apparent species differences in the amount of GnRH required to induce ovulation in the barbiturate-blocked animal. The hamster is most sensitive (0.11 μg/kg body wt) when compared with rabbit (0.88 μg/kg body wt) and rat (3.7 μg/kg body wt) (Humphrey et al., 1973).

2.3. The Brain

The necessity to include the brain in an explanation of the factors that regulate the ovulatory cycle originated with the realization that the number of hours of light during each 24-hr period determined whether or not the gonads remained active (Hammond, 1954). Later it was demonstrated that by changing the time of lights-on, one could change ovulation time (Wurtman, 1967). Since the gonadotropins necessary for follicle growth and ovulation are made and released from the pituitary, light acting via the brain must somehow be influencing the pituitary to release gonadotropins at a specific time and in a pattern necessary for follicle maturation and ovulation. This neural timing mechanism, whatever its basis, appeared to be present only in the female, because if ovaries are transplanted into a male, no ovulations occurred (Harris, 1964). Pituitary transplantation between the sexes indicated there was no sexual differentiation of the pituitary's ability to release gonadotropin (Harris and Jacobson, 1952). Therefore, the failure of the pituitary in the male to release gonadotropins in the cyclic pattern necessary to ovulation implies that the signals from brain to pituitary are different in male and female.

If hamsters are gonadectomized within the first few days after birth and allowed to reach adult size without gonads present, when ovaries are implanted ovulatory cycles occur regardless of the genetic sex of the individual (Swanson, 1970). Since the male that develops with gonads present is incapable of supporting cyclic ovarian function, it is likely that some secretion of the testes during the neonatal period is responsible for the sexually dimorphic ovulatory response pattern observed.

Neonatal injection of testosterone or estrogen results in acyclicity and ovaries lacking copora lutea (Swanson, 1966; Alleva et al., 1969). Prenatal injection of androgens (Nucci and Beach, 1971; Landauer et al., 1981) are much less effective than postnatal injections for induction of anovular sterility. Therefore, the period of maximum sensitivity to steroids for differentiation of the male (acyclic) pattern of gonadotropin secretion occurs after birth in the hamster. Thus, the hamster is a suitable species for examination of the neural sites of steroid action for masculinization of the pattern of gonadotropin release. Swanson and Brayshaw (1974) extruded pellets containing 0.9 μg testosterone from the tips of 26 gauge hypodermic tubing into the hypothalamus of female hamsters either on the day of birth (day zero) or day 1, 2, 3, or 4. Less than 20% of the animals implanted on day zero and about 40% of the animals implanted on day 1 or 2 showed estrous cycles when adult. No specific site in the hypothalamus proved more sensitive than another site. Hamilton *et al.* (1981) extruded pellets containing 3 μg testosterone, from 26 guage hypodermic tubing into the hypothalamus, or inserted pellets bilaterally, extruded from 28 gauge tubing with the total hormone content 3 μg testosterone, into females on day 3 after birth. The unilateral versus bilateral pellet placement produced similar effects. As shown in Table 1, implantation in the preoptic anterior hypothalamic area resulted in 50% of the animals being acyclic when adult. If the group size had been larger, the arcuate ventromedial hypothalamus may also have proved equally sensitive to testosterone during the neonatal period for disruption of cyclicity. These experiments support the hypothesis that exposure of specific brain sites to androgen during the neonatal period permanently affects the feedback systems that regulate gonadotropin release in a cyclic or tonic pattern. A study of the localization of androgen-retaining neurons employing ^{3}H-labeled autoradiography (Doherty and Sheridan, 1981) shows that the septal-preoptic region, anterior hypothalamus, ventromedial and arcuate nucleus of the hypothalamus and amygdala are all highly labeled. Thus, these regions are prime candidates for neonatal effects of steroids and, as observed in the previous studies, the sites at which neonatal implants of testosterone are effective in producing anovular females are regions of high androgen retention.

To determine which nuclear groups in the hypothalamus are necessary for maintaining the feedback relationships between gonads and brain, Lamperti and his colleagues have made use of monosodium glutamate which, if given during the neonatal

Table 1. Adult Vaginal Cyclicity of Female Hamsters Treated Neonatally with Subcutaneous TP Injections or Intrahypothalamic T Implants[a]

Treatment	Number with regular vaginal cycles/number in group	Number acyclic	% acyclic
100 μg TP injection	0/6	6	100[b]
Subcutaneous oil injection	10/11	1	9
3 μg T POA-AH	7/14	7	50[c]
Sham POA-AH	21/22	1	5
3 μg T VMH-ARC	4/6	2	33
Sham VMH-ARC	10/10	0	0

[a]From Hamilton *et al.*, 1981.
[b]$P \leq 0.001$.
[c]$P \leq 0.01$.

period, selectively destroys neurons in the hypothalamus (Tafelski and Lamperti, 1977; Lamperti and Baldwin, 1979; Lamperti and Blaha, 1980; Lamperti et al., 1980). Injection of monosodium glutamate (8 mg/g body wt) on day 8 resulted in destruction of 85% to 92% of all neurons in the arcuate nucleus. As adults, the animals are acyclic; there is an inhibition of follicle growth, uterine and pituitary weights are lower, and FSH secretion is depressed. Feedback responses in LH release following estrogen treatment still occurs; however, there is no longer any tonic regulation of FSH secretion. Thus, the arcuate neurons are important in regulation of gonadotropin release and particularly the FSH secretion necessary for follicle development. Examination of the median eminence region of these animals, employing the Falk-Hillarp histofluorescence technique, showed a greatly reduced fluorescence indicating the catecholaminergic pathway was at least partly destroyed.

Attempts to identify the location of the GnRH-producing neurons by means of immunohistochemistry have not resulted in complete agreement. Jennes and Stumpf (1980) localized most GnRH neurons in the septal-preoptic region, with a smaller number in the olfactory bulb and tubercle and the anterior and ventromedial hypothalamus. Silverman et al. (1979) localized GnRH neurons only in the arcuate, ventromedial, and premamillary nuclei of the hypothalamus.

In a study of the localization of estrogen-retaining cells by use of ^{3}H-labeled autoradiography, Krieger et al. (1976) found a distribution throughout the hypothalamus from the medial preoptic area to the ventral premamillary area. Thus, there is an overlapping array of estrogen-retaining cells and GnRH-producing cells throughout the hypothalamus, allowing the possibility of feedback relationships to be easily established at all levels of the hypothalamus.

Clearly, the mechanism controlling the release of gonadotropin by the pituitary becomes differentiated as a result of some event that occurs during neonatal development. If one gonadectomizes adult males and females, an injection of estrogen in the female results in a midafternoon surge of LH on each of the next two afternoons. Estrogen injection in the male does not result in a surge of LH or FSH (Buhl et al., 1978). Thus, the positive feedback response to estrogen is lacking in the male.

Almost no work has been done on the neuropharmacology of the transmitters necessary for the surge of LH release in the hamster. Injection of alpha-methyltyrosine, which inhibits catecholamine synthesis, blocks ovulation. By use of various drugs that modulate the availability of noradrenaline and serotonin, it appears that whether or not ovulation occurs depends on the relative availability of noradrenalin and serotonin (Lippmann, 1968).

3. FOLLICLE MATURATION AND OVULATION

3.1. Time of Ovulation

Ovulation occurs at a specific time that appears to be determined by the time of lights-on during each 24-hr period (Austin, 1955; Orsini, 1961). Animals maintained on a 12-hr L:12-hr D cycle in which half are exposed to light from 6:00 AM to 6:00 PM and half are exposed to light from 12:00 noon to 12:00 midnight show a 6 hr difference in ovulation time (Alleva *et al.,* 1968). When animals are switched from a 12-hr L : 12-hr D cycle to constant light, 9 to 14 consecutive four-day cycles occurred prior to the appearance of altered cycles (Kent *et al.,* 1968). Cycle length then increases to 6 to 32 days with the extended part of the cycle characteristic of high estrogen levels. After six months the animals show persistent vaginal cornification with sustained intervals of willingness to mate. The ovaries contain many antral follicles, much interstitial tissue, and no corpora lutea (Kent *et al.,* 1968). Constant light thus interferes with those stimuli necessary to ovulation, but does not block follicular development.

The predictability of the time of ovulation coupled with the ability to change the time of ovulation by changing the relative time of lights-on versus lights-off suggested that a neural "clock" mechanism entrained by light provided the signal necessary for ovulation. A neural timing mechanism that regulates ovulation time has been demonstrated for the rat (Everett *et al.,* 1949). Using centrally active drugs to block neural activity and hypophysectomy to remove the source of ovulating hormone, Alleva and Umberger (1966) were the first to demonstrate that stimuli necessary to ovulation occurred during a 2-hr period (the critical period) early on the afternoon of day 4. Further examination of the limits of the critical period by Greenwald (1971) using phenobarbital (6.5 mg/100 g body wt) indicated a period of 1.25 to 1.5 hr by which time the critical period had ended in all animals. Hypophysectomy experiments indicated that the ovulatory quota of gonadotropin was released within 1 hr of the end of the critical period (Greenwald, 1971). To further understand the limits of response of the neural clock, anesthesia was maintained with pentobarbital or ether, neither of which block ovulation as a result of anesthetization during the critical period. Anesthesia could be maintained for up to 5 hr with each of these agents without blocking ovulation. However, if anesthesia was extended to 6 hr, ovulation was delayed by 24 hr (Reid *et al.,* 1972). Thus, the neural mechanism necessary to trigger an ovulatory release of hormone can, under some conditions, remain capable of response for at least 5 hr on the afternoon of proestrus before the system is blocked and reset by 24 hr.

By combining radioimmunoassay for measurement of the serum levels of luteinizing hormone with phenobarbital to block the ovulatory surge of LH, it was possible to evaluate the effectiveness of a variety of agents in overcoming the block to LH release. Progesterone (Turgeon and Greenwald, 1972; Norman, 1975; Siegel *et al.,* 1976) given concurrently with phenobarbital restores LH release and about 75% of the animals release a normal number of ova (Siegel *et al.,* 1976). Concurrent injection of estrogen or testosterone with the phenobarbital does not restore ovulation (Siegel *et al.,* 1976). The preovulatory surge of LH release is well defined in the hamster, being completed in all animals within a 3-hr period (Turgeon and Greenwald, 1972). When phenobarbital and progesterone are given concurrently, the onset of the LH surge is delayed (Turgeon and Greenwald, 1972) or continues for a longer time (Norman, 1975).

In overiectomized animals, a daily surge in LH release occurs that is about 35% to 40% of the peak preovulatory surge level (Norman *et al.,* 1973b; Stetson *et al.,* 1978). This is accompanied by a daily surge of FSH that is 3 times the preovulatory surge level.

Ovariectomized animals show high elevations of pituitary stores of LH and FSH (Stetson et al., 1978). Twenty-four hours following estrogen injection, there is an LH surge that is quantitatively similar to the preovulatory surge and this continues for several days. These daily surges are blockable by phenobarbital injections (Norman et al., 1973a; Stetson et al., 1978).

Progesterone injection also increases the peak serum LH concentration in ovariectomized animals; however, progesterone is not as effective as estrogen in facilitating LH release following ovariectomy (Norman et al., 1973a). Estrogen is probably acting at the neural level to facilitate LH release because addition of estrogen does not increase the effectiveness of GnRH which acts on the pituitary to facilitate release of LH (Norman et al., 1973b).

Stetson and his colleagues have demonstrated that the preovulatory gonadotropin release is regulated by a neural clock with a 24-hr periodicity (Stetson and Watson-Whitmyre, 1977; Stetson et al., 1981). By giving phenobarbital injections daily for one, two, or three days and examining the time of the LH and FSH surges, it was demonstrated that no shift in the time of the surge occurred on subsequent days, which implies control of the initiation of the surge by an accurate timing mechanism. Terranova (1980) examined the fate of follicles delayed from ovulating as a result of daily phenobarbital injections. When ovulation had been blocked for one or two days, increased numbers of ova were found indicating that during the extra days more follicles were maturing and responded to the surge of gonadotropins. When ovulation was blocked for three days by phenobarbital injections, the original large follicles showed atresia and did not ovulate; however, ovulation occurred 1.5 days later as a result of a newly recruited set of follicles that reached maturity.

3.2. Ovarian Steroids and Ovulation

Estrous cycles cease during periods of continuous high-level progesterone secretion into the systemic circulation, e.g., pregnancy or pseudopregnancy. Therefore, the effect of progesterone injection on cycle length and ovulation are of interest. Reuter et al. (1970) showed that a single injection of 100 μg progesterone in oil, which results in serum levels of progesterone within the range occurring during the estrous cycle, will delay ovulation by 24 hr if given during day 3. Similar injections given earlier in the cycle did not affect cycle length (Fig. 2). However, supraphysiological dosages (2.5 to 5 mg progesterone) injected on day 1 will lengthen the cycle by two to three days (Greenwald, 1977). This is probably the result of the inability of the animal to quickly metabolize such a large quantity of hormone resulting in high levels of hormone remaining on day 3, which is probably the critical time for delaying the cycle. As little as 25 μg progesterone on day 3 can delay the cycle for one day in 33% to 75% of the animals, depending on time of injection and vehicle (Reuter et al., 1970; Greenwald, 1978). Progesterone injections given starting at lights-off on day 3, in addition to delaying the cycle, result within 2 to 4 hr of injection in the female becoming sexually receptive. Although these animals will readily mate, ovulation does not occur. However, the delayed ovulation is always accompanied by a second period of sexual receptivity

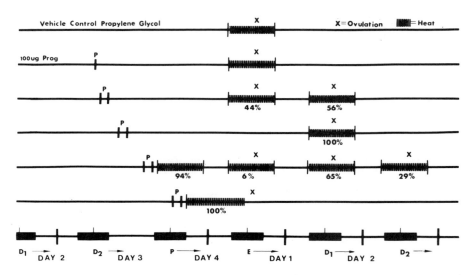

Figure 2. Effect of injection of 100 μg progesterone at various times during the estrous cycle on ovulation and sexual receptivity, based on Reuter *et al.* (1970). The numbers beneath the stripped bars indicate the percentage of the population responding on that day of the cycle. Progesterone results in an advance in sexual receptivity (heat) not accompanied by ovulation, or a delay by 24-hr increments in ovulation that is accompanied by sexual receptivity. See legend to Fig. 1 for details on the light cycle and designation of days of the cycle.

(Reuter *et al.,* 1970). Thus, early release of progesterone is disruptive to the estrous cycle both in terms of physiology and the linkage of behavioral receptivity to the period of ovulation.

Further examination of the events that follow injection of a physiological dosage (50 to 100 μg) of progesterone on day 3 of the cycle indicates that rapid fall (within 1 to 3 hr) in serum levels of estrogen occurs (Greenwald, 1978). Progesterone probably suppresses LH release from the pituitary which in turn decreases estrogen synthesis by the ovaries, resulting in lowered estrogen secretion. Until serum levels of estrogen have been restored for a sufficient time, the hypothalamic pituitary axis fails to be triggered to release the ovulatory surge of gonadotropins. Cycle length in the hamster therefore appears to be fixed by the duration of progesterone secretion during each cycle.

The effect of progesterone on the timing of ovulation is more complex than suggested by the preceding observations. Progesterone can result in an advance of ovulation by several hours as well as delaying ovulation by one or more days (Biphasic effect) (Reuter and Lisk, 1973a). This was demonstrated using progesterone pellets that were inserted subcutaneously and removed at varying times relative to the onset of the critical period. If the next expected critical period is during the first 10 hr of progesterone exposure, ovulation is advanced. If the next expected critical period is after the first 15 hr of continuous progesterone exposure, ovulation is blocked. If the progesterone exposure is 5 hr or less, no effect on ovulation is observed.

Long-term (14 to 18 days) suppression of ovulation can also be produced by a single injection of a long-acting estrogen (50 μg estradiol cyclopentylpropionate) on day 1 of the cycle (Greenwald, 1975). The high systemic estrogen levels block both LH and FSH secretion resulting in atresia of mature and developing follicles. As long as estrogen levels are high, FSH release will be blocked, thus inhibiting growth and maturation of new follicles with the result that new ovulations will not occur until FSH and LH release have recovered. Although the hypothalamic pituitary axis is very sensitive to estrogen levels under normal physiological conditions, continued high-level secretion of estrogen would not normally occur.

The necessity of the ovary as a source of hormones for regulation of estrous cycle events has been evaluated by use of timed ovariectomies followed by examination of acute changes in uterine weight, presence of the postovulatory vaginal discharge, and sexual receptivity. If the ovary remains *in situ* until 10 AM on day 4, uterine weight is increased indicating considerable estrogen secretion has already occurred (Brom and Schwartz, 1968). However, the ovaries must remain *in situ* until after the termination of the critical period on day 4 for sexual receptivity to occur. The period of sexual receptivity is always abbreviated even when ovariectomy is carried out after heat onset (Ciaccio and Lisk, 1971).

3.3. Role of the Nervous System in Ovulation

The demonstration of a critical period in which injection of certain anesthetic agents or neural blockers at a specific time of day inhibits ovulation for 24 hr (Alleva and Umberger, 1966; Terranova, 1980) implies that the nervous system is controlling the time of ovulation via a clock mechanism. Since removal of the pituitary (Greenwald, 1971) at the start of the critical period results in all animals failing to ovulate, whereas if the pituitary is not removed until 1 hr after the end of the critical period, all animals ovulate suggests that neural stimuli initiated during the critical period result in discharge from the pituitary of the gonadotropins necessary for ovulation.

The use of sensitive radioimmunoassay methods to measure serum levels of hormones demonstrated that gonadotropins remain at baseline levels throughout the estrous cycle until the afternoon of day 4 at which time a surge of LH and FSH resulted in these hormones rapidly increasing in the serum to 5 to 12 times the concentration found earlier in the cycle (Goldman and Porter, 1970; Varavudhi and Meites, 1974; Bast and Greenwald, 1974; Bex and Goldman, 1975). When serum LH and FSH levels were measured on the afternoon of day 4 in animals in which phenobarbital had been injected at a time known to be effective in blocking ovulation, the serum peak in LH and FSH was not found (Greenwald, 1971). This indicates that ovulation is blocked due to the lack of release of gonadotropins from the pituitary.

Anatomical localization of the neural sites responsible for the critical period events that result in a surge of gonadotropin release from the pituitary has involved use of lesioning and knife cuts (deafferentation) to localize inputs and control "centers" in the hypothalamus that make a connection with the pituitary via the neural stalk and the hypophyseal portal system. Sites of hormone feedback for regulation of gonadotropin release have been explored by direct application of hormone to the brain. Finally, drugs that affect DNA and RNA synthesis or neurotransmitter release have been employed in an attempt to determine the detailed mechanism that regulates the cyclicity of gonadotropin release necessary to ovulation.

Total deafferentation of the medial basal hypothalamus results in an immediate permanent block to ovulation (Norman et al., 1972). Extended frontal cuts produce a similar effect. However, frontal cuts of smaller radius block ovulation for two or three cycles only; following this, normal cycles resume. Posterior or lateral cuts have no effect on the ovulatory cycle. Animals with complete medial basal hypothalamic islands to which the pituitary is attached had decreased uterine weight, low serum LH, and high pituitary LH. This suggests no follicle development or estrogen secretion is occurring. In contrast, the animals with extended frontal cuts had increased uterine weight and ovaries full of large cystic follicles indicating complete follicular development with the consequence of high estrogen output. However, estrogen was not able to trigger the ovulatory release of gonadotropin.

To determine whether the positive feedback response to estrogen was absent, ovariectomized estrogen-primed females were employed to study the effect of neural deafferentation on the surge of LH release. Injection of estradiol into the ovariectomized female results in a surge of LH release 24 hr later. When a knife cut was made to interrupt the fibers between the medial preoptic area and the remainder of the hypothalamus, the estrogen-induced LH release no longer occurred. Small bilateral lesions in the medial preoptic area also blocked the estrogen-induced LH release. Therefore, the positive feedback of estrogen that triggers the LH surge requires intact connections between the medial preoptic area and the medial basal hypothalamus (Norman and Spies, 1974).

Follicle-stimulating hormone is released synchronously with LH during the critical period (afternoon of day 4) and manipulations that block the critical period also block the FSH surge. However, a second extended release of FSH not accompanied by the release of other gonadotropins occurs on the following morning (day 1). Study of the neural control of the day 1 release of FSH indicates that it is completely eliminated by electrocoagulation lesions of the arcuate nucleus median eminence region of the hypothalamus (Chappel et al., 1977). However, deafferentation of the medial preoptic area from the remainder of the hypothalamus, or complete deafferentation producing a hypothalamic pituitary island has no effect on FSH release on day 1. Thus, the mechanism for regulating the day 1 release of FSH must be contained within the medial basal hypothalamus.

To determine whether the release of FSH on day 1 was dependent on a specific neural event, frontal deafferentations were made at hourly intervals throughout the afternoon on day 4. It was possible to dissociate the day 4 surge of LH and FSH from the day 1 surge of FSH providing evidence that a neural mechanism is essential to the FSH release on day 1 (Chappel et al., 1979). To further examine the mechanism involved, phenobarbital was used to block the critical period surge of gonadotropins. Then, exogenous FSH was injected. This was followed by the normal day 1 peak of FSH. A similar effect was produced by injection of FSH directly into the third ventricle in the phenobarbital-blocked hamster, suggesting that a short loop feedback of the FSH surge during the critical period may trigger the day 1 surge of FSH (Coutifaris and Chappel, 1982).

Steroid hormones released into the general circulation by the ovaries exert both positive and negative feedback control on gonadotropin release. The short-term effect is positive resulting in increased gonadotropin output; however, if steroid levels remain elevated, they exert a negative effect and gonadotropin release from the pituitary is suppressed.

Injection of estrogen or progesterone in the cyclic hamster can alter the pattern of gonadotropin synthesis and release and block ovulation (Keever and Greenwald, 1967). Following unilateral ovariectomy there is an immediate increase in serum FSH (Bast and Greenwald, 1977). If both ovaries are removed, increases in serum LH and FSH are measurable in 3 to 6 hr (Vomachka and Greenwald, 1978). Within two days serum levels of FSH have peaked, whereas serum levels of LH continue to rise even 14 days after ovariectomy (Vomachka and Greenwald, 1980). Ovariectomy on all days of the cycle results in a rapid fall in serum levels or estrogen and progesterone (Vomachka and Greenwald, 1978). The increase in serum FSH following acute ovariectomy occurs following the decline in serum estrogen (Chappel et al., 1978). Addition of estrogen to the ovariectomized animal results in a rapid return of FSH to baseline levels and LH levels in serum are measurably lower within 3 hr (Vomachka and Greenwald, 1980). Progesterone is not as effective as estrogen in suppressing serum gonadotropins in chronically ovariectomized animals. Injection of GnRH results in FSH and LH levels in the serum rising even higher in chronically ovariectomized animals (Vomachka and Greenwald, 1980).

These findings of rapid increases in serum levels of LH and FSH following ovariectomy plus the fact that gonadotropin secretion can be depressed both in the intact and ovariectomized animals by injection of estrogen and progesterone indicate that a negative feedback relationship exists by which the gonadal steroids affect release of FSH and LH. Since GnRH can raise LH and FSH output still higher, a probable site for this negative feedback is at the hypothalamus where the GnRH neurons are located.

In a study employing the hormone implantation technique to localize neural sites of negative feedback for estrogen (Lisk and Ferguson, 1973), the inhibition of gonadotropin release was assayed by the presence or absence of the post ovulatory vaginal discharge. If the discharge was present, the oviduct was checked and the number of ova counted. Estrogen implant in and about the arcuate nucleus of the hypothalamus was found significantly more effective than all other implant locations for blocking ovulation (Table 2). Estrogen implants outside the basal medial hypothalamus resulted in a significantly larger percentage of the population showing normal cycles as compared with all other implant locations.

In a study using the hormone implantation technique to localize sites of negative feedback for progesterone (Fig. 3), implants of the smallest surface area blocked ovulation only when placed in the medial preoptic area (Reuter and Lisk, 1973b). Larger implants blocked ovulation when placed throughout the hypothalamus, but not when placed in the pituitary. Testosterone implants when placed in the preoptic area also

Table 2. Percentage of Animals Showing a Change in the Vaginal Cycle following Cranial Implantation of 30 Gauge Estradiol-containing Tubes or Control Implants (Empty or Testosterone-containing Tubes)[a]

Location	Number of animals	BLK[b]	INT[c]	IPS[d]	NOR[e]
Preoptic area exptl.	5	20	20	20	40
Control	0	—	—	—	—
Anterior hypothalamus exptl.	17	18	29	12	41
Control	4	50[f]	0	0	50
Arcuate nucleus exptl.	20	80[g]	5	0	15
Control	13	7	7	7	79
Mamillary bodys exptl.	9	11	44	—	34
Control	1	—	100	—	—
Outside basal exptl.	24	17	17	4	62[h]
Control	10	—	—	10	90

[a]From Lisk and Ferguson, 1973.
[b]BLK = cycle blocked continuous diestrus smears (day 2).
[c]INT = 4-day cycles interrupted by periods of diestrus.
[d]IPS = 4-day cycles interrupted by one or more periods of pseudopregnancy.
[e]NOR = normal continuation of 4-day estrous cycles.
[f]These two animals had testosterone implants.
[g]Percent of animals showing block is significantly different ($P = 0.001$) from all other cycle types and all other implant locations.
[h]Percent of animals showing normal cycles is significantly different ($P = 0.001$) from all other cycle types and all other implant locations.

blocked ovulation, but at a significantly lower rate than progesterone. The testosterone effect may be due to the similarity of the steroid molecules. If the effect is due to receptor-mediated transport of the hormone to the cell nucleus, it is probable that testosterone can also bind to the receptor and result in its transport to the cell nucleus. The negative feedback resulting from progesterone treatment differs from that produced by estrogen treatment. Following progesterone treatment, the system escapes after the lapse of one or more cycles and ovulation occurs again.

To determine the possible neural feedback site for progesterone facilitation of ovulation (positive feedback), phenobarbital was given to block the spontaneous LH surge (Norman *et al.*, 1973a). These animals do not show any change in sensitivity to GnRH for release of LH, suggesting phenobarbital is acting at the brain. Progesterone injection overcomes the phenobarbital block suggesting that the positive feedback action of progesterone is exerted at the neural level. If a frontal deafferentation is made in the phenobarbital-blocked animal just prior to the critical period, injection of progesterone does not result in a surge of LH. Thus, progesterone appears to act outside of the medial basal hypothalamus to trigger the LH surge in phenobarbital-blocked animals (Norman *et al.*, 1973a).

Nutritional status can also affect follicular development and ovulation. In a study of food-deprived versus well-fed animals, bilateral lesions of the mamillary bodies or brain stem reticular formation blocked ovulation (Printz and Greenwald, 1971). The authors suggest that starvation activates a center in the brain stem (ventral tegmental nuclei) that acts through the mamillary bodies to inhibit the FSH release necessary for follicle development. Since these effects were not observed in fed animals and posterior deafferentation of the hypothalamus has no effect on the ovulatory cycle of fed animals, this pathway would appear to be a specific mechanism activated only during starvation to inhibit the reproductive cycle during times that the female would be incapable of successfully producing a litter.

The critical period event, if blocked, occurs 24 hr later. This represents one example of circadian events such as locomotor rhythms, feeding patterns, or estrous cyclicity, all of which have in common their regularity of occurrence plus the ability to be entrained by the photoperiodic cycle on which the animal is maintained. A search for the biological clock that maintains these rhythms has resulted in the discovery that bilateral lesioning of the suprachiasmatic nuclei in the anterior hypothalamus abolishes all these rhythms and renders the animal insensitive to photoperiodic treatment (Stetson and Watson-Whitmyre, 1976). Animals remain reproductively mature irrespective of the photoperiodic cycle they are maintained under. Thus, the integration of light into

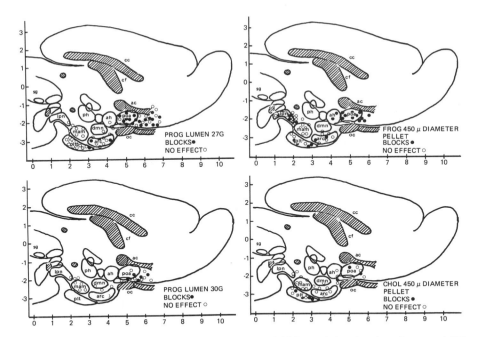

Figure 3. Neural loci at which progesterone implantation inhibits ovulation, based on Reuter and Lisk (1973b). As the surface area of hormone in contact with the brain is decreased to 200 μ diameter (27 gauge tube) and 150 μ diameter (30 gauge tube), ovulation is blocked only by implantation in and about the preoptic region. A large control implant (cholesterol 450 μ diameter pellet) blocks ovulation in a small percentage of the animals as a result of a nonspecific lesion effect. With smaller implant sizes, none of the animals with cholesterol showed an interruption of the estrous cycle.

the control mechanism for maintaining the reproductive cycle occurs within the hypothalamus. However, steroids do not appear to act directly on the suprachiasmatic nucleus, thus the biological clock does not appear to be directly regulated by steroids, but it in turn affects the neurons on which the steroids act.

Drugs that interfere with protein synthesis have been employed to determine whether production and release of new species of proteins are essential for ovulation. Ovulation can be inhibited by treatment with actinomycin-D, suggesting that new messenger RNA made in the follicles is necessary for ovulation (Barros and Austin, 1968). Synthesis inhibitors that act on DNA (cycloheximide, diptheria toxin) also block ovulation (Alleva *et al.*, 1979).

3.4. Linkage of Sexual Receptivity with Ovulation

Hamsters are nocturnal animals with most activity confined to the dark phase of the 24-hr photoperiod. The female becomes sexually receptive about the time of lights-off on the evening of day 4. If not mated, the receptive period lasts throughout the night and up to noon the next day (Ciaccio and Lisk, 1971), a period of 14 to 16 hr. Ovulation occurs just after the midpoint of the dark phase (about 1:00 AM). Thus, sexual receptivity and ovulation are normally linked, as shown in Fig. 1. Injection of progesterone early in the cycle can disrupt this linkage, resulting in a period of sexual receptivity without ovulation occurring (Reuter *et al.*, 1970). Mating at such times would not serve any biological utility to the species. Therefore, it is probable that a mechanism has evolved to ensure that receptivity and ovulation are linked together.

If the critical period events are blocked by injection of phenobarbital (Bosley and Leavitt, 1972), ovulation does not occur and sexual receptivity is not displayed during the night of day 4 unless exogenous progesterone is given. If phenobarbital-blocked animals are injected with exogenous FSH or LH in a dosage that induces ovulation, then sexual receptivity is also expressed. Therefore, the critical period surge of gonadotropins necessary for ovulation also results in ovarian secretion of progesterone which is necessary to facilitate sexual receptivity. Thus, the critical period events serve to link receptivity and ovulation together. This hypothesis is supported by examination of the effect of injection of antisera to the gonadotropins on proestrus (Goldman and Sheridan, 1974). Both ovulation and sexual receptivity were blocked. Exogenous progesterone restored receptivity but not ovulation. Luteinizing hormone restored both receptivity and ovulation. Estrogen had no effect.

4. THE LUTEAL PHASE

4.1. Pseudopregnancy

The hamster has a four-day estrous cycle that does not include a luteal phase. Sustained corpus luteum activity results from cervical stimulation that occurs at mating and results in either pseudopregnancy or pregnancy (see Chapter 3).

Following ovulation, the corpus luteum develops from the remaining granulosa cells of the follicle. During the first two days of the cycle, these cells continue to develop and the increasing level of progesterone in the serum correlates with this development (Leavitt et al., 1973). However, by day 3 regression of the corpora lutea occurs. There is a decrease in cell size and progesterone synthetic activity. By day 4 extensive phagocytic activity and cell autolysis are observed. Continued secretion of progesterone by the corpus luteum requires daily injection of FSH and prolactin (Greenwald, 1967; Choudary and Greenwald, 1967). Mating stimuli are necessary to initiate the required daily surge of prolactin. Breakdown of the corpora lutea (luteolysis) appears to result from estrogen action (Choudary and Greenwald, 1968).

5. SEASONAL ANESTRUS

The number of hours of light during each 24 hr has no effect on the age at which sexual maturity is reached nor does blinding of the animal (Darrow et al., 1980). Only after animals are more than seven weeks of age does light affect the reproductive cycle. Exposure of reproductively mature hamsters to short photoperiods results in acyclicity after several months (Seegal and Goldman, 1975). Blinding produces a similar effect (Reiter et al., 1976). The acyclic animals show diurnal fluctuations in serum gonadotropin. This daily surge of LH and FSH in the acyclic animal is replaced by a surge every four days—the four-day estrous cycle when the animals resume cyclicity. This can be reinduced by exposure to long days or will occur spontaneously even in the complete absence of light since animals eventually become refractory to the short-day stimulus (Stetson and Tate-Oshoff, 1981). Animals maintained on long days will continue to be reproductively active indefinitely. However, after 11 or more weeks exposure to long days, the animals are again sensitive to the short-day signal that will now result in gonadal regression. In the hamster the pineal is essential for expression of the photoperiodic effects. If melatonin injections are given, the short day effects can be prevented as can also the effects of blinding (Reiter et al., 1976, 1977). Thus, the pineal appears to mediate the photoperiodic effect on gonadotropin release via its regulation of melatonin synthesis and release.

In the hamster rendered acyclic as a result of exposure to short photoperiods, the diurnal surges of LH and FSH persist following ovariectomy and adrenalectomy (Bittman and Goldman, 1979). Thus, steroid feedback is probably not involved in the regulation of the daily gonadotropin surges. Prolactin is not detectable in the serum of acyclic hamsters (Widmaier and Campbell, 1981). Introduction of Silastic estrogen-containing capsules elevates serum prolactin in hamsters maintained on either short or long days. Thus, the loss of detectable prolactin in short days is due to the lack of estrogen secretion in the acyclic females.

6. CONCLUSIONS

Those factors that control the onset of sexual maturation and regulate the annual breeding season differ in some respects. Photoperiod is a major "zeitgeiber" for regulation of the length of the annual breeding season (> 12.5 hr light per day are required in the hamster). However, light does not influence the age at sexual maturation (Darrow *et al.*, 1980), nor does the absence of light block the onset of a breeding season. Under long-day conditions, hamsters breed throughout the year. Exposure to long-day length makes the hamster sensitive to short-day length (Stetson *et al.*, 1977), which will result in the cessation of estrous cycles. However, even if long-day length does not recur after five months, the gonads once again become active (Stetson and Tate-Oshoff, 1981). Therefore, the main function of light is to sensitize animals to short-day length that will terminate breeding prior to the onset of environmental conditions that are unsuitable for raising young.

The immature animal prior to sexual maturity, the mature animal during seasonal anestrus induced by short photoperiod, and the breeding animal during pregnancy or pseudopregnancy all show a common feature in that there is a daily blip in gonadotropin release (Smith *et al.*, 1980, 1982). During the breeding season there is a surge of gonadotropin release every fourth day, thus the estrous cycle differs from anestrus by there being a suppression of the daily blip in gonadotropin release and initiation of a much larger release every fourth day. This indicates that the sensitivities of some elements in the feedback control that regulates gonadotropin release change. Although the expression of the change, i.e., daily gonadotropin release versus gonadotropin release every fourth day, is similar for the immature animal, anestrus, pregnancy, and pseudopregnancy, no evidence is currently available to indicate whether or not there is a common basis for this change.

During the cycle, gonadotropin release is regulated via feedback effects of the gonadal steroids on the brain and pituitary (Keever and Greenwald, 1967; Bast and Greenwald, 1977; Vomachka and Greenwald, 1978, 1980; Chappel *et al.*, 1978). Ovariectomy of animals maintained on short photoperiods does not eliminate the daily surge of gonadotropins (Bittman and Goldman, 1979). Estrogen implants increase prolactin release in animals whether maintained on long or short photoperiods (Widmaier and Campbell, 1981). This suggests that short photoperiods do not simply block steroid feedback effects; however, more work needs to be done to determine whether changes in feedback sensitivity to steroids occur as a result of change in length of photoperiod.

Integration of light effects requires an intact suprachiasmatic nucleus (Stetson and Watson-Whitmyre, 1976). Although steroid-accumulating neurons (Kreiger *et al.*, 1976) and GnRH-producing neurons (Silverman *et al.*, 1979; Jennes and Stumpf, 1980) are found throughout the hypothalamus, neither appear to be present within the suprachiasmatic nucleus. Thus, the neurons of the suprachiasmatic nucleus appear to affect the function of neurons located elsewhere in the hypothalamus to regulate the release of gonadotropins from the pituitary.

The pineal is another organ that is essential to light-induced changes in the estrous cycle. Removal of the pineal prevents gonadal regression in hamsters exposed to short photoperiods (Hoffman and Reiter, 1965). Since the mammalian pineal appears to lack neural efferents, its influence on the reproductive cycle must be chemically mediated (Reiter, 1972). The pineal is innervated via the superior cervical ganglion, ablation of which produces the same effect as pineallectomy indicating that neural input is necessary for the pineal effect on reproduction.

One product of the pineal, an indoleamine, melatonin, can cause gonadal regression or recrudescence depending on dosage, time of injection, and the photoperiod on which the hamster is being maintained (Tamarkin et al., 1976, 1977; Turek and Campbell, 1979). Melatonin injections parallel the effects of photoperiod. In both cases the changes appear only after several weeks of treatment and in the female hamster include the appearance of daily afternoon peaks in LH and FSH secretion (Goldman et al., 1979). For the male hamster it has been demonstrated that if daily melatonin treatment is continued, after about 20 weeks the system escapes and gonadal recrudescence occurs, a situation similar to that found for photoperiod. Photorefractory hamsters do not respond to melatonin (Bittman, 1978).

Melatonin secretion and release follow a diurnal pattern with peak secretion at night (Klein and Weller, 1970; Rollag and Niswender, 1976). Exposure to light promptly decreases melatonin output. This suggests the melatonin rhythm might serve as an indicator of photoperiod; however, experiments in the hamster indicate that changes in photoperiod are not closely reflected by changes in melatonin secretion (Tamarkin et al., 1979; Panke et al., 1979). The site of action of melatonin is not known.

In the past 15 years much progress has been made in delineating those factors that regulate the estrous cycle. Most of the critical stimuli have been determined and their sites of action identified. Much less certainty exists concerning the mechanism that regulates the onset and termination of the annual reproductive season. Initiation of breeding does not require a photoperiodic cue. However, long photoperiod sensitizes the animals to short photoperiod so that breeding is terminated before the external environment becomes unsuitable for raising young.

A pineal indoleamine, melatonin, produces effects on the reproductive cycle on a similar time scale to the photoperiodic effects. Light affects the release of melatonin. Both short photoperiod and daily melatonin injections must be presented for several weeks before reproduction ceases. More work is required to identify whether the photoperiod changes the sensitivity of the system to hormonal stimuli and whether melatonin may serve as a masking agent through which the change in sensitivity to hormonal stimuli is affected.

Finally, the mechanism of action at the cellular level may involve change in numbers of receptors, affinity of receptors, or a whole host of other biochemical changes that affect the efficiency of transmission between cells. Mechanisms of action at the molecular level are ripe for exploitation once the necessary stimuli and their sites of action have been identified.

REFERENCES

Adams, T. E., and Spies, H. G., 1981a, Binding characteristics of gonadotropin-releasing hormone receptors throughout the estrous cycle of the hamster, *Endocrinology* 108:2245–2253.

Adams, T. E., and Spies, H. G., 1981b, GnRH-induced regulation of GnRH receptor concentration in the phenobarbital-blocked hamster, *Biol. Reprod.* 25:298–302.

Adler, S., 1948, Origin of the golden hamster, *Cricetus auratus* as a laboratory animal, *Nature (London)* 162:256–257.

Alleva, J. J., and Umberger, E. J., 1966, Evidence for neural control of the release of pituitary ovulating hormone in the golden Syrian hamster, *Endocrinology* 78:1125–1129.

Alleva, F. R., Alleva, J. J., and Umberger, E. J., 1969, Effect of a single prepubertal injection of testosterone propionate on later reproductive functions of the female golden hamster, *Endocrinology* 85:312–318.

Alleva, J. J., Waleski, M. V., Alleva, F. R., and Umberger, E. J., 1968, Synchronizing effect of photoperiodicity on ovulation in hamsters, *Endocrinology* 82:1227–1235.

Alleva, J. J., Bonventre, P. F., and Lamanna, C., 1979, Inhibition of ovulation in hamsters by the protein synthesis inhibitors Diptheria toxin and cycloheximide, *Proc. Soc. Exp. Biol. Med.* 162:170–174.

Arimura, A., Debeljuk, L., and Schally, A. V., 1972, LH release by LH-releasing hormone in golden hamsters at various stages of estrous cycle, *Proc. Soc. Exp. Biol. Med.* 140:609–612.

Austin, C. R., 1955, Ovulation, fertilization, and early cleavage in the hamster (*Mesocricetus auratus*), *J. Roy. Microbiol. Soc.* 73:141–154.

Baranczuk, R., and Greenwald, G. S., 1973, Peripheral levels of estrogen in the cyclic hamster, *Endocrinology* 92:805–812.

Barros, C., and Austin, C. R., 1968, Inhibition of ovulation by systematically administered actinomycin D in the hamster, *Endocrinology* 83:177–179.

Bast, J. D., 1979, A comprehensive profile of peripheral progesterone levels in cyclic hamsters sequentially bled at 2-hr intervals, *Proc. Soc. Exp. Biol. Med.* 162:199–201.

Bast, J. D., and Greenwald, G. S., 1974, Serum profiles of follicle-stimulating hormone, luteinizing hormone and prolaction during the estrous cycle in the hamster, *Endocrinology* 94:1295–1299.

Bast, J. D., and Greenwald, G. S., 1977, Acute and chronic elevation in serum levels of FSH after unilateral ovariectomy in the cyclic hamster, *Endocrinology* 100:955–966.

Bex, F. J., and Goldman, B. D., 1975, Serum gonadotropins and follicular development in the Syrian hamster, *Endocrinology* 96:928–933.

Bex, F. J., and Goldman, B. D., 1977, Serum gonadotropins associated with puberty in the female Syrian hamster, *Biol. Reprod.* 16:557–560.

Bittman, E. L., 1978, Hamster refractoriness: The role of insensitivity of pineal target tissues, *Science* 202:648–650.

Bittman, E. L., and Goldman, B. D., 1979, Serum levels of gonadotropins in hamsters exposed to short photoperiods: Effects of adrenalectomy and ovariectomy, *J. Endocrinol.* 83:113–118.

Bodemer, C. W., Rumery, R. E., and Blandau, R. L., 1959, Studies on induced ovulation in the intact immature hamster, *Fertil. Steril.* 10:350–360.

Bond, C. R., 1945, The golden hamster (*Cricetus auratus*) care, breeding and growth, *Physiol. Zool.* 18:52–59.

Bosley, C. G., and Leavitt, W. W., 1972, Dependence of preovulatory progesterone on critical period in the cyclic hamster, *Am. J. Physiol.* 222:129–133.

Brom, G. M., and Schwartz, N. B., 1968, Acute changes in the estrous cycle following ovariectomy in the golden hamster, *Neuroendocrinology* 3:366–377.

Buhl, A. E., Norman, R. L., and Resko, J. A., 1978, Sex differences in estrogen-induced gonadotropin release in hamsters, *Biol. Reprod.* 18:592–597.

Chappel, S. C., 1979, Cyclic fluctuations in ovarian FSH inhibiting material in golden hamsters, *Biol. Reprod.* 21:447–453.

Chappel, S. C., and Selker, F., 1979, Relation between the secretion of FSH during the periovulatory period and ovulation during the next cycle, *Biol. Reprod.* 21:347–352.

Chappel, S. C., Norman, R. L., and Spies, H. G., 1977, Regulation of the second (estrous) release of follicle-stimulating hormone in hamsters by the medial basal hypothalamus, *Endocrinology* 101:1339–1342.

Chappel, S. C., Norman, R. L., and Spies, H. G., 1978, Effects of estradiol on serum and pituitary gonadotropin concentrations during selective elevations of follicle stimulating hormone, *Biol. Reprod.* 19:159–166.

Chappel, S. C., Norman, R. L., and Spies, H. G., 1979, Evidence for a specific neural event that controls the estrous release of follicle-stimulating hormone in golden hamsters, *Endocrinology* **104**:169–174.

Choudary, J. B., and Greenwald, G. S., 1967, Effect of an ectopic pituitary gland on luteal maintenance in the hamster, *Endocrinology* **81**:542–552.

Choudary, J. B., and Greenwald, G. S., 1968, Comparison of the luteolytic action of LH and estrogen in the hamster, *Endocrinology* **83**:129–136.

Ciaccio, L. A., and Lisk, R. D., 1971, The hormonal control of cyclic estrus in the female hamster, *Am. J. Physiol.* **221**:936–942.

Coutifaris, C., and Chappel, S. C., 1982, Intraventricular injection of follicle-stimulating hormone (FSH) during proestrus stimulates the rise in serum FSH on entrus in phenobarbital-treated hamsters through a central nervous system-dependent mechanism, *Endocrinology* **110**:105–113.

Darrow, J. M., Davis, F. C., Elliott, J. A., Stetson, M. H., Turek, F. W., and Menàker, M., 1980, Influence of photoperiod on reproductive development in the golden hamster, *Biol. Reprod.* **22**:443–450.

Diamond, M., and Yanagimachi, R., 1970, Reproductive development in the female golden hamster in relation to spontaneous estrus, *Biol. Reprod.* **2**:223–229.

Doherty, P. C., and Sheridan, P. J., 1981, Uptake and retention of androgens in neurons of the brain of the golden hamster, *Brain Res.* **219**:327–334.

Everett, J. W., Sawyer, C. H., and Markee, J. E., 1949, A neurogenic timing factor in control of the ovulatory discharge of luteinizing hormone in the cyclic rat, *Endocrinology* **44**:234–250.

Goldman, B. D., and Porter, J. C., 1970, Serum LH levels in intact and castrated golden hamsters, *Endocrinology* **87**:676–679.

Goldman, B. D., and Sheridan, P. I., 1974, The ovulatory surge of gonadotropin and sexual receptivity in the female golden hamster, *Physiol. Behav.* **12**:991–995.

Goldman, B. D., Hall, V., Hollister, C., Roychoudbury, P., Tamarkin, L., and Westrom, W., 1979, Effects of melatonin on the reproductive system in Syrian hamsters maintained under various photoperiods, *Endocrinology* **104**:82–88.

Greenwald, G. S., 1960, The effects of unilateral ovariectomy on follicular maturation in the hamster, *Endocrinology* **66**:89–95.

Greenwald, G. S., 1961, Quantitative study of follicular development in the ovary of the intact or unilaterally ovariectomized hamster, *J. Reprod. Fertil.* **2**:351–361.

Greenwald, G. S., 1962a, Analysis of superovulation in the adult hamster, *Endocrinology* **71**:378–389.

Greenwald, G. S., 1962b, Temporal relationship between unilateral ovariectomy and the ovulatory response of the remaining ovary, *Endocrinology* **71**:664–666.

Greenwald, G. S., 1967, Luteotropic complex of the hamster, *Endocrinology* **80**:118–130.

Greenwald, G. S., 1971, Preovulatory changes in ovulating hormone in the cyclic hamster, *Endocrinology* **88**:671–677.

Greenwald, G. S., 1973, Effect of an anti-PMS serum on ovulation and estrogen secretion in the PMS-treated hamster, *Biol. Reprod.* **9**:437–446.

Greenwald, G. S., 1975, Proestrous hormone surges dissociated from ovulation in the estrogen-treated hamster, *Endocrinology* **97**:878–884.

Greenwald, G. S., 1977, Exogenous progesterone: Influence on ovulation and hormone levels in the cyclic hamster, *J. Endocrinol.* **73**:151–155.

Greenwald, G. S., 1978, Modification by exogenous progesterone of estrogen and gonadotropin secretion in the cyclic hamster, *Endocrinology* **103**:2315–2322.

Greenwald, G. S., and Papkoff, H., 1980, Induction of ovulation in the hypophysectomized proestrous hamster by purified FSH or LH, *Proc. Soc. Exp. Biol. Med.* **165**:391–393.

Greenwald, G. S., and Peppler, R. D., 1968, Prepubertal and pubertal changes in the hamster ovary, *Anat. Rec.* **161**:447–458.

Hamilton, M. A., Vomachka, A. J., Lisk, R. D., and Gòrski, R. A., 1981, Effect of neonatal intrahypothalamic testosterone implants on cyclicity and adult sexual behavior in the female hamster, *Neuroendocrinology* **32**:234–241.

Hammond, J., Jr., 1954, Light regulation of hormone secretion, *Vitam. Horm.* **12**:157–206.

Harris, G. W., 1964, Sex hormones, brain development and brain function, *Endocrinology* **75**:627–648.

Harris, G. W., and Jacobson, D., 1952, Functional grafts of the anterior pituitary gland, *Proc. Roy. Soc. (London) Ser., B* 139:263–276.

Hoffman, R. A., Robinson, P. F., and Magalhaes, H. (eds.), 1968, *The Golden Hamster, Its Biology and Use in Medical Research,* Iowa State University Press, Ames, Iowa.

Hoffman, R. H., and Reiter, R. J., 1965, Pineal gland: Influence on gonads of male hamsters, *Science* 148:1609–1611.

Humphrey, R. R., Dermody, W. C., Brink, H. O., Bousley, F. G., Schottin, N. H., Sakowski, R., Vaitkus, J. W., Veloso, H. T., and Reel, J. R., 1973, Induction of luteinizing hormone (LH) release and ovulation in rats, hamsters, and rabbits by synthetic luteinizing hormone-releasing factor (LRF), *Endocrinology* 92:1515–1525.

Jennes, L., and Stumpf, W. E., 1980, LHRH-systems in the brain of the golden hamster, *Cell Tissue Res.* 209:239–256.

Keever, J. E., and Greenwald, G. S., 1967, Effect of oestrogen and progesterone on pituitary gonadotrophic content of the cyclic hamster, *Acta Endocrinol.* 56:244–254.

Kent, G. C., Ridgway, P. M., and Strobel, E. F., 1968, Continual light and constant estrus in hamsters, *Endocrinology* 82:699–703.

Klein, D. C., and Weller, J., 1970, Indole metabolism in the pineal gland: A circadian rhythm in N-acetyltransferase, *Science* 169:1093–1095.

Knigge, K. M., and Leatham, J. H., 1956, Growth and atresia of follicles in the ovary of the hamster, *Anat. Rec.* 124:679–707.

Krieger, M. S., Morrell, J. I., and Pfaff, D. W., 1976, Autoradiographic localization of estradiol-concentrating cells in the female hamster brain, *Neuroendocrinology* 22:193–205.

Lamperti, A. A., and Baldwin, D. M., 1979, The effects of gonadal steroids on gonadotropin secretion in hamsters with a lesion of the arcuate nucleus of the hypothalamus, *Endocrinology* 104:1041–1045.

Lamperti, A., and Blaha, G., 1980, Further observation on the effects of neonatally administered monosodium glutamate on the reproductive axis of hamsters, *Biol. Reprod.* 22:687–693.

Lamperti, A., Pupa, L., and Tafelski, T., 1980, Time-related effects of monosodium glutamate on the reproductive neuroendocrine axis of the hamster, *Endocrinology* 106:553–558.

Landauer, M. R., Atlas, A. I., and Liu, S., 1981, Effects of prenatal and neonatal androgen on estrous cyclicity and attractiveness of female hamsters, *Physiol. Behav.* 27:419–424.

Leavitt, W. W., Bosley, G. G., and Blaha, G. C., 1971, Source of ovarian preovulatory progesterone, *Nature New Biol.* 234:283–284.

Leavitt, W. W., Basom, C. R., Bagwell, J. N., and Blaha, G. C., 1973, Structure and function of the hamster corpus luteum during the estrous cycle, *Am. J. Anat.* 136:235–249.

Lippmann, W., 1968, Relationship between hypothalamic norepinephrine and serotonin and gonadotrophin secretion in the hamster, *Nature (London)* 218:173–174.

Lisk, R. D., and Ferguson, D. S., 1973, Neural localization of estrogen-sensitive sites for inhibition of ovulation in the golden hamster *Mesocricetus auratus, Nueroendocrinology* 12:157–160.

Long, J. A., and Evans, H. M., 1922, The oestrous cycle of the rat and its associated phenomena, *Mem. Univ. Calif.* 6:1–148.

Nakano, A., 1960, Histological studies of the prenatal and postnatal development of the ovary of the golden hamster (*Cricetus auratus*), *Okajimas Folia Anat. Jpn.* 35:183–217.

Norman, R. L., 1975, Estrogen and progesterone effects on the neural control of the preovulatory LH release in the golden hamster, *Biol. Reprod.* 13:218–222.

Norman, R. L., and Greenwald, G. S., 1971, Effect of phenobarbital hypophysectomy and X-irradiation on preovulatory progesterone levels in the cyclic hamster, *Endocrinology* 89:598–605.

Norman, R. L., and Spies, H. G., 1974, Neural control of the estrogen-dependent twenty-four hour periodicity of LH release in the golden hamster, *Endocrinology* 95:1367–1372.

Norman, R. L., Blake, C. A., and Sawyer, C. H., 1972, Effects of hypothalamic deafferentation on LH secretion and the estrous cycle in the hamster, *Endocrinology* 91:95–100.

Norman, R. L., Blake, C. A., and Sawyer, C. H., 1973a, Evidence for neural sites of action of phenobarbital and progesterone on LH release in the hamster, *Biol. Reprod.* 8:83–86.

Norman, R. L., Blake, C. A., and Sawyer, C. H., 1973b, Estrogen-dependent, twenty-four hour periodicity in pituitary LH release in the female hamster, *Endocrinology* 93:965–970.

Nucci, L. P., and Beach, F. A., 1971, Effects of prenatal androgen treatment on mating behavior in female hamsters, *Endocrinology* 88:1514–1515.

Oritiz, E., 1947, The postnatal development of the reproductive system of the golden hamster (*Cricetus auratus*) and its reactivity to hormones, *Physiol. Zool.* 20:45–66.

Orsini, M. W., 1961, The external vaginal phenomena characterizing the stages of the estrous cycle, pregnancy, pseudopregnancy, lactation, and the anestrous hamster *Mesocricetus auratus* Waterhouse, *Proc. Anim. Care Panel* 11:193–206.

Oxberry, B. A., and Greenwald, G. S., 1982, An autoradiographic study of the binding of ^{125}I-labeled follicle-stimulating hormone, human chorionic gonadotropin and prolactin in the hamster ovary throughout the estrous cycle, *Biol. Reprod.* 27:505–516.

Pandergrass, P. B., and Reber, M., 1980, Scanning electron microscopy of the graafian follicle during ovulation in the golden hamster, *J. Reprod. Fertil.* 59:21–24.

Panke, E. S., Rollag, M. D., and Reiter, R. J., 1979, Pineal melatonin concentrations in the Syrian hamster, *Endocrinology* 104:194–197.

Printz, R. H., and Greenwald, G. S., 1971, A neural mechanism regulating follicular development in the hamster, *Neuroendocrinology* 7:171–182.

Ramaley, J. A., 1979, Development of gonadotropin regulation in the prepubertal mammal, *Biol. Reprod.* 20:1–31.

Rani, C. S. S., and Moudgal, N. R., 1977, Role of the proestrus surge of gonadotropins in the initiation of follicular maturation in the cyclic hamster: A study using antisera to follicle stimulating hormone and luteinizing hormone, *Endocrinology* 101:1484–1494.

Reid, L. N., Blake, C. A., and Sawyer, C. A., 1972, Delay of the proestrous ovulatory surge of LH in the hamster by pentobarbital or ether, *Endocrinology* 91:1025–1029.

Reiter, R. J., 1972, Surgical procedures involving the pineal gland which prevent gonadal degeneration in adult male hamsters, *Ann. Endocrinol.* 33:571–581.

Reiter, R. J., Rudeen, P. K., and Vaughan, M. K., 1976, Restoration of fertility in light-deprived female hamsters by chronic melatonin treatment, *J. Comp. Physiol.* 111:7–13.

Reiter, R. J., Vaughan, M. K. and Waring, P. J., 1977, Prevention by melatonin of short-day induced atropy of the reproductive systems of male and female hamsters, *Acta Endocrinol,* 84:410–418.

Reuter, L. A., and Lisk, R. D., 1973a, A biphasic effect of progesterone on ovulation in the hamster, *Fed.Proc.* 32:230.

Reuter, L. A., and Lisk, R. D., 1973b, Progesterone-sensitive loci for blockage of ovulation in the hamster, *Neuroendocrinology* 12:17–29.

Reuter, L. A., Ciaccio, L. A., and Lisk, R. D., 1970, Progesterone: Regulation of estrous cycle, ovulation and estrous behavior in the golden hamster, *Endocrinology* 86:1287–1297.

Ridley, K., and Greenwald, G. S., 1975, Progesterone levels measured every two hours in the cyclic hamster, *Proc. Soc. Exp. Biol. Med.* 149:10–12.

Rollag, M. D., and Niswender, G. D., 1976, Radioimmunoassay of serum concentrations of melatonin in sheep exposed to different lighting regimes, *Endocrinology* 98:482–489.

Saidapur, S. K., and Greenwald, G. S., 1978, Peripheral blood and ovarian levels of sex steroids in the cyclic hamster, *Biol. Reprod.* 18:401–408.

Saidapur, S. K., and Greenwald, G. S., 1979, Regulation of 17 β-estradiol synthesis in the proestrous hamster: Role of progesterone and luteinizing hormone, *Endocrinology* 105:1432–1439.

Schwartz, N. B., and Channing, C. P., 1977, Evidence for ovarian "inhibin": Suppression of the secondary rise in serum follicle stimulating hormone levels in proestrous rats by injection of porcine follicular fluid, *Proc. Natl. Acad. Sci. U.S.A.* 74:5721–5724.

Seegal, R. F., and Goldman, B. D., 1975, Effects of photoperiod on cyclicity and serum gonadotropins in the Syrian hamster, *Biol. Reprod.* 12:223–231.

Selle, R. M., 1945, Hamster sexually mature at twenty-eight days of age, *Science* 102:485–486.

Shander, D., and Goldman, B., 1978, Ovarian steroid modulation of gonadotropin secretion and pituitary responsiveness to luteinizing hormone releasing hormone in the female hamster, *Endocrinology* 103:883–893.

Shander, D., and Goldman, B., 1979, Serum gonadotropins and progesterone in proestrous hamsters compared to hamsters infused with LH-RH, *Life Sci.* 24:525–534.

Siegel, H. I., Bast, J. D., and Greenwald, G. S., 1976, The effects of phenobarbital and gonadal steroids on periovulatory serum levels of luteinizing hormone and follicle-stimulating hormone in the hamster, *Endocrinology* 98:48–55.

Silverman, A. J., Kray, L. C., and Zimmerman, E. A., 1979, A comparative study of the luteinizing hormone releasing hormone (LHRH) neuronal networks in mammals, *Biol. Reprod.* 20:98–110.

Smith III, S. G., Shalger, G., and Stetson, M. H., 1980, Maturation of the clock-timed gonadotropin release mechanism in hamsters: A key event in the pubertal process, *Endocrinology* 107:1334–1337.

Smith, III, S. G., Matt, K. S., Prestowitz, W. F., and Stetson, M. H., 1982, Regulation of tonic gonadotropin release in female hamsters, *Endocrinology* 110:1262–1267.

Stetson, M. H., and Tate-Oshoff, B., 1981, Hormonal regulation of the annual reproductive cycle of golden hamsters, *Gen. Comp. Endocrinol.* 45:329–344.

Stetson, M. H., and Watson-Whitmyre, M., 1976, Nucleus suprachiasmaticus: The biological clock in the hamster, *Science* 191:197–199.

Stetson, M. H., and Watson-Whitmyre, M., 1977, The neural clock regulating estrous cyclicity in hamsters: Gonadotropin release following barbiturate blockade, *Biol. Reprod.* 16:536–542.

Stetson, M. H., Watson-Whitmyre, M., and Matt, K. S., 1977, Termination of photorefractoriness in golden hamsters: Photoperiodic requirements, *J. Exp. Zool.* 202:81–87.

Stetson, M. H., Watson-Whitmyre, M., and Matt, K. S., 1978, Cyclic gonadotropin release in the presence and absence of estrogenic feedback in ovariectomized golden hamsters, *Biol. Reprod.* 19:40–50.

Stetson, M. H., Watson-Whitmyre, M., Pipinto, M. N., and Smith III, S. G., 1981, Daily luteinizing hormone release in ovariectomized hamsters: Effect of barbiturate blockade, *Biol. Reprod.* 24:139–144.

Swanson, H. H., 1966, Modification of the reproductive tract of hamsters of both sexes by neonatal administration of androgen or oestrogen, *J. Endocrinol.* 36:327–328.

Swanson, H. H., 1970, Effects of castration at birth in hamsters of both sexes on luteinization of ovarian implants, estrous cycles and sexual behaviour, *J. Reprod. Fertil.* 21:183–186.

Swanson, H. H., and Brayshaw, J. S., 1974, Effects of brain implants of testosterone propionate in newborn hamsters on sexual differentiation, *Adv. Biosci.* 13:119–137.

Tafelski, T. J., and Lamperti, A. A., 1977, The effects of a single injection of monosodium glutamate on the reproductive neuroendocrine axis of the female hamster, *Biol. Reprod.* 17:404–411.

Tamarkin, L., Westrom, W. K., Hamill, A. I., and Goldman, B. D., 1976, Effect of melatonin on the reproductive systems of male and female Syrian hamsters: A diurnal rhythm in sensitivity to melatonin, *Endocrinology* 99:1534–1541.

Tamarkin, L., Hollister, C. W., Lefebvre, N. G., and Goldman, B. D., 1977, Melatonin induction of gonadal quiescence in pinealectomized Syrian hamsters, *Science* 198:953–955.

Tamarkin, L., Reppert, S. M., and Klein, D. C., 1979, Regulation of pineal melatonin in the Syrian hamster, *Endocrinology* 104:385–389.

Terranova, P. F., 1980, Effects of phenobarbital-induced ovulatory delay on the follicular population and serum levels of steroids and gonadotropins in the hamster: A model for atresia, *Biol. Reprod.* 23:91–99.

Terranova, P. F., Connor, J. S., and Greenwald, G. S., 1978, *In vitro* steroidogenesis in corpora lutea and nonluteal ovarian tissues of the cyclic hamster, *Biol. Reprod.* 19:245–255.

Turek, F. W., and Campbell, C. S., 1979, Photoperiodic regulation of neuroendocrine-gonadal activity, *Biol. Reprod.* 20:32–50.

Turgeon, J., and Greenwald, G. S., 1972, Preovulatory levels of plasma LH in the cyclic hamster, *Endocrinology* 90:657–662.

VanVeen, T. H., Brackmann, M., and Moghimzadeh, E., 1978, Postnatal development of the pineal organ in the hamsters *Phodopus sungorus* and *Mesocricetus auratus*, *Cell Tissue Res.* 189:241–250.

Varavudhi, P., and Meites, J. 1974, Serum LH, FSH and prolactin levels during the estrous cycle in the hamster, *Proc. Soc. Exp. Biol. Med.* 145:571–573.

Vomachka, A. J., and Greenwald, G. S., 1978, Acute negative feedback effects of ovarian steroid removal and replacement in the cyclic female hamster, *Biol. Reprod.* **19:**1040–1045.

Vomachka, A. J., and Greenwald, G. S., 1979, The development of gonadotropin and steroid hormone patterns in male and female hamsters from birth to puberty, *Endocrinology* **105:**960–966.

Vomachka, A. J., and Greenwald, G. S., 1980, Negative feedback effects of ovarian steroids in the chronically ovariectomized hamster, *Biol. Reprod.* **22:**1127–1135.

Widmaier, E. P., and Campbell, C. S., 1981, The interaction of estradiol and daylength in modifying serum prolactin secretion in female hamsters, *Endocrinology* **108:**371–376.

Wurtman, R. J., 1967, Effects of light and visual stimuli on endocrine function, in: *Neuroendocrinology,* Volume 2 (L. Martini and W. F. Ganong, eds.), Academic Press, New York, pp. 19–59.

3

Endocrinology of the Pregnant Hamster

GILBERT S. GREENWALD

The objective of this chapter is to provide a comprehensive account of reproductive endocrinology of the pregnant hamster and to point out the major differences from the rat, which has been the most extensively studied rodent. As the chapter unfolds it should become clear that major differences exist between the two species.

1. COPULATORY BEHAVIOR IN THE ESTROUS HAMSTER

The signal that transforms the corpora lutea (CL) of the cycle to CL of pseudopregnancy or pregnancy is copulation on the evening of proestrus. One of the earliest observations of the unusual mating behavior of the hamster was provided by Chang and Sheaffer (1957), who reported that the male may mate as many as 50 times in an hour resulting in the depletion of 40% to 45% of sperm from the epididymis. The first five copulations usually result in no sperm or just a few deposited in the female reproductive tract with the highest sperm counts observed in the 11th to 30th copulations. Only 20% of females were pregnant after one ejaculatory series that consists of ten intromissions followed by ejaculation (Lanier *et al.,* 1975). In contrast, all females receiving at least seven ejaculations became pregnant. Conversely, the rat requires only one ejaculatory series to ensure pregnancy (Adler, 1969).

Mechanical stimulation of the vagina with an electric vibrator was most successful in inducing pseudopregnancy in the hamster when the female was stimulated 30 times over a total of 300 sec (Diamond and Yanagamachi, 1968). In a very interesting study, Diamond (1970) analyzed the "vaginal code" necessary to induce pseudopregnancy in mice; i.e., the duration and pattern of cervical stimulation required to transform the CL of the cycle to CL of pseudopregnancy. In the mouse, optimal results were obtained

GILBERT S. GREENWALD • Department of Physiology, University of Kansas Medical Center, Kansas City, Kansas 66103.

with three 20-sec insertions with 30 sec between insertions. In contrast, the formula for the hamster was 30 5-sec insertions with 20-sec intervals between each insertion. Obviously, the hamster requires a much different pattern of neural input via cervical afferent nerves than the rat or mouse to trigger the release of the initial luteotropic stimulus from the anterior pituitary.

Cyclic hamsters displaying behavioral estrus and that were artificially inseminated and injected with 5 or 10 mg of progesterone very rarely delivered at term in contrast to animals that received vaginal stimulation (per vibrator) plus artificial insemination (Diamond, 1972). It was concluded that vaginal stimulus is somehow remembered and translated at the end of gestation into a signal initiating parturition. A much more likely explanation for the failure of the former group of hamsters to maintain pregnancy is that the CL of the estrous cycle were not transformed into functional CL and therefore rapidly regressed. Hence, when the circulating titers of exogenous progesterone began to wane at midpregnancy, even if the placentae provided any luteotropic hormone(s), there were no CL capable of responding.

What is the luteotropic hormone that is released in response to the elaborate copulatory ritual? A clue is provided by comparing serum levels of follicle-stimulating hormone (FSH), luteinizing hormone (LH), and prolactin (PRL) between the four days of the estrous cycle (day 1 = morning of ovulation; day 4 afternoon = proestrus) and the first four days of pregnancy (Table 1). Throughout this chapter, day 1 of pregnancy is defined as the day on which spermatozoa are present in the vaginal lavage. It is evident from Table 1 that on day 2 of pregnancy serum FSH is approximately threefold higher than during the cycle, whereas the concentration of LH is reduced 40% to 50% from cyclic values. Injection of 50 μg progesterone on day 3 of the hamster cycle (when the hormone has normally dropped to undetectable levels) results in the four-day cycle extending to five days and is associated with a concomitant drop in serum LH and an increase in FSH (Greenwald, 1978) comparable to the changes shown in Table 1. Hence, sustained secretion of progesterone alters the ratio of circulating FSH and LH. Prolactin is the peptide hormone that differs most drastically between the first four days

Table 1. Comparison of FSH, LH, and PRL Levels in Cyclic and Pregnant Hamsters (Means ± S.E.M.)[a]

Hormone	Reproductive state	Hormone concentrations (ng/ml serum) at 09:00 hr on			
		Day 1	Day 2	Day 3	Day 4
FSH	Cyclic	965 ± 97(6)[b]	94 ± 12(6)	43 ± 5(5)	51 ± 6(6)
	Pregnant	1087 ± 59(6)	204 ± 32(6)	197 ± 10(6)	186 ± 16(6)
LH	Cyclic	100 ± 9	77 ± 7	87 ± 8	95 ± 6
	Pregnant	65 ± 10	52 ± 5	43 ± 4	46 ± 4
Prolactin	Cyclic	13.4 ± 2.2	9.2 ± 0.9	11.9 ± 1.5	17.3 ± 4.0
	Pregnant	27.9 ± 1.6	46.0 ± 3.9	51.3 ± 3.4	60.2 ± 6.2

[a]Data from Bast and Greenwald, 1974.
[b]Number of animals in which FSH, LH, and PRL levels were determined in the same animal, in parentheses.

of the cycle and gestation (Table 1) and therefore appears to be the initial luteotropic stimulus acting in conjunction with FSH and LH. There is a significant diurnal variation in serum PRL at least until day 8 of pregnancy with peak levels observed at 1900 hr in hamsters maintained on a lighting schedule of 14 hr light and 10 hr dark with lights on at 0500 hr (Terranova and Greenwald, 1979). In contrast, the pregnant rat shows two daily surges of PRL: during the light as well as in the dark period (for references, see Voogt *et al.*, 1982).

Two other lines of evidence point to PRL as the initial luteotropic stimulus in the hamster that converts the CL of the cycle to CL of pseudopregnancy or pregnancy. Transplanting a single pituitary to the kidney (PRL is secreted continuously and is the only hormone produced in appreciable quantities by the ectopic pituitary) results by the second cycle in the establishment of recurrent pseudopregnancies lasting eight to ten days (Choudary and Greenwald, 1967). Moreover, if the hamsters were mated at the end of the second pseudopregnant cycle, and the animals hypophysectomized on day 4 of pregnancy, daily injection of 50 or 200 μg FSH plus PRL (secreted by the transplanted pituitary) enabled pregnancy to continue, whereas it was interrupted in hamsters injected with 1 μg LH or uninjected hypophysectomized controls. Similarly, daily injection of 4 mg PRL (an enormous dose!) plus 200 μg FSH beginning on the day of ovulation results in 75% of cyclic hamsters becoming pseudopregnant. However, lower doses of PRL or FSH alone are ineffective in inducing pseudopregnancy (Grady and Greenwald, 1968).

Injection of 1 mg ergocryptine (ECR) that prevents PRL release on day 5 interrupts pregnancy in all treated hamsters (n = 7), whereas on day 6 similar treatment results in three of eight hamsters maintaining pregnancy on day 11. Moreover, administering ECR on day 7 or 8 is ineffective in affecting gestation (Ford and Yoshinaga, 1975). Daily injection of 300 μg PRL on days 5 and 6 reverses the abortifacient action of ECR given on day 5. Similarly, in pseudopregnant hamsters injected with ECR on days 1, 2, or 3, luteolysis is rapidly induced as evidenced by a drastic decline in serum progesterone (Harris and Murphy, 1981a). In hysterectomized, pseudopregnant hamsters, a single injection of 100 μg ECR on day 8 induces luteolysis: serum levels of progesterone are halved within 6 to 9 hr although serum PRL does not diverge from control values until 12 to 15 hr after treatment (Terranova and Greenwald, 1978). This suggests that dopaminergic compounds may have some direct luteolytic action.

2. PSEUDOPREGNANCY IN THE HAMSTER

In addition to mechanical stimulation of the cervix (Diamond, 1970) pseudopregnancy can be produced in hamsters by mating with vasectomized males or by using normal males but severing the uterus bilaterally at the utero-tubal junction on the morning that sperm are detected in the vaginal smear (Lukaszewska and Greenwald, 1969). The duration of pseudopregnancy was established by checking the females daily for the appearance of the conspicuous postovulatory vaginal discharge that defines day 1 of the next cycle. Under these circumstances, the duration of pseudopregnancy was 8.9

versus 10.0 days—a significant difference. Hysterectomy on day 1 of pregnancy delays ovulation until day 17 (Greenwald, 1974). On the other hand, deferring hysterectomy until days 8 to 13 of pregnancy (Klein, 1937) or day 8 of pregnancy (Duby et al., 1969) or day 9 of pregnancy (Greenwald, 1974) results in rapid structural luteolysis within three days as evidenced by a characteristic leukocytic infiltration of the CL and reduction in the cytoplasm: nucleus ratio of the luteal cells. Evidently, if hysterectomy is performed early enough, the pituitary continues to secrete luteotropic hormone(s) longer than usual. However, once the definitive placenta has formed and it assumes some luteotropic function, hysterectomy results in rapid involution of the CL.

In the intact pseudopregnant hamster (i.e., with the uterus in situ), peripheral levels of serum progesterone fall drastically between days 6 and 8 compared with pregnant animals (Greenwald and Bast, 1978) indicating that placental luteotropic function is normally established during this period. Another study measured steroids in ovarian venous plasma in pseudopregnant hamsters and found that the concentration of progesterone drops from 367 ng/ml on day 7 to 62 ng/ml on day 8 (Shaikh et al., 1973). In contrast, there is a sharp increase in estradiol in ovarian venous plasma by the morning of day 8 (5,698 pg/ml). Shaikh et al. (1973) also measured prostaglandin F in peripheral blood and the levels increased on the evening of day 4—three days before the fall in progesterone that marked the termination of pseudopregnancy. However, there was a dramatic increase in the PGF content of the uterus on day 8 of pseudopregnancy, whereas it was undetectable on day 5 (Lukaszewska et al., 1972).

3. PREGNANCY IN THE HAMSTER—GENERAL CONSIDERATIONS

The hamster has the shortest gestation period of any eutherian mammal. The mean gestation length for one-month-old mated hamsters is 373 hr and there is a statistically significant increase in the duration of pregnancy up to 402 hr in 14-month-old animals (Soderwall et al., 1960). There is a sharp drop in mating and deliveries after 14 months of age. Similarly, when classified according to parity, primiparous hamsters show a greater percentage of 16-day pregnancies (84.3%) than multiparous hamsters (71.1%) (Czyba et al., 1970). In our colony, hamsters normally deliver by 1000 hr on day 16 of gestation compared with 1200 hr in multiparious animals with maternal behavior first elicited 2.34 hr and 6.41 hr, respectively, prior to delivery (Siegel and Greenwald, 1975).

Gestation length in the hamster is about six days less than in the rat and for prepubertal females adjusting peak serum gonadotropins levels for postmating age makes the peaks coincide for the two species (Vomachka and Greenwald, 1979). It is tempting to speculate that maturational events are unfolding according to a definite time table related to fetal development at term. There is a seasonal variation in breeding success in France with an optimal period of reproduction in May (87% pregnancies) and a seasonal low from November to May (24%) (Czyba, 1968). Orsini (1961) was the first to report an anestrous period in some laboratory-housed hamsters. The females may be of almost any age, but it is more evident in winter and following lactation. The ovaries at this time lack antral follicles and corpora lutea and are reduced in size. One wonders whether in the wild, hibernation is initiated at this time of the year? This "acyclic" ovary is characteristic of a number of anovulatory states in hamsters: lactation (Greenwald, 1965); lesions of the mamillary bodies (Printz and Greenwald, 1971); starvation (Printz and Greenwald, 1970); and presumably reduction of illumination from 14 hr to 10 hr daily (Seegal and Goldman, 1975).

Following successful mating, the fertilized eggs on day 3 are three to eight cells, and on the morning of day 4 morulae or early blastocysts (Austin, 1956). Ova can no longer be flushed from the uterus by the late morning of day 4, which therefore marks the beginning of nidation as demonstrated by cleared uterine tracts (Orsini, 1962) in which opacities develop that are rich in glycogen (Foster *et al.*, 1963). After ovariectomy in the hamster, unlike the rat, implantation can be induced with progesterone alone; estrogen is not required (Harper *et al.*, 1969). It is of interest that protein synthesis, as demonstrated by incorporation of $[^{35}S]$methionine, occurs at a high rate in fertilized eggs throughout the preimplantation period in the hamster contrary to the mouse in which high levels of protein synthesis are evident only in the blastocyst (Weitlauf, 1971). Moreover, if pregnant hamsters are ovariectomized on day 3, the embryos continue to incorporate $[^{35}S]$methionine through day 4, whereas in the mouse, the embryos enter a stage of "suspended animation" and can survive for long periods of time. In contrast, the blastocysts degenerate by day 5 in the ovariectomized hamster (Weitlauf, 1971) indicating that the hamster does not possess the luxury of delayed implantation which is an option open to the mouse and rat.

An experimental substitute for implantation of ova as a stimulus for differentiation of uterine decidua can be produced in the hamster by inserting a thread into the uterine lumen (Czyba, 1963; Harper, 1970) or by injecting air into the uterus (Orsini, 1963). The resultant transformation of undifferentiated stromal cells into rich, plump secretory cells is referred to as the deciduomal response. In the hamster, the response can be elicited in ovariectomized animals by progesterone alone (Kehl *et al.*, 1962), but after traumatization, a better response is produced by the combination of progesterone and estrone (Harper, 1970). It is interesting that the duration of pseudopregnancy is not prolonged in the hamster beyond the normal eight to ten day range by induction of massive deciduoma (Kent and Atkins, 1959) contrary to the situation in the rat.

There are profound differences between the response of the reproductive tracts of the rat and hamster to exogenous estrogens or progestins. For example, in ovariectomized animals, estradiol and diethylstilbestrol are only 4% to 5% as active in the hamster as uterotrophic or contraceptive compounds compared with the rat (Giannina *et al.*, 1971). Perhaps this correlates with the much higher levels of circulating estrogen in the pregnant hamster than in the pregnant rat.

In contrast to its relative refractoriness to estrogen, the hamster is exquisitely sensitive to small amounts of progesterone compared with the rat. Thus, in hamsters castrated on day 2 of pregnancy, daily injection of 62 μg of progesterone results in nidation in 50% of treated hamsters (Orsini and Meyer, 1962). Conversely, in rats ovariectomized on day 3 of pregnancy, daily administration of 3 mg progesterone maintains viable free blastocysts and superimposition of 100 ng estradiol induces implantation (MacDonald, 1982). A deciduomal response is produced in ovariectomized hamsters by 75 μg progesterone per day (Czyba et al., 1962), whereas the rat requires a minimum of 0.5 mg of progesterone for a minimal response (Yochim and DeFeo, 1962). Still another species difference between the hamster and other rodents is that a ratio of estrogen:progesterone of 1 : 100 (mouse); 1 : 3000 (rat), 1 : 1000 (guinea pig) leads to antagonism in the experimental induction of deciduoma (for references, see Czyba, 1963). In contrast, in the ovariectomized hamster an estrogen:progesterone ratio of between 1 : 50 and 1 : 10 leads to synergism rather than antagonism of the decidual response (Czyba, 1963). Thus, the hamster uterus is much more resistant to doses of estrogen that would be deleterious in other species.

In part, this difference may be attributable to the appearance during pregnancy in the hamster of two steroid-binding proteins that bind testosterone, progesterone, and corticosterone, but not estradiol (Savu et al., 1977). Still another factor in the hamster that may be involved in the uterine refractoriness to estrogen and the sensitivity to progesterone may be attributable to the content and affinity of steroid-binding receptors. This is an area that has not been extensively explored in the hamster compared with the rat. In the cyclic hamster, large increases in uterine nuclear E_2 receptors occur when serum estradiol is increasing and serum progesterone is falling (West et al., 1978). It is especially interesting that on day 8 of pregnancy the hamster uterus contains very low concentration of total nuclear E_2 receptors: about 30 dpm[^{3}H]estradiol per μg DNA compared with 205 dpm per μg DNA on the morning of proestrus (West et al., 1978). There is also a specific progesterone receptor in the uterus of the cyclic hamster that increases as serum estradiol levels increase with a maximum uterine concentration on the morning of proestrus (Leavitt et al., 1974). The peak concentration of progesterone receptors on proestrus at 1000 hr is followed by a sharp decline by 2000 hr of the same day in response to either the dramatic increase in serum progesterone or the equally striking decrease in serum estradiol. In hamsters ovariectomized at proestrus, the concentration of progesterone binding sites declines slowly, but even two weeks after ovariectomy there is still significant retention of uterine receptors to levels approximating the amounts present on the first two days of the cycle. Estrogen treatment 14 days after ovariectomy results in a rapid return of uterine progesterone binding sites to the levels attained at proestrus (Leavitt et al., 1974). A subsequent study revealed that the preovulatory surge of progesterone is associated with translocation of the uterine progesterone receptor from the cytosol to the nucleus. (Evans et al., 1980). Moreover, hamsters ovariectomized at 0800 hr on day 3 of the cycle and given Silastic implants of estradiol maintained cytosol progesterone receptor synthesis at the level normally observed on the morning of cycle day 4. Removal of the estrogen implant did not change total uterine estrogen receptors over the next 8 hr. However, if 2.5 mg of progesterone were administered on day 4 to hamsters still retaining the estrogen implants, total concentration of uterine estrogen receptor (cytosol plus nuclear) fell rapidly within 2 hr to extremely low levels by 8 hr after progesterone treatment (Evans et al., 1980).

When a decidual reaction is induced on day 4 of pregnancy, hamsters killed on days 7 and 9 show that in deciduoma, compared with myometrium, there is an increase in cytoplasmic progesterone receptors and a concomitant fall in cytoplasmic and nuclear estrogen receptors (Leavitt *et al.*, 1979). The ratio of receptor progesterone : receptor estrogen in nuclei of deciduoma becomes correspondingly greater than the ratio in cytosol as pregnancy progresses, suggesting either selective retention of nuclear progesterone or loss of estrogen (Leavitt *et al.*, 1979).

The same general mechanisms regulating uterine estrogen and progesterone receptors in the hamster seem to apply to the rat. The critical question is whether there is any species differences in the physicochemical properties of the receptors or the number of receptors that can account for the physiological differences in sensitivity to estrogen and progesterone that exists between the rat and hamster. Unfortunately, at present, it is not possible to give a clear-cut answer to this question because of the differences in methodology used in different laboratories and the fact that few investigators have dealt with more than one species (for references, see Leavitt *et al.*, 1983). A possible clue is provided by comparing the number of nuclear progesterone receptors in the uterus during the estrous cycles of the rat and hamsters with the caveat that different methods were used in the nuclear exchange assays. On this basis, the nuclear progesterone receptors were fourfold to fivefold higher in the hamster (Leavitt *et al.*, 1979). Obviously, considerably more work on comparative aspects of "receptorology" of the uterus is in order.

Figure 1 summarizes various parameters of pregnancy in the hamster and is based on several studies from this laboratory.

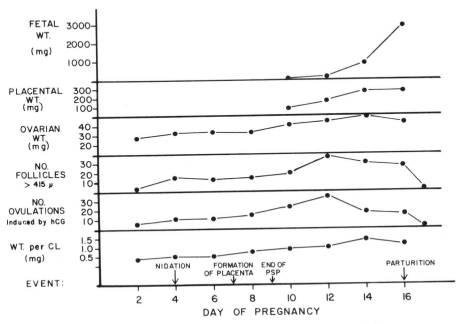

Figure 1. Temporal changes in various parameters during pregnancy in the hamster.

3.1. Fetal, Placental, and Ovarian Weights

On day 6 of pregnancy, discrete embryonic swellings are discernible (61 ± 6 mg) and there is an approximate doubling in the size of the swelling by day 8 (105±5 mg). Day 10 is the first day it is convenient to weigh the fetus (22 ± 3 mg). It is evident that most of the fetal growth occurs on days 14 and 16 (Lukasweska and Greenwald, 1970).

The placenta grows from 95 mg on day 10 to its maximal value of 290 mg on days 14 and 16 (Lukaszewksa and Greenwald, 1970).

The weight per paired ovaries is fairly constant until day 8 of pregnancy (33 ± 2 mg) followed by a sudden increase in size on day 10 with maximal ovarian weight attained by day 14 (52 ± 2 mg) (Lukaszewska and Greenwald, 1970).

3.2. Number of Follicles Greater than 415 μ in Diameter

In large part, the number of large healthy antral follicles parallels the changes in ovarian weight. Thus, until day 10 there are about 10 to 18 follicles less than 415 μ per pair of ovaries followed by the development of 36 follicles per ovaries on day 12. By the day of parturition (day 16), the number of antral follicles has dropped to 18 per animal and by the next day, only one or two scattered follicles have not degenerated (Greenwald, 1964a). The proliferation of numerous antral follicles during the second half of pregnancy in the hamster is markedly different from the pattern in the pregnant rat (Greenwald, 1966) wherein there are no really large follicles until day 20 and the follicles then grow rapidly in preparation for postpartum ovulation. This is again an example of an interesting species difference since postpartum ovulation does not take place in the hamster unlike in the rat and mouse. One of the most potent stimuli to produce delayed implantation in the latter two species is suckling, and in view of the inability of the hamster blastocyst to become quiescent (Weitlauf, 1971) the absence of postpartum ovulation in the hamster is understandable.

3.3. Induction of Ovulation in the Pregnant Hamster by 20 IU HCG

Human chorionic gonadotropin (HCG) has a number of biochemical and biological similarities to LH including its ability to act as an ovulating hormone. After the subcutaneous injection of HCG, at any day of hamster pregnancy, ovulation is induced by the next morning (Greenwald, 1967a). The number of induced ovulations parallels the number of antral follicles maturing and the increase in ovarian weight during the second half of pregnancy. Thus, the induction of ovulation on day 8 of pregnancy results in an average of 13.8 ± 1.3 ova per animal (the ovulation rate of the cyclic hamster) compared with 35.5 ± 3.7 ovulations when HCG is administered on day 12.

3.4. Growth and Regression of the Corpus Luteum

The mean weight per CL increases from 0.29 mg on day 2 to a maximum of 1.5 mg on day 15 which also accounts for some of the increase in ovarian weight during the second half of pregnancy in the hamster (Lukaszewska and Greenwald, 1970). In the rat following hypophysectomy the CL persist as definite histological entities for at least nine months (Smith, 1930), and moreover they secrete appreciable quantities of 20α-dihydroprogesterone in the absence of the pituitary gland (Taya and Greenwald, 1982). In contrast, the hamster CL of pregnancy is a much more labile structure that rapidly disappears in a few days after hypophysectomy (Greenwald, 1967b) or the deprivation of endogenous LH support for 12 hr as produced by the injection of a potent antiserum to ovine LH (Mukku et al., 1974). The latter study is especially instructive since animals were killed from 30 min to 8 hr after the injection of the antisera and CL excised and prepared for electron microscopy. The earliest regressive signs appear at 30 min in the form of increased heterochromatin in the nucleus, invasion of leukocytes by 1 hr, and the majority of mitochondria are affected by 2 hr. By 9 hr, luteolysis is advanced with more rough endoplasmic reticulum present than the normal smooth endoplasmic reticulum that is associated with steroidogenesis (Mukku et al., 1974). The ephemeral nature of the hamster corpus luteum is thus in marked contrast to the stability of the rat corpus luteum.

A factor possibly contributing to the lability of the hamster corpus luteum is that it contains—especially at term—a great deal of acid phosphatase that is a marker for lysosomal activity (Chatterjee and Greenwald, 1978). The ease with which various perturbations lead to irreversible luteolysis in the hamster may account for the fact that prostaglandin F_{2a} terminates pregnancy in the hamster at a much lower dose compared with the rat. When injected subcutaneously, 50 μg of PGF_{2a} from days 4 to 6 interrupts pregnancy in the hamster, whereas 20-fold more PGF_{2a} is required in the pregnant rat to cause fetal resorption (Labhsetwar, 1972).

There is still one more interesting difference between the CL of the hamster and rat that reflects different rates of structural luteolysis. In the cyclic hamster, the CL regress extremely rapidly so that by day 4 they have almost vanished from the ovary. The net result is that there is never more than one generation of CL in the hamster. In contrast, CL from several consecutive ovulations accumulate in the ovary of the cylic rat; consequently in the first half of gestation, the ovary of the rat contains CL of different ages.

4. PERIPHERAL STEROID AND PEPTIDE SERUM HORMONE LEVELS IN THE PREGNANT HAMSTER

Figure 2 is a composite profile of serum levels of the principal steroid and peptide hormones for the pregnant hamster. Two caveats are necessary: (1) the results represent hormone patterns between 0900 to 1100 hr and consequently significant circadian rhythms may be missed; and (2) different laboratories utilizing different procedures and antibodies for radioimmunoassays may show different actual values but the trends are similar.

4.1. Progestins

Progesterone is the major serum progestin in the hamster showing a slow steady climb until day 10 of pregnancy (10 to 20 ng/ml) followed by a sharp increase on day

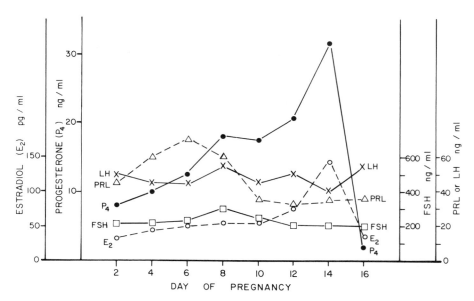

Figure 2. Serum steroid and peptide hormone levels in the pregnant hamster.

14 (30 ng/ml) and an equally abrupt fall to barely detectable levels by day 16 (Luka-szewska and Greenwald, 1970; Leavitt and Blaha, 1970; Gutknecht *et al.*, 1971; Baranczuk and Greenwald, 1974; Shaha and Greenwald, 1982). If the number of CL is increased to 40 or 50 by inducing superovulation with pregnant mare serum, the circulating levels of progesterone are threefold to fourfold higher than in control hamsters in which ovulation normally produces about ten CL (Greenwald, 1976).

Conversely, if all but one CL is extirpated by electrocautery on day 12 of hamster pregnancy, healthy young are delivered on day 16, whereas bilateral ovariectomy or enucleation of all CL on day 12 results in abortion on days 14 and 15, respectively (Weitlauf and Greenwald, 1967). At the time this experiment was performed, methods for measuring blood levels of progesterone were in their infancy, but the authors speculated that pregnancy could be maintained after day 12 by much lower serum levels of progesterone than are normally present. A much likelier explanation is that in pregnant hamsters with one CL, the uterine venous level of progesterone is considerably higher than the peripheral values suggesting that the placenta can contribute some progesterone in the second half of gestation (Frenkel and Greenwald, unpublished). Subsequently, Soares and Talamantes (1982) have shown that hamster placentae *in vitro* can synthesize progesterone but not testosterone or estradiol. This finding is consistent with the histochemical demonstration of Δ^5-3β-hydroxysteroid dehydrogenase in the hamster placenta (Legrand, 1977).

Peripheral levels and the pattern of progesterone during pregnancy are completely different between the rat and the hamster. In the rat, serum progesterone rises to 60 ng/ml between days 4 and 12 of pregnancy with a final plateau of 100 ng/ml on days 14 and 16 before the levels subside over the next six days (Taya and Greenwald, 1981). Similarly, in ovarian venous plasma during the first nine days of pregnancy, progesterone is fivefold higher in the rat compared with the hamster (Shaikh *et al.*, 1973).

In the pregnant hamster, 20α-dihydroprogesterone (20α-OHP) is a minor luteal metabolite with its serum levels never more than 2 ng/ml (Shaha and Greenwald, 1982). This differs from the rat where serum values of 20α-OHP run between 40 to 60 ng/ml through day 20 of pregnancy and then rise sharply to 100 ng/ml on day 22 (Taya and Greenwald, 1981). This species difference probably reflects the fact that in the hamster structural and functional luteolysis are concomitant events.

4.2. Estrogens

Serum estradiol levels are approximately 50 pg/ml until day 10 of pregnancy and then rise to a peak of 140 pg/ml on day 14, coinciding with the maximal values of progesterone (Baranczuk and Greenwald, 1974). The increase in peripheral levels of estradiol beginning on day 12 of pregnancy most likely represents the contribution of the massive proliferation of antral follicles. In the rat, serum estradiol runs at about 30 pg/ml on day 22 in preparation for postpartum ovulation (Taya and Greenwald, 1982). Similarly, high estradiol levels are found in ovarian venous plasma in the hamster versus the rat; for example, on day 4 of pregnancy at 1000 to 1200 hr, the ovarian venous effluent concentration of estradiol is 2856 ± 417 pg/ml (Shaikh *et al.*, 1973) compared with 163 ± 31 pg/ml in the rat (Shaikh, 1971).

Hamster follicles also produce estrone, but in lower amounts than estradiol. Thus, in ovarian venous blood during the first nine days of pregnancy, the levels of estradiol are about threefold higher than estrone (Shaikh *et al.*, 1973).

Although the antral follicles are presumably the major source of estrogen during pregnancy, the corpus luteum of the rat does produce limited amounts of estradiol (Taya and Greenwald, 1981). However, indirect evidence suggests that the hamster corpus luteum does not normally synthesize estrogen. After irradiation of an exteriorized ovary on day 1 of pregnancy (which destroys all follicles), the removal of the normal ovary on day 10 results in the interruption of pregnancy unless 1.0 μg of estrone is injected per day (Greenwald, 1972). Experiments currently in progress indicate that hamster CL *in vitro* fail to produce much estrogen unless exogenous testosterone is provided as a substrate (Greenwald, unpublished).

The only other class of steroids that has been measured in the pregnant hamster is plasma levels of cortisol, which is the major glucocorticoid in this species as opposed to corticosterone in the rat (Frenkel *et al.*, 1965). In the hamster, there is a significant increase in plasma cortisol in the second half of pregnancy with the levels about fivefold higher than in nonpregnant controls (Pellet *et al.*, 1970). It is of interest that after adrenalectomy, hamsters cannot survive on physiological saline, unlike the rat (Gaunt *et al.*, 1971). However, if adrenalectomized hamsters are injected with a long-acting mineralocorticoid for one month and treatment then discontinued, the males die and the females live because of the ultimate development of adrenal-like cells in the ovary (Frenkel, unpublished observations).

5. PEPTIDE HORMONES DURING HAMSTER PREGNANCY

Figure 2 summarizes the patterns of FSH, LH, and PRL measured at 0900 hr on each day of pregnancy (Bast and Greenwald, 1974). The serum concentration of FSH hovered at approximately 200 ng/ml throughout gestation, whereas LH was equally stable at approximately 40–50 ng/ml. This constant pattern of the gonadotropins (at least at 0900 hr) contrasts with the pregnant rat where serum levels of LH increase dramatically during the last two days of pregnancy (Taya and Greenwald, 1981).

A circadian pattern of PRL secretion exists in the pregnant hamster with a prolonged nocturnal elevation for about 16 hr and a peak at 2400 hr (Bast, 1978). This pattern disappears after day 11. In contrast, serum levels of LH are about 180° out of phase with PRL during the first eight days of pregnancy and thereafter there is no diurnal pattern of LH with the levels showing less and less oscillations as the time of delivery approaches (Bast, 1978). One wonders how these changes in gonadotropins are related to steroid profiles.

In the hamster, the abrupt fall between days 14 and 16 in progesterone, rather than estrogen, is the immediate trigger for parturition. The daily injection of 50 μg estradiol cyclopentylpropionate on days 13 to 15 of pregnancy fails to delay delivery by the afternoon of day 16, whereas 2 mg progesterone given on the same schedule significantly delays parturition and often causes maternal death with the fetuses retained *in utero* but separated from still attached placentae (Baranczuk and Greenwald, 1974). Injection of 0.1 mg progesterone at noon on day 15 delayed delivery by 10 hr, but 10 μg estradiol had no effect (Siegel and Greenwald, 1975). The demise of the CL does not correlate with the once-a-day determinations of the gonadotropins (Fig. 2). Whether prostaglandins affect the CL and thus play a role in parturtion in the hamster has not been explained, but, if so, what in turn regulates prostaglandin synthesis is equally a mystery in this species. It will be necessary to monitor levels of steroids and pituitary gonadotropins at frequent intervals in the last few days of gestation to see if any causal relationships exist in determining the onset of delivery.

However, two more peptide hormones may be involved as factors initiating parturition. As previously mentioned, several lines of evidence suggest that the hamster placenta assumes some endocrine activity as pregnancy progresses. Recent evidence indicates that the hamster placenta contains a factor that has the same elution profile by radioimmunoassays as human chorionic gonadotropin (Wide *et al.*, 1980). When hamster chorionic gonadotropin, which presumably has LH-like activity, appears and its peripheral levels at various times during pregnancy are intriguing subjects for future investigation.

The hamster placenta also appears to contain a PRL-like factor as measured by a radioreceptor assay (Soares and Talamantes, 1982). Fetal portions of the placentae from hamsters on days 8 to 16 of pregnancy were explanted and cultured for 12 hr. Placental content and release of PRL-like activity increased significantly from day 8 to 14. It is also of interest that there is a significant increase in PRL-like activity in serum on day 16 that *may* represent increased release from the placentae and not the pituitary (Soares and Talamantes, 1982). Evidence is therefore beginning to appear that the hamster placenta contains LH and PRL-like factors that fits previous findings about the existence of a luteotropic complex in this species.

6. LUTEOTROPIC COMPLEX OF THE PREGNANT HAMSTER

The purpose of this section is to relate pituitary gonadotropins to ovarian steroid production; specifically, what peptide hormones act as luteotropic hormones (i.e., are necessary to maintain a functional corpus luteum capable of synthesizing and releasing progesterone) and as folliculotropic hormones (produce large antral follicles secreting estradiol as their principal steroid product).

The Luteotropic Complex

Work from this laboratory was the first to establish that the interaction of PRL, FSH, and LH is necessary for optimal luteal function in the pregnant hamster (Greenwald, 1967b). In hamsters hypophysectomized on day 4 of pregnancy, the CL involute within a few days unless provided with PRL on a daily basis. Thus, PRL is necessary for the morphologic maintenance of the CL, but this treatment alone is insufficient to sustain pregnancy and moreover by day 8 peripheral levels of progesterone fall to nondetectable levels (Greenwald, 1973). The daily combination of PRL (0.1 to 1.0 mg) and FSH (100 to 200 µg) is the *minimal* luteotropic complex that maintains pregnancy and serum progesterone levels (Greenwald, 1967b, 1973). However, the embryonic swellings are smaller than usual, but they are restored to normal size if estrogen is provided along with PRL and FSH (but luteal weight is still reduced). This suggests that estrogen is required at the uterine level. The best treatment to restore all luteal parameters in the day 4 hypophysectomized hamster to day 8 control values is 1 mg PRL plus 1 IU pregnant mare serum (PMS). Pregnant mare serum is a "complete" gonadotropin possessing both FSH and LH activity. In addition, PMS has a much longer half-life than FSH, which also contributes to the greater efficacy of the former compound. Moreover, PRL and FSH never developed antral follicles in the hypophysectomized hamster, whereas substitution of PMS for FSH enabled large graafian follicles to mature (Greenwald, 1973). This seems to be another species difference in that in the hypophysectomized pregnant rat or mouse, we have found that 200 µg FSH (which is contaminated with some LH) enables large estrogen-secreting follicles to mature, whereas the same preparation of FSH is ineffective as a folliculotropin in the hamster. Evidently, development of antral follicles in the hamster requires exposure to tonic, high levels of FSH and/or LH.

When hamsters hypophysectomized at day 4 of pregnancy are injected daily with PRL and FSH plus various doses of LH, it was found that the addition of 0.1 to 0.25 µg LH was not deleterious to pregnancy but lowered plasma progesterone, whereas 1 to 50 µg LH along with the luteotropic complex interrupted pregnancy in most animals (Greenwald, 1973). It was therefore concluded that PRL plus FSH constituted the minimal luteotropic complex of the hamster and that LH served in a modulating role with low doses synergizing and high doses antagonizing the basic luteotropic complex (Greenwald, 1967b). It should be added that no combination of PRL and LH (1 to 10 μg daily) maintained pregnancy or plasma progesterone; PRL plus 50 μg LH at least maintained pregnancy in two or eight hypophysectomized hamsters and plasma progesterone levels of 4.1 ± 1.4 ng/ml (Greenwald, 1973). However, in comparison, PRL plus 200 μg FSH maintained pregnancy in 12 of 13 hypophysectomized hamsters with plasma progesterone concentration of 8.4 ± 1.3 ng/ml compared with 8.1 ± 0.57 ng/ml in intact, day 8 pregnant females. The ovine preparation used in these experiments was NIH-S5 and 200 μg FSH per day had 0.36 μg of LH as a contaminant.

The pregnant rat when hypophysectomized on day 12 maintains gestation to normal term without any problems indicating that the placenta has now assumed production of the necessary luteotropic hormone(s) and that consequently the pituitary is dispensable. In contrast, hamsters hypophysectomized at day 12 of pregnancy suffered a very heavy mortality and all animals aborted by day 16 (Greenwald, 1967b). Daily treatment with 200 μg FSH enabled seven of eight hypophysectomized hamsters to carry young to term, whereas 1 mg prolactin per day was ineffective. This suggested that the placenta had assumed the ability to produce both prolactin and LH-like hormones, but that the pituitary was still necessary for its secretion of FSH. The recent discovery of hamster placental lactogen and chorionic gonadotropin seems to fit this interpretation.

In a related study, hamsters were hypophysectomized on day 8 or 12 of pregnancy and various hormones were injected with the endpoint on day 14 or 15 being the number of animals pregnant and the percentage of live fetuses (Madhwa Raj et al., 1974). On day 8, a single injection of a potent LH antiserum interrupted pregnancy in intact controls. Hypophysectomized hamsters given a single injection of LH in bees wax sesame oil (which prolongs the half-life of the hormone) showed that 5 to 25 μg LH per day was ineffective, but 50 or 100 μg LH maintained pregnancy. These are rather enormous doses of LH. These same authors also found that in animals hypophysectomized at day 8, replacement with 25 μg LH + 2 mg prolactin was 100% effective in maintaining pregnancy (n = 10), whereas 200 μg FSH and 2 Mg PRL worked in only 4 of 15 animals (Madhwa Raj et al., 1974). They also found that after hypophysectomy at day 12, six of nine hamsters retained fetuses when autopsied on day 14 or 15. Differences in experimental design make it difficult to compare my results with those of Madhwa Raj et al. (1974). For example, after a single injection of 25 μg LH to hamsters on day 8 of pregnancy, 24 hr later there is an increase in progesterone in serum, luteal, and nonluteal tissues (Mukku and Moudgal, 1976).

The concept that on day 4 of gestation in the hamster a minimal luteotropic complex of PRL plus FSH is needed to sustain functional CL was greeted with some skepticism because a steroidogenic role of FSH on the corpus luteum was (and still is) usually not considered as a possibility: PRL and/or LH predominate in the literature as luteotropic hormones. It must also be borne in mind that LH has never been excluded as part of the luteotropic complex of the hamster and that part of the efficacy of 200 μg of ovine FSH is attributable to its LH contamination (Greenwald, 1967b).

Hamsters injected on day 1 of pregnancy with 50 μg or more of estradiol cyclopentylpropionate (ECP, a long-acting esterified estrogen) showed histological signs of luteal regression by day 4 that could be prevented by daily injection of human chorionic gonadotropin, PMS, or 100 to 200 μg FSH but not by LH (100 to 200 μg) or PRL (1 mg) (Greenwald, 1964b). A subsequent study revealed that structural luteolysis could be produced in pregnant hamsters not only by ECP but also by daily injection of 100 μg LH; the effects of ECP were reversible by daily injection of FSH and of LH by large doses of PRL (Choudary and Greenwald, 1968). At the time these experiments were performed, we could not explain how LH and PRL were interacting, but in light of recent studies on regulation of luteal peptide receptors (Richards and Midgley, 1976), a tentative explanation is now possible. That is, LH in large amounts "down regulates" its luteal receptors, whereas PRL increases the number of LH binding sites. Thus, large doses of PRL could counteract the luteolytic effects of daily administration of 100 μg LH by "protecting" luteal LH receptors against the desensitization effect of LH. In the case of the growing antral follicle, one of the actions of FSH is to increase the number of LH binding sites in the granulosa cells (Richards and Midgley, 1976). Is it possible that FSH may not exert some of its luteotropic effects on the hamster corpus luteum by a similar role?

With the above information as background, what evidence has accumulated over the past 15 years in favor of a luteotropic complex in the pregnant hamster and for a role of FSH? First, in hamsters hypophysectomized on day 6 of pregnancy and treated daily with 1 mg PRL and 5.0 μg of a highly purified ovine FSH preparation, pregnancy and a normal peripheral concentration of progesterone (9.6 ng/ml) were maintained in 100% of animals (Giannina et al., 1974). Second, a compound that interrupts pregnancy in the hamster is N,N-dimethylacetamide (DMA). When injected SC on day 4, DMA interrupted pregnancy in 100% of hamsters by day 8 (Miller et al., 1981). However, following DMA, daily treatment with 1 mg PRL plus 10 IU PMS on days 4 to 7 maintained pregnancy in all animals. However, injection of either PRL or PMS alone resulted in termination of 67% of the pregnancies (Miller et al., 1981). Third, pseudopregnant superovulated hamsters injected on day 5 at 1100 hr with 50 μg PGF_{2a} significantly reduced plasma progesterone by 6 hr and by 24 hr the levels were reduced to 0.2 the control value (Harris and Murphy, 1981b). In other groups, before injecting PGF[2a], hamsters were injected at 0700 hr of day 5 singly or in combination with LH (25 μg), PRL (100 μg), or FSH (200 μg). Administration of any of the hormones alone or a combination of any two failed to prevent luteolysis. It was only the combination of all three hormones that prevented luteolysis in 11 of 14 PGF_{2a}-treated hamsters.

What evidence has been cited against the concept of an luteotropic complex, which includes FSH, in the pregnant hamster? Organ culture of day 8 hamster CL for 20 hr in the presence of 1 μg LH increased the accumulation of progesterone in the medium to about ⅓ above the level of control incubated CL, whereas 0.5 μg FSH was ineffective (Behrman et al., 1974). The response to LH is not particularly impressive in this experiment, but again reflects species differences between the hamster and rat. As previously mentioned, the in vivo levels of serum progesterone are much greater in the rat than in the hamster. This difference also exists in vitro. When dispersed luteal cells on day 4 of pregnancy are incubated for 2 hr in the presence of 10 ng LH, the accumulation of progesterone by 5×10^4 luteal cells is 26 ng in the rat versus 4 ng in the hamster (Wada and Greenwald, unpublished). Moreover, the fact that FSH had no stimulatory effect on the hamster CL (Behrman et al., 1974) is not without precedent since PRL, which is a luteotropic hormone in the rat, fails to stimulate in vitro luteal steroidogenesis.

Another approach to determining whether various gonadotropins possess luteotropic activity has been to prepare monospecific potent antibodies against them and to test whether the antisera can affect gestation. Utilizing this approach, it has been found that a single injection of LH antiserum given between days 6 and 11 interrupted pregnancy in the hamster, whereas FSH antisera administered on either day 4 or 8 was wholly ineffective (Jagannahda Rao et al., 1972). Also, in hamsters injected on days 8 and 9, the luteolytic action of LH antiserum could not be reversed by concurrent injection of 200 μg FSH and 4 mg PRL, but luteolysis was prevented by administration of 100 μg LH (Jagannadha Rao et al., 1972).

We have also found that a potent LH antibody terminates pregnancy in the hamster when 0.1 to 0.2 ml is injected as a single dose at any time from day 4 through 10 and is partially effective on day 12 (Terranova and Greenwald, 1979). The same anti-LH serum is equally effective in interrupting pregnancy in the rat with 0.2 ml of the antibody being the minimal effective dose when administered on day 8 of pregnancy (Terranova and Greenwald, 1981). Hence, there does not seem to be any drastic difference in the requirements for LH between the two species. What is intriguing, however, is that with the resumption of estrus after administration of anti-LH to pregnant animals, the hamsters ovulate 29 ova at the first ovulation compared with 12.5 ova for the rat (Terranova and Greenwald, 1979, 1981).

The finding that an anti-LH serum terminates pregnancy in the hamster, whereas an anti-FSH serum is ineffective (Jagannadha Rao et al., 1972; Terranova and Greenwald, 1979) is not incompatible with the existence of a luteotropic complex that includes FSH. In all of my work, I have stressed that PRL plus FSH is the minimal luteotropic complex and that small doses of LH enhance the production of progesterone. Luteinizing hormone may well be the final luteotropic stimulus, but this does not negate the importance of PRL and FSH, which may still be essential in providing the necessary background for LH to exert its effects. The fact that anti-FSH is not effective in interrupting pregnancy merely indicates that its role is different from LH. Similarly, almost everyone will concede that pituitary PRL is luteotropic in the rat in the first part of gestation and yet, to my knowledge, no one has succeeded in interrupting pregnancy in the rat by administering PRL antiserum.

The previous discussion therefore substantiates that there is indeed a luteotropic complex in the hamster although the specific action of any of the hormones has not yet been resolved. The role of FSH is especially still controversial, but it may also be involved in other species as well as the hamster. For example, there are specific FSH-binding sites in the CL of the pseudopregnant rat (Torjesen and Aakvaag, 1977). A recent study has shown the presence of receptors for PRL and LH in the CL of pseudopregnant hamsters, but no attempt was made to determine whether there were also FSH-binding sites (Harris et al., 1981). Utilizing topical autoradiography, it has been recently reported that the newly formed CL of the cyclic hamster exhibit significant FSH binding (Oxberry and Greenwald, 1982); it will be interesting in the future to determine FSH binding sites in the CL of the pregnant hamster and other species.

With the availability of much purer FSH preparations, potent antisera to LH, radioreceptor assays and topical autoradiography for peptide hormones, and procedures for *in vitro* incubation of corpora lutea, the concept of the luteotropic complex is definitely ripe for renewed exploration.

REFERENCES

Adler, N. T., 1969, Effects of the males' copulatory behaviour on successful pregnancy of the female rat, *J. Comp. Physiol. Psychol.* **69**:613–622.

Austin, C. R., 1956, Ovulation, fertilization and early cleavage in the hamster, *J. Roy. Microsc. Soc.* **75**:141–154.

Baranczuk, R., and Greenwald, G. S., 1974, Plasma levels of oestrogen and progesterone in pregnant and lactating hamsters, *J. Endocrinol.* **63**:125–135.

Bast, J. D., 1978, Converse circadian patterns of luteinizing hormone and prolactin levels in the serum of pregnant hamsters, in: *11th Annual Meeting of the Society for the Study of Reproduction* Abstract 82, p. 43A, Society for the Study of Reproduction, Champaigne, Illinois.

Bast, J. D., and Greenwald, G. S., 1974, Daily concentrations of gonadotrophins and prolactin in the serum of pregnant or lactating hamsters, *J. Endocrinol.* **63**:527–532.

Behrman, H. R., Ng, T. S., and Orczyk, G. R., 1974, Interactions between prostaglandins and gonadotropins on corpus luteum function, in: *Gonadotropins and Gonadal Function* (N. R. Moudgal, ed.), Academic Press, New York, pp. 332–344.

Chang, M. C., and Sheaffer, D., 1957, Number of spermatozoa ejaculated at copulation, transported into the female tract, and present in the male tract of the golden hamster, *J. Hered.* **48**:107–109.

Chatterjee, S., and Greenwald, G. S., 1978, Biochemical changes in the corpora lutea of pregnant and lactating hamsters, *J. Endocrinol.* **78**:261–265.

Choudary, J. B., and Greenwald, G. S., 1967, Effect of an ectopic pituitary gland on luteal maintenance in the hamster, *Endocrinology* **81**:542–552.

Choudary, J. B., and Greenwald, G. S., 1968, Comparison of the luteolytic action of LH and estrogen in the hamster, *Endocrinology* **83**:129–136.

Czyba, J. C., 1963, Donnees numèriques sur les equilibres oestroprogestéroniques qui conditionnent le déciduome experimental chez le Hamster doré, *C.R. Seances Acad. Sci. Ser. III* **256**:5628–5629.

Czyba, J. C., 1968, Les fluctuations de la fécondité chez le Hamster doré (Mesocricetus auratus Waterhouse) au cours de l'annee. *C.R. Seances Soc. Biol. Ses Fil.* **162**:113–115.

Czyba, J. C., Chiris, M., and Duboi, P., 1962, La Période de sensibilité de l'endomètre au traumatisme dans la réalisation du deciduome expérimental chez le Hamster doré en phase lutéinique artificielle, *C.R. Seances Soc. Biol. Ses Fil.* **156**:2074–2076.

Czyba, J. C., Cams, R., and Laurent, J. L., 1970, Donnees sur l'évolution normale de la gestation du Hamster doré (Mesocricetus auratus) en laboratoire, *C.R. Seances Soc. Biol. Ses Fil.* **164**:2267–2269.

Diamond, M., 1970, Intromission pattern and species vaginal code in relation to induction of pseudopregnancy, *Science* **169**:995–997.

Diamond, M., 1972, Vaginal stimulation and progesterone in relation to pregnancy and parturition, *Biol. Reprod.* **6**:281–287.

Diamond, M., and Yanagimachi, R., 1968, Induction of pseudopregnancy in the golden hamster, *J. Reprod. Fertil.* **17**:165–168.

Duby, R. T., McDaniel, J. W., Spilman, C. H., and Black, D. L., 1969, Utero-ovarian relationships in the golden hamster, *Acta Endocrinol.* **60**:595–602.

Evans, R. E., Tong, J. C., Hendry, W. J., and Leavitt, W. W., 1980, Progesterone regulation of estrogen receptor in the hamster uterus during the estrous cycle, *Endocrinology* **107**:383–390.

Ford, J. J., and Yoshinaga, K., 1975, Ergocryptine and pregnancy maintenance in hamsters, *Proc. Soc. Exp. Biol. Med.* 150:425–427.

Foster, G. A., Orsini, M. W., and Strong, F. M., 1963, Localization of glycogen in the opacity characterizing decidualization in the cleared hamster uterus, *Proc. Soc. Exp. Biol. Med.* 113:262–265.

Frenkel, J. K., Cook, K., Grady, H. J., and Pendleton, S. K., 1965, Effects of hormones on adrenocortical secretion of golden hamsters, *Lab. Invest.* 14:142–156.

Gaunt, R., Grisoldi, E., and Smith, N., 1971, Refractoriness to renal effects of aldosterone in the golden hamster, *Endocrinology* 89:63–69.

Giannina, T., Butler, M., Popick, F., and Steinetz, B. G., 1971, Comparative effects of some steroidal and nonsteroidal antifertility agents in rats and hamsters, *Contraception* 3:347–359.

Giannina, T., Butler, M., Sawyer, W. K., and Steinetz, B. G., 1974, On the mechanism of prostaglandin F_{2}GA-induced abortion in hamsters, *Contraception* 9:507–522.

Grady, K. L., and Greenwald, G. S., 1968, Gonadotropic induction of pseudopregnancy in the cyclic hamster, *Endocrinology* 83:1173–1180.

Greenwald, G. S., 1964a, Ovarian follicular development in the pregnant hamster, *Anat. Rec.* 148:605–609.

Greenwald, G. S., 1964b, Luteolytic effect of estrogen on the corpora lutea of pregnancy of the hamster, *Endocrinology* 76:1213–1219.

Greenwald, G. S., 1965, Histologic transformation of the ovary of the lactating hamster, *Endocrinology* 77:641–650.

Greenwald, G. S., 1966, Ovarian follicular development and pituitary FSH and LH content in the pregnant rat, *Endocrinology* 79:572–578.

Greenwald, G. S., 1967a, Induction of ovulation in the pregnant hamster, *Am. J. Anat.* 121:249–258.

Greenwald, G. S., 1967b, Luteotropic complex of the hamster, *Endocrinology* 80:118–130.

Greenwald, G. S., 1972, Effects of x-irradiation on ovarian function in the pregnant hamster, *Endocrinology* 91:75–86.

Greenwald, G. S., 1973, Further evidence for a luteotropic complex in the hamster: Progesterone determinations of plasma and corpora lutea, *Endocrinology* 92:235–242.

Greenwald, G. S., 1974, Modifications in ovarian and pituitary function in the hysterectomized pregnant hamster, *J. Endocrinol.* 61:45–51.

Greenwald, G. S., 1976, Effects of superovulation on fetal development and hormone levels in the pregnant hamster, *J. Reprod. Fertil.* 48:313–316.

Greenwald, G. S., 1978, Modification by exogenous progesterone of estrogen and gonadotropin secretion in the cyclic hamster, *Endocrinology* 103:2315–2322.

Greenwald, G. S., and Bast, J. D., 1978, Hormone patterns in pregnant or pseudopregnant hamsters after unilateral ovariectomy or hysterectomy, *Biol. Reprod.* 18:658–662.

Gutknecht, G. D., Wyngarden, L. J., and Pharriss, B. B., 1971, The effect of prostaglandin $F_{2\alpha}$ on ovarian and plasma progesterone levels in the pregnant hamster, *Proc. Soc. Exp. Biol. Med.* 136:1151–1157.

Harper, M. J. K., 1970, Hormonal control of the deciduomal response of the golden hamster uterus, *Anat. Rec.* 167:225–230.

Harper, M. J. K., Dowd, D., and Elliott, A. S. W., 1969, Implantation and embryonic development in the ovariectomized-adrenalectomized hamster, *Biol. Reprod.* 1:253–257.

Harris, K. H., and Murphy, B. D., 1981a, Prolactin in maintenance of the corpus luteum of early pseudopregnancy in the golden hamster, *J. Endocrinol.* 90:145–150.

Harris, K. H., and Murphy, B. D., 1981b, Luteolysis in the hamster: Abrogation by gonadotropin and prolactin pretreatment, *Prostaglandins* 21:177–187.

Harris, K. H., Murphy, B. D., and Grinwich, D. L., 1981, Characteristics of luteal function in the superovulated, pseudopregnant hamster, *Biol. Reprod.* 25:699–707.

Jagannadha Rao, A., Madhwa Raj, H. G., and Moudgal, N. R., 1972, Effect of LH, FSH and their antisera on gestation in the hamster (*Mesocricetus auratus*), *J. Reprod. Fertil.* 29:239–249.

Kehl, R., Czyba, J. C., and Chiris, M., 1962, Maintien du deciduome experimental chez le Hamster dore, *C. R. Seances Soc. Biol. Ses Fil.* 156:674–676.

Kent, G. C., and Atkins, G., 1959, Duration of pseudopregnancy in normal and uterine-traumatized hamsters, *Proc. Soc. Exp. Biol. Med.* 101:106–107.

Klein, M., 1937, Relation between the uterus and the ovaries in the pregnant hamster. *Proc. R. Soc. (London) ser. B* 125:348–364.

Labhsetwar, A. P., 1972, Effects of prostaglandin $F_{2\alpha}$ on some reproductive processes of hamsters and rats, *J. Endocrinol.* 53:201–213.

Lanier, D. L., Estep, D. Q., and Dewsbury, D. A., 1975, Copulatory behavior of golden hamsters: Effects on pregnancy, *Physiol. Behav.* 15:209–212.

Leavitt, W. W., and Blaha, G. C., 1970, Circulating progesterone levels in the golden hamster during the estrous cycle, pregnancy and lactation, *Biol. Reprod.* 3:353–361.

Leavitt, W. W., Toft, D. O., Strott, C. A., and O'Malley, B. W., 1974, A specific progesterone receptor in the hamster uterus: Physiologic properties and regulation during the estrous cycle, *Endocrinology* 94:1041–1053.

Leavitt, W. W., Chen, J. J., and Evans, R. W., 1979, Regulation and function of estrogen and progesterone receptor systems, in: *Steroid Hormone Receptor Systems* (W. W. Leavitt, and J. H. Clark, eds.), Plenum Publishing Corp., New York, pp. 197–222.

Leavitt, W. W., MacDonald, R. G., and Okulicz, W. C., 1983, Hormonal regulation of estrogen and progesterone receptor systems, in: *Biochemical Actions of Hormones,* Volume 10 (G. Litwack, ed.), Academic Press, New York, pp. 324–356.

Legrand, C., 1977, Histochemical distribution of Δ5-3β-and 17β-hydroxysteroid dehydrogenases in hamster trophoblast, *J. Reprod. Fertil.* 51:405–408.

Lukaszewska, J. H., and Greenwald, G. S., 1969, Comparison of luteal function in pseudopregnant and pregnant hamsters, *J. Reprod. Fertil.* 20:185–187.

Lukaszewska, J. H., and Greenwald, G. S., 1970, Progesterone levels in the cyclic and pregnant hamster, *Endocrinology* 86:1–9.

Lukaszewska, J., Wilson, Jr., L., and Hansel, W., 1972, Luteotropic and luteolytic effects of prostaglandins in the hamster, *Proc. Soc. Exp. Biol. Med.* 140:1302–1307.

MacDonald, G. J., 1982, Maintenance of pregnancy in ovariectomized rats with steroid analogs and the reproductive ability of the progeny, *Biol. Reprod.* 27:261–267.

Madhwa Raj, H. G., MacDonald, G. J., and Greep, R. O., 1974, Involvement of luteinizing hormone as a luteotropin in the golden hamster, *Proc. Soc. Exp. Biol. Med.* 145:196–199.

Miller, W. L., Frank, D. W., and Sutton, M. J., 1981, Antifertility activity of DMA in hamsters: Protection with a luteotropic complex, *Proc. Soc. Exp. Biol. Med.* 166:199–204.

Mukku, V., and Moudgal, N. R., 1976, Relative sensitivity of the corpus luteum of different days of pregnancy to LH-deprivation in the rat and hamster, *Mol. Cell. Endocrinol.* 6:71–80.

Mukku, V., Anand Kumar, T. C., Kumar, K., Jagannadha Rao, A., and Moudgal, N. R., 1974, Ultrastructural changes in the corpus luteum of pregnancy in the golden hamster following LH deprival, in: *Gonadotropins and Gonadal Function* (N. R. Moudgal, ed.), Academic Press, New York, pp. 281–291.

Orsini, M. W., 1961, The external vaginal phenomena characterizing the stages of the estrous cycle, pregnancy, pseudopregnancy, lactation, and the anestrous hamster, *Mesocricetus auratus* Waterhouse, *Proc. Anim. Care Panel* 11:193–206.

Orsini, M. W., 1962, Study of ovo-implantation in the hamster, rat, mouse, guinea pig and rabbit in cleared uterine tracts, *J. Reprod. Fertil.* 3:288–293.

Orsini, M. W., 1963, Induction of deciduomata in hamster and rat by injected air, *J. Endocrinol.* 28:119–121.

Orsini, M. W., and Meyer, R. K., 1962, Effect of varying doses of progesterone on implantation in the ovariectomized hamster, *Proc. Soc. Exp. Biol. Med.* 110:713–715.

Oxberry, B. A., and Greenwald, G. S., 1982, An autoradiographic study of the binding of ^{125}I-labelled FSH, hCG and prolactin to the hamster ovary throughout the estrous cycle, *Biol. Reprod.* 27:505–516.

Pellet, H., Czyba, J. C., Rollet, J., and Cottinet, D., 1970, Variations des stéroides plasmatiques chez le Hamster doré pendant la gestation, *C.R. Seances Soc. Biol. Ses Fil.* 164:2255–2257.

Printz, R. H., and Greenwald, G. S., 1970, Effects of starvation on follicular development in the cyclic hamster, *Endocrinology* 80:290–295.

Printz, R. H., and Greenwald, G. S., 1971, A neural mechanism regulating follicular development in the hamster, *Neuroendocrinology* 7:171–182.

Richards, J. S., and Midgley, A. R., 1976, Protein hormone action: A key to understanding ovarian follicular and luteal development, *Biol. Reprod.* 14:82–94.

Savu, L., Nunez, E. A., and Jayle, M. F., 1977, High affinity testosterone, corticosterone and progesterone binding activities in pregnant hamster serum, *Endocrinology* 101:369–377.

Seegal, R. F., and Goldman, B. D., 1975, Effects of photoperiod on cyclicity and serum gonadotropins in the Syrian hamster, *Biol. Reprod.* 12:223–231.

Shaha, C., and Greenwald, G. S., 1982, In vivo and in vitro production of progestins by the corpus luteum of pregnancy of the hamster, *Biol. Reprod.* 26:854–860.

Shaikh, A. A., 1971, Estrone and estradiol levels in the ovarian venous blood from rats during the estrous cycle and pregnancy, *Biol. Reprod.* 5:297–307.

Shaikh, A. A., Birchall, K., and Saksena, S. K., 1973, Steroids in the ovarian venous plasma and F prostaglandins in the peripheral plasma during pseudopregnancy and days 1–9 of pregnancy in the golden hamster, *Prostaglandins* 4:17–30.

Siegel, H. L., and Greenwald, G. S., 1975, Prepartum onset of maternal behavior in hamsters and the effects of estrogen and progesterone, *Horm. Behav.* 6:237–245.

Smith, P. E., 1930, Hypophysectomy and a replacement therapy in the rat, *Am. J. Anat.* 45:205–273.

Soares, M. J., and Talamantes, F., 1982, Placental and serum hormone changes during the second-half of pregnancy in the hamster, *Biol. Reprod.* 27:523–529.

Soderwall, A. L., Kent, Jr., H. A., Turbyfill, M. S., and Britenbaker, A. L., 1960, Variation in gestation length and litter size of the golden hamster, *Mesocricetus auratus, J. Gerontol.* 15:246–248.

Taya, K., and Greenwald, G. S., 1981, In vivo and in vitro ovarian steroidogenesis in the pregnant rat, *Biol. Reprod.* 25:683–691.

Taya, K., and Greenwald, G. S., 1982, In vivo and in vitro ovarian steroidogenesis in the long term hypophysectomized rat, *Endocrinology* 110:390–397.

Terranova, P. F., and Greenwald, G. S., 1978, Steroid and gonadotropin responses to ergocryptine-suppressed prolactin secretion in hysterectomized, pseudopregnant hamster, *Endocrinology* 103:845–853.

Terranova, P. F., and Greenwald, G. S., 1979, Antiluteinizing hormone: Chronic influence on steroid and gonadotropin levels and superovulation in the pregnant hamster, *Endocrinology* 104:1013–1019.

Terranova, P. F., and Greenwald, G. S., 1981, Effects of an antiserum to luteinizing hormone and steroid replacement therapy on maintenance of pregnancy in the rat: Serum and luteal levels of progesterone, testosterone and oestradiol, *J. Endocrinol.* 90:9-18.

Torjesen, P. A., and Aakvaag, A., 1977, The serum levels of progesterone and 20α-dihydroprogesterone and the ovarian LH, FSH and PRL binding during luteolysis of the superovulated rat ovary, *Acta Endocrinol.* 86:162–172.

Vomachka, A. J., and Greenwald, G. S., 1979, The development of gonadotropin and steroid hormone patterns in male and female hamsters from birth to puberty, *Endocrinology* 105:960–966.

Voogt, J., Robertson, M., and Friesen, H., 1982, Inverse relationship of prolactin and rat placental lactogen during pregnancy, *Biol. Reprod.* 26:800–805.

Weitlauf, H. M., 1971, Protein synthesis in vivo by preimplantation hamster embryos, *Am. J. Anat.* 132:103–107.

Weitlauf, H. M., and Greenwald, G. S., 1967, Maintenance of pregnancy in the hamster by a single corpus luteum from day 12, *J. Reprod. Fertil.* 14:489–491.

West, N. B., Norman, R. L., Sandow, B. A., and Brenner, R. M., 1978, Hormonal control of nuclear estradiol receptor content and the luminal epithelium in the uterus of the golden hamster, *Endocrinology* 103:1732–1741.

Wide, L., Hobson, B., and Wide, M., 1980, Chorionic gonadotropin in rodents, in: *Chorionic Gonadotropin* (S. J. Segal, ed.), Plenum Press, New York, pp. 37–51.

Yochim, J. M., and DeFeo, V. J., 1962, Control of decidual growth in the rat by steroid hormones of the ovary, *Endocrinology* 71:134–142.

Yoshinaga, K., and Ford, J. J., 1975, Ergocryptine and pregnancy maintenance in hamsters, *Proc. Soc. Exp. Biol. Med.* 150:425–427.

4

Male Hamster Reproductive Endocrinology

ANDRZEJ BARTKE

Male reproductive functions are regulated by hormones secreted by the pituitary and by the testis. The anterior pituitary controls testicular growth and function primarily by secreting two glycoprotein hormones: luteinizing hormone (LH) and follicle-stimulating hormone (FSH). Luteinizing hormone binds to Leydig cells in the testicular interstitium and stimulates several processes related to steroid production. This results in increased secretion of androgenic steroids, primarily testosterone, by the testis. Follicle-stimulating hormone binds to Sertoli cells in the seminiferous tubules of the testis and stimulates their secretory activity, which is essential for the normal progression of multiplication and differentiation of germ cells, leading to the production of spermatozoa. In rodents, pituitary control of the testis also involves prolactin (PRL), a protein hormone that binds to Leydig cells and regulates the number of LH receptors. The principal androgenic steroid produced by the mammalian testis is testosterone. Testosterone plays a pivotal role in the regulation of male reproductive physiology since it controls virtually all aspects of reproductive and sexual functions, including spermatogenesis, maturation of spermatozoa in the epididymis, growth of male accessory reproductive glands and their ability to produce seminal plasma, as well as territorial, aggressive, and copulatory behavior.

The golden (Syrian) hamster is a seasonal breeder and consequently the function of the pituitary-testicular axis undergoes cyclic variations consisting of activation during sexual maturation and at the beginning of each annual breeding season, and suppression coinciding with the end of this season and with preparation for the period of hibernation.

Seasonal differences in reproductive and endocrine function are readily apparent in laboratory populations of golden hamsters exposed to natural photoperiod and can be

ANDRZEJ BARTKE • Department of Obstetrics and Gynecology, The University of Texas Health Science Center at San Antonio, San Antonio, Texas 78284. *Present Address:* Department of Physiology, Southern Illinois University, School of Medicine, Carbondale, Illinois 62901.

elicited at any time of the year by appropriate alterations of artificial photoperiod. Thus golden hamsters, unlike most commonly used laboratory rodents, are ideally suited for studies of the seasonal pattern of reproduction. It should be emphasized that in the natural habitat this pattern of gonadal and sexual function is characteristic of the overwhelming majority of animal species.

This chapter will examine changes in the release of LH, FSH, and PRL and in the endocrine function of the testis during sexual maturation and during the alternating sequence of the periods of seasonal quiescence and activity. Several parameters of hypothalamic function that are belived to be responsible for the control of gonadotropin and PRL release will also be discussed. This will be followed by a listing of typical hormone values in reproductively active and gonadally regressed adult hamsters, as reported by different laboratories.

1. SEXUAL MATURATION

According to Festing (1972), golden hamsters reach puberty at 45 to 60 days of age and their minimal breeding age is 42 to 56 days. Although this appears not to have been systematically examined, the males tend to mature later than females and thus the males probably become fertile at the age of 50 to 55 days.

A study by Vomachka and Greenwald (1979) provides detailed information on serum levels of LH, FSH, PRL, and androgens during the sexual maturation of the male hamster. These investigators examined hormone levels in 16 groups of hamsters ranging in age from one to three days to 60 days, using from five to 17 animals per group. Serum LH and FSH levels (Fig. 1) show very similar patterns of changes with age, with low concentrations during the first three to four weeks of postnatal life, followed by a pronounced linear increase between the ages of 25 to 27 and 40 days. The increase from the low prepubertal levels to the peak levels at 40 days of age is approximately fivefold for LH and approximately threefold for FSH. After the age of 40 days, serum gonadotropin levels gradually decline. The age-related changes in serum PRL levels (Fig. 1) are more pronounced and follow a different time course. A gradual increase during the first 18 days of life is followed by a steep rise between the ages of 16 to 18 days and 20 to 24 days. The concentration of PRL achieved at this time is more than sevenfold higher than the levels recorded in neonatal males. Serum PRL levels remain high between the ages of 22 and 30 days and thereafter decline slightly, yet rise to another peak at the age of 55 days. Androgens were measured using an antiserum to testosterone that crossreacts with dihydrotestosterone and to a lesser extent with androstenedione. For adult animals, the results of these studies, as well as of other studies that will be discussed further in this chapter, can probably be considered as equivalent to serum testosterone levels because the adult hamster testis appears to produce primarily testosterone and little, if any, dihydrotestosterone (Lau et al., 1978; Terada et al., 1980). Androgen levels (Fig. 1) are low and relatively constant until the animals are 28 to 30 days old and thereafter rise steeply to a maximum at 50 days of age. This is followed by a slight decline in serum androgen levels. These age-related changes in serum levels of pituitary and testicular hormones in the developing male hamster differ

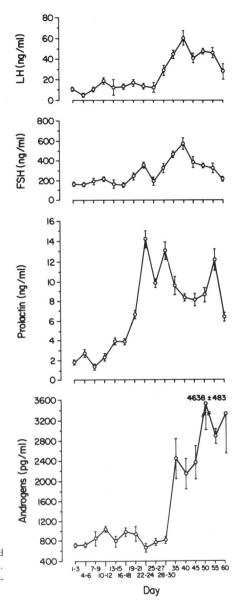

Figure 1. Serum levels of gonadotropins, PRL, and androgens in developing male hamsters. Means ± SE. Age of animals is indicated on the abscissa. (Reproduced from Vomachka and Greenwald, 1979.)

from the changes described in other species of laboratory rodents. In comparison with the rat (Steinberger and Ficher, 1969; Swerdloff *et al.*, 1971; Negro-Vilar *et al.*, 1973; Döhler and Wuttke, 1975; MacKinnon *et al.*, 1976; Corbier *et al.*, 1978), hamsters show a more pronounced age-related increase in serum LH levels, earlier onset of rapid changes in serum PRL levels and a conspicuous absence of the perinatal elevation of serum testosterone levels. In comparison with the mouse (Sinha *et al.*, 1974; Bartke and Dalterio, 1975; Selmanoff *et al.*, 1977; Barkley, 1979), hamsters exhibit a much greater and steeper pubertal increase in serum PRL combined with delayed and relatively modest increases in serum gonadotropin and androgen levels. However, it is difficult to make valid comparisons between hamsters and mice because of the major strain differences in plasma hormone levels and in their changes during development in the latter species.

Effects of sexual maturation on the ability of testicular tissue to metabolize labeled progesterone *in vitro* were examined by Terada *et al.* (1980). Their results indicated that testes from 8-day-old hamsters produce primarily 17-hydroxyprogesterone. Testes from 21-day-old animals convert a greater proportion of added progesterone, but produce very little testosterone or other androgens. Significant amounts of testosterone (corresponding to 14% and 18% of added precursor, respectively) were detected in incubations of testes from 30- and 36-day-old animals. In addition, testes of 30-day-old hamsters converted 15% of added progesterone to 5β-C_{19} steroids, primarily 5β-androstane-$3\alpha,17\beta$-diol. These biologically inactive steroids are derived from androstenedione and testosterone, indicating that this metabolic pathway may effectively lower the amounts of testosterone released from immature testes. Surprisingly, testes from older animals are much less active in converting the substrate and produce primarily 17-hydroxyprogesterone. Increased formation of androstenedione plus testosterone could be obtained by incubating greater amounts of tissue from very young hamsters or by increasing the time of incubation of the testes from 44-day-old animals.

2. ENDOCRINE FUNCTION IN THE ADULT MALE

Following sexual maturation, the concentration of testosterone in the peripheral circulation of male hamsters maintained under standard laboratory conditions fluctuates over a fairly wide range, but rarely declines below 0.75 ng/ml or exceeds 6 ng/ml. The variability of plasma testosterone levels in the golden hamster is much greater than that observed in man (Judd, 1979), but comparable to the fluctuations of plasma testosterone in the rat (Bartke *et al.*, 1973; Ellis and Desjardins, 1982) and considerably smaller than the fluctuations of testosterone in the mouse (Bartke *et al.*, 1973; Bartke and Dalterio, 1975). The wide fluctuations of peripheral androgen levels in rodents result from a pulsatile pattern of testosterone release by the testis and from the absence of high-affinity androgen-binding components in the serum (Corvol and Bardin, 1973; Ellis and Desjardins, 1982). Experiments involving castration and testosterone replacement therapy provide evidence that in the golden hamster, as in other vertebrate species, testicular androgens stimulate the growth and secretory function of male accessory reproductive glands (Ellis and Turek, 1979), maintain the functional integrity of the epididymis as assessed from maturation and survival of spermatozoa (Lubicz-Nawrocki, 1974), exert a regulatory influence of the secretion of LH and FSH from the anterior pituitary (Tamarkin *et al.*, 1976), maintain secondary sexual characteristics (Drickamer *et al.*, 1973), and are necessary for the expression of copulatory (Beach and

Pauker, 1949; Campbell *et al.*, 1978), territorial, and aggressive behaviors. There is much indirect evidence that testicular testosterone is also the principal hormonal regulator of spermatogenesis in the golden hamster, as would be expected from results obtained in other species (Boccabella, 1963; Clermont and Harvey, 1967). However, the ability of testosterone to induce and/or maintain spermatogenesis in hypophysectomized hamsters remains to be demonstrated.

Results reported by Ewing and his collaborators (1979) allow comparing the testes of hamsters to those of four other mammalian species with respect to capacity for testosterone production. Dog, rabbit, guinea pig, rat, and hamster testes were perfused *in vitro* with artificial media containing maximally stimulating amounts of ovine LH. Under these conditions, testes of hamsters produce much less testosterone than do the testes of other species. This is not due to differences in the size of the testes or in the mass of the Leydig cells, because testosterone production per gram of testicular weight and per gram of Leydig cells is also markedly lower in the hamster than in any of the other species examined. These results imply that the capacity of individual Leydig cells to produce testosterone is much lower in adult golden hamsters than in adult male rats, rabbits, guinea pigs, or dogs.

A subsequent study from the same group of investigators (Zirkin *et al.*, 1980) provided evidence that hamster Leydig cells contain an unusually low amount of smooth endoplasmic reticulum, a cellular organelle containing many of the enzymes required for biosynthesis of testosterone. It was also shown that species differences in the capacity to secrete testosterone in response to maximal LH stimulation are closely related to the volume and surface density of smooth endoplasmic reticulum in the Leydig cells. In the five species examined, these parameters of Leydig cell ultrastructure were positively and linearly correlated to testosterone output with a correlation coefficient of 0.99.

In comparison to other common laboratory species, male hamsters have several unique behavioral characteristics. They exhibit very active marking behavior consisting of depositing the secretions of the so-called flank organs on any unfamiliar objects, including the sides of a newly cleaned cage. Flank organs consist of aggregations of specialized sebaceous glands located on each side of the animal, approximately equidistant from the shoulder and the hip. Flank organs are androgen dependent and measurement of their diameter has been used as a way of assessing peripheral androgen levels in the male hamster (Drickamer *et al.*, 1973). Another interesting behavioral characteristic of the hamster is the relatively low level of aggressive behavior in the male. Aggressive interaction of male hamsters rarely leads to wounding or continued attacks on the submissive animal. This is in marked contrast to the behavior of adult female hamsters which, except for the periods of sexual receptivity, are very aggressive toward conspecifics of either sex. Behavioral characteristics of the golden hamster are discussed in some detail in other chapters of this volume (Chapters 6 to 12).

Plasma testosterone levels in adult male hamsters rise in response to the presence of a receptive female (Macrides *et al.*, 1974). This effect appears to be mediated via olfactory perception of the odor of the vaginal discharge of estrous females (Macrides *et al.*, 1974). Extrapolation of the results obtained in mice (Macrides *et al.*, 1975; Maruniak and Bronson, 1976) and in other species suggests that the increase in peripheral testosterone levels in response to exteroceptive stimuli from female conspecifics is caused by enhanced release of LH from the pituitary in response to pheromones or other clues to the presence of a female.

3. AGING

The effect of aging on testicular weight and plasma LH and PRL levels in male hamsters was examined by Chen (1981). One of eight aged (22- to 26-month-old) hamsters had atrophic testes. The average testicular weight in 22- to 26-month-old animals was slightly but significantly lower than in 4- to 5-month-old animals. A decrease in relative testicular weight in old animals was more pronounced because of an increase in body weight with age. In old males, plasma LH seemed lower and plasma PRL higher than the corresponding values in young animals, but neither of these apparent differences was statistically significant.

More recently, the effects of aging on various parameters of pituitary and testicular function in the golden hamster were examined by Swanson et al. (1982). There were no age-related changes in body weight, in the weights of the testes or the prostate, or in serum levels of LH, FSH, or testosterone in animals sacrificed at 4, 11, 24, or 31 months. The weights of the seminal vesicles and the epididymides were significantly reduced between 24 and 31 months of age. In the same study, the animals were tested for copulatory behavior at monthly intervals between the ages of 12 and 24 months. Apart from a gradual increase in the ejaculation latency, no age-related changes in sexual behavior were detected and over 80% of the animals achieved ejaculation during each of the tests. These results suggest that function of the pituitary–gonadal axis in male golden hamsters continues with little or no change to a very advanced age and probably for the duration of the natural life span. However, in hamsters that died from unknown "natural" causes between the ages of 16 and 30 months, testicular weight was greatly reduced. Only 29% of these animals had testes weighing over 2 g (normal value is approximately 3 g), and in some animals the gonads were completely regressed. This could indicate that the male reproductive system undergoes fairly rapid atrophy during the period immediately preceding death. However, a far more likely explanation is that the reproductive failure in some aging hamsters is due to disease, the incidence of which might increase with age, rather than to the process of aging per se. A similar conclusion was reached earlier from studies in mice (Nelson et al., 1976) and men (Harman, 1980).

4. SEASONAL CHANGES

4.1. Photoperiod-induced Testicular Regression

In male hamsters maintained under standard laboratory conditions including artificial illumination for 14 hr per day, secretion of gonadotropins, PRL, and testosterone continues with little change throughout the adult life. However, when the animals are exposed to natural illumination, the male reproductive system undergoes atrophy in the fall and recrudescence in the spring (Mogler, 1958; Czyba *et al.*, 1971; Vendrely *et al.*, 1971; Reiter, 1973/74). Thus, hamsters exposed to seasonal changes in day length retain their natural pattern of breeding activity in the spring and the summer, followed by reproductive quiescence in the fall and the winter. Subsequent studies of the response of the male reproductive system to different photoperiods resulted in the demonstration that exposure of adult hamsters at any time of the year to complete darkness or to day lengths of less than 12.5 hr results in testicular atrophy (Gaston and Menaker, 1967). It is of interest that postpubertal testicular atrophy was described also in hamsters homozygous for the Anophthalmic white (Wh) gene, which are blind due to congenital absence of the eyes (James *et al.*, 1980). This response to short photoperiod or darkness can be completely eliminated by surgical removal of the pineal gland (Reiter *et al.*, 1970; Reiter, 1980). The role of the pineal and its secretory products in mediating the effects of photoperiod on reproductive function will be discussed in detail in Chapter 5 (this volume). Therefore, the responses of the hypothalamic-pituitary-gonadal system to photoperiod will be discussed in this chapter without further reference to the fact that these phenomena are pineal mediated or that comparable endocrine changes can be elicited by administration of a pineal product, melatonin.

Short photoperiod-induced testicular regression is accompanied by major changes in pituitary hormone release, in the endocrine function of the testis, and in the mutual regulatory relationship between the adenohypophysis and the gonads. Exposure to short photoperiods for two to three months leads to a significant reduction in the concentrations of LH, FSH, PRL and testosterone in peripheral circulation (Turek *et al.*, 1975a; Bex *et al.*, 1978) and in the number of LH, FSH, and PRL receptors in the testis (Bex and Bartke, 1977; Tamarkin *et al.*, 1981; Bartke *et al.*, 1982; Klemcke *et al.*, 1982). Concomitant with these changes is complete suppression of spermatogenesis (Desjardins *et al.*, 1971; Gravis and Weaker, 1977) and a precipitous decline in testicular weight (review in Reiter, 1980). Gravis (1977) examined the ultrastructure of Leydig cells six weeks after blinding, a procedure believed to cause changes equivalent to those observed after transferring the animals to a short photoperiod. In blind animals, the diameter of Leydig cells was reduced, the nuclei were shrunken and infolded, lipid droplets were absent, the size of the Golgi complex was reduced, and there was less smooth endoplasmic reticulum. These findings indicate that biosynthesis of steroid hormones in the Leydig cells of blinded hamsters is severely suppressed. Severe androgen deficiency in animals maintained for two to three months in short photoperiod is evident from atrophy of the seminal vesicles and other accessory reproductive glands (Desjardins *et al.*, 1971) and loss of libido (Morin *et al.*, 1977). Short photoperiod very consistently produces suppression of testicular weight, plasma testosterone levels, and other indices of testicular function, but in some experiments, no change in peripheral LH or FSH levels can be detected. The unexpected coexistence of severely reduced plasma testosterone levels and unaltered plasma gonadotropin levels in an occasional group of animals has been reported by several laboratories and does not appear to be related to any particular characteristic of the animals, geographic location, assay system used to measure LH and FSH, or experimental protocol (Bex *et al.*, 1978; Reiter, 1980).

However, it should be emphasized that the usual pattern of changes in endocrine function in response to short photoperiod certainly includes a significant reduction in peripheral LH and FSH levels.

Studies of temporal patterns of the various endocrine responses to short photoperiod reveal that suppression of LH and FSH levels usually precedes the decline in testosterone levels and in testicular weight (Fig. 2) (Berndtson and Desjardins, 1974; Turek *et al.*, 1975a) and that reductions in peripheral PRL levels (Fig. 3) and in the number of testicular LH receptors occur either concomitantly with or before the changes in plasma LH and FSH levels (Bartke *et al.*, 1980c; Goldman *et al.*, 1981; Tamarkin *et al.*, 1981). These observations and the well-established role of gonadotropins and PRL in the control of testicular function (Bartke, 1980) indicate that suppression of endocrine and spermatogenic functions of the testes in short photoperiod is a consequence of reduced release of LH, FSH, and PRL. Since testicular regression in response to short photoperiod is accompanied also by significant reductions in the numbers of LH, FSH, and PRL receptors in the testes, it is very likely that the effects of gonadotropin and PRL deficiency are combined with the effects of reduced responsiveness of the testes to adenohypophyseal hormones. However, results of detailed studies in the rat suggest that the ability of LH to stimulate testosterone production by the testis may be limited by factors other than the number of LH receptors (Zipf *et al.*, 1978; Huhtaniemi and Catt, 1981). Thus the functional significance of a partial depletion of testicular LH, FSH, or PRL receptors remains to be thoroughly characterized. Although the total amount of testosterone produced by the atrophic testes of short-photoperiod animals in response to large doses of human chorionic gonadotropin (hCG) *in vivo* and *in vitro* is drastically reduced, the relative increase in testosterone output in response to low doses of hCG *in vitro* in these animals is augmented rather than decreased (Bartke *et al.*,

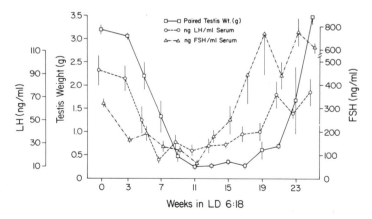

Figure 2. Effects of a short photoperiod (6 : 18) on testicular weight and on serum LH and FSH levels in adult male hamsters transferred from a long photoperiod on day 0. Means ± SE. During both regression and spontaneous recrudescence of the testes, changes in serum gonadotropins (particularly FSH) precede changes in testicular weight. (Reproduced from Turek *et al.*, 1975a.)

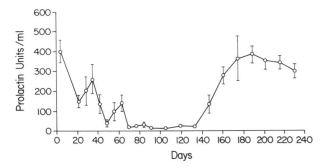

Figure 3. Effects of exposure to short photoperiod (10 : 14) on serum PRL levels in adult male hamsters. Means ± SE. Number of days in 10 : 14 is indicated on the abscissa. Serum PRL levels decline during testicular regression (days 0 to 70) and increase during recrudescence (after 140 days in 10 : 14). (Constructed from data in Goldman *et al.,* 1981.)

1980c, and unpublished observations). The reduced ability of the regressed testes to respond to large doses of LH (or hCG) *in vivo* or *in vitro* (Bartke *et al.,* 1980c, and unpublished observations) and to convert endogenous or exogenous precursors to testosterone (Desjardins *et al.,* 1971; Berndtson and Desjardins, 1974; Bartke *et al.,* 1981b) is almost certainly due to chronic suppression of LH, FSH, and PRL release and the resulting loss of Leydig cell enzymes that control biosynthesis of androgenic steroids.

The analysis of data on the effects of photoperiod on testicular hormone binding in the golden hamster is complicated by changes in the cellular composition of the testes during photoperiod-induced regression. After transfer of animals to a short photoperiod, there is an initial decline in both the total content (number per testis) and the concentration (number per mg of protein in a membrane preparation from testicular homogenate) of LH and PRL receptors. Subsequently the total content of receptors remains suppressed, but concentration often rises and may significantly exceed the values observed in the control, long-photoperiod animals (Bartke *et al.,* 1980c, and unpublished observations). This is also true for FSH receptors. Divergent effects of short photoperiod-induced testicular regression on the concentration and total content of gonadotropin binding sites in the testes are presumably related to increases in the percent of testicular volume occupied by Sertoli cells, which bind FSH, and Leydig cells, which bind LH. Suppression of spermatogenesis during photoperiod-induced testicular regression and repopulation of the seminiferous tubules during gonadal recrudescence lead to enormous shifts in testicular volume, whereas the number of Sertoli cells almost certainly remains stable and the number of Leydig cells probably fluctuates only very slightly. Detailed morphometric studies are needed to relate changes in hormone binding to changes in the number, volume, surface, and surface characteristics of target cells in the testis.

To determine if decreases in the peripheral levels of LH, FSH, and PRL are causally related to loss of gonadal function, several investigators have examined the effects of hormone replacement therapy on the responses of male golden hamsters to short photoperiod. Regression of the testes can be completely or nearly completely prevented if, during the exposure to short photoperiod, the animals are treated with combinations of PRL and luteinizing hormone-releasing hormone (LHRH) or PRL and hCG (Chen and Reiter, 1980; Bartke *et al.,* 1980a). In these experiments, PRL treatment consisted of implantation of one or two homologous pituitaries under the renal capsule, a procedure known to cause chronic elevation of peripheral PRL levels (Bex *et al.,* 1978). Injections of LHRH were administered in order to stimulate the release of LH and FSH, whereas hCG treatment was intended to substitute for endogenous LH. Since treatment with any of these hormones alone was only partially effective, the results indicate that testicular atrophy in short photoperiod is due to suppression of both gonadotropins and PRL release.

In animals that have undergone gonadal regression, a significant increase in testicular weight can be obtained by treatment with pituitary grafts, ovine PRL, sulpiride (a drug known to stimulate PRL release), ovine growth hormone, hCG, pregnant mare serum gonadotropin (PMSG), or testosterone (Czyba *et al.,* 1971; Bex *et al.,* 1978; Bartke *et al.,* 1981b). Prolonged treatment with ectopic pituitary grafts (Fig. 4) or sulpiride results in complete reversal of testicular atrophy that, in the case of the grafts, was shown to be accompanied by normalization of plasma testosterone levels and reappearance of fertility (Bex *et al.,* 1978; Bartke *et al.,* 1979). Treatment with PRL or grafts was also effective in increasing the number of LH receptors in the testes (Bex and Bartke, 1977; Bex *et al.,* 1978). Somewhat unexpectedly, treatment of gonadally regressed animals with LHRH, ovine LH, or ovine FSH has little, if any, effect on testicular function (Reiter *et al.,* 1975; Bex and Bartke, 1977). These results suggest that the ability of the testis to respond to LH and FSH may be compromised. This could be related to the loss of gonadotropin receptors. Excellent responses to exogenous PRL and to sulpiride-induced stimulation of PRL release (Czyba *et al.,* 1971; Bex *et al.,* 1978) emphasize the importance of changes in serum PRL levels in mediating the responses of the testes to short photoperiod. Increases in testicular weight and LH binding and in plasma testosterone levels were observed also in animals injected with ovine growth hormone (GH). However, the physiological meaning of these responses remains uncertain, since serum levels of endogenous GH apparently do not change in response to a short photoperiod (Bartke *et al.,* 1980b).

Changes in peripheral levels of gonadotropins and testosterone in animals exposed to a short photoperiod are accompanied by and almost certainly related to changes in feedback regulation of hypothalamic-pituitary function by the testis. An important change in the sensitivity of the hypothalamic–pituitary axis to inhibition by gonadal steroids, the so-called "set point" of pituitary-testicular feedback, is evident from a number of observations. First, as was already mentioned, both gonadotropin and testosterone levels decline, indicating that the usual compensatory increase of LH and FSH release in response to reduced testicular androgen output fails to occur in short photoperiod. Second, the castration-induced rise in serum LH and FSH levels is severely attenuated in animals exposed to a short photoperiod (Fig. 5) (Turek *et al.,* 1975b; Tate-Ostroff and Stetson, 1981). Third, much smaller doses of testosterone, dihydrotestosterone, or estradiol are required to suppress plasma LH and FSH levels in

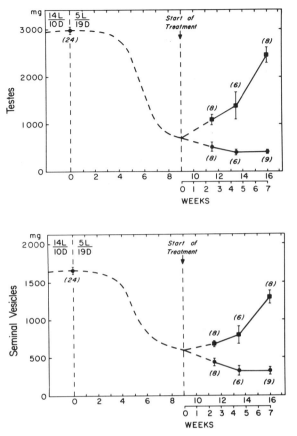

Figure 4. Effects of treatment with PRL (300 μg/day) for 2.5 weeks or with PRL-producing pituitary transplants for 4.5 or 7 weeks on the weights of the testes (top panel) and the seminal vesicles (bottom panel) in gonadally regressed hamsters. Treated animals: ■—■; Controls: ●—●. Means ± SE. The number of weeks in short photoperiod and the number of weeks after the start of the treatment are indicated on the abscissa. It is evident that the weights of the testes and the seminal vesicles in treated animals gradually return toward the values observed in gonadally active long-photoperiod controls. (Reproduced from Bex *et al.*, 1978.)

castrated animals exposed to short photoperiod than in castrated animals maintained in long photoperiod (Fig. 6) (Tamarkin *et al.*, 1976; Turek, 1977). It should be emphasized that changes in the sensitivity of the hypothalamic-pituitary system to suppression by gonadal steroids play an important role in mediating the effects of photoperiod on the function of the gonads also in other seasonally breeding species (Davis and Meyer, 1972, 1973a,b; Pelletier and Ortavant, 1975; Legan *et al.*, 1977; Lincoln and Short, 1980). Changes in sensitivity to gonadal feedback have also been proposed to explain the increase in gonadotropin secretion during pubertal development (McCann *et al.*, 1974; Odell and Swerdloff, 1976).

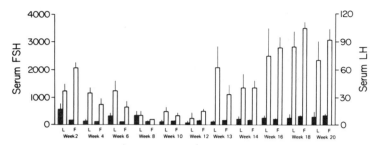

Figure 5. Effects of exposure to short photoperiod (6 : 18) for 2 to 20 weeks on serum gonadotropin response to castration. Serum levels of LH (L) and FSH (F) were measured in samples collected just prior to castration (solid bars) and in samples collected one week later (open bars). Means ± SE; values in ng/ml. The LH and FSH response to castration is gradually lost during the first eight weeks of exposure to 6 : 18 and regained after 13 weeks in short photoperiod. (Reproduced from Tate-Ostroff and Stetson, 1981.)

In contrast to the effects of testosterone on gonadotropin release, the ability of exogenous testosterone to activate copulatory behavior in castrated hamsters is reduced rather than increased by exposure to a short photoperiod (Morin *et al.,* 1977; Campbell *et al.,* 1978). Thus, increased responsiveness of the hypothalamic-pituitary system to feedback inhibition by testicular steroids is not a part of a more general increase in the responsiveness of the central nervous system to gonadal hormones, but rather a specific response of the neurons involved in the control of gonadotropin release.

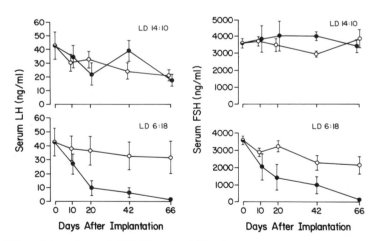

Figure 6. Effects of short photoperiod (6 : 18) on the ability of testosterone to suppress serum gonadotropin levels in castrated male hamsters. Sixty days after castration, the animals were implanted with empty (open circles) or 4 mm testosterone-filled (closed circles) Silastic® capsules and either remained in long photoperiod (top panels) or were transferred to a short photoperiod (bottom panels). Means ± SE. These implants were completely ineffective in animals exposed to long photoperiod, but suppressed serum LH and FSH to nondetectable levels in animals exposed to short photoperiods. (Reproduced from Ellis and Turek, 1979).

In addition to causing profound changes in the function of the pituitary-testicular axis, blinding or prolonged exposure to a short photoperiod reduces serum levels of thyroid-stimulating hormone (TSH) and thyroxine as well as free thyroxine, estimated from the uptake of thyroxine and 3,5,3'triiodothyronine (Vriend, 1981). These changes in pituitary-thyroid function could be related to gonadal atrophy as a consequence and/or as a contributing factor. Changes in serum thyroxine and particularly in free thyroxine could be secondary to the effects of altered gonadal function on the production of thyroxine-binding globulin. This possibility is addressed by Vriend *et al.* (1979), who conclude that suppression of thyroid function in short photoperiod appears to be due mainly to changes in the hypothalamic regulation of TSH release. Regardless of its cause, suppression of thyroid function could contribute to testicular failure in hamsters exposed to short day lengths. Both TSH and thyroid hormones are suspected of exerting direct stimulatory effects on the testes (Oppenheimer *et al.*, 1974; Amir *et al.*, 1978; Hutson and Stocco, 1978). Thyroid deficiency can interfere with gonadal function in experimental animals and hypothyroidism is a known contributor to male infertility in the human.

4.2. *Photoperiod-induced and Spontaneous Testicular Recrudescence*

Gonadally regressed hamsters respond to long photoperiod by an increase in testicular weight with eventual complete restoration of endocrine and gametogenic functions of the testes (Reiter, 1972; Berndtson and Desjardins, 1974). Stimulation of testicular growth and function is preceded by increases in serum LH and FSH levels (Turek *et al.*, 1975a; Matt and Stetson, 1980), whereas serum PRL levels appear to increase concomitantly with changes in testicular weight (Matt and Stetson, 1980). The effects of exposure of gonadally regressed animals to a long photoperiod on testicular gonadotropin and PRL binding remain to be examined.

It is an interesting feature of seasonal changes in reproductive functions in the golden hamster that restoration of gonadal activity in the spring does not depend on the increased day length, but occurs "spontaneously" at a predetermined time. The nature of the mechanism(s) regulating this phenomenon remains unknown, but its biological function can be deduced from recent results obtained in a closely related species, the Turkish hamster (*Mesocricetus brandti*). In Turkish hamsters, hibernation does not occur unless testicular function is suppressed; hibernation can be terminated by treatment with testosterone or dihydrotestosterone, but not with estradiol, and arousal from hibernation can be drastically delayed, or perhaps even prevented, by removal of the testes (Hall and Goldman, 1980). These results would imply that an increase in peripheral testosterone levels in early spring, i.e., toward the end of hibernation, is necessary for the subsequent arousal from the hibernating condition. However, other physiological mechanisms, including changes in the activity of the pituitary–thyroid axis (Hudson and Wang, 1979) and possibly also in the levels of endogenous opioids (Kromer, 1980), appear to be involved in control of the onset and duration of hibernation.

Endocrine changes associated with spontaneous testicular recrudescence in the golden hamster were examined in several laboratories. When golden hamsters are continuously exposed to short photoperiod, the testes remain fully regressed for approximately two months (between 10 and 19 weeks of exposure) and thereafter undergo recrudescence that is complete within approximately six weeks. The increase in testicular weight is preceded by an increase in serum gonadotropin levels (Figs. 2, 7). Matt and Stetson (1979) and Turek *et al.* (1975b) reported a significant increase in serum FSH levels after 13 to 15 weeks of exposure to short photoperiod (Fig. 7). Thereafter, serum FSH levels began to decline, whereas concentrations of LH, PRL, and testosterone in peripheral circulation gradually increased (Matt and Stetson, 1979). In animals examined by Klemcke *et al.* (1981), serum levels of FSH, LH, PRL, and testosterone remained suppressed somewhat longer and a significant increase in their concentrations was first evident between 18 and 19 weeks of exposure to short photoperiod. An increase in the number of testicular LH binding sites occurred simultaneously with changes in peripheral hormone levels (Fig. 8). Moreover, the temporal sequence of changes in endocrine function was not identical in different groups of animals. Additional experiments performed by Tamarkin *et al.* (1976), Matt and Stetson (1979), and Ellis *et al.* (1979) in intact and in castrated animals provide evidence for a gradual decrease in the sensitivity of testicular-pituitary feedback during the period in which intact animals undergo recrudescence of the testes. Collectively, these studies indicate

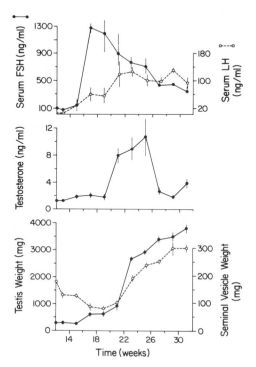

Figure 7. Changes in serum LH, FSH, and testosterone levels, and in the weights of the testes and seminal vesicles during spontaneous gonadal recrudescence, between 14 and 30 weeks of exposure to short photoperiods (6 : 18). Means ± SE. A sharp increase in serum FSH levels precedes other hormonal changes. (Reproduced from Matt and Stetson, 1979.)

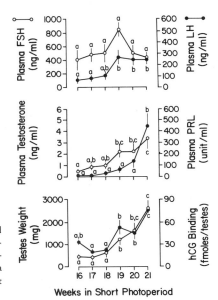

Figure 8. Changes in plasma gonadotropins, PRL, and testoterone levels and in testicular hCG (LH) binding during testicular recrudescence in adult male hamsters exposed to short photoperiod (5 : 19). Means ± SE. Data points with different superscripts are significantly different ($P < 0.05$). (Modified from Klemcke *et al.*, 1981.)

that restoration of gonadal activity is preceded by a number of changes in the hypothalamic-pituitary system and in the endocrine function of the testes. Since these changes occur either simultaneously or in rapid sequence, it is difficult to decide which of them are primary. It is also unclear whether the function of the hypothalamo-pituitary unit can be adequately evaluated from measurements of hormone levels in single blood samples. Most hormones are secreted in an episodic (pulsatile) fashion; therefore, hormone levels in the circulation normally fluctuate over a fairly wide range. In the sheep, there are seasonal changes in both amplitude and frequency of LH pulses (Lincoln *et al.*, 1977; Legan and Karsch, 1979) and these changes appear to constitute the main stimulus for the transition from nonbreeding to breeding condition (review in Karsch, 1980).

4.3. Hormonal Mechanisms of Seasonal Changes in Reproductive Function

The temporal pattern of changes in serum hormone levels and in testicular binding of gonadotropins and PRL during testicular regression and recrudescence have been described in preceding sections of this chapter. The results provide clear evidence that changes in testicular size are preceded by changes in the concentrations of gonadotropins, PRL, and testosterone in peripheral circulation. In addition to demonstrating these not unexpected relationships, studies of the temporal sequence of endocrine events suggest that the decrease in serum PRL and the consequent loss of testicular LH binding sites may be important in causing testicular regression, whereas the increase in serum FSH levels is an important, if not the primary, trigger for testicular recrudescence. Differences between the results obtained in different groups of animals and limitations of the conclusions that can be drawn from this type of studies have already been discussed.

Results of hormonal replacement therapy were also discussed in an earlier section of this chapter. They provide strong evidence for the importance of PRL in mediating the effects of photoperiod on the testes and suggest that regression of the testes in short photoperiod is caused by the decline in the release of both gonadotropins and PRL. Although the effects of LH and FSH on the function of Leydig and Sertoli cells, respectively, have been studied in considerable detail, the mechanism of PRL action on the testis is not fully understood. The presence of specific PRL receptors in testicular interstitium has been demonstrated in the rat (Charreau et al., 1977) and the ability of PRL to exert direct effects on the testis has been conclusively demonstrated in rats and mice (Bartke, 1980). The increase in the concentration of LH binding sites in the testes of hypophysectomized hamsters treated with ectopic pituitary homografts (Bartke et al., 1982) indicates that PRL can act directly on the hamster testis and confirms the ability of PRL to stimulate testicular LH binding. Prolactin also stimulates the release of pituitary FSH in the golden hamster (Bartke et al., 1981a). This would be expected to lead to stimulation of spermatogenesis and could also contribute to the increase in testicular LH binding in PRL-treated animals. Thus, PRL appears to exert both direct and pituitary-mediated stimulatory effects on testicular function. The ability of sulpiride to reverse autumnal regression of the testes (Czyba et al., 1971) provides important evidence for a critical role of PRL in the control of seasonal changes in gonadal function. However, our attempts to induce testicular regression by selective suppression of PRL release with bromocriptine (2-Br-α-ergocriptine, CB-154) were not successful. Chronic treatment with bromocriptine in adult male hamsters maintained in a long photoperiod reduced plasma PRL levels to nondetectable, but caused a very modest decrease in the weights of the testes and the seminal vesicles (Bex et al., 1978). This decrease, although statistically significant, was in no way comparable to the drastic reduction in testicular and accessory gland weights in animals exposed to short photoperiod. Thus, inducing isolated PRL deficiency did not mimic the effects of short photoperiod. It was subsequently demonstrated that treatment with identical daily doses of bromocriptine can significantly accelerate gonadal regression in a short photoperiod (Roychoudhury et al., 1980) and significantly delay the response of gonadally regressed animals to a long photoperiod (Bartke et al., 1980a), but it fails to prevent the recrudescence of the testis from eventually taking place (Bartke et al., 1980a). These results reinforce the conclusion that changes in testicular function occur in response to alterations in the release of both gonadotropins and PRL.

The following picture of the endocrine control of testicular function under natural conditions emerges from results available to date: Male hamsters mature sexually at the age of six to eight weeks, regardless of the season in which they are born or of the photoperiod conditions prevailing during this period of their development (Darrow et al., 1980). This is in contrast to the situation in other seasonally breeding rodents in which photoperiod influences the rate of sexual development and animals born toward the end of the breeding season often fail to reach sexual maturation until the spring of the following year (Foster and Ryan, 1979). Endocrine correlates of sexual maturation in male golden hamsters were described earlier in this chapter.

The photoperiod-related endocrine changes in the adult golden hamster can be summarized as follows: In the fall, when the length of the day decreases to less than 12.5 hr, the absence of light during the intrinsically regulated daily period of photosensitivity (Stetson et al., 1975) leads to inhibition of LH and FSH release from the pituitary. The decrease in stimulatory input from the pituitary leads to inhibition of Sertoli cell and Leydig cell function. The suppression of pituitary PRL release and the resulting loss of testicular LH receptors contribute to the reduction in the capacity of the testis to produce testosterone in response to LH stimulation. In parallel with these changes, sensitivity to the inhibitory action of testosterone on those hypothalamic centers regulating gonadotropin release gradually increases and thus suppression of LH and FSH secretion is maintained in spite of reduced serum testosterone levels.

The precipitous decline in testicular and peripheral androgen levels leads to suppression of spermatogenesis, loss of testicular weight, structural and functional involution of accessory reproductive glands, and loss of sexual behavior. In the case of sexual behavior, the consequences of reduced peripheral androgen levels are augmented by the concomitant decrease in the ability of testosterone to activate sexual behavior (Morin et al., 1977).

Approximately 16 weeks after the onset of these endocrine changes, an unknown internal stimulus or stimuli lead to a reversal of this process: the release of LH, FSH, and PRL is stimulated and sustained by a concomitant decrease in hypothalamic sensitivity to testosterone; the testicular capacity to bind LH and to produce testosterone gradually increases; continued stimulation by testosterone and FSH restores sperm production; and the animals return to breeding condition.

The changes in testicular function that occur in response to natural or artificial changes in photoperiod are unquestionably due to changes in pituitary hormone release. In an attempt to obtain some information on the regulation of pituitary function during these changes, we have recently begun to study the content and turnover rates of hypothalamic neurotransmitters during testicular regression and recrudescence in short photoperiod and after exposure of gonadally regressed hamsters to a long photoperiod (Steger et al., 1982, and unpublished observations). Turnover rates of norepinephrine and dopamine were estimated from the rate of their disappearance from the hypothalamic tissue 1 hr after intraperitoneal administration of α-methyl-para-tyrosine, which blocks catecholamine synthesis by inhibiting the activity of tyrosine hydroxylase. The rate of catecholamine decay is believed to be strictly comparable to the rate of its synthesis and thus to represent a useful measure of the functional state of catecholaminergic neurons. Results obtained to date indicate that short photoperiod induces an increase in hypothalamic LHRH concentration, a decrease in hypothalamic dopamine concentration (Fig. 9), and a decrease in turnover rates of both dopamine and norepinephrine in the hypothalamus. Spontaneous recrudescence of the testes is preceded by an increase in norepinephrine and dopamine turnover rates, an increase in hypothalamic norepinephrine content, and depletion of hypothalamic LHRH (Fig. 9). Pickard and Silverman (1979) have previously demonstrated that photoperiod does not

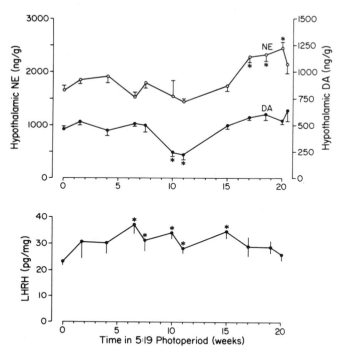

Figure 9. Effects of exposure to short photoperiod (5 : 19) on the concentration of norepinephrine and dopamine (top panel) and LHRH (bottom panel) in the hamster hypothalamus. Means ± SE. Asterisks denote values significantly different from the control value (time 0). (Reproduced from Steger *et al.*, 1982.)

modify the responsiveness of the hamster pituitary to LHRH stimulation. These results suggest that exposure of adult male golden hamsters to a short photoperiod leads to reduced activity of adrenergic and dopaminergic neurons in the hypothalamus and inhibition of LHRH release into the portal circulation. Eventual increases in the turn-over rate and the content of norepinephrine in the hypothalamus trigger recrudescence of the testes by stimulating LHRH and, consequently, LH and FSH release. We are uncertain whether the changes in the dopaminergic tone of the hypothalamus represent a cause or a consequence of changes in PRL release. There is experimental evidence to support both possibilities and we are presently attempting to elucidate this point by studying the effects of experimental alterations of peripheral PRL levels on hypothalamic neurotransmitters. Orstead and Benson (1980) reported that the reduction in pituitary and serum PRL levels in hamsters exposed to a short photoperiod is accompanied by reduced ability of the pituitary to produce PRL *in vitro* and by reduced release of PRL *in vivo* in response to urethane administration. It is not known whether these changes represent consequences of long-term suppression of PRL synthesis and release by endogenous dopamine or are due to alterations in mechanism(s) involved in stimulation of PRL release.

The hypothesis that the effects of photoperiod on the release of anterior pituitary hormones are mediated by changes in the synthesis of catecholamines in hypothalamic neurons is supported by our recent demonstration that significant changes in turnover rates of both norepinephrine and dopamine in the hamster hypothalamus are evident as early as one day after transferring gonadally regressed hamsters from a short to a long photoperiod (Steger and Bartke, unpublished observations). However, it is unlikely that photoperception directly regulates the activity of secretory neurons in the hypothalamus. There is a considerable amount of evidence to suggest that the primary site of action of photoperiod on neuroendocrine function is the pineal gland. The role of the pineal in mediating the effects of photoperiod on the pituitary and ultimately on the gonads will be discussed in Chapter 5.

5. TYPICAL HORMONE VALUES

Concentrations of LH, FSH, PRL, and testosterone in the peripheral circulation of adult male hamsters examined in different laboratories are listed in Tables 1 to 4. Some animals were maintained in a long (stimulatory) photoperiod of 14 : 10 or 16 : 8, whereas others were blinded or exposed to short photoperiod for periods ranging from 8 to 15 weeks. It is evident that the average hormone levels in different groups of control (i.e., gonadally active) animals can be quite different. These differences are evident both between and within different laboratories, even when the same preparation is used as standard reference in the radioimmunoassay system. Therefore, it would appear that there are substantial differences in the concentrations of plasma LH, FSH, and testosterone between different batches of adult male hamsters. Alternatively, differences in the procedures used for blood collection and/or the pulsatile nature of hormone release

Table 1. *Effects of Short Photoperiod or Blinding on LH Levels in Peripheral Circulation in Adult Male Golden Hamsters (Means ± SE)*

Treatment	Duration (weeks)	LH/ml of serum or plasma		Units	Reference
		Short photoperiod or blinded	Long photoperiod intact controls		
6 : 18	10	N.D.[a]	54 ± 29	ng RP-1	Stetson *et al.*, 1977
10 : 14	10	24 ± 2	146 ± 41	ng RP-1	Benson and Matthews, 1980
5 : 19	11.5	49 ± 22	56 ± 15	ng RP-1	Bex *et al.*, 1978
6 : 18	13	52 ± 8	129 ± 15	ng RP-1	Pickard and Silverman, 1979
5 : 19	13.5	46 ± 8	106 ± 24	ng RP-1	Bartke *et al.*, 1980b
6 : 18	13–14	N.D.[a]	50 ± 24	ng RP-1	Stetson *et al.*, 1977
Blinded	8	24 ± 6	146 ± 41	ng RP-1	Benson and Matthews, 1980
Blinded	12	1.7 ± 0.1	2.6 ± 0.2	ng NIAMDD[b]	Chen and Reiter, 1980

[a]N.D. = not detectable.
[b]NIAMDD LH preparation for iodination was used as standard in this study; its biological activity equals 33.3 × RP-1.

Table 2. Effects of Short Photoperiod or Blinding on FSH Levels in Peripheral Circulation in Adult Male Golden Hamsters (Means ± SE)

Treatment	Duration (weeks)	FSH/ml of serum or plasma		Units	Reference
		Short photoperiod or blinded	Long photoperiod intact controls		
6 : 18	10	103 ± 8	460 ± 62	ng RP-1	Stetson *et al.*, 1977
5 : 19	10	180 ± 18	292 ± 35	ng RP-1	Steger *et al.*, 1982
5 : 19	12	181 ± 31	179 ± 10	ng RP-1	Bartke *et al.*, 1980b
6 : 18	13–14	230 ± 73	239 ± 25	ng RP-1	Stetson *et al.*, 1977
Blinded	12	2.2 ± 0.1	3.2 ± 0.2	ng NIAMDD[a]	Chen and Reiter, 1980

[a]NIAMDD FSH preparation for iodination was used as standard in this study; its biological activity equals $47.6 \times$ RP-1.

could also account for these differences. It is difficult to determine whether similar differences exist also for plasma PRL, because different laboratories use different standard preparations.

In addition, data presented in Tables 1 to 4 illustrate the effects of short photoperiod and blinding on the pituitary–testicular axis. After two to three months, these treatments produce a precipitous decline in PRL levels (Table 3) and a marked decrease in testosterone levels (Table 4), as well as somewhat less pronounced but fairly consistent reductions in the levels of LH and FSH (Tables 1 and 2).

Table 3. Effects of Short Photoperiod or Blinding on PRL Levels in Peripheral Circulation in Adult Male Hamsters (Means ± SE)

Treatment	Duration (weeks)	PRL/ml of serum or plasma		Units	Reference
		Short photoperiod or blinded	Long photoperiod intact controls		
5 : 19	10	<8	58 ± 11	μl serum pool	Steger *et al.*, 1982
5 : 19	11.5	10 ± 2	52 ± 9	μl serum pool	Bex *et al.*, 1978
5 : 19	13.5	3 ± 1	46 ± 11	μl serum pool	Bartke *et al.*, 1980b
10 : 14	15	0.2 ± 0.1	10.6 ± 1.9	ng[a]	Borer *et al.*, 1982
Blinded	7	12 ± 0.5	22 ± 2	ng RP-1	Matthews *et al.*, 1978
Blinded	8	11 ± 0.7	24 ± 3	ng RP-1	Benson and Matthews, 1980
Blinded	10	8 ± 0.7	24 ± 3	ng RP-1	Benson and Matthews, 1980
Blinded	12	2.1 ± 0.1	3.1 ± 0.2	μg SHAP[b]	Chen and Reiter, 1980

[a]Results of this study were based on homologous hamster PRL radioimmunoassay; all other results were based partially or totally on reagents for measurement of rat PRL.
[b]SHAP = Standard hamster anterior pituitary.

Table 4. Effects of Short Photoperiod on Testosterone Levels in Peripheral Circulation in Adult Male Hamsters (ng/ml of Serum or Plasma; Means ± SE)

Treatment	Duration (weeks)	Testosterone		Reference
		Short photoperiod	Long photoperiod	
5 : 19	11.5	0.71 ± 0.15	1.59 ± 0.31	Bartke *et al.*, 1975
5 : 19	11.5	0.27 ± 0.05	1.05 ± 0.20	Bex and Bartke, 1977
5 : 19	12	0.47 ± 0.13	3.61 ± 0.24	Bartke *et al.*, 1981b
6 : 18	13	0.25 ± 0.07	1.95 ± 0.43	Pickard and Silverman, 1979
5 : 19	13.5	0.29 ± 0.05	1.85 ± 0.43	Bartke *et al.*, 1980b

ACKNOWLEDGMENTS. Unpublished studies referred to in this chapter were supported by NIH through grant HD 12642. I would like to express my gratitude to Dr. K. Matt for her helpful comments on the text of this chapter and to Ms. L. Rudloff and Ms. G. Small for their help in preparing the manuscript. I would also like to apologize to those investigators whose work related to this subject but were not discussed or quoted because of the limitations of space.

REFERENCES

Amir, S. M., Sullivan, R. C., and Ingbar, S. H., 1978, Binding of bovine thyrotropin to receptors in rat testis and its interaction with gonadotropins, *Endocrinology* 103:101–111.

Barkley, M. S., 1979, Serum prolactin in the male mouse from birth to maturity, *J. Endocrinol.* 83:31–33.

Bartke, A., 1980, Role of prolactin in reproduction in male mammals, *Fed. Proc.* 39:2577–2591.

Bartke, A., and Dalterio, S., 1975, Evidence for episodic secretion of testosterone in laboratory mice, *Steroids* 26:749–756.

Bartke, A., Steele, R. E., Musto, N., and Caldwell, B. V., 1973, Fluctuations in plasma testosterone levels in adult male rats and mice, *Endocrinology* 92:1223–1228.

Bartke, A., Croft, B. T., and Dalterio, S., 1975, Prolactin restores plasma testosterone levels and stimulates testicular growth in hamsters exposed to short day-length, *Endocrinology* 97:1601–1604.

Bartke, A., Smith, M. S., and Dalterio, S., 1979, Reversal of short photoperiod-induced sterility in male hamsters by ectopic pituitary homografts, *Int. J. Andrology* 2:257–262.

Bartke, A., Hogan, M. P., and Cutty, G. B., 1980a, Effects of human chorionic gonadotropin, prolactin, and bromocriptine on photoperiod-induced testicular regression and recrudescence in golden hamsters, *Int. J. Andrology* 1:115–120.

Bartke, A., Goldman, B. D., Bex, F. J., Kelch, R. P., Smith, M. S., Dalterio, S., and Doherty, P. C., 1980b, Effects of prolactin on testicular regression and recrudescence in the golden hamster, *Endocrinology* 106:167–172.

Bartke, A., Goldman, B. D., Klemcke, H. G., Bex, F. J., and Amador, A. G., i980c, Effects of photoperiod on pituitary and testicular function in seasonally breeding species, in: *Functional Correlates of Hormone Receptors in Reproduction* (V. Mahesh, T. G. Muldoon, B. B. Saxena, and W. A. Sadler, eds.), Elsevier North-Holland, New York, pp. 171–185.

Bartke, A., Siler-Khodr, T. M., Hogan, M. P., and Roychoudhury, P., 1981a, Ectopic pituitary transplants stimulate synthesis and release of follicle-stimulating hormone in golden hamsters, *Endocrinology* 108:133–139.

Bartke, A., Klemcke, H., and Amador, A., 1981b, Effects of testosterone, progesterone and cortisol on pituitary and testicular function in male golden hamsters with gonadal atrophy induced by short photoperiods, *J. Endocrinol.* 90:97–102.

Bartke, A., Klemcke, H. G., Amador, A., and Van Sickle, M., 1982, Photoperiod and regulation of gonadotropin receptors, in: *The Cell Biology of the Testis* (C. W. Bardin and R. J. Sherins, eds.), *Annals of the New York Academy of Sciences, Volume 383,* The New York Academy of Sciences, New York, pp. 122–134.

Beach, F. A., and Pauker, R. S., 1949, Effects of castration and subsequent androgen administration upon mating behavior in the male hamster, *Endocrinology* 45:211–221.

Benson, B., and Matthews, M. J., 1980, Possible role of prolactin and pineal prolactin-regulating substances in pineal-mediated gonadal atrophy in hamsters, *Horm. Res.* 12:137–148.

Berndtson, W. E., and Desjardins, C., 1974, Circulating LH and FSH levels and testicular function in hamsters during light deprivation and subsequent photoperiodic stimulation, *Endocrinology* 95:195–205.

Bex, F. J., and Bartke, A., 1977, Testicular LH binding in the hamster: Modification by photoperiod and prolactin, *Endocrinology* 100:1223–1226.

Bex, F., Bartke, A., Goldman, B. D., and Dalterio, S., 1978, Prolactin, growth hormone, luteinizing hormone receptors, and seasonal changes in testicular activity in the golden hamster, *Endocrinology* 103:2069–2080.

Boccabella, A. V., 1963, Reinitiation and restoration of spermatogenesis with testosterone propionate and other hormones after a long-term post-hypophysectomy regression period, *Endocrinology* 72:787–798.

Borer, K. T., Kelch, R. P., and Corley, K., 1982, Hamster prolactin: Physiological changes in blood and pituitary concentrations as measured by a homologous radioimmunoassay, *Neuroendocrinology* 35:13–21.

Campbell, C. S., Finkelstein, J. S., and Turek, F. W., 1978, The interaction of photoperiod and testosterone on the development of copulatory behavior in castrated male hamsters, *Physiol. Behav.* 21:409–415.

Charreau, E. H., Attramadal, A., Torjesen, P. A., Purvis, K., Calandra, R., and Hansson, V., 1977, Prolactin binding in rat testis: Specific receptors in interstitial cells, *Mol. Cell. Endocrinol.* 6:303–307.

Chen, H. J., 1981, Spontaneous and melatonin-induced testicular regression in male golden hamsters: Augmented sensitivity of the old male to melatonin inhibition, *Neuroendocrinology* 33:43–46.

Chen, H. J., and Reiter, R. J., 1980, The combination of twice daily luteinizing hormone-releasing factor administration and renal pituitary homografts restores normal reproductive organ size in male hamsters with pineal-mediated gonadal atrophy, *Endocrinology* 106:1382–1385.

Clermont, Y., and Harvey, S. C., 1967, Effects of hormones on spermatogenesis in the rat, *Ciba Found. Colloq. Endocrinol.* 16:173–189.

Corbier, P., Kerdelhue, B., Picon, R., and Roffi, J., 1978, Changes in testicular weight and serum gonadotropin and testosterone levels before, during and after birth in the perinatal rat, *Endocrinology* 103:1985–1991.

Corvol, P., and Bardin, C. W., 1973, Species distribution of testosterone-binding globulin, *Biol. Reprod.* 8:277–282.

Czyba, J. C., Cottinet, D., Dams, R., and Curé, M., 1971, Effets du sulpiride sur l'appareil genital du Hamster doré mâle en phase d'involution hivernale, *C.R. Seances Soc. Biol. Ses Fil.* 165:1624–1626.

Darrow, J. M., Davis, F. C., Elliott, J. A., Stetson, M. H., Turek, F. W. and Menaker, M., 1980, Influence of photoperiod on reproductive development in the golden hamster, *Biol. Reprod.* 22:443–450.

Davis, G. J., and Meyer, R. K., 1972, The effect of daylength on pituitary FSH and LH and gonadal development of snowshoe hares, *Biol. Reprod.* 6:264–269.

Davis, G. J., and Meyer, R. K., 1973a, FSH and LH of the pituitary gland of gonadectomized snowshoe hares, *Endocrinology* 92:340–344.

Davis, G. J., and Meyer, R. K., 1973b, Seasonal variation in LH and FSH of bilaterally castrated snowshoe hares, *Gen. Comp. Endocrinol.* 20:61–68.

Desjardins, C., Ewing, L. L., and Johnson, B. H., 1971, Effects of light deprivation upon the spermatogenic and steroidogenic elements of hamster testes, *Endocrinology* 89:791–800.

Döhler, K. D., and Wuttke, W., 1975, Changes with age in levels of serum gonadotropins, prolactin, and gonadal steroids in prepubertal male and female rats, *Endocrinology* 97:898–907.

Drickamer, L. C., Vandenburgh, J. G., and Colby, D. R., 1973, Predictors of dominance in the male golden hamster (*Mesocricetus auratus*), *Anim. Behav.* 21:557–563.

Ellis, G. B., and Desjardins, C., 1982, Male rats secrete luteinizing hormone and testosterone episodically, *Endocrinology* 110:1618–1627.

Ellis, G. B., and Turek, F. W., 1979, Time course of the photoperiod-induced change in sensitivity of the hypothalamic-pituitary axis to testosterone feedback in castrated male hamsters, *Endocrinology* 104:625–630.

Ellis, G. B., Losee, S. H., and Turek, F. W., 1979, Prolonged exposure of castrated male hamsters to a nonstimulatory photoperiod: Spontaneous change in sensitivity of the hypothalamic-pituitary axis to testosterone feedback, *Endocrinology* 104:631–635.

Ewing, L. L., Zirkin, B. R., Cochran, R. C., Kromann, N., Peters, C., and Ruiz-Bravo, N., 1979, Testosterone secretion by rat, rabbit, guinea pig, dog and hamster testes perfused *in vitro:* Correlation with Leydig cell mass, *Endocrinology* 105:1135–1142.

Festing, M. F. W., 1972, Hamsters, in: *The UFAW Handbook on the Care and Management of Laboratory Animals,* 4th Edition (Universities Federation for Animal Welfare, ed.), Williams and Wilkins, Baltimore, Maryland, pp. 242–256.

Foster, D. L., and Ryan, K. D., 1979, Endocrine mechanisms governing transition into adulthood: A marked decrease in inhibitory feedback action of estradiol on tonic secretion of luteinizing hormone in the lamb during puberty, *Endocrinology* 105:896–904.

Gaston, S., and Menaker, M., 1967, Photoperiodic control of hamster testis, *Science* 158:925–928.

Goldman, B. D., Matt, K. S., Roychoudhury, P., and Stetson, M. H., 1981, Prolactin release in golden hamsters: Photoperiod and gonadal influences, *Biol. Reprod.* 24:287–292.

Gravis, C. J., 1977, Testicular involution following optic enucleation: The Leydig cell, *Cell Tiss. Res.* 185:303–313.

Gravis, C. J., and Weaker, F. J., 1977, Testicular involution following optic enucleation: An ultrastructural and cytochemical study, *Cell Tiss. Res.* 184:67–77.

Hall, V., and Goldman, B., 1980, Effects of gonadal steroid hormones on hibernation in the Turkish hamster (*Mesocricetus brandti*), *J. Comp. Physiol.* 135:107–114.

Harman, S. M., 1980, Reproductive hormones in aging men, *J. Clin. Endocrinol. Metab.* 51:35–40.

Hudson, J. W., and Wang, L. C. H., 1979, Hibernation: Endocrinologic aspects, *Annu. Rev. Physiol.* 41:287–303.

Huhtaniemi, I. T., and Catt, K. J., 1981, Induction and maintenance of gonadotropin and lactogen receptors in hypoprolactinemic rats, *Endocrinology* 109:483–490.

Hutson, J. C., and Stocco, D. M., 1978, Specificity of hormone-induced responses of testicular cells in culture, *Biol. Reprod.* 19:768–772.

James, S. C., Hooper, G., and Asher, J. H., Jr., 1980, Effects of the gene *Wh* on reproduction in the Syrian hamster, *Mesocricetus auratus, J. Exp. Zool.* 214:261–275.

Judd, H. L., 1979, Biorhythms of gonadotropins and testicular hormone secretion, in: *Endocrine Rhythms* (D. T. Krieger, ed.), Raven Press, New York, pp. 299–321.

Karsch, F., 1980, Seasonal reproduction: A saga of reversible fertility, *Physiologist* 23:29–38.

Klemcke, H. G., Bartke, A., and Goldman, B. D., 1981, Plasma prolactin concentrations and testicular human chorionic gonadotropin binding sites during short photoperiod-induced testicular regression and recrudescence in the golden hamster, *Biol. Reprod.* 25:536–548.

Klemcke, H., Bartke, A., and Borer, K., 1982, Testicular PRL and hCG binding, and serum GH in male golden hamsters: Effects of photoperiod and time of day, *Int. J. Andrology* 3:20–21.

Kromer, W., 1980, Naltrexone influence on hibernation, *Experientia* 36:581.

Lau, I. F., Saksena, S. K., Dahlgren, L., and Chang, M. C., 1978, Steroids in the blood serum and testes of cadmium choloride treated hamsters, *Biol. Reprod.* 19:886–889.

Legan, S. J., and Karsch, F. J., 1979, Neuroendocrine regulation of the estrous cycle and seasonal breeding in the ewe, *Biol. Reprod.* 20:74–85.

Legan, S. J., Karsch, F. J., and Foster, D. L., 1977, The endocrine control of seasonal reproductive function in the ewe: A marked change in response to negative feedback action of estradiol on luteinizing hormone secretion, *Endocrinology* 101:818–824.

Lincoln, G. A., and Short, R. V., 1980, Seasonal breeding: Nature's contraceptive, *Recent Prog. Horm. Res.* 36:1–52.

Lincoln, G. A., Peet, M. J., and Cunningham, R. A., 1977, Seasonal and circadian changes in the episodic release of follicle-stimulating hormone, luteinizing hormone and testosterone in rams exposed to artificial photoperiods, *J. Endocrinol.* 72:337–349.

Lubicz-Nawrocki, C. M., 1974, The inhibition of fertilizing ability of epididymal spermatozoa by the administration of oestradiol benzoate to testosterone-maintained hypophysectomized or castrated hamsters, *J. Endocrinol.* 61:133–138.

McCann, S. M., Ojeda, S., and Negro-Vilar, A., 1974, Sex steroid, pituitary and hypothalamic hormones during puberty in experimental animals, in: *Control of the Onset of Puberty* (M. M. Grumback, G. D. Grave, and F. E. Mayer, eds.), John Wiley and Sons, New York, pp. 1–31.

MacKinnon, P. C. B., Mattock, J. M., and Ter Haar, M. B., 1976, Serum gonadotropin levels during development in male, female and androgenized female rats and the effect of general disturbance on high luteinizing hormone levels, *J. Endocrinol.* 70:361–371.

Macrides, F., Bartke, A., Fernandez, F., and D'Angelo, W., 1974, Effects of exposure to vaginal odor and receptive females on plasma testosterone in the male hamster, *Neuroendocrinology* 15:355–364.

Macrides, F., Bartke, A., and Dalterio, S., 1975, Strange females increase plasma testosterone levels in male mice, *Science* 189:1104–1106.

Maruniak, J. A., and Bronson, F. H., 1976, Gonadotropic responses of male mice to female urine, *Endocrinology* 99:963–969.

Matt, K. S., and Stetson, M. H., 1979, Hypothalamic-pituitary-gonadal interactions during spontaneous testicular recrudescence in the golden hamster (*Mesocricetus auratus*), *Biol. Reprod.* 20:739–746.

Matt, K. S., and Stetson, M. H., 1980, A compraison of serum hormone titers in golden hamsters during testicular growth induced by pinealectomy and photoperiodic stimulation, *Biol. Reprod.* 23:893–898.

Matthews, M. J., Benson, B., and Richardson, D. L., 1978, Partial maintenance of testes and accessory organs in blinded hamsters by homoplastic anterior pituitary grafts or exogenous prolactin, *Life Sci.* 23:1131–1138.

Mogler, R., 1958, Das endokrine System des Syrischen Goldhamsters unter Berücksichtigung des natürlichen Winterschlafs, *Z. Morph. Oekol. Tiere* 47:267–308.

Morin, L. P., Fitzgerald, K. M., Rusak, B., and Zucker, I., 1977, Circadian organization and neural mediation of hamster reproductive rhythms, *Psychoneuroendocrinology* 2:73–98.

Negro-Vilar, A., Krulich, L., and McCann, S. M., 1973, Changes in serum prolactin and gonadotropins during sexual development of the male rat, *Endocrinology* 93:660–664.

Nelson, J. F., Latham, K. R., and Finch, C. E., 1976, Plasma testosterone levels in C57BL/6J male mice: Effects of age and disease, *Acta Endocrinol.* 80:744–752.

Odell, W. D., and Swerdloff, R. S., 1976, Etiologies of sexual maturation: A model system based on the sexually maturing rat, *Recent Prog. Horm. Res.* 32:245–288.

Oppenheimer, J. H., Schwartz, H. L., and Surks, M. I., 1974, Tissue differences in the concentration of triiodothyronine nuclear binding sites in the rat: Liver, kidney, pituitary, heart, spleen and testis, *Endocrinology* 95:897–903.

Orstead, M., and Benson, B., 1980, Evidence for decreased release of PRL in hamsters placed in short photoperiod, *Anat. Rec.* 196:141A.

Pelletier, J., and Ortavant, R., 1975, Photoperiodic control of LH release in the ram, II. Light-androgens interaction, *Acta Endocrinol.* 78:442–450.

Pickard, G. E., and Silverman, A. J., 1979, Effects of photoperiod on hypothalamic luteinizing hormone releasing hormone in the male hamster, *J. Endocrinol.* 83:421–428.

Reiter, R. J., 1972, Evidence for refractoriness of the pituitary-gonadal axis to the pineal gland in golden hamsters and its possible implications in annual reproductive rhythms, *Anat. Rec.* 173:365–372.

Reiter, R. J., 1973/74, Influence of pinealectomy on the breeding capability of hamsters maintained under natural photoperiodic and temperature conditions, *Neuroendocrinology* 13:366–370.

Reiter, R. J., 1980, The pineal and its hormones in the control of reproduction in mammals, *Endocr. Rev.* 1:109–131.

Reiter, R. J., Sorrentino, S., Jr., and Hoffman, R. A., 1970, Early photoperiodic conditions and pineal. Antigonadal function in male hamsters, *Int. J. Fertil.* 15:163–170.

Reiter, J. R., Vaughan, M. K., Blask, D. E., and Johnson, L. Y., 1975, Pineal methoxyindoles: New evidence concerning their function in the control of pineal-mediated changes in the reproductive physiology of male golden hamsters, *Endocrinology* 96:206–213.

Roychoudhury, P., Cutting, R., and Goldman, B. D., 1980, Prolactin involvement in the testicular cycle in the Syrian hamster, in: *Program and Abstracts of the 62nd Annual Meeting of the Endocrine Society*, Washington, D.C., p. 253.

Selmanoff, M. K., Goldman, B. D., and Ginsburg, B. E., 1977, Developmental changes in serum luteinizing hormone, follicle stimulating hormone and androgen levels in males of two inbred mouse strains, *Endocrinology* 100:122–127.

Sinha, Y. N., Selby, F. W., and Vanderlaan, W. P., 1974, The natural history of prolactin and GH secretion in mice with high and low incidence of mammary tumors, *Endocrinology* 94:757–764.

Steger, R. W., Bartke, A., and Goldman, B. D., 1982, Alterations in neuroendocrine function during photoperiod induced testicular atrophy and recrudescence in the golden hamster, *Biol. Reprod.* 26:437–444.

Steinberger, E., and Ficher, M., 1969, Differentiation of steroid biosynthetic pathways in developing testes, *Biol. Reprod.* 1:119–133.

Stetson, M. H., Elliott, J. A., and Menaker, M., 1975, Photoperiodic regulation of hamster testes: Circadian sensitivity to the effects of light, *Biol. Reprod.* 13:329–339.

Stetson, M. H., Watson-Whitmyre, M., and Matt, K. S., 1977, Termination of photorefractoriness in golden hamsters—photoperiodic requirements, *J. Exp. Zool.* 202:81–87.

Swanson, L. J., Desjardins, C., and Turek, F. W., 1982, Aging of the reproductive system in the male hamster: Behavioral and endocrine patterns, *Biol. Reprod.* 26:791–799.

Swerdloff, R. S., Walsh, P. C., Jacobs, H. S., and Odell, W. D., 1971, Serum LH and FSH during sexual maturation in the male rat: Effect of castration and cryptorchidism, *Endocrinology* 88:120–128.

Tamarkin, L., Hutchison, J. S., and Goldman, B. D., 1976, Regulation of serum gonadotropins by photoperiod and testicular hormone in the Syrian hamster, *Endocrinology* 99:1528–1533.

Tamarkin, L., Katikinemi, M., Chan, V., Yellon, S., Klein, D., and Goldman, B., 1981, Photoperiod and pineal regulation of testis function in the Syrian hamster mediated by prolactin and by testicular prolactin and LH receptors, in: *Program and Abstracts of the 63rd Annual Meeting of the Endocrine Society*, Cincinnati, Ohio, p. 245.

Tate-Ostroff, B., and Stetson, M. H., 1981, Correlative changes in response to castration and the onset of refractoriness in male golden hamsters, *Neuroscience* 32:325–329.

Terada, N., Sato, B., and Matsumoto, K., 1980, Formation of 5β- and 5α-products as major C_{19}-steroids from progesterone *in vitro* in immature golden hamster testes, *Endocrinology* 106:1554–1561.

Turek, F. W., 1977, The interaction of photoperiod and testosterone in regulating serum gonadotropin levels in castrated male hamsters, *Endocrinology* 101:1210–1215.

Turek, F. W., Elliott, J. A., Alvis, J. D., and Menaker, M., 1975a, Effect of prolonged exposure to nonstimulatory photoperiods on the activity of the neuroendocrine-testicular axis of golden hamsters, *Biol. Reprod.* 13:475–481.

Turek, F. W., Elliott, J. A., Alvis, J. D., and Menaker, M., 1975b, The interaction of castration and photoperiod in the regulation of hypophyseal and serum gonadotropin levels in male golden hamsters, *Endocrinology* 96:854–860.

Vendrely, E., Guerillot, C., Basseville, C., and DaLage, C., 1971, Poids testiculaire et spermatogenese du Hamster dore au cours du cycle saisonniel, *C. R. Seances Soc. Biol. Ses Fil.* 165:1562–1565.

Vomachka, A. J., and Greenwald, G. S., 1979, The development of gonadotropin and steroid hormone patterns in male and female hamsters from birth to puberty, *Endocrinology* 105:960–966.

Vriend, J., 1981, The pineal and melatonin in the regulation of pituitary-thyroid axis, *Life Sci.* 29:1929–1936.

Vriend, J., Reiter, R. J., and Anderson, G. R., 1979, Effects of the pineal and melatonin on thyroid activity of male golden hamsters, *Gen. Comp. Endocrinol.* 38:189–195.

Zipf, W. B., Payne, A. H., and Kelch, R. P., 1978, Prolactin, growth hormone and luteinizing hormone in the maintenance of testicular luteinizing hormone receptors, *Endocrinology* 103:595–600.

Zirkin, B. R., Ewing, L. L., Kromann, N., and Cochran, R. C., 1980, Testosterone secretion by rat, rabbit, guinea pig, dog and hamster testes perfused *in vitro:* Correlation with Leydig cell ultrastructure, *Endocrinology* 107:1867–1874.

5

Pineal–Reproductive Interactions

RUSSEL J. REITER

1. INTRODUCTION

The Syrian hamster has been an extremely valuable experimental animal for clarifying the role of the pineal gland in the control of reproductive physiology. It was the first mammalian species in which a functional link between the pineal gland and the reproductive system was unequivocally established. The first of these reports appeared in the mid-1960s (Czyba *et al.*, 1964; Hoffman and Reiter, 1965a, 1966). In order to appreciate how little was known of the pineal–reproductive interactions only 15 years ago, the reader is referred to a chapter (Reiter and Hoffman, 1968) that appeared in the predecessor of the present book. In this review only four papers concerned with the influence of the pineal gland on sexual physiology are cited. In the intervening years numerous workers have utilized the Syrian hamster to define not only the interaction of the pineal secretory products with the reproductive system, but the mechanisms involved as well (Reiter, 1980a,b, 1981b; Stetson and Tate-Ostroff, 1981). It is the advances within the last 15 years that will receive primary consideration in the present resume.

2. PINEAL MORPHOLOGY

In the Syrian hamster the pineal consists of two portions, i.e., a superficial and a deep pineal gland (Sheridan and Reiter, 1970a). The subdivisions are ultrastructurally similar (Sheridan and Reiter, 1970b) and derive from the same embryonic anlage. The deep pineal is relatively small, being about 1/20th the size of the superficial gland. Although during the technique of superficial pinealectomy (Hoffman and Reiter,

RUSSEL J. REITER • Department of Cellular and Structural Biology, The University of Texas Health Science Center at San Antonio, San Antonio, Texas 78284.

1965b) only the superficial pineal is removed, the deep pineal is sympathetically denervated (Reiter and Hedlund, 1976) thereby rendering it nonfunctional. In a similar manner, bilateral superior cervical ganglionectomy interrupts the functional innervation to both the superficial and deep pineal gland and thus destroys their endocrine activity (Reiter and Hester, 1966). An intact peripheral sympathetic nervous system is known to be required for the normal functioning of the pineal gland of mammals.

Although it is generally conceded that the primary route of secretion of pineal hormones is into the capillaries that pass through the gland (Quay, 1974; Rollag et al., 1977), the epithalamus of the Syrian hamster exhibits modifications that may allow for the discharge of pineal secretory products directly into the cerebrospinal fluid (CSF) of the third ventricle. The superficial pineal is located directly under the confluence of the superior sagittal and transverse sinuses; this arrangement seemed to preclude the secretion of superficial pineal constitutents into the ventricular CSF until it was discovered (Sheridan et al., 1968) that an evagination of the third ventricle actually extends to the superficial gland and directly abuts it. This modification of the ventricle has been referred to as the suprahabenular recess and it would allow for the escape of pineal hormones directly into the third ventricle. Relative to the deep pineal gland of the Syrian hamster, there is also a modification that may allow for this portion of the pineal complex to discharge its products into the ventricle. The ependymal lining of the pineal recess, which is surrounded by the deep portion of the gland, is discontinuous thereby allowing some pinealocytes to come in direct contact with the CSF (Hewing, 1978). These cells could discharge their secretory products into the ventricular cavities; however, whether they do remains to be established.

The advantages of an intraventricular route of secretion have been summarized elsewhere (Reiter et al., 1975a). Although a CSF route of secretion is favored by some (Pavel, 1973), it is generally considered that the normal secretory route involves the peripheral circulation.

3. PINEAL MELATONIN SYNTHESIS

The major endocrine product of the pineal gland in mammals appears to be melatonin (Fig. 1). It is produced in the gland from the amino acid tryptophan (Axelrod et al., 1969). After the amino acid is taken up by the pinealocytes from the blood vascular system, it is metabolized to serotonin; the concentration of this constituent in the pineal gland exceeds that of any other organ (Giarman et al., 1960). As in other mammals, the concentration of serotonin in the hamster pineal gland exhibits a 24-hr rhythm with lowest levels being measured during the daily period of darkness (S. A. Matthews and R. J. Reiter, unpublished). Melatonin is synthesized from serotonin in a two-step process that involves the enzymes N-acetyltransferase (NAT) and hydroxy-indole-O-methyltransferase (HIOMT). First serotonin is N-acetylated to N-acetyl-serotonin (Klein and Weller, 1970) with this product being O-methylated to melatonin (Axelrod and Weissbach, 1960). Besides melatonin, other indole products are formed in the pineal gland of mammals. Of those that have been tested as to their reproductive

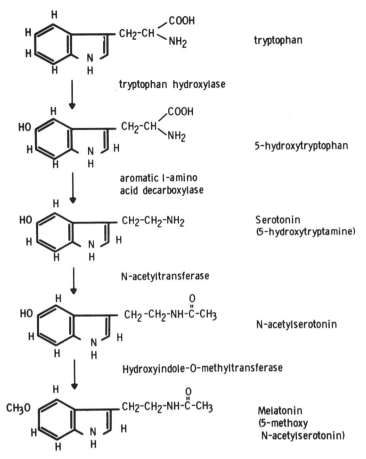

Figure 1. Conversion of tryptophan to melatonin as it has been shown to occur in the mammalian pineal gland.

consequences in hamsters (Reiter and Richardson, 1980), the one with the most consistent effects on sexual physiology undoubtedly is melatonin (Reiter, 1980b). Additionally, 5-methoxytryptamine may also induce changes in the reproductive physiology of hamsters (Pevet *et al.*, 1981); however, it seems to have only about 10% the inhibitory activity of melatonin (Rollag and Stetson, 1982).

3.1. Photoperiodic Influences

Well before assays were available to quantitate melatonin, the activity of the alleged rate-limiting enzyme, i.e., NAT, in melatonin production (Klein and Weller, 1970) was measured and found to have a rhythm in the pineal gland of the hamster

(Rudeen *et al.*, 1975). Under a light : dark cycle of 14 : 10, pineal NAT activity was found to be low at all time points during the light; during darkness, however, a threefold to fivefold increase in NAT activity was measured. From this it was assumed that pineal melatonin levels fluctuated accordingly. When measured three years later, it was indeed found that in both prepubertal and adult hamsters (Panke *et al.*, 1978) pineal melatonin concentrations are greatest late in the dark period and are usually closely correlated with the rhythm in the activity of the acetylating enzyme (Fig. 2). The fluctuations in pineal melatonin are of similar duration under a variety of different photoperiodic conditions (Panke *et al.*, 1979b, 1980; Tamarkin *et al.*, 1979; Reiter, 1981a), although Tamarkin and colleagues (1979) feel the peak may be significantly prolonged when the period of darkness is extended. It has been calculated that the hamster pineal gland produces about 18.6 ng melatonin per 24-hr period (Rollag *et al.*, 1980b).

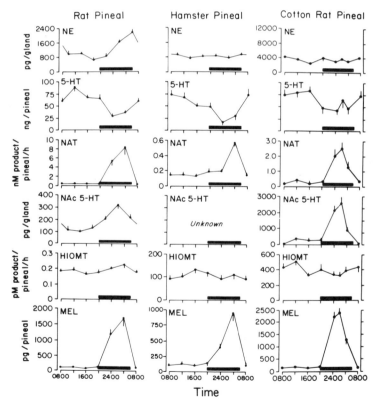

Figure 2. Changes in various constituents in the pineal gland of three species. The dark bar represents the period of darkness. NE = norepinephrine; 5-HT = serotonin; NAT = *N*-acetyltransferase; NAc 5-HT = *N*-acetylserotonin; HIOMT = hydroxyindole-*O*-methyltransferase; MEL = melatonin.

Since melatonin is believed not to be stored within the pineal gland in any appreciable quantity, plasma melatonin titers usually follow closely the production of the indole in the pineal gland. In the case of the hamster, the concentration of melatonin in the blood is greater at night than during the day (Brown *et al.*, 1981; Reiter, 1982; Reiter *et al.*, 1982).

The light : dark cycle is clearly the major overriding factor in determining melatonin production in the pineal gland of the hamster. Hence, if light is extended into the normal dark period, pineal melatonin levels remain low (Tamarkin *et al.*, 1979). Likewise, if hamsters are acutely exposed to light during the night when their pineal melatonin levels are elevated, the melatonin content of the gland is rapidly diminished (Rollag *et al.*, 1980b) with the half-time being on the order of 10 min. Very low irradiances of light are sufficient to depress nocturnal pineal melatonin in the domesticated hamster. According to Brainard and co-workers (1982b), an irradiance of 0.186 μW/cm^2 (1.08 lux) is sufficient to cause a maximal suppression of pineal melatonin content.

When hamsters are maintained under natural photoperiodic and temperature conditions throughout the year, they experience a waxing and waning of their sexual capability with the shortest days of the fall and winter being accompanied by involution of the reproductive organs. This seasonal regression is known to be mediated by the pineal gland, probably through the hormonal mediator melatonin (Reiter, 1980a,b). When pineal melatonin production in hamsters kept under natural environmental conditions was measured over five consecutive seasons (at the solstices and equinoxes), the indole fluctuated accordingly with the lowest levels of melatonin being measured in the summer and highest values being present during the winter months (Brainard *et al.*, 1982a). However, when the annual sexual cycle of the hamster was duplicated in the laboratory by appropriate photoperiodic manipulation, pineal melatonin levels did not vary significantly throughout the various phases of the circannual sexual cycle (Rollag *et al.*, 1980a). The differences in pineal indole production observed in these two experiments may be related to the fact that one group of animals (those outdoors) were also exposed to seasonal temperature fluctuations, whereas the indoor animals were maintained under a constant temperature throughout the various phases of the circannual cycle.

3.2. *Endocrine Influences*

Although in rats the synthesis of melatonin may be influenced by hormones from other endocrine glands, especially the gonads (Cardinali, 1981), in hamsters the evidence for such interactions is sparse. Indeed, when 24-hr pineal melatonin concentrations were measured during the various stages of the estrous cycle, it remained uniform throughout (Rollag *et al.*, 1979). In contrast, a similar study in rats revealed lowest levels of pineal melatonin during the estrous phase of the cycle (Johnson *et al.*, 1982). Removal of the harderian gland from male hamsters reportedly blunts the nocturnal rise in pineal melatonin concentration (Panke *et al.*, 1979a); this may be a hormonally mediated change.

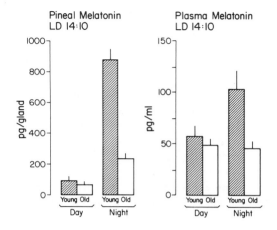

Figure 3. Pineal and plasma melatonin levels in young and old male hamsters killed at night or during the day.

3.3. Influence of Age

On the basis of minimal data from the human, it was assumed for years that the function of the pineal gland may decrease with age. A careful study of pineal melatonin levels in young and old male and female Syrian hamsters did, in fact, reveal a marked reduction in the nocturnal melatonin peak in the older hamsters (Reiter *et al.*, 1980a,b) (Fig. 3). Whereas in 2-month-old hamsters a nighttime peak of 800 to 900 pg melatonin/gland was reached, in 18-month-old animals the nocturnal values were 150 to 200 pg melatonin/gland. Both groups of hamsters had similar daytime levels (75 to 125 pg melatonin/pineal). Besides the levels of melatonin in the pineal, plasma titers of the indole also fall as hamsters age (Reiter *et al.*, 1982). There is still a question as to whether the reduction in pineal melatonin in aging hamsters is associated with a concomitant loss of function of the gland. It has been speculated that old hamsters may exhibit an increased sensitivity to melatonin (Reiter *et al.*, 1982).

4. PHOTOPERIOD, PINEAL, AND REPRODUCTION

The reproductive system of Syrian hamsters is exquisitely sensitive to changes in the length of the daily photoperiod. Thus, if hamsters are exposed to anything less than 12.5 hr of light daily, reproductive involution ensues (Elliot, 1976). The significance of this is that seasonal variations in photoperiod, because of the actions on the pineal gland, are important in determining circannual fluctuations in reproductive potential (Reiter, 1973a).

4.1. Males

When male Syrian hamsters are either totally deprived of light by blinding or if they are placed under short-day conditions (< 12.5 hr light daily), their reproductive organs (testes and accessory sex organs) totally involute unless the animals are either surgically pinealectomized (Hoffman and Reiter, 1965a) or have their pineal gland sympathetically denervated (Reiter and Hester, 1966). In hamsters kept under short days for a period of 8 to 10 weeks, the testes regress to 1/6 their original size (Hoffman and Reiter, 1965a), spermatogenesis is completely suppressed (Reiter, 1968a), and the epithelial cells lining the seminal vesicles become atrophic (Reiter, 1967). The structural changes in the testes and adnexa that accompany short-day exposure have been repeatedly confirmed (Reiter, 1980a). Without exception, the inhibitory effects of short daily photoperiods on the morphology of the reproductive organs of male hamsters are completely obviated either by pinealectomy or by its sympathetic denervation.

Besides the morphological changes, short-day exposure also alters circulating levels of reproductively related hormones. Hence, when measured by specific radioimmunoassays, plasma titers of luteinizing hormone (LH) (Berndtson and Desjardins, 1974; Tamarkin et al., 1976a), follicle-stimulating hormone (FSH) (Berndtson and Desjardins, 1974; Turek et al., 1976), prolactin (Bartke et al., 1977), and testosterone (Berndtson and Desjardins, 1974) are normally depressed (Fig. 4). Somewhat surprisingly, the depression in LH and FSH levels in the blood is not as consistent as might be predicted on the basis of the atrophic gonads. In the final analysis, it appears that prolactin may be very important in determining the functional status of the testes in male hamsters (Bartke, 1980). It generally appears that LH and FSH by themselves, in the absence of prolactin, may be only minimally capable of stimulating testicular function. When short-day-exposed male hamsters are treated with luteinizing hormone-releasing hormone (LH-RH), which releases both LH and FSH, and given intrarenal pituitary transplants, which release prolactin, only then is maximal testicular function maintained (Chen and Reiter, 1980). Neither treatment by itself, i.e., LH-RH administration or pituitary homografts, promote maximal testicular spermatogenesis and steroidogenesis. As with the structural changes, pineal removal reverses the inhibitory effects of short days on circulating hormone levels (Reiter, 1980b).

As with plasma hormone levels, short-day exposure leads to diminished pituitary levels of LH, FSH, and prolactin with the latter hormone exhibiting the most profound and consistent change (Berndtson and Desjardins, 1974; Reiter and Johnson, 1974a). Again, short days are ineffective in mediating these changes in pinealectomized male hamsters.

4.2. Females

The end organ responses of female hamsters to restricted daylengths are similar to those in males. Thus, prolonged short-day exposure causes ovarian involution, failure of ovulation, regression of the uterus, and cessation of cyclic vaginal phenomena (Hoffman and Reiter, 1966; Reiter, 1968b, 1969a; Sorrentino and Reiter, 1970). Unlike in males, however, the gonadal weights actually increase; this is primarily a result of the marked proliferation of the ovarian interstitial tissue (Reiter, 1968b). Unquestionably, female hamsters kept under short days are reproductively incompetent unless they are pinealectomized (Reiter, 1973a, 1980a).

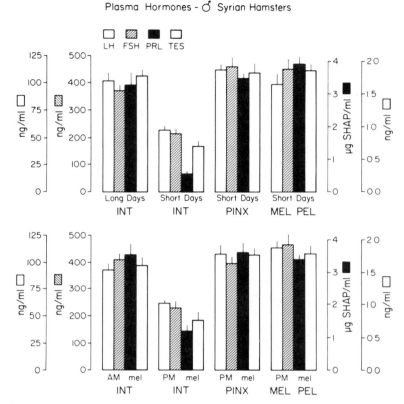

Figure 4. Plasma levels of LH, FSH, prolactin (PRL), and testosterone (TES) in male hamsters subjected to various treatments. INT = intact animals; PINX = pinealectomized animals; MEL PEL = hamsters bearing a subcutaneous implant of melatonin; AM mel = hamsters that received an early morning injection of melatonin; PM mel = animals that received an afternoon injection of melatonin.

The blood hormonal patterns of reproductively active substances in light-restricted females respond in what seems to be an unusual manner. Rather than being depressed as in males, they are actually elevated more frequently than normal. Reproductively competent females kept under long-day conditions experience an afternoon pulse of LH and FSH every fourth day. However, when the animals are experiencing pineal-mediated reproductive involution, the surge of LH and FSH occurs every afternoon (Bridges and Goldman, 1975; Seegal and Goldman, 1975). The significance of the seemingly paradoxical daily rise in the gonadotropins in the presence of involuted end organs remains unknown. Pinealectomy, besides reversing the morphological changes in the end organs induced by darkness, also prevents short days from causing the daily hormonal surges.

Pituitary levels of the gonadotropins in female hamsters respond in a similar manner when the animals are maintained under naturally or artificially shortened daylengths. When measured by radioimmunoassay, LH and FSH levels in the anterior pituitary gland increase significantly (Reiter and Johnson, 1974b,c). On the other hand, hypophyseal prolactin concentrations drop substantially (Reiter and Johnson, 1974b).

5. PINEAL HORMONES

Although the pineal gland unquestionably regulates, by hormonal means, the reproductive physiology of the male and female Syrian hamsters, the nature of the hormonal envoy(s) continues to be debated. The two general categories of potential hormones, i.e., the indoles and polypeptides, will be discussed separately.

5.1. Indoles

The pineal gland synthesizes a gamut of indoles anyone of which could mediate the effects of the pineal gland on the reproductive system. The data to date are overwhelmingly in favor of melatonin being the hormonal envoy (Reiter and Vaughan, 1977).

5.1.1. Antigonadotrophic Actions

Melatonin, when properly administered, can duplicate the effects of the stimulated pineal gland on reproductive physiology in both male and female hamsters. A critical aspect of its administration, however, is the timing of the injections. Many studies prior to the mid-1970s were unsuccessful in illustrating melatonin's antigonadotropic action because it was either injected at the improper time of the day (early in the light phase of the light : dark cycle) or it was administered by improper means (via subcutaneously placed pellets). The failure to demonstrate a neuroendocrine–reproductive action of melatonin confounded investigators since the reproductive system of the hamster was known to be exquisitely sensitive to the activated pineal gland and yet did not ostensibly respond to melatonin, the presumed pineal antigonadotropic envoy; the results of these early investigations have been summarized by Reiter (1974a).

In 1976, however, a revolutionary observation was made. Tamarkin and colleagues (1976b) reported that indeed the reproductive organs of hamsters do regress in response to melatonin provided it is injected daily late in the light phase; the dosage of melatonin used in this study was 25 μg daily. The injection of equivalent amounts of the indole early in the light period was totally ineffective in inhibiting sexual physiology. This was soon confirmed and furthermore it was shown that afternoon melatonin injections are only effective in suppressing the function of the neuroendocrine–reproductive system if the pineal gland is intact and sympathetically innervated (Reiter et al., 1976). This suggested that perhaps exogenously administered melatonin may be acting in concert with the pineal gland to induce gonadal atrophy.

The specific time interval during which the reproductive system of hamsters is sensitive to melatonin was further demarcated by the daily injection of the indole at various intervals after the onset of the light period. In hamsters kept under a light : dark cycle of 14 : 10, it was found that melatonin given between 6.5 and 13.75 hr following the dark to light transition was effective in suppressing reproductive physiology in male and female hamsters (Tamarkin et al., 1976b). Subsequently, Rollag and co-workers (1980b) established that as little as 1.6 μg melatonin daily in the afternoon totally suppresses the function of the neuroendocrine—reproductive system in the hamster. Finally, melatonin has also been shown to inhibit the gonads in pinealectomized hamsters, but the frequency of the melatonin injections must be increased to thrice daily (Tamarkin et al., 1977; Goldman et al., 1979).

As far as they have been examined, the gonadal regression that accompanies daily afternoon melatonin administration is accompanied by the same hormonal imbalances seen in hamsters with pineal-mediated reproductive collapse (Reiter, 1981b) (Fig. 4). Information in this area is, however, relatively scanty.

Besides melatonin, 5-methoxytryptamine has received some recent investigative effort as to its antigonadotropic potential. Hence, when injected in the afternoon, it, too, induces involution of the reproductive organs of hamsters (Pevet et al., 1981; Rollag and Stetson, 1982). According to the latter authors, however, 5-methoxytryptamine has only about 1/10 the activity of melatonin. The hormonal patterns in hamsters after they have been treated with 5-methoxytryptamine have yet to be investigated.

5.1.2. Counter Antigonadotropic Actions

Two years before it was discovered that afternoon melatonin injections cause reproductive atrophy in the Syrian hamster, it had been demonstrated that melatonin, given by means of a continuous-release subcutaneous pellet, actually prevented pineal-mediated reproductive regression (Reiter et al., 1974, 1975b). This action of melatonin was referred to as the counter antigonadotropic effect. At the time this observation was made, it raised serious questions about the ability of the indole to suppress sexual physiology. Soon after it was revealed that daily afternoon melatonin injections induce gonadal involution in the Syrian hamster, Reiter and colleagues (1977) examined the status of the reproductive organs in hamsters that possessed subcutaneous implants of melatonin, but were also given melatonin injections each afternoon. As the subcutaneous reservoirs of melatonin had prevented darkness-induced reproductive involution (Reiter et al., 1974, 1975b), so they also overcame the inhibitory actions of afternoon melatonin injections, i.e., continually available melatonin inhibited the action of daily melatonin administration (Fig. 4). Despite the seemingly paradoxical nature of these results, it provided the best evidence to that time that melatonin was in fact the pineal agent normally responsible for gonadal involution. The rationale for this judgement is as follows: Since subcutaneous melatonin depots prevented both darkness- and melatonin-induced gonadal regression, it was assumed that the atrophic responses in these two situations were due to the same factor, namely melatonin.

In 1978, a hypothesis was formulated to explain the complex actions of melatonin in reference to the reproductive system of the Syrian hamster (Reiter et al., 1978). This theory was further embellished and tested in subsequent years (Reiter, 1980a; Cardinali, 1981). Theoretically, melatonin has the ability to down regulate (either decrease their number or their sensitivity) its own receptors. Thus, when melatonin is available, it presumbaly interacts with its receptors and having done so, it renders them transiently insensitive to additional melatonin for a finite period of time. This theory explains the majority, if not all, of the known actions of melatonin in the Syrian hamster. For example, when hamsters are kept under light : dark cycles of 14 : 10, pineal melatonin is released late in the dark period, the indole circulates, is taken up by cells at its yet unknown site of action, and down regulates its receptors. When the lights come on in the morning, the melatonin receptors are still in a down-regulated state; this explains why the injection of melatonin early in the light phase does not cause gonadal regression. By late in the afternoon, however, the sensitivity of the melatonin receptors have been re-established and daily melatonin administration at this time causes involution of the neuroendocrine–reproductive axis (Fig. 5).

Obviously, when subcutaneous melatonin reserves are continually releasing the indole, the receptors are persistently occupied by melatonin and as long as this is so, the receptors are in a down-regulated state. This readily explains how melatonin pellets placed under the skin are able to prevent gonadal regression due to either short-day exposure or afternoon melatonin injections.

The theory, as outlined, has been further tested using another experimental paradigm. In this case it was reasoned that, if the theory is valid, perhaps a large injection of melatonin in the morning, although not causing gonadal regression, may prolong the receptors in a down-regulated state and thereby prevent the reproductive inhibitory actions of afternoon melatonin injections. When Chen et al. (1980) and Richardson and colleagues (1981) conducted the appropriate experiments, they obtained the results predicted by the theory. Thus, a 1-mg daily melatonin injection in the morning did not cause reproductive involution, but it completely overcame the action of 25 μg melatonin given each afternoon (Fig. 5). In essence, the 1-mg melatonin injection in the morning is equivalent to a continual-release subcutaneous melatonin reserve. These relations hold both for males and females of the species.

Relatively little is known concerning melatonin receptors themselves. The few studies that have been conducted indicate that in the hamster there are large numbers of hypothalamic melatonin receptors late in the light period as opposed to early in the light phase of the light : dark cycle (Vacas and Cardinali, 1979). This observation is consistent with the down regulation theory and suggests that perhaps it is a change in number rather than an alteration in sensitivity of the receptors to melatonin.

Attempts to disprove the down regulation theory of melatonin action to date have failed. Nevertheless, there may yet be other explanations for the complex actions of melatonin on the reproductive system. Finally, the reader is reminded that the theory was formulated to explain the actions of melatonin in the Syrian hamster and does not necessarily apply to other species.

5.2. Polypeptides

Besides the indoles, various polypeptides have been proposed as important pineal constituents in some species; in the hamster the effects of these compounds on reproduc-

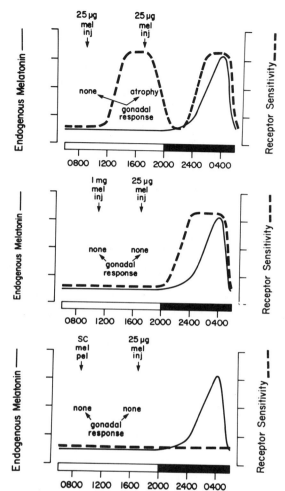

Figure 5. Theoretical explanation of the actions of melatonin on the reproductive system of the Syrian hamster. When melatonin acts, it may alter the sensitivity of its receptors. The dark bar is the period of darkness.

tion have either not been adequately tested or the experiments have yielded equivocal results (Vaughan, 1981). In one study where one of the polypeptides (arginine vasotocin or AVT) was used, it slowed gonadal and accessory organ growth in male hamsters (Vaughan *et al.*, 1974).

6. PINEAL AND SEASONAL REPRODUCTION

Inasmuch as daylength changes on an annual basis, pineal function is also altered accordingly. At virtually all latitudes and certainly in the area at which Syrian hamsters are normally found in the Middle East, during a portion of the year daylength is less than the minimal (i.e., 12.5 hr daily) required to maintain the gonads in a functionally mature state. Hence, in photosensitive species such as the hamster the changing photoperiod, because of its action on the pineal gland and associated structures, determines seasonal reproductive fluctuations. The phase of the annual cycle of reproduction as well as the role of the pineal gland in this cycle have been carefully outlined; the following phases have been defined: inhibition phase, sexually quiescent phase, restoration phase, and sexually active phase (Reiter, 1974b, 1975a) (Fig. 6).

6.1. Inhibition Phase

When Syrian hamsters are actually exposed to natural fluctuations in daylength, the fall of the year is associated with reproductive collapse (Reiter, 1973a, 1974b). This relates to the fact that at this time of the year daylengths fall below the necessary 12.5 hr required for maintenance of the reproductive organs (Elliott, 1976). The gonadal and accessory sex organ changes are identical to those described above for hamsters exposed to short daily photoperiods in the laboratory. In males, spermatogenic arrest is observed, the seminal vesicles and coagulating glands involute, and the hormonal patterns are altered accordingly (Reiter, 1973b). In females, the interstitial elements of the ovary proliferate, ovulation is interrupted, the uteri atrophy, and vaginal cyclicity ceases (Reiter, 1974c). If animals are pinealectomized during this interval, the decreasing days of the fall are incapable of inducing reproductive involution (Reiter 1973b, 1974b,c).

The period of time required for the gonads to regress, i.e., the inhibition phase, is believed to be approximately 8 weeks, although it may be shorter in hamsters exposed

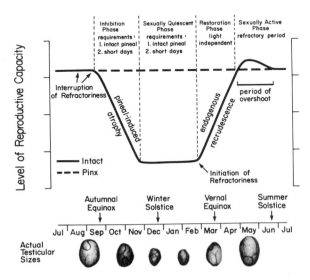

Figure 6. Presumed interrelationships of the photoperiod, the pineal gland, and seasonal reproduction in the hamster.

to naturally decreasing daylengths than in animals exposed to artificially shortened photoperiods of the laboratory. At the end of the inhibition phase, the hamsters are reproductively incompetent and physiologically prepared for hibernation (Reiter, 1978; Hall and Goldman, 1980) (Fig. 6).

6.2. Sexually Quiescent Phase

Following the collapse of the reproductive system, the animals enter a period of prolonged sexual dormancy (Reiter, 1973a,b). This normally coincides with the short days of the winter; at this time the animals may hibernate if the circumstances permit. Both males and females probably enter this phase at about the same time. Since the animals are sexually incompetent, failure of successful reproduction is ensured. This is important since young born during the winter months would be subjected to less than ideal environmental conditions and young born at this time would probably not survive (Reiter, 1973/74). If during the sexually quiescent phase hamsters are either pinealectomized or if they are moved into the long photoperiods of the laboratory, the gonads of these animals begin to regrow and reach their maximal size and function within about eight weeks.

The sexually quiescent phase of the annual cycle extends from the fall to at least midwinter. At some point after the winter solstice and by mechanisms that are not understood, the gonads begin to recrudesce. This happens even if hamsters are in short days and have an intact pineal gland. The neuroendocrine–reproductive axis merely becomes refractory to the inhibitory influence of the pineal hormone(s) (Reiter, 1975b). At the first point of spontaneous regeneration, by definition, the sexually quiescent phase is terminated (Fig. 6).

6.3. Restoration Phase

As noted above, the restoration phase begins and is probably completed while the animals are in the latter stages of hibernation and in the total darkness of their burrows (Reiter, 1975c). Indeed, the regrowth of the gonads with the resultant secretion of gonadal steroids is probably what terminates hibernation (Hall and Goldman, 1980). Gonadal recrudescence is a light-independent phenomenon that develops, as noted above, due to refractoriness of the reproductive system to the inhibitory influence of the pineal gland. Once the phase is initiated, gonadal recrudescence goes to completion and cannot be prevented with even total darkness due to blinding (Reiter, 1969a, 1975c). The restoration of the morphological and functional status of the reproductive system coincides with the emergence of hamsters above ground in the spring of the year; this concludes the restoration phase of the cycle (Fig. 6).

6.4. Sexually Active Phase

Very soon (perhaps within hours) after their emergence from hibernation in the spring, hamsters begin to mate. This initiates the sexually active phase. Considering the short gestation period (16 to 17 days) of the hamster, the first litters are born soon thereafter. It is likely that each female has several litters during this phase of the cycle.

The onset of refractoriness that initiated the restoration phase obviously extends into the phase of sexual activity since if in the spring or early summer hamsters are exposed to darkness, reproductive collapse does not ensue (Reiter, 1975b,c). Inasmuch as the reproductive system is refractory, the long days of the summer are not actually required to maintain the gonads in a functionally mature state; however, the long days play a very important role in the annual reproductive cycle, i.e., they eventually serve to interrupt the refractory condition thereby allowing the hypothalamopituitary–gonadal axis to become sensitive to the short days of the subsequent fall (Reiter, 1975b; Stetson et al., 1976).

In the final analysis, both long and short days are critical in determining the annual cycle of reproduction in the Syrian hamster. This circannual rhythm ensures that the young are born at the seasons that will favor their survival. In pinealectomized hamsters, the circannual fluctuation in reproduction is lost; thus they remain continually reproductively active (Reiter, 1973/74).

7. MELATONIN AND THE SEASONAL REPRODUCTIVE CYCLE

Despite the ease with which the seasonal cycle of reproduction is demonstrated when hamsters are kept under natural photoperiodic conditions, the nature of the pineal hormone that mediates the annual cycle remains contested. However, by far the strongest candidate for the endocrine mediator of these events is melatonin. When properly administered, melatonin can duplicate the seasonal cycle of reproduction in the laboratory. Certainly, when administered by means of a daily afternoon injection, it causes the same degree of reproductive involution as that seen in animals kept under short days (Tamarkin et al., 1976a,b; Reiter, 1973a). Once the gonads are atrophic, both short-day exposure and melatonin are capable of maintaining the sexual organs in an involuted condition for a finite period of time (Tamarkin et al., 1976; Reiter, 1969b). Thereafter, the system becomes refractory to both darkness and melatonin treatment (Reiter, 1975b; Bittman, 1978; Reiter et al., 1979). Finally, the refractoriness to both treatments is interrupted by long-day exposure (Reiter, 1975b; Stetson et al., 1976; Reiter et al., 1979). It is therefore obvious that it is not necessary to envoke any other pineal constituent in the mediation of the annual reproductive cycle. This does not mean that other pineal substances may not have important physiological consequences.

8. SITES OF ACTION OF PINEAL HORMONES

At which level of the neuroendocrine–reproductive axis the active pineal constituent intervenes to control reproduction is not agreed upon, although the preponderance of evidence is for a neural site (Blask, 1981).

8.1. Neuroendocrine Axis

Considering that melatonin receptors have been at least tentatively identified in the hypothalamus of the hamster (Vacas and Cardinali, 1979), a neural site of action of melatonin certainly seems likely. Although the actual data seem to be in conflict, measurement of the hypothalamic content of LH-RH in hamsters also suggests that the pineal of short-day-exposed hamsters acts within the central nervous system (Blask et al., 1979; Pickard and Silverman, 1979).

The claim has been made that melatonin specifically interacts with the suprachiasmatic nuclei, to regulate the reproductive state (Rusak, 1980). Another more conservative speculation merely suggests that melatonin may act in the anterior hypothalamic area (Reiter et al., 1981). Unlike in newborn rats, melatonin appears to be incapable of interfering with the action of LH-RH on LH release from incubated hamster pituitary glands (Bacon et al., 1981).

8.2. Gonads and Accessory Sex Glands

Although there is rather extensive literature indicating that various pineal constituents may control reproductive physiology by acting directly on the gonads or their adnexa (Blask, 1981), these findings have come from species other than the hamster. It appears, however, that the active pineal constituent cannot prevent the combined action of LH, FSH, and prolactin of the testes of Syrian hamsters (Chen and Reiter, 1980).

9. CONCLUDING REMARKS

Clearly, the Syrian hamster has been an extremely valuable species for identifying the interactions of the photoperiod and the pineal gland with the reproductive system. Indeed, the majority of what is known of these relationships have either been derived directly from experiments on the hamster or from studies that were designed according to previously published successful experiments that used this species. Despite all that is known, there is a dearth of information on the mechanisms involved. For example, refractoriness, the nature and location of the melatonin receptors along with the concept of their down regulation, and sites of action of melatonin and other pineal constituents require clarification. These will undoubtedly be fields of active investigation within the next decade.

REFERENCES

Axelrod, J., and Weissbach, H., 1960, Enzymatic O-methylation of N-acetylserotonin to melatonin, *Science* **131**:1312.

Axelrod, J., Shein, H. M., and Wurtman, R. J., 1969, Stimulation of ^{14}C-melatonin synthesis from ^{14}C-tryptophan by noradrenaline in rat pineal in organ culture, *Proc. Natl. Acad. Sci. U.S.A.* **62**:655–549.

Bacon, A., Sattler, C., and Martin, J. E., 1981, Melatonin effect on the hamster pituitary response to LHRH, *Biol. Reprod.* **24**:993–999.

Bartke, A., 1980, Role of prolactin in reproduction in male mammals, *Fed. Proc.* 39:2577–2581.

Bartke, A., Goldman, B. D., Bex, F. J., and Dalterio, S., 1977, Mechanism of reversible loss of reproductive capacity in a seasonally breeding mammal, *Int. J. Andrology* (suppl.) 2:345–353.

Berndtson, W. E., and Desjardins, C., 1974, Circulating LH and FSH levels and testicular function in hamsters during light deprivation and subsequent photoperiodic stimulation, *Endocrinology* 95:195–205.

Bittman, E. L., 1978, Hamster refractoriness: Role of insensitivity of pineal target tissues, *Science* 202:648–649.

Blask, D. E., 1981, Potential sites of action of pineal hormones within the neuroendocrine-reproductive axis, in: *The Pineal Gland,* Volume II: *Reproductive Effects* (R. J. Reiter, ed.), CRC Press, Boca Raton, pp. 189–216.

Blask, D. E., Reiter, R. J., Vaughan, M. K., and Johnson, L. Y., 1979, Differential effects of the pineal gland on LH-RH and FSH-RH activity in the medial basal hypothalamus of the male golden hamsters, *Neuroendocrinology* 23:36–43.

Brainard, G. C., Petterborg, L. J., Richardson, B. A., and Reiter, R. J., 1982a, Pineal melatonin in Syrian hamsters: Circadian and seasonal rhythms in animals maintained under laboratory and natural conditions, *Neuroendocrinology* 35:342–348.

Brainard, G. C., Richardson, B. A., Petterborg, L. J., and Reiter, R. J., 1982b, The effect of different light intensities on pineal melatonin content, *Brain Res.* 233:75–81.

Bridges, R. S., and Goldman, B. D., 1975, Diurnal rhythms in gonadotropins and progesterone in lactating and photoperiod induced acyclic hamster, *Biol. Reprod.* 13:613–622.

Brown, G., Grota, L., and Niles, L., 1981, Melatonin: Origin, control of circadian rhythm and site of action, in: *Melatonin—Current Status and Perspectives* (N. Birau and W. Schloot, eds.), Pergamon, New York, pp. 193–199.

Cardinali, D. P., 1981, Melatonin. A mammalian pineal hormone. *Endocr. Rev.* 2:237–250.

Chen, H. J., and Reiter, R. J., 1980, The combination of twice daily luteinizing hormone-releasing factor administration and renal pituitary homografts restores normal reproductive organ size in male hamsters with pineal-mediated gonadal atrophy, *Endocrinology* 106:1382–1385.

Chen, H. J., Brainard, G. C., and Reiter, R. J., 1980, Melatonin given in the morning prevents the suppressive action on the reproductive system of melatonin given in the late afternoon, *Neuronendocrinology* 31:129–132.

Czyba, J. C., Girod, C., et Durand, N., 1964, Sur l'antagonisme épiphysohypophysaire et les variations saisonnières chez le hamster doré (*Mesocricetus auratus*), *C. R. Seances Soc. Biol. Ses Fil.* 158:742–745.

Elliott, J., 1976, Circadian rhythms and photoperiodic time measurement in mammals, *Fed. Proc.* 35:2339–2346.

Giarman, N. J., Freedman, D. X., and Picard-Ami, L., 1960, Serotonin content of the pineal glands of man and monkey, *Nature (London)* 186:480–481.

Goldman, B., Hall, V., Hollister, C., Roychoudbury, P., Tamarkin, L., and Westrom, W., 1979, Effects of melatonin on the reproductive system in intact and pinealectomized male hamsters maintained under various photoperiods, *Endocrinology* 104:82–88.

Hall, V., and Goldman, B., 1980, Effects of gonadal steroid hormones on hibernation in the Turkish hamster (*Mesocricetus brandti*), *J. Comp. Physiol.* 135:107–114.

Hewing, M., 1978, A liquor contacting area in the pineal recess of the golden hamster (*Mesocricetus auratus*), *Anat. Embryol.* 153:295–304.

Hoffman, R. A., and Reiter, R. J., 1965a, Pineal gland: Influence of gonads of male hamsters, *Science* 148:1609–1611.

Hoffman, R. A., and Reiter, R. J., 1965b, Rapid pinealectomy in hamsters and other small rodents, *Anat. Rec.* 153:19–22.

Hoffman, R. A., and Reiter, R. J., 1966, Responses of some endocrine organs of female hamsters to pinealectomy and light, *Life Sci.* 5:1147–1151.

Johnson, L. Y., Vaughan, M. K., Richardson, B. A., and Reiter, R. J., 1982, Variation in pineal melatonin content during the estrous cycle of the rat, *Proc. Soc. Exp. Biol. Med.* 169:416–419.

Klein, D. C., and Weller, J. C., 1970, Indole metabolism in the pineal gland: A circadian rhythm in N-acetyltransferase, *Science* 169:1093–1094.

Panke, E. S., Reiter, R. J., Rollag, M. D., and Panke, T. S., 1978, Pineal serotonin N-acetyltransferase activity and melatonin concentrations in prepubertal and adult Syrian hamsters exposed to short daily photo-periods, *Endocr. Res. Commun.* 5:311–324.

Panke, E. S., Reiter, R. J., and Rollag, M. D., 1979a, Effect of removal of the Harderian glands on pineal melatonin concentrations in the Syrian hamster, *Experientia* 35:1405–1406.

Panke, E. S., Rollag, M. D., and Reiter, R. J., 1979b, Pineal melatonin concentrations in the Syrian hamster, *Endocrinology* 104:194–197.

Panke, E. S., Rollag, M. D., and Reiter, R. J., 1980, Effects of photoperiod on hamster pineal melatonin concentrations, *Comp. Biochem. Physiol.* 66A:691–693.

Pavel, S., 1973, Arginine vasotocin release into cerebrospinal fluid of cats induced by melatonin, *Nature (London)* 246:183–184.

Pevet, P., Haldar-Misra, C., Öcal, T., 1981, The independency of an intact pineal gland of the inhibition by 5-methoxytryptamine of the reproductive organs in the male hamster, *J. Neural Transm.* 52:95–106.

Pickard, G., and Silverman, A. J., 1979, Effects of photoperiod on hypothalamic luteinizing hormone releasing hormone in the male hamster, *J. Endocrinol.* 83:421–428.

Quay, W. B., 1974, *Pineal Chemistry*, C. C. Thomas, Springfield, Illinois.

Reiter, R. J., 1967, The effect of pinealectomy, pineal grafts, and denervation of the pineal gland on the reproductive organs of male hamsters, *Neuroendocrinology* 2:138–146.

Reiter, R. J., 1968a, Morphological studies on the reproductive organs of blinded male hamsters and the effects of pinealectomy or superior cervical ganglionectomy, *Anat. Rec.* 160:13–24.

Reiter, R. J., 1968b, Changes in the reproductive organs of cold-exposed and light-deprived female hamsters, *J. Reprod. Fertil.* 16:217–222.

Reiter, R. J., 1969a, Pineal function in long term blinded male and female golden hamsters, *Gen. Comp. Endocrinol.* 12:460–468.

Reiter, R. J., 1969b, Pineal-gonadal relationship in male rodents, in: *Progress in Endocrinology* (C. Gual, ed.), Excerpta Medica, Amsterdam, pp. 631–636.

Reiter, R. J., 1973a, Comparative physiology: Pineal gland, *Annu. Rev. Physiol.* 35:305–328.

Reiter, R. J., 1973b, Pineal control of a seasonal reproductive rhythm in male golden hamsters exposed to natural daylight and temperature, *Endocrinology* 92:423–430.

Reiter, R. J., 1973/74, Influence of pinealectomy on the breeding capability of hamsters maintained under natural photoperiodic and temperature conditions, *Neuroendocrinology* 13:366–370.

Reiter, R. J., 1974a, Pineal regulation of the hypothalamo-pituitary axis: Gonadotrophins, in: *Handbook of Physiology, Endocrinology IV*, Part 2 (E. Knobil and W. H. Sawyer, eds.), American Physiological Society, Washington, D.C., pp. 519–550.

Reiter, R. J., 1974b, Circannual reproductive rhythms in mammals related to photoperiod and pineal function: A review, *Chronobiologia* 1:365–395.

Reiter, R. J., 1974c, Pineal-mediated regression of the reproductive organs of female hamsters exposed to natural photoperiods during the winter months, *Am. J. Obstet. Gynecol.* 118:878–880.

Reiter, R. J., 1975a, The pineal gland and seasonal reproductive adjustments, *Int. J. Biometeorol.* 19:282–288.

Reiter, R. J., 1975b, Evidence for refractoriness of the pituitary-gonadal axis to the pineal gland in golden hamsters and its possible implications in annual reproductive rhythms, *Anat. Rec.* 173:365–372.

Reiter, R. J., 1975c, Exogenous and endogenous control of the annual reproductive cycle in the male golden hamster: Participation of the pineal gland, *J. Exp. Zool.* 191:111–120.

Reiter, R. J., 1978, Interaction of photoperiod, pineal and seasonal reproduction as exemplified by findings in the hamster, in: *The Pineal and Reproduction* (R. J. Reiter, ed.), Karger, Basel, pp. 169–190.

Reiter, R. J., 1980a, The pineal and its hormones in the control of reproduction in mammals, *Endocr. Rev.* 1:109–131.

Reiter, R. J., 1980b, The pineal gland: A regulator of regulators, *Prog. Psychobiol. Physiol. Psychol.* 9:323–356.

Reiter, R. J., 1981a, Chronobiological aspects of the mammalian pineal gland, in: *Biological Rhythms in*

Structure and Function (H. V. Mayersbach, L. E. Schewing, and J. E. Pauly, eds.), Alan R. Liss, New York, pp. 223–233.

Reiter, R. J., 1981b, Reproductive effects of the pineal gland and pineal indoles in the Syrian hamster and the albino rat, in: *The Pineal Gland,* Volume II: *Reproductive Effects* (R. J. Reiter, ed.), CRC Press, Boca Raton, Florida, pp. 45–81.

Reiter, R. J., 1982, Neuroendocrine effects of the pineal gland and of melatonin, in: *Frontiers in Neuroendocrinology,* Volume 7 (W. F. Ganong and L. Martini, eds.), Academic, New York, pp. 287–316.

Reiter, R. J., and Hedlund, L., 1976, Peripheral sympathetic innervation of the deep pineal gland of the golden hamster, *Experientia* 32:1071–1072.

Reiter, R. J., and Hester, R. J., 1966, Interrelationships of the pineal gland, the superior cervical ganglia and the photoperiod in the regulation of the endocrine systems of hamsters, *Endocrinology* 79:1168–1170.

Reiter, R. J., and Hoffman, R. A., 1968, The endocrine system, in: *The Golden Hamster* (R. A. Hoffman, P. F. Robinson, and H. Magalhaes, eds.), Iowa State Press, Ames, pp. 139–155.

Reiter, R. J., and Johnson, L. Y., 1974a, Depressant action of the pineal gland, the superior cervical ganglia and the photoperiod on pituitary luteinizing hormone and prolactin in male hamsters, *Horm. Res.* 5:311–320.

Reiter, R. J., and Johnson, L. Y., 1974b, Pineal regulation of immunoreactive luteinizing hormone and prolactin in light-deprived female hamsters, *Fertil. Steril.* 25:958–964.

Reiter, R. J., and Johnson, L. Y., 1974c, Elevated pituitary LH and depressed pituitary prolactin levels in female hamsters with pineal-induced gonadal atrophy and the effects of chronic treatment with synthetic LRF, *Neuroendocrinology* 14: 310–322.

Reiter, R. J., and Richardson, B. A., 1980, The physiology of melatonin, in: *Biochemical and Medical Aspects of Tryptophan Metabolism* (O. Hayaishi, Y. Ishimura, and R. Kido, eds.), Elsevier/North Holland, Amsterdam, pp. 247–256.

Reiter, R. J., and Vaughan, M. K., 1977, Pineal antigonadotrophic substances: Polypeptides and indoles, *Life Sci.* 21:159–172.

Reiter, R. J., Vaughan, M. K., Blask, D. E., and Johnson, L. Y., 1974, Melatonin: Its inhibition of pineal antigonadotrophic activity in male hamsters, *Science* 185:1169–1171.

Reiter, R. J., Vaughan, M. K., and Blask, D. E., 1975a, Possible role of the cerebrospinal fluid in the transport of pineal hormones in mammals, in: *Brain-Endocrine Interaction II* (K. M. Knigge, D. E. Scott, H. Kobayashi, and S. Ishii, eds.), Karger, Basal, pp. 337–354.

Reiter, R. J., Vaughan, M. K., Blask, D. E., and Johnson, L. Y., 1975b, Pineal methoxyindoles: New evidence concerning their function in the control of pineal-mediated changes in the reproductive physiology of male golden hamsters, *Endocrinology* 96:206–213.

Reiter, R. J., Blask, D. E., Johnson, L. Y., Rudeen, P. K., Vaughan, M. K., and Waring, P. J., 1976, Melatonin inhibition of reproduction in the male hamster: Its dependency on time of administration and on an intact and sympathetically innervated pineal gland, *Neuroendocrinology* 22:107–116.

Reiter, R. J., Rudeen, P. K., Sackman, J. W., Vaughan, M. K., Johnson, L. Y., and Little, J. C., 1977, Subcutaneous melatonin implants inhibit reproductive atrophy in male hamsters induced by daily melatonin injections, *Endocr. Res. Commun.* 4:35–44.

Reiter, R. J., Rollag, M. D., Panke, E. S., and Banks, A. F., 1978, Melatonin: Reproductive effects, *J. Neural Transm. Suppl.* 13:209–223.

Reiter, R. J., Petterborg, L. J., and Philo, R. C., 1979, Refractoriness to the antigonadotrophic effects of melatonin in male hamsters and its interruption by exposure of the animals to long daily photoperiods, *Life Sci.* 25:1571–1576.

Reiter, R. J., Johnson, L. Y., Steger, R. W., Richardson, B. A., and Petterborg, L. J., 1980a, Pineal biosynthetic activity and neuroendocrine physiology in the aging hamster and gerbil, *Peptides* (suppl. 1) 1:69–77.

Reiter, R. J., Richardson, B. A., Johnson, L. Y., Ferguson, B. N., and Dinh, D. T., 1980b, Pineal melatonin rhythm: Reduction in aging Syrian hamsters, *Science* 210:1372–1373.

Reiter, R. J., Dinh, D. T., de los Santos, R., and Guerra, J. C., 1981, Hypothalamic cuts suggest a brain site for the antigonadotrophic action of melatonin in the Syrian hamster, *Neurosci. Lett.* 23:315–318.

Reiter, R. J., Vriend, J., Brainard, G. C., Matthews, S. A., and Craft, C. M., 1982, Reduced pineal and plasma melatonin levels and gonadal atrophy in old hamsters kept under winter photoperiods, *Exp. Aging Res.* **8**:27–30.

Richardson, B. A., Vaughan, M. K., Brainard, G. C., Huerter, J. J., de los Santos, R., and Reiter, R. J., 1981, Influence of morning melatonin injections on the antigonadotrophic effects of afternoon melatonin administration in male and female hamsters, *Neuroendocrinology* **33**:112–117.

Rollag, M. D., and Stetson, M. H., 1982, Melatonin injection into Syrian hamsters, in:*The Pineal and Its Hormones* (R. J. Reiter, ed.), Alan R. Liss, New York, pp. 143–151.

Rollag, M. D., Morgan, R. J., and Niswender, G. D., 1977, Route of melatonin secretion in sheep, *Biomed. Sci. Instrum.* **13**:111–117.

Rollag, M. D., Chen, H. J., Ferguson, B. N., and Reiter, R. J., 1979, Pineal melatonin content throughout the hamster estrous cycle, *Proc. Soc. Exp. Biol. Med.* **162**:211–213.

Rollag, M. D., Panke, E. S., and Reiter, R. J., 1980a, Pineal melatonin content in male hamsters throughout the seasonal reproductive cycle, *Proc. Soc. Exp. Biol. Med.* **165**:330–334.

Rollag, M. D., Panke, E. S., Trakulrungsi, W., Trakulrungsi, C., and Reiter, R. J., 1980b, Quantification of daily melatonin synthesis in the hamster pineal gland, *Endocrinology* **106**:231–236.

Rudeen, P. K., Reiter, R. J., and Vaughan, M. K., 1975, Pineal serotonin-N-acetyltransferase activity in four mammalian species, *Neurosci. Lett.* **1**:225–229.

Rusak, B., 1980, Suprachiasmatic lesions prevent an antigonadal effect of melatonin, *Biol. Reprod.* **22**:148–154.

Seegal, R. F., and Goldman, B. D., 1975, Effects of photoperiod on cyclicity and serum gonadotrophins in the Syrian hamster, *Biol. Reprod.* **12**:223–231.

Sheridan, M. N., and Reiter, R. J., 1970a, Observations on the pineal system in the hamster. I. Relations of the superificial and deep pineal to the epithalamus, *J. Morphol.* **131**:153–162.

Sheridan, M. N., and Reiter, R. J., 1970b, Observations on the pineal system in the hamster. II. Fine structure of the deep pineal, *J. Morphol.* **131**:163–178.

Sheridan, M. N., Reiter, R. J., and Jacobs, J. J., 1968, An interesting anatomical relationship between the hamster pineal gland and the ventricular system of the brain, *J. Endocrinol.* **45**:131–132.

Sorrentino, S., Jr., and Reiter, R. J., 1970, Pineal-induced alteration of estrous cycles in blinded hamsters, *Gen. Comp. Endocrinol.* **15**:39–42.

Stetson, M. H., and Tate-Ostroff, B., 1981, Hormonal regulation of the annual reproductive cycle of golden hamsters, *Gen. Comp. Endocrinol.* **45**:329–344.

Stetson, M. H., Matt, K. S., and Watson-Whitmyre, M., 1976, Photoperiodism and reproduction in golden hamsters: Circadian organization and the termination of photorefractoriness. *Biol. Reprod.* **14**:531–539.

Tamarkin, L., Hutchinson, J. S., and Goldman, B. D., 1976a, Regulation of serum gonadotropins by photoperiod and testicular hormone in the Syrian hamster, *Endocrinology* **99**:1528–1533.

Tamarkin, L., Westrom, W. K., Hamill, A. I., and Goldman, B. D., 1976b, Effect of melatonin on the reproductive systems of male and female Syrian hamsters: A diurnal rhythm in sensitivity to melatonin, *Endocrinology* **99**:1534–1541.

Tamarkin, L., Hollister, C. W., Lefebvre, N. G., and Goldman, B. D., 1977, Melatonin induction of gonadal quiescence in pinealectomized Syrian hamsters, *Science* **198**:953–955.

Tamarkin, L., Reppert, S. M., and Klein, D. C., 1979, Regulation of pineal melatonin in the Syrian hamster, *Endocrinology* **104**:385–389.

Turek, F. W., Alvis, J. D., Elliott, J. A., and Menaker, M., 1976, Temporal distribution of serum levels of LH and FSH in adult male golden hamsters exposed to long or short days, *Biol. Reprod.* **14**:630–637.

Vacas, M. I., and Cardinali, D. P., 1979, Diurnal changes in melatonin binding sites of hamster and rat brains. Correlation with neuroendocrine responsiveness to melatonin, *Neurosci. Lett.* **15**:259–263.

Vaughan, M. K., 1981, The pineal gland—A survey of its antigonadotrophic substances and their actions, in: *International Review of Physiology,* Volume 24: *Endocrine Physiology III* (S. M. McCann, ed.), University Park, Baltimore, pp. 41–95.

Vaughan, M. K., Vaughan, G. M., and Klein, D. C., 1974, Arginine vasotocin: Effect on development of reproductive organs, *Science* **186**:938–939.

III

Social Behaviors

6

Communication

ROBERT E. JOHNSTON

1. INTRODUCTION

In this chapter, I concentrate on the analysis of specific signals and their behavioral functions. More integrative treatments concerning sexual and aggressive behavior and their endocrine mechanisms can be found in other chapters of this volume. Section 2, on postures used during social interactions, should be useful for anyone beginning behavioral studies of hamsters. In Sections 3, 4, and 5 I have summarized and evaluated the more extensive literature on auditory and chemical communication.

2. POSTURES USED DURING SOCIAL INTERACTIONS

In this section I describe some of the most obvious and typical postures that hamsters use during social interactions. The information is distilled from personal observations and a few relevant references (Dieterlen, 1959; Grant and Mackintosh, 1963; Johnston, 1976; Floody and Pfaff, 1977). Although the postures of hamsters are rarely stereotyped in the sense that many displays of birds or reptiles are—being nearly invariant in their form, timing, rhythm, or repetition rate—hamsters do perform a few actions that can easily be distinguished and seem to have communication functions. However, except for the analysis of upright postures (Johnston, 1976), there has not been any research to quantitatively demonstrate such functions.

First I describe postures that would typically be seen during the course of agonistic interactions, and then I describe aspects of sexual interactions. The purpose is not to provide a quantitative analysis of the sequences of interactions (compare Johnston, 1976; Floody and Pfaff, 1977c), but to indicate how and when the postures are used.

Nose-to-nose. (Fig. 1) Most social interactions begin with two individuals opposing their noses, apparently sniffing one another. This quickly changes to a brief mutual

ROBERT E. JOHNSTON • Department of Psychology, Cornell University, Ithaca, New York 14853.

Figure 1. Postures during social interactions between hamsters (two males). (A) Nose-to-nose. Two males that have fought previously; subordinate is on the right, displaying greater "tension" in the rigidity of his stance, more erect ears, more widely opened eyes. (B) Mutual investigation in head region. Both males are also in a Paw-up posture, in which one paw is raised to fend off the other male. (C) Upright. Subordinate male on left assumes upright as dominant male approaches. (D) On Back. Male partially hidden from view has frozen in this posture, leading to a temporary cessation of a rolling fight.

investigation of each other's face and head (Fig. 1). Investigation in the facial region may provide hamsters with information such as sexual identity and possibly species and individual identity, since functionally relevant responses may follow immediately. For example, estrous females may turn away from a male, engage in the solicitation walk, and then go into lordosis; a male that has been beaten in a fight by another may briefly sniff him and then flee.

Circle, Mutual Sniff. In an anti-parallel position, two individuals circle one another while sniffing the other's body, particularly the genital region. The attempt to sniff the genitals may also lead to a Head Under posture, in which one of the pair pushes its head under the body of the other, usually from a position to the side of or anti-parallel to the one investigated (e.g., see Fig. 2).

At any time one individual may Walk Away, and the other may or may not Follow, two activities that are important parts of interactions that are not particularly stereotyped or distinguished in any way.

Sparring or Mutual Agonistic Postures, including Upright and Sideways Postures. During both mild and intense agonistic interactions hamsters often engage in one or more series of "sparring" interactions in which one animal attempts to press the attack by making

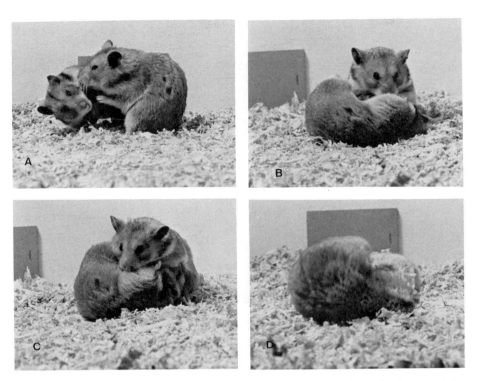

Figure 2. Attack sequence between two males. (A) Dominant aggressor on left, subordinate on right. Dominant is beginning attack by attempting to get at subordinate's ventrum. Subordinate in Upright. (B) Attack continues. 0.6 sec. later, dominant moves under subordinate; this male still in upright posture. (C) Attack continues. 0.3 sec later, dominant has grabbed a mouthful of fur. (D) 0.3 sec later, Rolling Fight is well underway.

contact with the mouth on the other's ventral surface while the second animal attempts to keep the first at bay or fend him off. One type of posture seen in these interactions is the Upright (Figs. 1 and 2), in which the hamster has both forepaws off the substrate, its body angled upward and usually facing its opponent. Grant and Mackintosh (1963) claimed that there are two distinct types of Uprights, a Defensive Upright and an Offensive Upright, but during agonistic interactions individuals constantly jockey for position and modify the angle of their bodies, extension of their forelegs and paws, and their orientation to one another; for the majority of cases I think it would be difficult or impossible to correctly identify the aggressor or defender on the basis of the posture alone (Johnston, 1976). However, if two hamsters with a clear difference in status are interacting with one another and both are simultaneously in upright postures, the subordinate will be standing more erect, with its paws held higher and more extended from the body, the digits more extended and the mouth more open (Fig. 1) (Johnston,

Figure 3. Attack and Chase sequence. [(All approximately 0.3 sec intervals, except (F) follows by 0.6 sec.)] (A) As dominant (left) Attacks, subordinate executes a Fly Away. (B) Both have recovered, dominant presses attack while subordinate flees. (C) Subordinate escapes but runs into corner and so as dominant approaches, (D) subordinate leaps away, (E) leaving dominant temporarily empty-handed. (F) But not for long; dominant has followed subordinate to the other side of the enclosure and launches a vicious, leaping attack.

1976). From the illustrations presented in Grant and Mackintosh (1963) one would probably conclude the opposite. The most extreme form of the upright posture as performed by subordinate males can be quite rigid and held for long periods of time. This rigidity, duration, degree of uprightness, and limb and digit extension reliably distinguishes it from actions a dominant male would perform. Furthermore, when such subordinates adjust their position they do so in a rapid, rather jerky manner, whereas a dominant male moves more slowly and smoothly. Maintaining an upright posture is effective in forestalling attacks by dominants (Johnston, 1976). The same type of upright posture is observed in response to predators or their odors in both golden hamsters (Dieterlen, 1959) and European hamsters (Eibl-Eibesfeldt, 1953); the upright in response to predators appears to closely resemble the extreme form of upright posture displayed by subordinate males.

Another common posture that occurs during this phase of agonistic interactions is Paw Up, in which the hamster lifts one paw and turns its head slightly away from an approaching individual (Fig. 1). This seems to me to be the same as the "defensive sideways" of Grant and Mackintosh (1963). It is characteristically performed when one hamster approaches another from the side, and may be effective in fending off an attack.

Attack can be initiated from either an upright or a standing (on all four paws) position, and can be directed at an individual in either position. If the attackee is in an upright position, the attacker rotates and turns its head and upper body so that it can bite the attackee at about the junction between the dark fur of the dorsum and the light fur of the underside. One or both forepaws are placed on the opponents side (Fig. 2) (see also Floody and Pfaff, 1977). If the opponent is on all fours the attacker approaches from the side, rotates and turns it head and upper body so that it can push under the opponent, and again directs this action at the midsection of the opponent. In either case the attackee is knocked off his feet and the two hamsters engage in a Rolling Fight (Fig. 2), in which the two are usually wrapped around each other at right angles, and roll about like an animated ball. Both grab mouthfuls of fur and skin. At first such a fight may be relatively slow paced and not too vigorous. During this phase (or immediately after Attack) there may be occasional pauses in the action, apparently due to one of the pair freezing in the On Back position (Fig. 1). Sometimes the animal in the On Back may be able to get up and move away without an immediate resumption of the Rolling Fight, but on other occasions the aggressor attacks again and the Rolling Fight resumes.

Often there is a point at which there is a qualitative change in the nature of the Rolling Fight. Squeaking is heard and the animals' movements become extremely vigorous and rapid; to the observer it now appears to be really serious and vicious. During this type of fight severe wounds may be inflicted. This phase of the fight can be at least temporarily terminated by Fly Away, an explosive movement that allows one individual to disentangle itself from the other (Fig. 3) (see also Dieterlen, 1959; Floody and Pfaff, 1977). Fly Away is performed by individuals that have lost a fight: after this maneuver such an individual repeatedly flees from the other, although this still may not be insurance against vicious attacks during a chase (Fig. 3). In high-intensity fights the fleeing animal seems panicked: it dashes about, crashing into the walls of enclosures,

Figure 4. Various postures. (A) Tail-up Freeze. Subordinate male (left) fled to the top of a wooden platform and remained immobile in the Tail-up posture; dominant male then approached. Note the arched back of the subordinate. (B) Subordinate male, in Tail-up Freeze, is mounted by dominant male. (C) Lordosis by female (left) while male sniffs and licks genital region and the vaginal secretions extruded by female.

and leaping wildly. The pursuer, although avid, usually moves more slowly. In a confined space, all of this leaping about often has the effect of slowing down the pursuit, sometimes just because the pursuer gets left behind, but other times because he may be knocked over by a violent action.

The loser of a fight (or an animal that is subordinate to another) may perform a Tail-up Move or the Tail-up Freeze (Fig. 4) first described by Dieterlen (1959) and also reported by Eibl-Eibesfeldt (1953) for the European hamster. The tail is raised and the back is arched upward convexly (opposite to that of the lordosis posture) in both the Tail-up Move and Tail-up Freeze. When walking in this posture a hamster often proceeds with a peculiar, stiff-legged gait. In Tail-up Freeze the hamster remains immobile. Both Tail-up postures are quite effective in turning off further attacks; in the case of a male aggressor this may be because these postures resemble lordosis, and tend to elicit sexual tendencies incompatible with fighting. Occasionally one male will mount another that is in Tail-up Freeze posture (Fig. 4). Although males use this tail-up posture when severely beaten by a female, females have not been observed to mount males in response (Dieterlen, 1959). I have not observed females use this posture during fights with males or other females, nor has Floody (personal communication). Females, however, use what is apparently the same posture as a means of sexual solicitation before they are sexually receptive (Ciaccio *et al.*, 1979a). Finally, Dieterlen (1959) noted that young hamsters that were introduced into a strange nest used the tail-up posture, again apparently having the function of forestalling attacks.

During sexual interactions most of these agonistic activities are not observed. After an initial nose-to-nose exchange, females indicate their readiness to copulate by turning away and shuffling off in a Solicitation Walk and by extruding vaginal secretions (see Section 4.2). The characteristics of this walk cannot be captured in a photograph or drawing because its uniqueness seems to lie in the type of movement, quick little steps ending in the assumption of the Lordosis posture (Fig. 4). At present it is not clear if this walk is unique and distinctive by itself or not. Lordosis itself is unusual in hamsters because of its duration and the ease with which it is elicited in estrous females. Females may remain immobile in this posture for five to ten minutes at a time, and even for several minutes when the male is not present. Lordosis can be elicited by tactile stimulation of the genital region or back, by the smell and ultrasonic calls of males (Floody and Pfaff, 1977b), or even by being followed by a female hamster. While females maintain lordosis and before the males mount, the males sniff and lick the genital region (Fig. 4), consuming relatively large quantities of vaginal secretions extruded by the females. Males also spend a good deal of time sniffing the head region, possibly investigating harderian gland or ear gland scent, but the special significance of this interest in the head has not been determined (see Sections 4 and 5).

The postures described above seem to be the most stereotyped of those that hamsters engage in and they all seem to have a clear significance in the course of interactions. This is not to say that other, less obvious behaviors may not also have communicative functions. In male–female interactions, for example, approaches by females may act as a threat to males (Steel, 1980). As another example, vaginal scent marking is often performed by nonestrous females when being followed by males, and it is possible that the sight of a female marking has some significance for males.

2.1. Do the Dark Chest Bars Have a Communicatory Significance?

Hamsters of the genus *Mesocricetus* and the European hamster, *Cricetus cricetus* have an area of dark fur underneath their forelegs that contrasts sharply with the otherwise light pelage of the ventrum. This dark and light pattern is exposed when an animal stands up on its hind legs, and is more evident the more upright an animal is and the more it extends its forelegs. Grant *et al.*, (1970) suggested that this dark and light pattern might serve as a visual threat stimulus since, they claimed, it was exposed more completely by aggressive, dominant individuals than by subordinates. Johnston (1976), however, showed that subordinates actually exposed the dark chest pattern more completely than dominants did, and suggested that at best it might serve as a defensive threat. European hamsters engage in this type of posture when opposing predators (Eibl-Eibesfeldt, 1953) and it may be useful in this context for *Mesocricetus* species as well.

In a second line of evidence, Grant *et al.* (1970) dyed the chests of some hamsters so that the dark area was both larger and darker, and found that males with artificially enhanced chest patches became dominant over unaltered males when they were allowed to interact. There were several flaws in their experimental design, most notably that the pairs of males were tested first in the cages of the males that had been dyed, thus giving these males a home cage advantage. Johnston (1976) repeated the experiment using weight-matched pairs that encountered each other in a familiar but neutral arena and found that 11 of 14 black-chested males won their encounter with the sham-dyed normal males ($P < 0.029$, one-tailed binomial test). However, in a subsequent study (D. Sherman, B. Finlay, and R. Johnston, unpublished observations) undertaken to determine the effects of various visual system lesions on this effect, we were unable to replicate this finding. We used a total of 30 weight-matched pairs of males, one member of each having an enlarged and darkened chest patch and the other having a normal chest patch (sham dyed with water). Fourteen of the males with black chests won their initial encounter. It thus appears unlikely that alteration of the chest patches has an effect on the outcome of agonistic encounters; combining the data from the two experiments done in my laboratory yields no significant difference in fight outcome, and the results obtained by Grant *et al.* (1970) can be explained on other grounds (Johnston, 1976).

3. AUDITORY COMMUNICATION

The only sounds produced by hamsters that have been analyzed in any detail are ultrasonic ones that occur primarily in sexual contexts (see below). Hamsters do make other sounds, some of which probably have communicatory functions.

3.1. Sonic Signals

Adult hamsters produce noisy, audible calls in a variety of situations. During Rolling Fights loud squeaks are often heard by observers, usually just prior to termination of the fight by Fly Away and Chase. When hamsters are startled and assume an upright posture (similar to that seen in agonistic encounters), they sometimes produce a noisy sound. When they are picked up and held firmly by humans they may also produce a loud squeak, almost always in association with sudden, vigorous explosive movement (probably the same as the Fly-Away maneuver). Since none of these adult calls have been analyzed spectrographically, it is not clear how similar they are in structure. They do all occur in conditions of extremely high arousal, suggesting similar causal mechanisms.

One other sound that probably has communicative functions is teeth chattering. The sound produced is a rapid staccato one as if the teeth were repeatedly struck together. According to Dieterlen (1959) the hamster's lower jaw is protruded when it is making this sound so that the lower incisors are in front of the upper ones. Dieterlen observed pairs of hamsters in the home cage of one member of the pair and found that teeth chattering occurred most often in interactions between two males (in 92% of the encounters), much less often in female–female encounters (39%), and least in encounters between males and females (5%). The severity of fighting between pairs of animals follows the same rank order (Johnston, 1975a, 1977a), suggesting that teeth chattering is directly related to the degree of agonistic tension or arousal. In Dieterlen's observations, the intruding individual's teeth chattered more often than the home cage animal's. In encounters between like-sex individuals in which teeth chattering occurred, the intruder was the first to teeth chatter in 75% of the pairings of males and in 83% of the encounters between females, suggesting that teeth chattering is more closely related to defensive tendencies and "fear" than to aggressive or attack tendencies. Although observations of my own agree with those of Dieterlen, I would caution that it is often difficult to tell which of the two animals is the one producing the sound.

3.2. Ultrasonic Calling and Communication

Infant hamsters produce a noisy, wide-band sound that may sometimes be purely ultrasonic or may include sonic components. Presumably these sounds elicit maternal attention and care, as they seem to do in mice and rats (Allin and Banks, 1972; Smotherman et al., 1974), but I am not aware of any studies that prove such functions in hamsters.

Adult hamsters of both sexes produce 60 to 170 msec calls in the frequency range of 20 to 60 kHz, the average dominant frequency being 32 to 42 kHz (Floody and Pfaff, 1977a). Some of these calls are relatively constant in frequency, whereas others show considerable frequency modulation. The intensity of the calls of females has been measured to be 48 to 62 dB with the microphone 5 to 15 cm from the animal. Floody and Pfaff (1977a) estimate that under ideal conditions in open air, the effective range of such calls would be 11 to 23 m, but in nature it is likely to be less due to sound scattering by obstacles, background noise, etc. These calls are used primarily in sexual contexts, and based on a number of intersecting lines of evidence it has been suggested that they function in the attraction and location of a sexual partner and in the facilitation of copulatory activity (Floody and Pfaff, 1977b; Floody, 1979). The evidence on which these conclusions are based is summarized briefly in the paragraphs that follow.

First, the frequency of calling by female hamsters is correlated with their reproductive condition. During estrous cycles females call most frequently when in estrus but rarely call when diestrus (Floody et al., 1977). Females also rarely call when pregnant or lactating, or when they are acyclic due to exposure to short days (Johnston, unpublished observations). High rates of calling are dependent on high levels of estrogen followed by a surge of progesterone, the same sequence of hormonal events underlying lordosis and receptivity (Floody et al., 1979a). Thus females are most likely to call when they are sexually receptive, a correlation that strongly suggests a sexual-advertisement function. Calling by male hamsters is also strongly influenced by gonadal hormones. Castrated males display very low rates of calling, whereas castrates injected with 200 μg/day testosterone propionate called at rates comparable to their precastration levels (Floody et al., 1979b).

Second, the social signals that influence ultrasonic calling suggest sexual functions. Both male and female hamsters are stimulated to call by contact with animals of the opposite sex or their odors (Floody *et al.*, 1977). Although informal observations indicate that odors of same sex individuals do not stimulate calling, no experimental proof has been published. The olfactory basis of this effect has been further documented in female hamsters. Elimination of the sense of smell by olfactory bulbectomy drastically reduces calling frequency (Kairys *et al.*, 1980). The sources of male odors that stimulate calling by females are unknown, although preliminary experiments in my laboratory indicate that cues are androgen dependent, since females call more in the presence of odors of intact males than castrated males. The olfactory cues that stimulate calling by males are primarily contained in the vaginal secretions of females. Male hamsters call frequently in areas that are vaginally marked by intact females, but rarely call in areas marked by vaginectomized females (Johnston and Kwan, 1984).

Third, the timing of calls during actual interactions suggests that hamsters use calls to facilitate sexual interactions. When a male and an estrous female first meet and begin to interact, they produce calls; presumably both the male and female call at such times, but this has not been definitely established. Once the female assumes the lordosis posture (remaining immobile for minutes at a time), she apparently does not call (Floody and Pfaff, 1977a). If the male wanders off or if he is removed by the experimentor, females, on coming out of lordosis, begin to move about and to call, as if they were attempting to locate and/or attract the male back. Males will likewise call after removal of a female.

Males also call when females are in lordosis, especially during intermount intervals (Floody and Pfaff, 1977b), and these calls help to keep the female immobile. Estrous females exposed to synthetic ultrasounds after removal of male partners maintain the lordosis posture much longer than females not exposed to ultrasounds. Females occasionally assume the lordosis posture in the presence of male odors when ultrasounds are played to them (Floody and Pfaff, 1977b). Ultrasounds thus seem to facilitate lordosis, and calling is probably one means by which males stimulate this receptive posture by females and maintain them in it. It would be interesting to know more about the specificity of this response; is lordosis facilitated by any sounds that might be made by a hamster (such as those made while moving about) or is facilitation specific to ultrasounds of male hamsters?

Finally, hamsters show several other responses to hearing ultrasounds that are consistent with the proposed sexual functions. Floody and Pfaff (1977b) found that both males and females were more attracted to the arm of a Y maze from which synthetic ultrasounds were emanating than to the arm with no ultrasounds. Another common response to conspecific ultrasounds is to call in return. By mutually calling and approaching the source of another animal's call, hamsters could quickly locate one another (Floody and Pfaff, 1977b).

Some of the most interesting questions that remain at the behavioral level of analysis concern the kinds of information that may be provided by the ultrasonic calls. Do the calls allow a hamster to identify the sex of the caller? Floody and Pfaff (1977a) have shown that there appear to be some consistent differences in the structure of the calls of males and females, most notably in that female calls have more frequency changes and a greater proportion of female calls have silent gaps of 5 msec or longer. However, at the present time we have no indication that hamsters use this information. The fact that males and females readily respond by calling to simplified, synthesized calls and to crude imitations by humans (Floody and Pfaff, 1977b; Johnston, unpublished observations) suggests that they may not respond differently to male and female calls, which differ in relatively subtle ways.

Do hamster calls provide information about changes in motivation of the caller? That is, do the calls produced vary consistently in different social contexts to provide different messages, or can the variation that is observed in call structure be attributed to individual differences and random variation? Floody and Pfaff (1977a) showed that calls produced by males after brief encounters with estrous females tended to be shorter in duration and had narrower bandwidths than those produced by males during contact with a female in lordosis. Whether these differences have any functional significance is unknown. Analysis of calls in a greater variety of situations might reveal other interesting differences in call structure. It is curious that despite the fact that many species of rodents seem to use ultrasounds for purposes of communication, no one has yet shown that any rodent has an extensive repertoire of ultrasonic signals.

Another type of information that could be encoded in hamster ultrasounds is species identity. Both Turkish hamsters (*M. brandti*) and Djungarian hamsters (*Phodopus sungorus*) produce ultrasounds in sexual contexts (Frank and Johnston, 1981; Levin and Johnston, unpublished observations). The causation of calls by female Turkish hamsters appears to be quite similar to that observed for Syrian hamsters, suggesting similar functions (Frank and Johnston, 1981). Although we have not yet undertaken a thorough, quantitative analysis, the calls of these three species appear to differ. The calls of *Phodopus sungorus* are qualitatively distinct, with a range of frequencies of 56 to 72 kHz compared with the range for *M. auratus* of 32 to 42 kHz (Levin and Johnston, unpublished observations).

3.3. Ultrasounds and Echolocation

Some small terrestrial mammals, such as shrews, use ultrasounds as an aid in exploration and navigation in their environment (Gould *et al.*, 1964; Sales and Pye, 1974; Buchler, 1976). A brief note in the literature claims that hamsters also do (Kahmann and Ostermann, 1951). In an extensive but as yet unpublished series of experiments in my laboratory, we have failed to confirm this claim. Initially our results were encouraging: When male hamsters were placed on a platform in the light or in the dark, the latency to jump off of a platform was directly proportional to the height of the platform. This finding was repeated several times with both golden hamsters, *M. auratus*, and Turkish hamsters, *M. brandti*. Once we obtained equipment to monitor and record ultrasounds, however, we were not able to detect any ultrasounds in this situation. Attempts to train thirsty hamsters to jump from a large platform to a nearby small one for a water reward were unsuccessful, as were attempts to train them to go to the long side of a tunnel maze for a water reward. Finally, when we tried to determine the effects of plugging the ears of hamsters on our basic latency-to-jump phenomenon, we failed to replicate it. Thus it is highly unlikely that hamsters regularly use ultra-sound for the purposes of echolocation, although it is possible that under some circumstances they could make use of such information, much as humans can use sonic cues if necessary.

4. CHEMICAL COMMUNICATION: SCENT SIGNALS AND SCENT MARKING

By far the greatest quantity of research on communication in hamsters has concerned communication by chemical signals. In this section, the sources of scent signals and the mechanisms of scent-marking behavior are discussed; in Section 5, the functions of odor signals are examined.

4.1. Flank Glands and Flank Marking

The flank glands are two round to slightly oblong regions of darkly pigmented sebaceous glands on the dorsal flank. They occur in the three species of *Mesocricetus* that I have examined (*M. auratus, M. brandti,* and *M. newtoni*), as well as in the European hamster (*Cricetus cricetus;* Eibl-Eibesfeldt, 1953). These glands are sexually dimorphic in secretory activity and size, measuring 7 to 10 mm in length in males and 2 to 4 mm in females (Vandenbergh, 1973). This sex difference is androgen dependent: Castration greatly reduces the size and activity of the flank glands of males, and replacement therapy with testosterone propionate or other androgens restores them (Hamilton and Montagna, 1950; Vandenbergh, 1973, 1977). The flank glands of females can be made to resemble those of males by giving females injections of androgens (Frost *et al.,* 1973). This sexual difference occurs at puberty and is correlated with the growth and higher secretory activity of the testes (Algard *et al.,* 1966; Miller *et al.,* 1977). The flank glands produce a thick, slightly yellowish and greasy secretion that appears on the surface in small, irregularly shaped particles. Compared with adjacent skin, the glandular area contains greatly enlarged subaceous glands filled with many lipid droplets, greatly enlarged hair follicles, and dendritic cells containing dark brown granules similar to those of melanocytes (Montagna and Hamilton, 1949; Hamilton and Montagna, 1959; Frost *et al.,* 1973).

Hamsters deposit the secretion of the flank glands by use of a specialized scent marking behavior known as flank marking, in which the animal arches its back as if to expose the gland and rubs its side against vertical surfaces such as the walls of its cage or burrow, or the entrances to tunnels, nest boxes, and so on (see Fig. 5) (Dieterlen, 1959; Johnston, 1975a). Flank gland secretions may also be deposited on the ground or floor of an enclosure from an animal's paws, since it it likely that flank gland secretions are transferred to the paws during grooming, especially when hamsters scratch the flanks with the hind paws. Secretions from other sources (such as the ear glands and the harderian glands) may be mixed with those of the flank gland when an animal marks, since the whole side of the body usually makes contact with the substrate.

Figure 5. Scent-marking behaviors. (A) Male hamster flank marking the walls of enclosure by arching the back out toward wall and vigorously rubbing his entire side against it. (B) Female hamster vaginal marking by pressing the genital region against the substrate and moving forward a few inches. The short tail is almost always slightly raised.

The causation of flank marking has been investigated in considerable detail (Dieterlen, 1959; Johnston, 1975a,b,c 1977a). Hamsters flank mark in both social and nonsocial contexts. They mark with low frequency and intensity in nonsocial contexts, such as just before or after grooming, when exploring the environment, or as they enter or leave their nest area. However, the most frequent and vigorous marking is observed in social contexts. Hamsters are stimulated to mark by odors of conspecifics and by contact with conspecifics (Johnston, 1975a, 1977a). At least for males, the odor of another male's flank mark seems to be an important component of the stimulatory odors (Johnston, 1975b,c). Odors of another species in the same genus are not nearly as stimulating, further suggesting that flank marking behavior serves communicative functions within the species (Johnston and Brenner, 1982).

Flank marking seems to be an aspect of agonistic behavior, and to be on the aggressive side of the aggressive-to-submissive motivational continuum. In pairs of either males or females that have established a dominance relationship the dominant individual flank marks much more frequently than the subordinate (Johnston, 1975a, 1977a), and in groups of four males or four females the dominant individual flank marks significantly more than any of the others (Drickamer and Vandenbergh, 1973; Drickamer et al., 1973). In a study of pairs of males in laboratory enclosures in which they could interact at will, subordinate males rarely marked and when they did it was always in or immediately adjacent to their nest compartment, often in the context of defending the nest against entry by the dominant male (Johnston, 1975c, and unpublished observations). Hamsters rarely mark during actual interactions with another individual. Usually they mark after contact has been discontinued, although occasionally they seem to break off interactions in order to move a little distance away and mark, and I have occasionally observed one female flank mark a second female that was in lordosis (Johnston, 1977a).

Among male hamsters, flank-marking frequency is greatly influenced by androgen levels. The frequency of marking in the presence of odors from another male is greatly reduced after castration but is restored to precastration levels with testosterone propionate therapy (Johnston, 1979). Marking is not entirely dependent on the presence of the testes, however, since castrated males that engage in frequent aggressive interactions sometimes mark as frequently as intact males (Whitsett, 1975). Hormonal influences on flank marking by females have not been explicitly studied, although in experiments investigating hormones and aggression it was found that ovariectomy reduced flank-marking frequency and replacement therapy with either testosterone propionate or estradiol benzoate increased marking levels above those seen in ovariectomized females (Vandenbergh, 1971).

Sexual motivation inhibits flank marking. Males mark less frequently after encounters with estrous as opposed to diestrous females, they flank mark less in the empty home cages of estrous as opposed to diestrous females, and their marking frequency is suppressed by exposure to vaginal secretions of females (Johnston, 1975a, 1980). Females flank mark at high rates throughout their estrous cycle in the presence of odors from other females, but in the presence of odors from males, flank marking is drastically reduced on their estrous days (that is, when they are receptive to males). The suppression of flank marking by sexual tendencies tends to reinforce the notion that this type of marking is primarily an agonistic behavior, since encounters of adult individuals seem to be limited to being either sexual or agonistic.

The information obtained so far suggests that flank marking may on the one hand have general broadcast functions, advertising to other animals that, for example, a particular individual male *M. auratus* in reproductive condition is occupying an area. The specific function of such marks will depend on the social context, what other individual perceives them, and so forth. On the other hand, the correlations of high marking frequency with high agonistic tension also suggest a specific role for flank marking in defense of a burrow, nest, or home area. Several reports have addressed these issues and are discussed in Section 5.

4.2. Vaginal Secretions and Vaginal Marking

Hamsters have especially copious vaginal secretions that change in both volume and consistency with their reproductive state. These changes are so consistent that they can be used with virtually 100% reliability to monitor estrous cycles (Orsini, 1961).

Morphological specializations of the vagina may be involved in the production and deployment of the secretion. There are two lateral pouches of the vagina near the vulva; the epithelium in these pouches is continually growing and sloughing off, and the sloughed cells remain in the pouches and form two leaflike masses (Deanesly, 1938). The pouches first appear about 17 days postpartum and then undergo dramatic enlargement on days 27 to 29, coincident with the growth of the vagina, uterus, ovary, and fallopian tubes just prior to the first ovulation at about 30 days. In ovariectomized females, the lateral pouches appear at 17 days of age but do not enlarge (LaVelle, 1951). The upper vaginal epithelium contains mucus-secreting cells that undergo cyclic growth during the estrous cycle, reaching a peak during estrus, and shedding the following day (Deanesly, 1938). Urethral and preputial (or clitoral) glands are also present, but whether these glands contribute active components to the vaginal secretions is unknown. Dieterlen (1959) claimed that the secretions expressed from the vagina were produced in the clitoral gland, but the reasons for his belief were not explained. Female hamsters presumably have some muscular control over the vaginal walls and the lateral pouches, since they leave a thin streak of quickly drying secretion when vaginal marking yet extrude large quantities of secretion, including accumulated cellular debris, during precopulatory interactions with males.

Hamster vaginal secretions contain appreciable amounts of at least several hundred chemical compounds. Among the volatile components, positive identification has been made of three sulfur-containing compounds, eight fatty acids, and four alcohols (Singer et al., 1976, 1983; O'Connell et al., 1978). One of these compounds, dimethyl disulfide, has been identified as a major component of the attractive properties of vaginal secretions, but so far little behavioral role has been found for any other components (see Section 5). Much of the sexual aphrodisiac effects seem to be associated with the nonvolatile fraction (Macrides et al., 1977; Frey, 1978). Two compounds that were characterized as steroids were found to be behaviorally inactive (Watson, Frey, Meutterties, and Johnston, unpublished results).

Female hamsters deposit vaginal secretions in the environment by a scent-marking behavior known as vaginal marking (Dieterlen, 1959; Johnston, 1977a). By pressing their genital region against the substrate and moving forward, they deposit streaks of clear secretion. These marks dry within a few minutes, leaving a thin, translucent film that can be easily seen on a smooth, clean surface (Johnston and Schmidt, 1979). In addition to information on the effects of vaginal secretions (see Section 5), several lines of evidence suggest that vaginal marking is a major means of sexual advertisement for female hamsters, important in advertising an approaching estrus and aiding in attraction of males to a female's home area.

First, females direct their vaginal marks toward males. They mark more frequently in an area containing the odors of males than in a clean area, whereas they mark less in areas containing another female's odors than in clean areas, suggesting that vaginal marks are directed at males or areas scent marked by males (Fig. 6) (Johnston, 1977a).

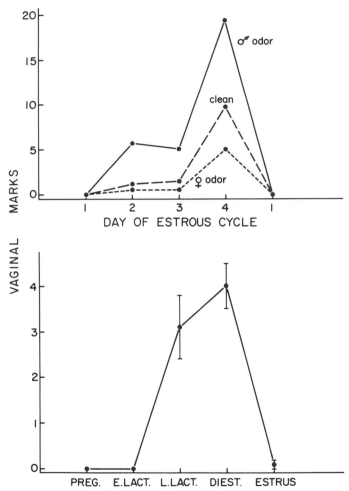

Fiure 6. Vaginal marking by females in different reproductive states. Upper panel: mean number of marks by females during the 10-min trial throughout the estrous cycle in the presence of male odors, or no conspecific odors (from Johnston, 1977a). Lower panel: mean number of marks by females during 4-min trials in a variety of odor conditions when the females were in different reproductive states: pregnancy, early lactation, late lactation, diestrus, estrus. (Redrawn from Johnston, 1979.)

Females often vaginal mark as males follow closely behind; males then stop, sniff, and lick the secretions deposited (Dieterlen, 1959; Johnston, 1977a). Vaginal marking is more frequent in the presence of odors of conspecific males than heterospecific males (Johnston and Brenner, 1982). Species specificity is especially appropriate for signals with sexual advertising functions, since such signals may serve as a species-isolating mechanisms.

Second, vaginal marks aid in location of females by males. In laboratory enclosures, Lisk et al. (1983) have noted that females tend to vaginal mark most frequently in areas that lie in a broad path between the female's nest and that of a male, suggesting that in nature vaginal marks may be distributed so that trails of marks lead to a female's home burrow. In a two-arm, Y-maze situation, males tend to approach, enter, and traverse an arm marked by an intact female more quickly than a clean arm, but do not do so if the scented arm was marked by a female lacking vaginal secretions (Johnston and Kwan, 1984). Thus, males may follow trails of vaginal marks deposited by females.

Another way in which vaginal marks may aid in location of a male is that the secretions deposited seem to be the primary olfactory stimulus that elicits ultrasonic calling from males (Johnston and Kwan, 1984). Males call more frequently in an arena that has been marked by an intact female than in an area marked by females lacking a vagina and vaginal secretions. As discussed in Section 3.2, calling stimulates calling and attraction in response, and thus is no doubt an important link in the location of a partner.

Third, the frequency of vaginal marking is highly correlated with reproductive state (Fig. 6). Females rarely mark when pregnant or during the early stages of lactation, but as lactation proceeds, the frequency of vaginal marking increases (Leonard, 1972; Johnston, 1979, and unpublished observations). During the two diestrous days of the estrous cycle, females vaginal mark at approximately this same moderate level, but then mark much more frequently on day 4, the active period 24 hr before estrus (Johnston, 1977a). This period corresponds to the time when females are actively soliciting males and attempting to lead them back into their burrows (Ciacco et al., 1979b; Lisk et al., 1983). Relatively high levels of vaginal marking can still be observed several hours before females become receptive (Lisk et al., 1983). Thus the frequency of vaginal marking peaks prior to receptivity, when females are actively soliciting males, and is virtually absent during pregnancy and early lactation, when the females are furthest from a receptive phase. Vaginal marking is also virtually absent when females are reproductively quiescent due to exposure to short days (Johnston, unpublished observations).

Perhaps surprising is that female hamsters rarely vaginal mark when they are receptive. If females usually solicit and sequester a male before they become receptive, as the studies from Lisk's laboratory suggest (Ciacco et al., 1979; Lisk et al., 1983), then marking during estrus would be superfluous. Females are probably not always successful in this attempt, but even if they find themselves receptive with no male, it may be a better strategy to remain in the general vicinity of their own nest and wait for males than to venture out refreshing their scent trails. Vaginal marks are quite long-lasting (Johnston and Schmidt, 1979), and males in the vicinity are sure to know the location of females' burrows and to check them regularly, especially if the density of vaginal marks in an area has been on the increase.

4.3. The Harderian Gland

This gland, described in 1694 by Harder in deer, occurs in a wide variety of vertebrates and is especially prominent in rodents. It is a relatively large, often bilobate gland located within the orbit immediately behind the eyeball (e.g., Christensen and Damm, 1953). In hamsters, the weight of the two glands per 100 g of body weight averages 195 mg (Payne, 1977). Traditional theories about its functions include lubrication for the eye, antibacterial action, and as an extraretinal photoreceptive device; more recently it has been implicated in maintenance of the pelage, temperature regulation, and as a source of pheromones (Thiessen et al., 1976; Thiessen and Kittrel, 1980). The gland has also been implicated in pineal function and melatonin synthesis (Clabough and Norvell, 1973; Panke et al., 1979). In gerbils, Thiessen et al. (1976) have shown that this gland secretes at least some of its products to the external nares via the harderian-lacrimal tract. By observing gerbils under UV illumination, it was possible for them to observe the fluorescent harderian material as it was spread around the facial region during grooming. This behavior in turn caused investigation by other gerbils.

In hamsters the harderian gland is sexually dimorphic. The glands of males are larger than those of females and contain two types of secretory cells, whereas the glands of females have only one type of secretory cell but contain porphryn pigments that fluoresce bright orange under UV illumination. These sexual dimorphisms are dependent on androgens. Harderian glands of castrated males have the characteristics of the glands of females, but return to malelike glands when androgen therapy is administered (Clabough and Norvell, 1973; Payne et al., 1977). A preliminary report of the lipid content of male and female harderian glands suggests a quantitative dimorphism in the amount of two lipids (Lin and Nadakavukaren, 1981). It is also of some interest that the harderian glands of the hamster are innervated and the nerve endings are associated with secretory cells, myoepithelial cells, and blood vessels (Bucana and Nadakavukaren, 1972).

The sexual dimorphism suggests communication functions. Although no specific scentmarking behavior has been observed, hamsters distribute harderian secretion around the face and head when grooming (Thiessen et al., 1976). Males, at least, spend a large amount of time investigating the head region of females during social interactions. Examination of males' responses to crushed harderian glands reveals that they spend more time sniffing preparations of female glands than those of male glands and more time investigating male glands than control substances (Payne, 1979). The aggressive responses of experimental males are less when stimulus males are scented with harderian gland secretions from females than when they are scented with harderian secretions from other males or with distilled water (Payne, 1977). Observations from my laboratory also suggest a role for harderian secretions in the stimulation of male sexual behavior (Johnston, 1983b). Thus it seems likely that the harderian gland provides information that is sufficient for sexual recognition and that may influence a variety of social behaviors, but the specifics of these functions are yet to be discovered (see also Section 5).

4.4. Other Glands, Other Marking Behaviors

Although there is relatively little work on any other source of scent that might have a role in communication, there are a few intriguing hints of additional sources of communicatory odors.

The most likely addition to the list are the ear glands. On the ventral surfaces of the pinna, hamsters have sebaceous glands that show a marked sexual dimorphism in size and activity. The glands of females can be stimulated to grow by injecting them with testosterone, and male glands can be reduced in size by injecting males with the antiandrogen cyproterone acetate (Plewig and Luderschmidt, 1977). Male hamsters spend more time sniffing at the odors of ear tissue from females than that from males in a two-choice preference test, suggesting that the ear glands may provide sexually distinguishable chemical cues (Landauer et al., 1980). This result is less compelling than it might be, however, since males are also more attracted to harderian gland material from females than that from males (Payne, 1979), and the ear tissues used in the Landauer et al. experiments were probably contaminated with harderian secretions.

Preputial glands apparently do exist in golden hamsters. LaVelle (1951) shows small preputial glands attached to the side of the glans penis, but compared with the size of the glands in house mice, for example, they are very small. Preputial glands are common in cricetid and murid rodents in general, and the European hamster does have them (Brown and Williams, 1972). There are no behavioral studies suggesting the use of this gland by hamsters in communication, but a few observations may be relevant. Female hamsters will often attempt to poke their heads underneath a male, usually from the side, as if to sniff the genital region. Some form of anogenital scent marking is common in rodents and although it is not conspicuous, I have occasionally observed males touching their ventrum to the ground as if they might be marking. At the end of one experiment in which males were investigating odors placed on glass floors, I found spots and streaks of additional material that resembled vaginal marks deposited by females, even though we had not observed the males engaging in any obvious marking behaviors.

Other discrete sources of odors that are potential sources of communicative signals are urine, feces, and saliva. Despite the use of urine in many other rodent species, attempts to demonstrate potential communicative functions in hamsters have been negative. Male hamsters are more interested in investigating a clean bottle placed into their cage than a bottle containing urine from females (Johnston, 1974), and do not show a preference for the odor of female versus male urine (Landauer, 1975). Hamsters tend to urinate in just one or two places in their living area, and the scent of the urine may tend to elicit urination, perhaps especially in young pups (Dieterlen, 1959). The only suggestion that feces may be used in scent marking is the observation that a pile or two of feces can often be observed on or near a new food hoard (Johnston, 1972, 1977b). Saliva has recently been implicated in social communication among house mice (Block *et al.*, 1981), and the salivary glands of this species show dramatic sexual dimorphism. Saliva is also important in nipple attachment and nursing among rats (Blass and Teicher, 1980). I am not aware of any investigations on its possible functions in hamsters, but even if saliva does not contain chemical signals itself, it may facilitate distribution of other signals. The flank gland area is licked repeatedly during grooming, often to the extent of becoming wet and matted. Likewise, harderian and/or ear gland secretions are no doubt mixed with saliva during grooming.

5. FUNCTIONS OF SCENT SIGNALS

In this section I review what is known about the information communicated by hamster scent signals.

5.1. Species Recognition

Both male and female hamsters behave differently toward conspecifics as opposed to heterospecifics, and odor cues seem to be of primary importance.

Murphy (1977) tested estrous females of three congeneric species (*Mesocricetus auratus, M. brandti,* and *M. newtoni*) in a preference task in which a female could approach and investigate two different males confined behind wire screens. Females of all three species showed strong preferences for investigating conspecific males over either of the heterospecific males (e.g., 75% to 97% of investigation time). Perhaps more significantly, two of the three species also engaged in sexual presenting to the males and did so almost exclusively to males of their own species, indicating that the preferences observed were clearly sexual preferences. The females of one species, *M. brandti,* were also shown to have clear mating preferences; these females mated with conspecific males, but attacked heterospecific males (Murphy, 1978). Visual, auditory, and chemical cues were all available to the hamsters in these experiments, so the degree to which each contributed to the preferences for conspecifics is unknown. However, females apparently do show preferences for conspecific males on the basis of odors alone (Murphy, personal communication). Furthermore, female Syrian hamsters vaginal mark more frequently in response to the odors of conspecific males than to those of heterospecific males (Johnston and Brenner, 1982; see Section 4.2).

Male hamsters of two species (*M. auratus* and *M. brandti*) have been shown to have similar sexual preferences for conspecific females (Murphy, 1980). Males spend more time investigating and mounting anesthetized, conspecific, estrous females than heterospecific, estrous females. In this case visual and chemical cues were available, but when ablations were made of the olfactory and vomeronasal systems, the conspecific preferences were eliminated. Furthermore, intact males of both species demonstrated strong preferences for investigating vaginal secretions from conspecific females over those from heterospecific females. Sexual experience is not necessary for these preferences to be manifested. Sexually experienced and naive males both demonstrated strong preferences for conspecific females or their odors. Cross-fostering experiments with these two species showed that preferences of males for conspecific females were reduced or eliminated if a male was raised by a female of the other species, but that preferences for the foster species over the conspecific species were not induced. Thus, the preferences are due in part to genetically controlled mechanisms and in part to experiences of males with their mothers (Murphy, 1980). Male hamsters have also been shown to react differently to conspecific and heterospecific males, as shown by a greater frequency of flank marking toward the odors of conspecific males (Johnston and Brenner, 1982; see Section 4.1).

5.2. Sexual Recognition and Attraction

Male and female hamsters behave differently toward one another than toward another hamster of the same sex, and again this discrimination seems to be based primarily on chemical cues. Behavioral cues are also important during actual interactions, but differences in behaviors must be initiated by some kind of sexual recognition process.

Male hamsters are more strongly attracted to the odors of females than those of other males. Likewise, they are preferentially attracted to and direct more mounting attempts toward anesthetized females than anesthetized males (Landauer et al., 1978). These preferences are based on two classes of chemical cues, those in vaginal secretions and others that are androgen dependent.

Many different investigators, using different techniques, have demonstrated the powerful attractive effects of vaginal secretions on male hamsters (e.g., Murphy, 1973; Johnston, 1974; Powers et al., 1979; Singer et al., 1976; O'Connell et al., 1979). The odors of females that lack vaginal secretions are less attractive to males than the odors of intact females (Kwan and Johnston, 1980; Johnston and Kwan, 1984).

The importance of androgen-dependent cues has been demonstrated in several kinds of preference tests. Males direct more sniffing investigation and more mounting attempts toward anesthetized, castrated males than toward similar intact males or castrated males receiving testosterone propionate injections. In addition, males are more attracted to ovariectomized females than ovariectomized females receiving testosterone injections (Landauer et al., 1977). That nothing other than androgen-dependent odors and vaginal secretions are involved in males' preferences for females is shown by the finding that males are equally attracted to the odors of vaginectomized females and castrated males; that is, females that lack vaginal secretions and males that lack high concentrations of androgens are equally attractive (Kwan and Johnston, 1980).

Several androgen-dependent scent glands may be sufficient for sexual discrimination, including the harderian glands, the ear glands, and the flank glands (see Section

3). I am not aware of any investigations that demonstrate which of these are necessary for sexual discrimination or whether some may have slightly different uses than others, such as being useful for discrimination at a distance versus close range.

The attraction of males to odors of vaginal secretions is dependent on the presence of testicular androgens (Gregory et al., 1974). Castrated males spend an equal amount of time investigating a bottle containing vaginal secretions and a clean bottle, whereas intact males and castrated males given testosterone propionate show a strong preference for vaginal secretions. Adult females show little attraction to vaginal secretions of other females, and the sexual difference between males and females appears at puberty (Johnston and Coplin, 1979). Immature males and females can be induced to show attraction to the secretion by giving them testosterone propionate injections, suggesting that this aspect of sexual behavior does not depend on neonatal androgens for the organization of the appropriate neural substrates (Johnston and Coplin, 1979). Pups of both sexes demonstrate a period of strong interest in many different odors between about 10 and 15 days of age, and they are especially strongly attracted to vaginal secretions. Since this is a time when pups have become quite mobile but before their eyes have opened, the secretion could have functions as a "maternal pheromone." This brief period of interest in olfactory stimuli may also be a sensitive period for the effects of experience on responses to the vaginal secretion (Johnston and Coplin, 1979).

A great deal of progress has been made in the chemical analysis of components of the vaginal secretion involved in sexual attraction. Singer et al. (1976) identified dimethyl disulfide as one component that elicited prolonged investigatory responses from male hamsters in an assay procedure in which stimulus jars were attached to ports underneath standard plastic cages. Macrides et al. (1977) found that when stimuli were presented in glass jars inside an animal's cage, dimethyl disulfide was more attractive to males than vanillin, a moderately attractive control substance. Furthermore, placing either dimethyl disulfide or vaginal secretions on the genital regions of anesthetized males made this area more attractive to experimental males than if no scent was added. In both the Singer et al. (1976) and Macrides et al. (1977) studies it was noted that whole vaginal secretions were significantly more attractive than dimethyl disulfide alone. For example, in the Singer et al. study, all males responded to vaginal secretion, but only 5 of 12 males responded to dimethyl disulfide, depending on the amount used. Of the males that did respond to dimethyl disulfide, the duration of response averaged 71% of the duration of the response to vaginal secretion.

In one study the role of dimethyl disulfide as an attractant has been questioned (Johnston, 1981). I showed that the two types of attraction tests used as assays for dimethyl disulfide tend to elicit relatively high levels of response from a broad range of odorants, and thus the methods might yield a high number of false positives. Although the attractive properties of dimethyl disulfide should be verified using other methods, it seems likely that dimethyl disulfide is an important component eliciting attraction of males to females. First, male hamsters do not respond to the alcohols or aliphatic acids that were tested in the same assays (O'Connell et al., 1978), indicating some specificity of the response to dimethyl disulfide. Second, dimethyl disulfide is attractive in the assays employed in extremely small quantities, on the order of 500 fg (O'Connell *et al.*, 1979). It seems unlikely that such extreme sensitivity would be exhibited to a substance that lacked biological significance. Third, the concentration of dimethyl disulfide in the vaginal secretion reaches a peak during the time that females are receptive (O'Connell *et al.*, 1981). Any one of these three findings might not provide a strong case by itself, but the three of them together lend much weight to the argument that dimethyl disulfide has a special role in sexual attraction. Other evidence that would greatly strengthen this argument would be to demonstrate the androgen dependence of attraction to dimethyl disulfide, since the attraction to whole vaginal secretions is androgen dependent (Gregory *et al.*, 1975; Johnston and Coplin, 1979), but attraction to other odorants, such as peanut oil, is not (Johnston, unpublished observations).

In an attempt to find other components of vaginal secretions that might be attractive to males, O'Connell *et al.* (1978) tested mixtures of components identified from the secretions. These were predominantly short-chain fatty acids and alcohols and were presented in the proportions in which they occur in vaginal secretions. They found that these substances did not elicit much investigation from males, and neither enhanced nor reduced responses to dimethyl disulfide.

It may be that there are other compounds that are attractive by themselves that have not yet been identified. Alternatively, the function of some volatile components may depend on being bound to larger molecules and released from them in a way that is different from the way they are released when separated by chemical extraction. Using a vacuum distillation method that was probably a less thorough method for separating the volatile and nonvolatile fractions than the one used by Singer *et al.* (1976), we found that the "nonvolatile" residue still had a strong odor to humans and elicited considerable interest from males (Johnston, 1977b; Frey, 1978). Yet a third possibility is that most individual components have no attractive properties by themselves, but only acquire this capacity when they are mixed in the proper ratios with other compounds. In other words, the meaning of the signal (e.g., "female Syrian hamster") may be dependent on a particular odor quality or gestalt that can only be obtained when most or all of the individual components are present.

Much less is known about sexual recognition and attraction among female hamsters. Estrous females are more attracted to males than to other females (Landauer and Banks, 1973; Beach *et al.*, 1976), and both estrous and diestrous females are more attracted to the odors emanating from intact males than those from other females (Johnston, 1979). Androgen-dependent odors from males seem to be involved, since females show a strong preference for intact males over castrated males, but are not differentially attracted to castrated males versus diestrous females (Johnston, 1981). The sources of the relevant androgen-dependent odors have not been determined. The flank gland cannot be the only source, however, since females were strongly attracted to both intact and flank glandectomized males (Johnston, 1979).

A female's reproductive condition influences her attraction to odors of males. Estrous females show the strongest degree of attraction to the odors of intact males, and both diestrous and lactating females also tend to show high levels of interest, although responses of such females are more variable. Pregnant females show the lowest level of interest in the odors of males and show no indication of a preference for intact over castrated males (Johnston, 1979).

5.3. Reproductive State of Females

Females in a variety of mammalian species produce special scent signals during estrus that signal their receptive condition and render them much more highly attractive to males at this time (Johnston, 1983b). In contrast, the odors of female hamsters do not seem to be more attractive to males when they are estrus as opposed to diestrus (Landauer et al., 1978; Johnston, 1980). Males will attempt to mount females throughout the estrous cycle, and when tested with anesthetized females, there is no difference in attraction or mounting attempts directed at estrous and diestrous females (Landauer et al., 1978). The odors of female hamsters do, however, vary in attractiveness across other reproductive conditions. The odors of both pregnant and lactating females are less attractive than the odors of estrous and diestrous females (Johnston, 1980). The contrast between these states may be more relevant than the estrous and diestrous states, since in the wild females would be pregnant or lactating most of the time and would only rarely be going through repeated estrous cycles. Especially since hamsters live solitarily, one might expect them to become more attractive to males several days before they become receptive (i.e., in a hormonal state like that of the nonestrous days of the cycle) to ensure the presence of one or more males as partners (see Section 4.2).

The preceding should not be taken to mean that females do not provide any cues or signals to males that would allow them to distinguish between estrous and diestrous females. Many aspects of female behavior change over the estrous cycle. Ultrasonic calling peaks during estrus (Section 3) and vaginal marking peaks the day before (Section 4.2) along with a submissive "prelordosis" behavior pattern apparently identical to the Tail-up posture (Section 2). The aggressive tendencies of females also change across the cycle, as do females' attraction to males (Section 4).

It is also possible that odors deposited in the environment may provide some information about estrus versus diestrus. Males are slightly more attracted to the odors of bedding material from the cages of estrous as opposed to diestrous females, and they flank mark less in the empty cages of estrous females than in those of non-estrous females (Johnston, 1980). These later two effects are due to vaginal secretions, since they do not occur if the stimulus females have been vaginectomized. Since neither the attractiveness nor the aphrodisiac properties of vaginal secretions seem to vary with reproductive conditions (estrus, diestrus, lactating and pregnant) (Lisk et al., 1972; Johnston, 1974; Darby et al., 1975; Macrides et al., 1977), I interpreted these results to mean that the changes observed were due to differences in the quantity of vaginal secretions that were deposited by females. However, several chemical changes have been discovered that correlate with the estrous cycle. The concentration of dimethyl disulfide in vaginal secretions is maximal during estrus, but since the quantity of secretion is maximal the day after estrus the absolute amount of dimethyl disulfide is greatest the day after estrus (O'Connell et al., 1981). Recently another compound, methyl thiobutyrate, has been identified in hamster vaginal secretions. This compound is a much more reliable indicator of estrus, existing at very low levels during most of the cycle and then showing a rapid rise coinciding with estrus (Singer *et al.,* 1983). Unfortunately it has not been possible to demonstrate any effects of this compound on the behavior of males that could not be accounted for by trace amounts of dimethyl disulfide, so it is not clear whether it has any behavioral functions.

5.4. Sexual Arousal and Aphrodisiac Odors

A diversity of cues probably influences a male's tendency to engage in copulatory behavior, including the female's Solicitation Walk and Lordosis posture, her ultrasonic calls, and her odors. It is the olfactory signals and their effects that have been analyzed in most detail, since a functional olfactory system is apparently necessary for male sexual behavior (Murphy and Schneider, 1970; Devor and Murphy, 1973).

The vaginal secretion of females has a powerful, sexually arousing effect on males. Males attempt to mount a variety of stimulus animals (ovariectomized females, intact males, castrated males) that have been scented with vaginal secretions, whereas without this scenting treatment very few if any males would attempt to copulate. The stimulus animals may be awake or anesthetized, but in both cases addition of vaginal secretions leads to an increase in attempted copulations (Lisk *et al.,* 1972; Murphy, 1973; Johnston, 1975d). Correlated with this heightened sexual arousal is a reduction in aggressive tendencies of males (Murphy, 1973; Johnston, 1975d).

Several lines of evidence suggest at least partially separate signals and mechanisms for the attractive and aphrodisiac effects of vaginal secretions. Whereas attraction seems to be a function of the volatile fraction, as would seem necessary, sexual stimulation may be primarily accomplished through one or more compounds found in the non-volatile fraction (Macrides *et al.,* 1975; Singer *et al.,* 1980). Although it would certainly not be necessary that an aphrodisiac be relatively involatile, it does make sense for a female to restrict such effects to her immediate vicinity and not have that kind of a stimulus wafting off on the slightest breeze. With regard to neural mechanisms, experiments with selective lesions of the olfactory and vomeronasal systems of hamsters indicate that the olfactory system may be primarily responsible for the attraction of males to vaginal secretions, whereas the vomeronasal system has little or no role in this function. Both systems, however, seem to be involved in facilitating copulatory behavior (Powers and Winans, 1975; Winans and Powers, 1977; Powers *et al.,* 1979; Meredith, 1980; Johnston, 1983a).

Sexual experience is not necessary for vaginal secretions to have aphrodisiac effects. Although sexually experienced males usually engage in more copulatory attempts toward stimulus animals than sexually naive males, vaginal secretions stimulate copulatory attempts in both types of male (Lisk *et al.,* 1972; Murphy, 1973; Johnston, 1975d). Thus, the response mechanism is probably a result of an interaction between genetic biases and experience of the pups with the mother prior to weaning. Nonetheless, learning experiences can have a dramatic effect on responses to vaginal secretions. In research that was initiated for entirely different reasons, we demonstrated that hamsters would form a so-called taste aversion to vaginal secretions when ingestion of the secretion was paired with an injection of LiCl (Johnston and Zahorik, 1975; Zahorik and Johnston, 1976). Furthermore, males with taste aversions to vaginal secretions show dramatic reductions in sexual behavior compared with control males (Johnston *et al.*, 1978). When male hamsters that were averted to an arbitrary odorant were tested with receptive females that were scented with that odorant, no depression of sexual activity was noted (Emmerick and Snowdon, 1976). Thus we suggested that the reason that we saw reductions in male sexual performance was because we had altered the meaning of the vaginal secretion for the experimental males (Johnston *et al.*, 1978).

Despite the powerful effects that vaginal secretions exert on the copulatory tendencies of male hamsters, these substances do not fit the notion of a sex pheromone as a releaser in the classical ethological sense. Vaginal secretions will not elicit copulatory attempts toward inanimate models of hamsters (Darby *et al.*, 1975), indicating that some other cues from hamsters are required as well (Johnston, 1977b). Furthermore, vaginal secretions are not necessary for mounting to occur: In one experiment, females lacking a vagina and vaginal secretions were mounted just as readily and just as often as intact females (Kwan and Johnston, 1980). In more recent and more extensive studies we have shown that there is a quantitative deficit in copulatory behavior toward vaginectomized females compared with that toward intact females or vaginectomized females scented with vaginal secretions (Johnston, 1983b). Further deficits in copulatory performance can be produced by eliminating the flank glands and the harderian glands of vaginectomized females. A prior claim that the presence of a female's flank glands is necessary for male copulatory behavior (Lipkow, 1954) is clearly false. A more recent report suggesting deficits in mating behavior on removal of the female's flank glands is difficult to interpret because of parallel deficits in the receptivity of females (Florez-Lozano *et al.*, 1980). Nonetheless, even the olfactory influences on males are complex and diversified and not a simple one-stimulus, one-response mechanism.

Among female hamsters, a functional olfactory system is not necessary for receptivity (Carter, 1973). Olfactory stimuli may still be important for sexual interactions, particularly courtship activities (Johnston, 1979). It may be important for females to identify males of the proper species (see Section 5.1) or males in reproductive condition (see Section 5.2), and females may have preferences for particular individuals (see Section 5.5); this kind of information may have more robust effects on the behavior of females in more natural settings.

5.5. Discrimination of Individuals

Hamsters clearly distinguish between individuals on the basis of olfactory cues (Johnston, 1983a; Johnston and Rasmussen, 1984). Male hamsters display the so-called Coolidge effect, in which a male that has mated to satiation is rearoused when placed with a new female (Bunnell *et al.*, 1977; Dewsbury, 1981). We have shown that this arousal is dependent on discrimination of the odors of the females, that it is dependent on olfactory but not vomeronasal function, and that the female's flank gland is an important source of individually discriminable cues (Johnston and Rasmussen, 1984).

In another situation we have also observed discrimination of individuals—in agonistic situations, males react differently to familiar as opposed to unfamiliar opponents (Johnston, unpublished observations).

5.6. Defense of the Home Burrow

Hamsters apparently live solitarily (Murphy, 1977), and in laboratory enclosures defend a nest box or home burrow (Johnston, 1975c, and unpublished observations; Lisk et al., 1983). Odors may help an individual defend its burrow. Since (1) flank marking is often associated with agonistic interactions, (2) hamsters mark frequently in their nest area and at the entrance to it, and (3) individuals sometimes mark while defending their nest from another (Johnston, 1975a,c; 1977a), it seems likely that the flank gland is involved in burrow defense. Indeed, Alderson and Johnston (1975) found that males were more hesitant to enter an area recently occupied by a normal male than a clean area or an area occupied by a flank-glandectomized male. These experiments should not be interpreted to mean that the flank gland odor necessarily functions as a threat. In some situations males may spend considerable time investigating the odors of other males, even preferring such odors to odors of clean areas (Solomon and Glickman, 1977). What the marks probably do is advertise the presence of another individual hamster of a particular sex, and they may indicate how recently that individual has been in the area (Johnston and Lee, 1976; Johnston and Schmidt, 1979). Other individuals will respond in different ways depending on their sex, reproductive status, past history, and so forth. They should be interested in investigating such odors to determine what they can about the animal that deposited the scent, but in some circumstances flank marks of another should also make them hesitant or induce them to turn back or not enter a burrow.

6. DIRECTIONS FOR FUTURE RESEARCH

There are numerous obvious questions that remain about virtually all of the communication signals discussed. In addition, some intriguing gaps in our knowledge include the following.

6.1. Interaction of Signals

In what ways are different signals used to orchestrate social interactions? Are there close functional ties between some signals in different modalities, such as the suggested complementary use of vaginal marks and ultrasonic calls for attraction of sexual partners (Floody and Pfaff, 1977b)? What functions are common to several scent signals and what functions are specific to one secretion?

6.2. Neural and Neuroendocrine Mechanisms

Given the relatively detailed knowledge about the stimulus and situational control of ultrasonic calling and the two scent-marking behaviors, it would be valuable to investigate the neural mechanisms underlying these behaviors. Floody has published some relevant lesion studies on ultrasonic calls, but there is little else (Floody and O'Donohue, 1980). Since all three behaviors have some courtship functions, such studies would be particularly valuable if done in contrast to the mechanisms of copulatory behavior. It is clear, for example, that vaginal marking and other aspects of female courtship behaviors have different hormonal substrates than ultrasonic calling and lordosis (Johnston, 1979), but the actual hormonal mechanisms are unknown. Likewise, the perceptual and response mechanisms related to ultrasonic calls and odor signals should prove fertile ground. On the neural side, a more thorough analysis of the functions of the main olfactory and vomeronasal systems should prove rewarding. In terms of hormone mechanisms, an intriguing observation is that responsiveness to vaginal secretions apparently does not depend on perinatal androgens (Johnston and Coplin, 1979), but male copulatory behavior does (Eaton, 1970; Swanson, 1971; Noble, 1973).

6.3. Development of Responses to Olfactory Signals and the Related Neural Substrates

Johnston and Coplin (1979) showed that hamsters were particularly responsive to many olfactory stimuli (especially vaginal secretions) for a brief period between 11 and 15 days of age, suggesting a sensitive period for development of responses to odors. This period correlates rather closely with the time that functional synapses are being formed in the most caudal portion of the olfactory pathway (Leonard, 1975), and it could be that this correlation is more than just circumstantial. In his cross-fostering studies, Murphy (1980) demonstrated clear effects of rearing on species-specific preferences, indicating that experience with the mother can have important effects on adult behavior.

6.4. Comparative Studies

Comparative studies can help to answer a variety of questions about the species specificity of the signals, the important chemical or acoustic components of signals, and the differing functions that the communication system may have in species with different social organizations. *Mesocricetus auratus*, *M. brandti*, and *M. newtoni*, as well as *Cricetus cricetus*, all appear to be solitary, but the Djungarian hamster, *Sungorus sungorus*, is apparently more gregarious. Unfortunately very little is known of the ecology of behavioral ecology of any of these species (Murphy, 1977a), and the prospects for research on these species in the field seem dim.

REFERENCES

Alderson, J., and Johnston, R. E., 1975, Responses of male golden hamsters (*Mesocricetus auratus*) to clean and male scented areas, *Behav. Biol.* 15:505–510.

Algard, F. T., Dodge, A. H., and Kirkman, H., 1966, Development of the flank organ (scent gland) of the Syrian hamster, II. Postnatal development, *Am. J. Anat.* 118:317–325.

Allin, J. T., and Banks, E. M., 1972, Functional aspects of ultrasound production by infant albino rats (*Rattus norvegicus*), *Anim. Behav.* **20**:175–185.

Beach, F. A., Stern, B., Carmichael, M., and Ranson, E., 1976, Comparisons of sexual receptivity and proceptivity in female hamsters, *Behav. Biol.* **18**:473–487.

Blass, E. M., and Teicher, M. J., 1980, Suckling, *Science* **210**:15–22.

Block, M. L., Volpe, L. C., and Hayes, M. J., 1981, Saliva as a chemical cue in the development of social behavior, *Science* **211**:1062–1064.

Brown, J. C., and Williams, J. D., 1972, The rodent preputial gland, *Mammal Rev.* **2**:105–147.

Bucana, C. D., and Nadakavukaren, M. J., 1972, Innervation of the hamster Harderian gland, *Science* **175**:205–206.

Buchler, E. R., 1976, The use of echolocation by the wandering shrew (*Sorex vagrans*), *Anim. Behav.* **24**:858–873.

Bunnell, B., Boland, B. D., and Dewsbury, D. A., 1977, Copulatory behavior of golden hamsters, *Mesocricetus auratus, Behavior* **61**:180–206.

Carter, C. S., 1973, Olfaction and sexual receptivity in the female golden hamster, *Physiol. Behav.* **10**:47–51.

Christensen, F., and Dam, H., 1953, A sexual dimorphism of the Harderian glands in hamsters, *Acta Physiol. Scand.* **27**:333–336.

Ciaccio, L. A., Lisk, R. D., and Reuter, L. A., 1979a, Prelordotic behavior in the hamster: A hormonally modulated transition from aggression to sexual receptivity, *J. Comp. Physiol. Psychol.* **93**:771–780.

Ciaccio, L. A., Cantanzaro, C., and Lisk, R. D., 1979b, Social behavior of the golden hamster under seminatural conditions. Paper presented at Animal Behavior Society meetings, New Orleans.

Clabough, J. W., and Norvell, J. E., 1973, Effects of castration, blinding, and the pineal gland on the Harderian glands of the male golden hamster, *Neuroendocrinology* **12**:344–353.

Darby, E. M., Devor, M., and Chorover, S. L., 1975, A presumptive sex pheromone in the hamster: Some behavioral effects, *J. Comp. Physiol. Psychol.* **88**:496–502.

Deanesly, R., 1938, The reproductive cycle of the golden hamster (*Cricetus auratus*), *Proc. Zool. Soc. London* **108** (Ser. A):31–37.

Devor, M., and Murphy, M. R., 1973, The effect of peripheral olfactory blockade on the social behavior of the male golden hamster, *Behav. Biol.* **9**:31–42.

Dewsbury, D. A., 1981, Effects of novelty on copulatory behavior: The Collidge Effect and related phenomena, *Psych. Bull.* **89**:464–482.

Dieterlen, F., 1959, Das Verhalten des syrischen Goldhamsters (*Mesocricetus auratus* Waterhouse), *Z. Tierpsychol.* **16**:47–103.

Drickamer, L. C., and Vandenbergh, J. G., 1973, Predictors of social dominance in the adult female golden hamster (*Mesocricetus auratus*), *Anim. Behav.* **21**:564–570.

Drickamer, L. C., Vandenbergh, J. G., and Colby, D. R., 1973, Predictors of dominance in the male golden hamster (*Mesocricetus auratus*), *Anim. Behav.* **21**:557–563.

Eaton, G., 1970, Effect of a single prepubertal injection of testosterone propionate on adult bisexual behavior of male hamsters castrated at birth, *Endocrinology* **87**:934–940.

Eibl-Eibesfeldt, I., 1953, Zur ethologie des hamsters (*Cricetus cricetus* L.), *Z. Tierpsychol.* **10**:204–254.

Emmerick, J. J., and Snowdon, C. T., 1976, Failure to show modification of male golden hamster mating behavior through taste/odor of aversion learning, *J. Comp. Physiol. Psychol.* **90**:857–869.

Floody, O. R., and O'Donohue, T. L., 1980, Lesions of the mesencephalic central gray depress ultrasound production and lordosis by female hamsters, *Physiol. Behav.* **24**:79–85.

Floody, O. R., 1979, Behavioral and physiological analysis of ultrasound production by female hamsters (*Mesocricetus auratus*), *Amer. Zool.* **19**:443–455.

Floody, O. R., and Pfaff, D. W., 1977a, Communication among hamsters by high-frequency acoustic signals: I. Physical characteristics of hamster calls, *J. Comp. Physiol. Psychol.* **91**:794–806.

Floody, O. R., and Pfaff, D. W., 1977b, Communication among hamsters by high-frequency acoustic signals: III. Responses evoked by natural and synthetic ultrasounds, *J. Comp. Physiol. Psychol.* **91**:820–829.

Floody, O. R., and Pfaff, D. W., 1977c, Aggressive behavior in female hamsters: The hormonal basis for fluctuations in female aggressiveness correlated with estrous state, *J. Comp. Physiol. Psychol.* **91**:443–464.

Floody, O. R., Pfaff, D. W., and Lewis, C. D., 1977, Communication among hamsters by high-frequency acoustic signals: II. Determinants of calling by females and males, *J. Comp. Physiol. Psychol.* 91:807–819.

Floody, O. R., Merkel, D. A., Cahill, T. J., and Shopp, G. M., 1979a, Gonadal hormones stimulate ultrasound production by female hamsters, *Horm. Behav.* 12:172–184.

Floody, O. R., Walsh, C. H., and Flanagan, M. T., 1979b, Testosterone stimulates ultrasound production by male hamsters, *Horm. Behav.* 12:164–171.

Florez-Lozano, J. A., Menendez-Patterson, A., and Marin, B., 1980, Effects of flank glandectomy of the female hamster (*Mesocricetus auratus*) upon the sexual behavior of the male hamster, *Behav. Neural Biol.* 29:399–404.

Frank, D. H., and Johnston, R. E., 1981, Determinants of scent marking and ultrasonic calling by female Turkish hamsters, *Mesocricetus brandti*, *Behav. Neural Biol.* 33:514–518.

Frey, K. F., 1978, Behavioral responses of male hamsters to relatively volatile and involatile fractions of the vaginal secretion, Masters Dissertation, Cornell University, New York.

Frost, P., Giegel, J. L., Weinstein, G. D., and Gomex, E. C., 1973, Biodynamic studies of hamster flank organ growth: Hormonal influences, *J. Invest. Dermatol.* 61:159–167.

Grant, E. C., and Mackintosh, J. H., 1963, A comparison of the social postures of some common laboratory rodents, *Behaviour.* 21:246–259.

Grant, E. C., Mackintosh, J. H., and Lerwill, C. J., 1970, The effect of a visual stimulus on the agonistic behaviour of the golden hamster, *Z. Tierpsychol.* 27:73–77.

Gould, E., Negus, N. C., and Novick, A., 1964, Evidence for echolocation in shrews, *J. Exp. Zool.* 156:19–38.

Gregory, E., Engel, K., and Pfaff, D., 1974, Male hamster preference for odors of female hamster vaginal secretions: Studies of experimental and hormonal determinants, *J. Comp. Physiol. Psychol.* 89:442–446.

Hamilton, J. B., and Montagna, W., 1950, The sebaceous glands of the hamster: I. Morphological effects of androgens on integumentary structures, *Am. J. Anat.* 86:191–233.

Johnston, R. E., 1972, Scent marking, olfactory communication and social behavior in the golden hamster, *Mesocricetus auratus, Diss. Abstr. Int.* 32 (University Microfilms No. 72–12, 666).

Johnston, R. E, 1974, Sexual attraction function of golden hamster vaginal secretion, *Behav. Biol.* 12:111–117.

Johnston, R. E., 1975a, Scent marking by male hamsters: I. Effects of odors and social encounters, *Z. Tierpsychol.* 37:75–98.

Johnston, R. E., 1975b, Scent marking by male hamsters: II. The role of flank gland odor in the causation of marking, *Z. Tierpsychol.* 37:138–144.

Johnston, R. E., 1975c, Scent marking by male hamsters: III. Behavior in a semi-natural environment, *Z. Tierpsychol.* 37:213–221.

Johnston, R. E., 1975d, Sexual excitation function of hamster vaginal secretion, *Anim. Learn. Behav.* 3:161–166.

Johnston, R. E., 1976, The role of dark chest patches and upright postures in the agonistic behavior of male hamsters, *Mesocricetus auratus, Behav. Biol.* 17:161–176.

Johnston, R. E., 1977a, The causation of two scent marking behaviors in female golden hamsters, *Anim. Behav.* 25:317–327.

Johnston, R. E., 1977b, Sex pheromones in golden hamsters, in: *Chemical Signals in Vertebrates* (D. Muller-Schwarze and M. M. Mozell, eds.), Plenum Press, New York, pp. 225–249.

Johnston, R. E., 1979, Olfactory preferences, scent marking and "proceptivity" in female hamsters, *Horm. Behav.* 13:21–39.

Johnston, R. E., 1980, Responses of male hamsters to the odors of females in different reproductive states, *J. Comp. Physiol. Psychol.* 94:894–904.

Johnston, R. E., 1981, Attraction to odors in hamsters: An evaluation of methods, *J. Comp. Physiol. Psychol.* 95:951–960.

Johnston, R. E., 1983a, Mechanisms of individual discrimination in hamsters, in: *Chemical Signals in Vertebrates*, 3rd International Symposium (D. Muller-Schwarze and R. M. Silverstein, eds.), Plenum Press, New York, pp. 245–258.

Johnston, R. E., 1983b, Chemical signals and reproductive behavior, in: *Pheromones and Reproduction in Mammals* (J. G. Vandenbergh, ed.), Academic Press, New York, pp. 3–37.

Johnston, R. E., and Brenner, D., 1982, Species specificity of scent marking in hamsters, *Behav. Neural Biol.* 35:46–55.

Johnston, R. E., and Coplin, B., 1979, Development of responses to vaginal secretion and other substances in golden hamsters, *Behav. Neural Biol.* 25:473–489.

Johnston, R. E., and Kwan, M., 1984, Vaginal scent marking: Effects on ultrasonic calling and attraction of male golden hamsters, *Behav. Neural. Biol.,* in press.

Johnston, R. E., and Rasmussen, K., 1984, Individual recognition of female hamsters by males: Role of chemical cues and of the olfactory and vomeronasal systems, *Physiol. and Behav.* 33:95–104.

Johnston, R. E., and Lee, N. A., 1976, Persistence of the odor deposited by two functionally distinct scent marking behaviors of golden hamsters, *Behav. Biol.* 16:199–210.

Johnston, R. E., and Schmidt, T., 1979, Responses of hamsters to scent marks of different ages, *Behav. Neural Biol.* 26:64–75.

Johnston, R. E., and Zahorik, D. M., 1975, Taste aversions to sexual attractants, *Science* 189:893–894.

Johnston, R. E., Zahorik, D. M., Immler, K., and Zakon, H., 1978, Alterations of male sexual behavior by learned aversions to hamster vaginal secretion, *J. Comp. Physiol. Psychol.* 92:85–93.

Kahmann, H., and Ostermann, K., 1951, Wahrnehmen und Herrorbringen hoher Tone bei Kleinen Saugetieren, *Experientia* 7:268–269.

Kairys, D. J., Magalhaes, H., and Floody, O. R., 1980, Olfactory bulbectomy depresses ultrasound production and scent marking by female hamsters, *Physiol. Behav.* 25:143–6.

Kwan, M., and Johnston, R. E., 1980, The role of vaginal secretion in hamster sexual behavior: Males' responses to normal and vaginectomized females and their odors, *J. Comp. Physiol. Psychol.* 94:905–913.

Landauer, M. R., 1975, Sexual and olfactory preferences of male hamsters (*Mesocricetus auratus* Waterhouse) for conspecifics in different hormonal conditions, *Diss. Abstr.* 36:2106B–2107B.

Landauer, M. R., and Banks, E. M., 1973, Olfactory preferences of male and female golden hamsters, *Bull. Ecol. Soc. Am.* 54:44.

Landauer, M. R., Banks, E. M., and Carter, C. S., 1977, Sexual preferences of male hamsters (*Mesocricetus auratus*) for conspecifics in different endocrine conditions, *Horm. Behav.* 9:193–202.

Landauer, M. R., Banks, E. M., and Carter, C. S., 1978, Sexual and olfactory preferences of naive and experienced male hamsters, *Anim. Behav.* 26:611–621.

Landauer, M. R., Liu, S., and Goldberg, N., 1980, Responses of male hamsters to the ear gland secretions of conspecifics, *Physiol. Behav.* 24:1023–1026.

LaVelle, F. W., 1951, A study of hormonal factors in the early sex development of the golden hamster, *Carnegie Contr. Embryol.* 356(223):220–230.

Leonard, C. M., 1972, Effects of neonatal (day 10) olfactory bulb lesions on social behavior of female golden hamsters (*Mesocricetus auratus*), *J. Comp. Physiol. Psychol.* 80:208–215.

Leonard, C. M., 1975, Developmental changes in olfactory bulb projections revealed by degeneration argyrophilia, *J. Comp. Neurol.* 162:467–486.

Lin, W.-L., and Nadakavukaren, M. J., 1981, Harderian gland lipids of male and female golden hamsters, *Comp. Biochem. Physiol.* 70B:627–630.

Lipkow, J., 1954, Uber das Seitenorgan des Goldhamsters, *Z. Morph. U. Okal Tiere* 42:333–372.

Lisk, R. D., Zeiss, J., and Ciaccio, L. A., 1972, The influence of olfaction on sexual behavior in the male golden hamster (*Mesocricetus auratus*), *J. Exp. Zool.* 181:69–78.

Lisk, R. D., Ciaccio, L. A., and Catanzaro, C., 1983, Mating behavior of the golden hamster under seminatural conditions, *Anim. Behav.* 310:659–666.

Macrides, F., Johnson, P. A., and Schneider, S. P., 1977, Responses of the male golden hamster to vaginal secretion and dimethyl disulfide: Attraction versus sexual behavior, *Behav. Biol.* 20:377–386.

Meredith, M., 1980, The vomeronasal organ and accessory olfactory system in the hamster, in: *Chemical Signals in Vertebrates and Aquatic Invertebrates* (D. Muller-Schwarze and R. Silverstein, ed.), Plenum Press, New York, pp. 303–326.

Miller, L. L., Whitsett, J. M., Vandenbergh, J. G., and Colby, D. R., 1976, Physical and behavioral aspects of sexual maturation in the male golden hamster, *J. Comp. Physiol. Psychol.* 91:245–259.

Montagna, W., and Hamilton, J. B., 1949, The sebaceous glands of the hamster: II. Some cytochemical studies in normal and experimental animals, *Am. J. Anat.* 84:365–395.

Murphy, M. R., 1973, Effects of female hamster vaginal discharge on the behavior of male hamsters, *Behav. Biol.* 9:367–375.

Murphy, M. R., 1977, Intraspecific sexual preferences of female hamsters, *J. Comp. Physiol. Psychol.* 91:1337–1346.

Murphy, M. R., 1978, Oestrous Turkish hamsters display lordosis toward conspecific males but attack heterospecific males, *Anim. Behav.* 26:311–312.

Murphy, M. R., 1980, Sexual preferences of male hamsters: Importance of preweaning and adult experience, vaginal secretion, and olfactory or vomeronasal sensation, *Behav. Neural Biol.* 30:323–340.

Murphy, M. R., and Schneider, G. E., 1970, Olfactory bulb removal eliminates mating behavior in the male golden hamster, *Science* 167:302–304.

Noble, R. G., 1973, The effects of castration at different intervals after birth on the copulatory behavior of male hamsters (*Mesocricetus auratus*), *Horm. Behav.* 4:45–52.

O'Connell, R. J., Singer, A. G., Macrides, F., Pfaffman, C., and Agosta, W. C., 1978, Responses of the male golden hamster to mixtures of odorants identified from the vaginal discharge, *Behav. Biol.* 24:244–255.

O'Connell, R. J., Singer, A. G., Pfaffmann, C., and Agosta, W. C., 1979, Pheromones of hamster vaginal discharge: Attraction to femtogram amounts of dimethyl disulfide and to mixtures of volatile components, *J. Chem. Ecol.* 5:575–585.

O'Connell, R. J., Singer, A. G., Stern, F. L., Jesmajian, S., and Agosta, W. C., 1981, Cyclic variations in the concentration of sex attractant pheromone in hamster vaginal discharge, *Behav. Neural Biol.* 31:457–464.

Orisini, M. W., 1961, The external vaginal phenomena characterizing the stages of the estrous cycle, pregnancy, psuedopregnancy, lactation, and the anestrous hamster, (*Mesocricetus auratus* Waterhouse), *Proc. Anim. Care Panel* 11:193–206.

Panke, E. S., Reiter, R. J., and Rollag, M. D., 1979, Effect of removal of the Harderian glands on pineal melatonin concentrations in the Syrian hamster, *Experientia* 35:1405.

Payne, A. P., 1977, Pheromonal effects of Harderian gland homogenates on aggressive behaviour in the hamster, *J. Endocrinol.* 73:191–192.

Payne, A. P., 1979, The attractiveness of Harderian gland smears to sexually naive and experienced male golden hamsters, *Anim. Behav.* 27:897–904.

Payne, A. P., McGadey, J., Moore, M. R., and Thompson, G., 1977, Androgenic control of the Harderian gland in the male golden hamster, *J. Endocrinol.* 75:73–82.

Plewig, G., and Luderschmidt, C., 1977, Hamster ear model for sebaceous glands, *J. Invest. Dermatol.* 68:171–176.

Powers, J. B., and Winans, S. S., 1975, Vomeronasal organ: Critical role in mediating sexual behavior of the male hamster, *Science* 187:961–963.

Powers, J. B., Fields, R. B., and Winans, S. S., 1979, Olfactory and vomeronasal system participation in male hamsters' attraction to female vaginal secretions, *Physiol. Behav.* 22:77–84.

Sales, G., and Pye, D., 1974, *Ultrasonic Communication by Animals*, Chapman & Hall, London.

Singer, A. G., Agosta, W. C., O'Connell, R. J., Pfaffman, C., Bowen, D. V., and Field, F. H., 1976, Dimethyl disulfide: An attractant pheromone in hamster vaginal secretion, *Science* 191:948–950.

Singer, A. G., Macrides, F., and Agosta, W. C., 1980, Chemical studies of hamster reproductive pheromones, in: *Chemical Signals in Vertebrates and Aquatic Invertebrates* (D. Muller-Schwarze and R. M. Silverstein, ed.), Plenum Press, New York, pp. 365–375.

Singer, A. G., O'Connell, R. J., Macrides, F., Bencsath, A. F., and Agosta, W. C., 1983, Methyl thiobutyrate: A reliable correlate of estrus in the golden hamster, *Physiol. Behav.* 30: 139–143.

Smotherman, W. P., Bell, R. W., Starzec, J., Elias, J., and Zachman, T. A., 1974, Maternal responses to infant vocalizations and olfactory cues in rats and mice, *Behav. Biol.* 12:55–66.

Solomon, J. A., and Glickman, S. E., 1977, Attraction of male golden hamsters (*Mesocricetus auratus*) to the odors of male conspecifics, *Behav. Biol.* 20:367–376.

Steel, E., 1980, Changes in female attractivity and proceptivity throughout the oestrus cycle of the Syrian hamster (*Mesocricetus auratus*), *Anim. Behav.* **28**:256–265.

Swanson, H. H., 1971, Determination of the sex role in hamsters by the action of sex hormones in infancy, *Proceedings of the International Society of Psychoneuroendocrinology* (D. H. Ford, ed.), Karger, Basel, pp. 424–440.

Thiessen, D. D., and Kittrell, M. W., 1980, The Harderian gland and thermoregulation in the gerbil (*Meriones unguiculatus*), *Physiol. Behav.* **24**:417–424.

Thiessen, D. D., Clancy, A., and Goodwin, M., 1976, Harderian gland pheromone in the Mongolian gerbil, *Meriones unguiculatus, J. Chem. Ecol.* **2**:231–238.

Vandenbergh, J. G., 1971, The effects of gonadal hormones on the aggressive behavior of adult golden hamsters (*Mesocricetus auratus*), *Anim. Behav.* **19**:589–594.

Vandenbergh, J. G., 1973, Effects of gonadal hormones on the flank gland of the golden hamster, *Horm. Res.* **4**:28–33.

Vandenbergh, J. G., 1977, Reproductive coordination in the golden hamster: Female influences on the male, *Horm. Behav.* **9**:264–275.

Winans, S. S., and Powers, J. B., 1977, Olfactory and vomeronasal deafferentation of male hamsters: Histological and behavioral analyses, *Brain Res.* **126**:325–344.

Whitsett, J. M., 1975, The development of aggressive and marking behavior in intact and castrated male hamsters, *Hormones and Behavior.* **6**:47–57.

Zahorik, D. M., and Johnston, R. E., 1976, Taste aversions to food flavors and vaginal secretion in golden hamsters, *J. Comp. Physiol. Psychol.* **90**:57–66.

7

Sexual Differentiation and Development

LYNWOOD G. CLEMENS and JEFFREY A. WITCHER

1. INTRODUCTION

The golden hamster provides an important model for the study of development. With a short gestation of only 16 days, the investigator is provided with a rapidly developing organism. In some cases, processes that take place prenatally in other animals occur postnatally in the hamster.

One of the major problems with the hamster as a model for developmental work is the tendency of the mother to cannibalize her young. Even in undisturbed situations, as many as 70% of hamster mothers may show some cannibalism (Day and Galef, 1977). Often this behavior appears as an attempt by the mother to reduce large litters, but it has also been the experience of some investigators that when the pups are disturbed by experimental treatment, mothers may destroy the entire litter. Hence, use of the hamster requires planning to accommodate relatively high loss of young animals.

In the present chapter, we will review current literature relevant to morphological, physiological, and behavioral development. In each instance emphasis has been placed on reproductive functions. Consequently, not all aspects of development will be considered.

2. MORPHOLOGICAL DIFFERENTIATION

The work of Ortiz (1945, 1947) and White (1947) provided much of the ground work for developmental reproductive morphology in the golden hamster. Although the gestation period of 16 days for the golden hamster is relatively short, when compared with 19 and 22 days for the mouse and rat, the stage of development for the urogenital system in the hamster at birth is comparable to newborn mice and rats (Ortiz, 1945). In

LYNWOOD G. CLEMENS and JEFFREY A. WITCHER • Department of Zoology, Michigan State University, East Lansing, Michigan 48823.

the hamster, one sees a more rapid embryonic differentiation of the reproductive system.

2.1. Differentiation of the Gonad

For the first ten days of gestation, fetal male and female hamsters can not be distinguished by morphological characteristics. The ten-day gonad is composed of gonia, mesonephric-derived cells, and coelomic epithelial cells, and is said to be indifferent. The indifferent gonad forms on the ventral edge of the cranial portion of the mesonephros. In the hamster the mesonephric glomeruli do not function, freeing cells from the mesonephric tubules to migrate and form a connection to the gonadal ridge. Ultimately, this connection will give rise to the testis and ovary (Byskov and Peters, 1981; Mossman and Duke, 1973). After 12 days of gestation, the ovary is still indifferent, consisting of an epithelial mass surrounded by germinal epithelium. By day 13, the ovary has begun to collect into cell clusters interspersed by connective tissue. By the 15th day the ovary descends just posterior to the kidney with little additional morphological development. At about this time meiosis is initiated and continues until primary oocytes reach the diplotene stage approximately nine days following birth (Challoner, 1974). The initiation of meiosis in female hamsters does not appear to be triggered exclusively by gonadotropins or steroids (Challoner, 1975). It has been suggested that a portion of the ovary is necessary to induce meiosis. The rete ovarii is the suggested candidate (Byskov, 1974; O and Baker, 1976). Whether initiation is accomplished through a diffusable substance or direct cellular interaction is not known. At birth the differentiated ovary contains distinct cell clusters and is entirely encased within the ovarian capsule. By day 12, sex can be determined through microscopic inspection of the gonad, with the future testis forming a surrounding tunica and the future ovary remaining indifferent. During the 13th day the epithelial cells begin to migrate into testis cords and are well formed by day 14. The testis forms the rete and efferent ducts (although both are solid at this point). The testis begins to descend toward the bladder during this period. The perinatal hamster testis contains solid rete and efferent ducts that begin to hollow by postnatal day 6. At six days the rete testis has a large lumen that contains efferent ducts that connect to the epididymus. Primary sex determination, the process of reading genetic sex into gonadal sex is dependent on the presence or absence of an antigen known as the H-Y antigen. In the presence of the H-Y antigen, the testis develops, whereas in the absence of H-Y antigen, an ovary develops (Wachtel et al., 1975; Ohno, 1979). For the importance of the H-Y antigen in sexual differentiation of mammalian gonads, the reader is referred to several excellent reviews (Muller and Schindler, 1983; Ohno, 1979; Wachtel, 1979).

2.2. Ductal Differentiation

Both wolffian and müllerian ducts are present in the 12-day hamster embryo. The wolffian ducts give rise to the epididymus and ductus deferens, and the müllerian ducts give rise to portions of the oviducts, uterus, and vagina. At day 12, the ducts are not differentiated by sex. However, by the 13th day occlusion of the müllerian duct begins to occur in the male. As with other mammalian species, degeneration of the müllerian duct is not prompted by the presence of androgens (Bruner and Witschi, 1946; White, 1947; Swanson, 1966) but is thought to occur in response to a fetal testicular-derived glycoprotein: Müllerian-inhibiting factor (Josso et al., 1980). In the absence of fetal testes, the müllerian duct develops further in the female with initial degeneration of wolffian ducts around the 14th day of gestation. Degeneration of the wolffian ducts occurs in the absence of androgenic stimulation from the fetal testes. Accessory reproductive organs begin to form on the 14th day. Bulbourethral glands develop in both sexes as outgrowths from the urethra with disappearance occurring in the female soon afterward. Seminal vesicles begin to form from the wolffian ducts in males. Through day 15, accessory reproductive organs continue to develop in males with the appearance of the prostate and coagulating glands. In general, the reproductive tract is as well developed in the hamster at birth as it is in longer gestation species such as the mouse and rat. However, at parturition the müllerian ducts have not yet fused to form the uterovaginal canal, but do so soon after birth.

2.3. Brain Differentiation

Sexual differentiation of brain morphology is poorly understood having been studied in only a few mammalian species (rat, guinea pig, squirrel monkey, hamster, and human). However, differences between sexes in the adult have been found. Many investigators trace the differences in the brain morphology to the differential hormone environment during the period of sexual differentiation. It has been speculated that quantitative differences in behavior between sexes may be traced to differential neuronal connections (Greenough et al., 1977). In some areas of the hypothalamus, greater dendritic density was noted in males when compared with female hamsters (Greenough et al., 1977).

In support of sexual differences in neuronal organization are the observed differences of the hypothalamic–pituitary–gonadal axis output in male and female hamsters. Female hamsters are known to have three- to five-day cycles in hormone production that support ovulation. In contrast, males support spermatogenesis by hormonal production that varies within days, but between-day differences are minimized. Male hamsters are said to be acyclic in gonadotropin release when compared with females.

2.4. Sexual Dimorphism in Reproductive Parameters

Sexual differences in reproductive physiology are observed throughout development in the hamster from prenatal life to senescence. The fetal testis produces measurable amounts of steroid hormone, whereas the fetal ovary is thought to be quiescent in regard to steroid hormone production. From birth through postnatal day 12, serum gonadotropin titers do not vary significantly between the sexes. However, at day 12 and 15, both follicle-stimulating hormone (FSH) and luteinizing hormone (LH) titers, respectively, reach levels that are higher in the female hamsters than in the male. This sex difference remains until hamsters reach 30 days of age (Vomachka and Greenwald, 1979). In addition to the differential release of gonadotropins that support ovulation or spermatogenesis depending on sex, other dimorphisms exist such as age at which reproductive competence begins to falter. Female hamsters begin to exhibit decrements in reproductive capabilities at about the age of 40 weeks (Blaha, 1964). These decrements include cyclic irregularity, reduction in the number of ova released, and an inability to maintain pregnancy. Males, on the other hand, display no serious impairments in reproductive function through the age of 31 months (Swanson et al., 1982).

2.5. Sex Ratio

Genetic sex determination occurs at fertilization in hamsters. As in other mammals, two classes of sperm are routinely produced: one carrying an X-accessory chromosome resulting in female offspring and one carrying the Y-accessory chromosome needed for a male. Since both classes of sperm are produced in equal numbers, an equal number of male and female hamster zygotes should occur at fertilization (primary sex ratio). However, at birth, hamsters display a sex ratio (secondary sex ratio) biased in favor of males, 106 : 100 (Sundell, 1962). This suggests that either the primary sex ratio is not 1:1 or that differential mortality occurs by sex *in utero*. To investigate the skewed ratio, hamster blastocysts were sexed 3.5 days after mating. Out of 98 embryos sexed, a sex ratio favoring males emerged 1.8 : 1 (Lindahl and Sundell, 1958; Sundell, 1962). Whether or not a primary sex ratio differing from 1:1 is indicated remains uncertain. The uncertainty arises since fertilization could occur at random with some mechanism operating to selectively prevent the development of female zygotes. After blastocyst formation, data strongly suggest male hamsters incur higher rates of *in utero* mortality than their littermates. A higher male *in utero* mortality rate has often been observed in other mammalian species. Whatever mechanism provides a skewed primary and secondary sex ratio in favor of male hamsters, no evidence exists for a sex ratio different from 1 : 1 at puberty (tertiary sex ratio) in hamsters.

2.5.1. Dosage Compensation in the Golden Hamster

The golden hamster has a diploid number of 44 chromosomes (Ohno and Weiler, 1961). Of the 44 chromosomes, 21 pair are autosomal with one pair comprising the sex chromosomes. The hamster, similar to other mammals utilizing an X-Y mechanism for sex determination, inactivates all but one X chromosome in each cell early in embryonic development. This process, known as *dosage compensation,* assures that each individual will have equal amounts of expressed X-chromatin material regardless of whether the individual is female (XX) or male (XY). The inactivated X chromosome forms into a condensed packet of chromatin known as a Barr body. Therefore, males, although they only posesses one X chromosome, and females both contain the same amount of expressed X-chromatin material in each cell.

A comparative study utilizing seven different species of mammals suggested a constant ratio of X : autosomal chromatin exists between species. In six of the seven species, the mass of X chromatin was 5% that of the total autosomal mass found within a diploid cell. In the seventh, the golden hamster, this figure was 10%, which could be explained by the relative larger size of the golden hamster's X chromosome (Ohno *et al.*, 1964). On careful examination, the hamster is found not be at variance with the other mammalian species. Since one X chromosome is inactivated in female mammal cells and only one is present in the males cells, the mass of active X-chromatin material when compared with autosomal mass is actually 2.5%. When active (expressed) X-chromatin mass is determined in hamsters, this, too, is 2.5% of the autosomal chromatin mass present. This outcome results from the fact that inactivation of X chromosome in hamsters occurs to a greater extent than in the other species investigated. In female hamsters, not only is one X chromosome converted to a Barr body, but so is approximately half the noncondensed X chromosome. In male hamsters the X chromosome also displays partial inactivation. Through this process the hamster achieves a 2.5% X chromatin : autosome ratio (Saksela and Moorhead, 1962).

3. MATURATION OF SEXUAL BEHAVIOR

3.1. The Female

The female hamster reaches sexual maturity around 35 days of age. This is the time when several features of the female's reproductive system become fully synchronized. Vaginal opening is the earlier reproductive feature to develop, occurring between 10 and 15 days of age. Spontaneous behavioral estrus has been reported as early as day 26, but the average age was 34 days (Diamond and Yanagimachi, 1970). Vaginal secretions that are typical of the estrous cycle generally preceded behavioral estrus by three or four days. Ovulation in young females was not always accompanied by behavioral estrus (silent heat). However, by day 35 there was a reliable coordination between ovulation and the occurrence of sexual receptivity (Diamond and Yanagimachi, 1970).

The substrates that mediate sexual behavior and stimulation develop earlier than 35 days. When young females were treated with estrogen and progesterone, it was found that lordosis responses could be observed as early as 18 days of age (Diamond *et al.*, 1974). Ovulation can be induced as early as 28 days of age (Greenwald and Peppler, 1968).

3.2. The Male

Male hamsters begin to show copulatory responses at about 35 days of age (Miller *et al.*, 1977). Associated with this is the onset of the male's attraction to vaginal secretions beginning around 31 days of age (Johnston and Coplin, 1979). Male attraction to the female and the males increased sexual behavior are preceded by a rise in plasma testosterone levels beginning around 28 days of age (Fig. 1) (Miller *et al.*, 1977; Johnston and Coplin, 1979). It would appear that the rise in plasma testosterone initiates the behavioral changes that occur during this phase of development.

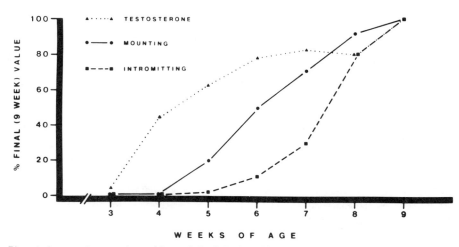

Figure 1. Increases in mounting and intromission behavior with age in relation to plasma testosterone levels (after Miller *et al.*, 1977).

Olfactory bulb lesions as well as chemically induced anosmia abolish male sexual behavior both in males (Devor and Murphy, 1973; Murphy and Schneider, 1970) and in neonatally androgenized females (Doty *et al.*, 1971) indicating that masculine sexual behavior in the hamster is heavily dependent on olfactory cues. However, it is not clear that the attraction of vaginal secretions is dependent on androgens neonatally (Johnston and Coplin, 1979; Gregory and Pritchard, 1983). It should also be mentioned that vaginal secretions are probably not the olfactory cue that is essential for masculine sexual behavior (Kwan and Johnston, 1980). For a discussion of the role of experience and vomeronasal sensations in male preferences for vaginal secretions, see Murphy, (1980).

4. BEHAVIORAL DIFFERENTIATION

In addition to sexual differentiation of morphological structures, the golden hamster displays sexual dimorphism in both reproductive and nonreproductive behaviors. Sexual dimorphism of behavior is seldom an all or none phenomenon. Instead, both the male and the female display the same behavior, but one or the other sex will show the behavior to a greater or lesser extent than the other. Some of the behaviors in which sexual dimorphism have been demonstrated are sexual behavior, aggressive behavior, parental behavior, and open-field behaviors. In this section we will review some of the more recent work on these behaviors and the factors that control their expression in the adult.

4.1. Sexual Behavior

Of the sexually dimorphic behaviors that have been studied from a developmental perspective, sexual behavior has perhaps been studied in the most detail. This behavior is sexually dimorphic in the sense that whereas male hamsters will readily show mounting behavior toward a receptive female, the female does not show mounting behavior as readily as the male. Normal, adult female hamsters will show mounting behavior only after they have been treated for prolonged periods of time with gonadal hormones (Noble, 1974, 1977; DeBold and Clemens, 1978).

With regard to female reproductive behavior, i.e., the display of lordosis, both the male and female hamsters show lordosis, but the duration of lordosis is normally much longer in female hamsters than in male hamsters, even when both individuals have been given the ovarian hormones estrogen and progesterone (Swanson and Crosley, 1971; Tiefer, 1970; Noble and Alsum, 1975; Carter *et al.,* 1972). Thus, for both mounting and lordosis, the dimorphism is quantitative rather than qualitative. Further details of hormonal control of adult sexual behavior in the adult are discussed in Chapters 2 to 5.

4.1.1. Influence of Hormones in Development

The expression of adult sexual behavior in both the male and the female is dependent on the presence of gonadal hormones. The enhancement of sexual behavior following exposure to such hormones in the adult is referred to as *activational* or *concurrent* hormone influences. In addition, gonadal hormones also influence sexual behavior during development. The presence or absence of gonadal hormones during development will determine, to a great extent, the probability of particular sexual behaviors occurring in the adult. This action of hormones on behavior during development is referred to as the *developmental* or *organizational* influences of the hormone (Phoenix *et al.,* 1959; Beach, 1971).

In the hamster, the occurrence of mounting in the adult appears to require exposure to gonadal hormones during early neonatal development (Eaton, 1970; Swanson, 1970; Swanson and Crosley, 1971; Carter *et al.,* 1972; Noble, 1973). Testicular hormones, present for a brief period neonatally, increase the probability of mounting in the adult. Castration of the male hamster on the day of birth resulted in failure of the adult to show mounting behavior, even when testosterone is given to the adult. However, castration on the 5th or 6th day of postnatal life did not abolish mounting in the adult male, but castration on day 6 did reduce the probability of intromission (Carter *et al.,* 1972; Noble, 1973). Consistent with this finding is the finding that the induction of the potential for mounting by administering exogenous testosterone neonatally either to males castrated on day 1 or to normal female hamsters was more effective when the testosterone was given during the first five days following birth as compared with treatment with testosterone during the time period from day 6 to day 0 postnatally (Coniglio and Clemens, 1976). It has been reported that continued treatment of normal adult females with testosterone will eventually result in mounting behavior even though the female did not receive androgens neonatally (Noble, 1977). Thus, although the potential for mounting behavior appears to be present in normal females, its expression requires prolonged hormone treatment in the adult if androgens were not given neonatally. Neonatal treatments appear to enhance the expression of mounting in the adult.

Development of the potential for feminine sexual behavior, lordosis, requires that the hamster, during the first few days of postnatal life, develops in the absence of exposure to gonadal steroids (Carter *et al.*, 1972; Swanson and Crosley, 1971; Coniglio and Clemens, 1976; Noble, 1973). Male hamsters castrated on the day of birth show lordosis of relatively long duration in response to estrogen-progesterone treatment (Eaton, 1970; Carter *et al.*, 1972). However, when castration was delayed until day 6, the adult male showed lordosis responses of relatively short duration following treatment as an adult with estrogen and progesterone (Carter *et al.*, 1972; Noble, 1973). Thus the expression of lordosis in the adult is inhibited by the presence of the testis during the first few days of postnatal life.

4.1.2. Prenatal Determinants of Sexual Behavior

Development of the potential for masculine behavior appears to be largely a postnatal process in the hamster (Nucci and Beach, 1971). When pregnant hamsters were treated with testosterone propionate, it was found that there was no influence on feminine or masculine sexual responses in female offspring (Nucci and Beach, 1971). There is limited evidence that the development of the potential for lordosis in hamsters may begin prenatally. Experiments in the authors' laboratory have demonstrated that exposure of hamster fetuses to the antiandrogen, flutamide, increased the potential for lordosis in male offspring when they became adults. Although this would support the idea that the potential for feminine behavior begins prenatally in the hamster and extends into early postnatal development, such a conclusion must be qualified by the findings that treatment with testosterone during prenatal development does not clearly inhibit the development of feminine behavior in female hamsters. Prenatal treatments have, however, been shown to modify other aspects of the female. Females that were exposed to testosterone propionate prenatally, in addition to showing disruption of the estrous cycle when they reach adulthood, were found to be less attractive to sexually experienced males than were normal control females (Landauer *et al.*, 1981). That is, sexually experienced males spent more time investigating control females than either prenatally or neonatally androgenized females. These results seem to indicate that the sexual attractiveness of the female hamster is altered by exposure to androgens prenatally.

A number of observations are consistent with the concept that the potential for feminine sexual responses may begin differentiating prenatally in the hamster. In addition to the finding that prenatal exposure to antiandrogens enhance lordosis in the male, there is also the observation that castration of the male on day 1 does not result in the male having the same potential for lordosis as seen in normal females (Carter *et al.*, 1972; Gerall and Thiel, 1975). The lordosis duration of day-1 castrate males was significantly shorter than that of normal females. If the process of feminization were entirely a postnatal process, one would expect that castration on day 1 would result in the male achieving a lordosis duration similar to that of the female. Also consistent with the concept of a prenatal process for organization of feminine behavior are the findings of DeBold and Whalen (1975). They found that less testosterone is needed during postnatal development to induce the potential for mounting than is required for suppression of lordosis in female hamsters. If behavior masculinization occurs postnatally, whereas the process of feminization begins prenatally, one would expect postnatal treatments to have a greater influence on masculine behavior.

There is also evidence to suggest that the processes of masculinization and feminization, regardless of their time courses, are independent processes. DeBold and Whalen (1975) report that 1 μg of testosterone propionate neonatally was sufficient to induce the potential for mounting in females. However, lordosis duration was not inhibited until the dose of testosterone propionate was 0 μg or more. Thus it is possible to have the potential for both feminine and masculine behaviors reside in one animal. Payne (1976) has reported that treatment with testosterone propionate neonatally facilitated mounting and reduced lordosis, whereas a single injection of 300 μg of free testosterone on the day of birth failed to suppress lordosis, but enhanced the potential for mounting behavior in females. This study using the alcohol form of testosterone and the longer-acting testosterone propionate again demonstrates that one can observe alterations in masculine behavior independent of alterations in feminine behavior suggesting that the processes controlling these two behavior patterns are independent.

4.1.3. Specificity of Steroid Action during Development

Although testosterone has both the ability to induce the potential for masculine behavior as well as inhibit development of the potential for feminine behavior, it is not clear that testosterone itself is the active molecule. There is, in fact, considerable evidence to suggest that testosterone must first be aromatized to estradiol and that estradiol is the active steroid molecule responsible for the changes in behavior potential. In the hamster, testosterone can be converted to estradiol by cells within the central nervous system (Callard et al., 1979).

Treatment of neonatal females or day-1 castrated males with testosterone, other aromatizable androgens, estradiol, or with the synthetic estrogen diethylstilbestrol resulted in both masculinization and defeminization (Coniglio et al., 1973a,b; Paup et al., 1972, 1974; Tiefer and Johnson, 1971; Whalen and Etgen, 1978; Payne, 1976, 1979). In general, nonaromatizable androgens have been found to be less effective in achieving masculinization and defeminization than those capable of being converted to estrogen (Gerall et al., 1975; Payne, 1976; Coniglio et al., 1973a,b; Paup et al., 1972).

Consistent with the idea that estrogen metabolites of testosterone are the active steroids in masculinization and defeminization is the finding that estrogens are often more potent than testosterone in altering the potential sexual behavior of the hamster (Paup et al., 1974; Coniglio et al., 1973a,b; Whalen and Etgen, 1978).

4.1.4. Influence of Aromatase Inhibitors and Antiestrogens

If the action of testosterone is dependent on its conversion to estrogen, the interruption of this process should alter the potential for sexual behavior in adult hamsters. The steroid 1,4,6-androstatriene-3,17-dione (ATD) interferes with the aromatization (conversion) of testosterone to estrogen. When ATD was administered along with

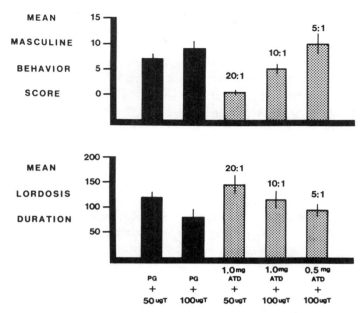

Figure 2. Neonatal administration of the aromatase inhibitor ATD (see text) reduced the masculinizing (upper panel) and defeminizing (lower panel) effects of testosterone in female hamsters (Clemens *et al.*, unpublished data).

testosterone to neonatal females, the masculinizing and defeminizing effects of the testosterone were blocked. ATD failed to block the defeminizing effects of simultaneous estrogen administration (Ruppert and Clemens, 1981) (Fig. 2). Other investigators using ATD have failed to inhibit the masculinizing and defeminization effects of testosterone propionate (Hazzard and Etgen, 1982). The varied results may be due to the different doses of testosterone that were utilized and from the fact that the long-acting propionate form of testosterone was used. These two factors could serve to effectively raise the amount of active testosterone available. This is important since ATD and testosterone actively compete for the same enzymes. The study by Hazzard and Etgen (1982) did provide support for the aromatization hypothesis since neonatal treatment with ATD resulted in increased mount latency for males and a reduction of mounting at the same doses of ATD.

Antiestrogens do not block the masculinizing and defeminizing effects of testosterone. When the antiestrogen MER 25 was administered simultaneously with testosterone propionate, no attenuation of action was noted (Gottlieb *et al.*, 1973). When the antiestrogen nafoxidine was given alone to neonatal female hamsters, a defeminization without concomitant masculinization was observed (Etgen and Whalen, 1979). However, when nafoxidine was given to males or females treated concomitantly with estrogen, nafoxidine had a demasculinizing action. Another antiestrogen, tamoxifen, resulted in both defeminization and masculinization (Etgen, 1981). Interpretation of the results obtained from the use of antiestrogens is difficult since all of the antiestrogens mentioned have been shown to have estrogenic as well as antiestrogenic properties. In summary, however, Etgen (1981, p. 299) suggests that "The data support the hypothesis that estrogen derived via aromatization from androgens plays an important role in both masculinization and defeminization, at least in hamsters."

4.2. Aggressive Behavior

The golden hamster is unique with regard to aggressive behavior in that both the male and the female are quite aggressive. In contrast to many other species, the female hamster is generally dominant over the male (Guhl, 1961; Payne and Swanson, 1970). Aggressive behavior of the female is triggered by an intruder coming into her living area whether it be a male, female, or young. The male differs from the female in that aggression is normally triggered only by intrusion of another male (Rowell, 1961; Marques and Valenstein, 1976).

Levels of aggression in both male and female adult hamsters are strongly influenced by caging conditions (Grelk et al., 1974; Brain, 1972). When males or females were housed individually there was an increase in the amount of aggression compared with those housed in groups.

The influence of gonadal hormones on aggression in adult hamsters is unclear. Although some report a reduction in aggression of males following castration (Payne and Swanson, 1970, 1973; Grelk et al., 1974; Payne, 1974a), others have not observed this (Tiefer, 1970; Vandenberg, 1971; Whitsett, 1975). A similar situation pertains in the female with some reports indicating a reduction in aggression following ovariectomy (Payne and Swanson, 1971b, 1972b) and others indicating no changes in aggression following ovariectomy (Vandenbergh, 1971; Floody and Pfaff, 1977; Tiefer, 1970). One investigator found that females housed in groups showed a reduction in aggression following ovariectomy, whereas females housed individually did not show a reduction in aggression following ovariectomy (Grelk et al., 1974). Such findings may help to explain some of the variation in aggression for female hamsters since those reports indicating no influence of ovariectomy on aggression also report the females were housed individually. However, even this variable does not explain all of the variation because Kislak and Beach (1955) used animals that were individually housed but still found a reduction in the frequency of fights following ovariectomy.

The uncertainty concerning the factors that control aggression in the adult is summarized here as a note of caution in the interpretation of data concerning the development of aggression. For the most part, studies of early hormone treatments in relation to aggression in the adult have been carried out with group-housed animals.

Aggression developed rapidly in male hamsters that were housed singly following weaning at 21 days of age (Whitsett, 1975). Mean number of attacks rose rapidly between 30 and 40 days of age (Fig. 3). In this particular study, males castrated at day 24 of age also showed a parallel increase in aggression.

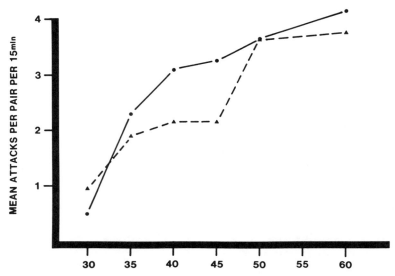

Figure 3. Increases in aggressiveness of male hamsters was not influenced by castration at 24 days of age (after Whitsett, 1975). ●, Intact; ▲, castrate.

4.2.1. Influence of Early Hormone Treatment

Treatment of the hamster neonatally with testosterone propionate (day 1 or 2 after birth) results in an increased level of aggression in the adult. Although male hamsters are normally less aggressive than female hamsters, when males were treated neonatally with testosterone propionate, the males, as adults, showed increased levels of aggression equal to or greater than that of females (Payne and Swanson, 1972; Carter and Landauer, 1975). Treatment of the female hamster with testosterone propionate on the day after birth also resulted in a level of aggression that was greater than that shown by control females. However, ovariectomy of the androgenized females reduced their level of aggression to that of the control ovariectomized female (Payne, 1974b).

When females were treated neonatally, either with 300 μg of the alcohol form of testosterone or with androstenedione, they showed increased levels of aggression as adults (Payne, 1976, 1977). Treatment of the neonatal female with dihydrotestosterone (300 μg on the day after birth) failed to alter levels of aggression in the adult. It is difficult to draw conclusions concerning the role of dihydrotestosterone in this case, since only one dose level was administered. Failure to see an effect in an adult could result from the fact that dihydrotestosterone may not be active or it is possible that the level used was not sufficient.

In any event, it does seem clear from the studies that have been carried out that treatment of neonatal hamsters with testosterone or androstenedione, will increase the amount of fighting shown by males or females when they reach adulthood. All the testing summarized above deals with tests in which males are paired with females.

4.3. Parental Behavior

Parental behaviors such as nest building, retrieving, crouching over young and licking young have been reported for both male and nulliparous female hamsters (Rowell, 1961; Swanson and Campbell, 1979; Marques and Valenstein, 1976). However, in contrast to other species, male hamsters are more apt to pick up and carry a pup than are females. Females (about 50% in one study) are more apt to cannibalize the pup (Marques and Valenstein, 1976) than are male hamsters.

Both male and female hamsters can be induced to display maternal behavior by constant exposure to young, a process termed *sensitization*. In the hamster, sensitization was found to require a shorter priming period in nulliparous females than in virgin males (Swanson and Campbell, 1979). In addition to the longer period required for sensitization of males, there were other differences between male and female parental behavior. In the naive males, retrieval by the male was often incomplete in that the male did not take the pup to the nest, whereas in females, if they did not eat the pup, they were more apt to take it to the nest. Although males seldom ate the pup, the male injured the pup more frequently than did a retrieving female. Also, males did not show the lactation posture toward pups in the nest: "laying in the nest on top of the pups with paws spread, giving pups easy access to all nipples." (Swanson and Campbell, 1979).

In female hamsters, nest building increases with advancing pregnancy or following treatment with estrogen and progesterone (Richards, 1969). In the male, estrogen and progesterone did not increase nest building in either intact or castrated males (Richards, 1969).

It is not clear to what extent these sex differences in maternal behavior may be due to differential exposure to gonadal hormones during development. Castration of neonatal males between days 2 and 4 did not enhance nest building; administering 500 μg testosterone propionate to females at the same age did not decrease nest-building behavior (Richards, 1969).

Similarly, treatment of neonatal female hamsters with 300 μg testosterone propionate on days 6,8, or 10 of neonatal life resulted in deficiencies in reproductive success such as low birth weight and higher litter losses. There was no detectable effect on retrieval and crouching behavior (Stern and Strait, 1983).

Thus, although there are marked sex differences in the parental behaviors of male and female hamsters, at the present time it is unclear what, if any, developmental events might be associated with these sex differences. With only a limited number of studies available on this topic, the lack of a definitive statement here reflects the current need for work in this area.

5. SUMMARY AND CONCLUSIONS

The golden hamster provides an important model for the study of developmental and reproductive phenomena. With a short gestation time, it undergoes a rapid development and differentiation. For the behaviorist, the hamster provides a species in which much of the behavioral differentiation occurs postnatally. Basic studies are available on a variety of sexually dimorphic characters and the endocrinology of the hamster has received considerable attention. Although cannibalism is higher in the hamster than in many other species, careful planning of experiments can accommodate this drawback and make effective use of this unique species.

ACKNOWLEDGMENTS. We thank Dave Brigham for his assistance and we are grateful to Ruth Ann Reynolds for her secretarial help in all phases of this manuscript preparation. This work was supported in part by Public Health Service Grant No. HD06760.

REFERENCES

Beach, F. A., 1971, Hormonal factors controlling the differentiation, development, and display of compulatory behavior in the ramstergig and related species, in: *The Biopsychology of Development* (E. Tobach, L. Aronson, and E. Shae, eds.), Academic Press, New York, pp. 249–296.

Blaha, G. C., 1964, Reproductive senescence in the female golden hamster, *Anat. Rec.* 150:405–412.

Brain, P. F., 1972, Effects of isolation grouping on endocrine function and fighting behavior in male and female golden hamsters (*Mesocricetus auratus* Waterhouse), *Behav. Biol.* 7:349–357.

Bruner, J. A., and Witschi, E., 1946, Testosterone-induced modifications of sex development in female hamsters, *Am. J. Anat.* 79:293–320.

Byskov, A. G., 1974, Does the rete ovarii act as a trigger for the onset of meiosis? *Nature (London)* 252:396–397.

Byskov, A. G., and Peters, H., 1981, *Development and Function of Reproductive Organs,* Excerpta Medica, Amsterdam, Oxford, Princeton.

Callard, G. V., Hoffman, R. A., Petro, Z., and Ryan, K. J., 1979, In vitro aromatization and other androgen transformations in the brain of the hamster (*Mesocricetus auratus*), *Biol. Reprod.* 21:33–38.

Carter, C. S., and Landauer, M. R., 1975, Neonatal hormone experience and adult lordosis and fighting in the golden hamster, *Physiol. Behav.* 14:1–6.

Carter, C. S., Clemens, L. G., and Hoekema, D. J., 1972, Neonatal androgen and adult sexual behavior in the golden hamster, *Physiol. Behav.* 9:89–95.

Challoner, S., 1974, Studies of oogenesis and follicular development in the golden hamster 1: A quantitative study of meiotic prophase *in vitro, J. Anat.* 117:373–383.

Challoner, S., 1975, Studies of oogenesis and follicular development in the golden hamster 2: Initiation and control of meiosis *in vitro, J. Anat.* 119:149–156.

Coniglio, L. P., and Clemens, L. G., 1976, Period of maximal susceptibility to behavioral modification by testosterone in the golden hamster, *Horm. Behav.* 7:267–282.

Coniglio, L. P., Paup, D. C., and Clemens, L. G., 1973a, Hormonal factors controlling the development of sexual behavior in the male golden hamster, *Physiol. Behav.* 10:1087–1094.

Coniglio, L. P., Paup, D. C., and Clemens, L. G., 1973b, Hormonal specificity in the suppression of sexual receptivity of the female golden hamster, *J. Endocrinol.* 57:55–61.

Day, C. S. D., and Galef, B. G., Jr., 1977, Pup cannibalism: One aspect of maternal behavior in golden hamsters, *J. Comp. Physiol. Psychol.* 91:1179–1189.

DeBold, J. F., and Clemens, L. G., 1978, Aromatization and the induction of male sexual behavior in male, female, and androgenized female hamsters, *Horm. Behav.* 11:401–413.

DeBold, J. F., and Whalen, R. E., 1975, Differential sensitivity of mounting and lordosis control systems to early androgen treatment in male and female hamsters, *Horm. Behav.* 6:197–209.

Devor, M., and Murphy, M. R., 1973, The effect of peripheral olfactory blockage on the social behavior of the male golden hamster, *Behav. Biol.* 9:31–42.

Diamond, M., and Yanagimachi, R., 1970, Reproductive development in the female golden hamster in relation to spontaneous estrus, *Biol. Reprod.* 2:223–229.

Diamond, M., Mast, M., and Yanagimachi, R., 1974, Reproductive development and induced estrous in the prepubertal hamster, *Horm. Behav.* 5:129–133.

Doty, R. L., Carter, C. S., and Clemens, L. G., 1971, Olfactory control of sexual behavior in the male and early androgenized female hamster, *Horm. Behav.* 2:325–335.

Eaton, G., 1970, Effect of a single prepubertal injection of testosterone propionate on adult bisexual behavior of male hamsters castrated at birth, *Endocrinology* 87:934–940.

Etgen, A. M., 1981, Differential effects of two estrogen antagonists on the development of masculine and feminine sexual behavior in hamsters, *Horm. Behav.* 15:299–311.

Etgen, A. M., and Whalen, R. E., 1979, Masculinization and defeminization induced in female hamsters by neonatal treatment with estradiol, RU-2858, and nafoxidine, *Horm. Behav.* 12:211–217.

Floody, O. R., and Pfaff, D. W., 1977, Aggressive behavior in female hamsters: The hormonal basis for fluctuations in female aggressiveness correlated with estrous state, *J. Comp. Physiol. Psychol.* 91:443–462.

Gerall, A. A., and Thiel, A. R., 1975, Effect of perinatal gonadal secretions on parameters of receptivity and weight gain in hamsters, *J. Comp. Physiol. Psychol.* 89:580–589.

Gerall, A. A., McMurray, M. M., and Farell, A., 1975, Suppression of the development of female hamster behaviour by implants of testosterone and non-aromatizable androgens administered neonatally, *J. Endocrinol.* 67:439–445.

Gottlieb, H., Gerall, A. A., and Thiel, A., 1973, Receptivity in female hamsters following neonatal testosterone, testosterone propionate, and MER-25, *Physiol. Behav.* 12:61–68.

Greenough, G. C., Carter, C. S., Steerman, C., and DeVoogd, T., 1977, Sex differences in dendritic patterns in hamster preoptic area, *Brain Res.* 126:63–72.

Greenwald, G., and Peppler, R. D., 1968, Prepubertal and pubertal changes in the hamster ovary, *Anat. Rec.* 161:447–458.

Gregory, E., and Pritchard, W. S., 1983, The effects of neonatal androgenization of female hamsters on adult preference for female hamster vaginal discharge, *Physiol. Behav.* 31:861–864.

Grelk, D. F., Papson, B. A., Cole, J. E., and Rowl, R. A., 1974, The influence of caging conditions and hormone treatments on fighting in male and female hamsters, *Horm. Behav.* 5:355–366.

Guhl, A. M., 1961, Gonadal hormones and social behavior in infrahuman vertebrates, in: *Sex and Internal Secretions* (W. C. Young, ed.), Vol. 2, Williams and Wilkins, Baltimore, Maryland, pp. 1240–1267.

Hazzard, M. D., and Etgen, A., 1982, Effects of an aromatase inhibitor on masculinization and defeminization in hamsters, *Conference on Reproductive Behavior Abstract,* p. 65.

Johnston, R. E., and Coplin, B., 1979, Development of responses to vaginal secretion and other substances in golden hamsters, *Behav. Neural Biol.* 25:473–489.

Josso, N., Picard, J. Y., and Tran, D., 1980, A new testicular glycoprotein-centimullerian hormone, in: *Testicular Development, Structure and Function* (A. Steinberger and E. Steinberger, eds.), Raven Press, New York, p. 21.

Kislak, J. W., and Beach, F. A., 1955, Inhibition of aggressiveness by ovarian hormones, *Endocrinology* 56:684–692.

Kwan, M., and Johnston, R. E., 1980, The role of vaginal secretion in hamster sexual behavior: Male's responses to normal and vaginectomized females and their odors, *J. Comp. Physiol. Psychol.* 94:905–913.

Landauer, M. R., Attas, A. I., and Lui, S., 1981, Effects of prenatal and neonatal androgen on estrous cyclicity and attractiveness of female hamsters, *Physiol. Behav.* 27: 419–424.

Lindahl, P. E., and Sundell, G., 1958, Sex ratio in the golden hamster before uterine implantation, *Nature (London)* 182:1392.

Marques, D. M., and Valenstein, E. S., 1976, Another hamster paradox: More males carry pups and fewer kill and cannibalize young than do females, *J. Comp. Physiol. Psychol.* 7:653–657.

Miller, L. L., Whitsett, J. M., Vandenbergh, J. G., and Colby, D. R., 1977, Physical and behavioral aspects of sexual maturation in male golden hamsters, *J. Comp. Physiol. Psychol.* 91:245–259.

Mossman, H. W., and Duke, K. L., 1973, *Comparative Morphology of the Mammalian Ovary,* University of Wisconsin Press, Madison.

Muller, U., and Schindler, H., 1983, Testicular differentiation—A developmental cascade, *Differentiation* 23 (suppl):S99–S103.

Murphy, M. R., 1980, Sexual preferences of male hamsters: Importance of preweaning and adult experience, vaginal secretion, and olfactory or bomeronasal sensation, *Behav. Neural Biol.* 30:323–340.

Murphy, M. R., and Schneider, G. E., 1970, Olfactory bulb removal eliminates mating behavior in the male golden hamster, *Science* 167:302–304.

Noble, R. G., 1973, The effects of castration at different intervals after birth on the copulatory behavior of male hamsters (*Mesocricetus auratus*), *Horm. Behav.* 4:45–52.

Noble, R. G., 1974, Estrogen plus androgen induced mounting in adult female hamsters, *Horm. Behav.* 5:227–234.

Noble, R. G., 1977, Mounting in female hamsters: Effects of different hormone regimens, *Physiol. Behav.* 19:519–526.

Noble, R. G., and Alsum, P. B., 1975, Hormone dependent sex dimorphisms in the golden hamster (*Mesocricetus auratus*), *Physiol. Behav.* 14:567–574.

Nucci, L. P., and Beach, F., 1971, Effects of prenatal androgen treatment on mating behavior in female hamsters, *Endocrinology* 88:1514–1515.

O, W. S., and Baker, T. G., 1976, Initiation and control of meiosis in hamster gonads *in vitro, J. Reprod. Fertil.* 48:399–401.

Payne, A. P., and Swanson, H. H., 1971, Hormonal control of aggressive dominance in the female hamster, *Physiol. Behav.* 6:355–357.

Payne, A. P., and Swanson, H. H., 1972a, The effect of sex hormones on the aggressive behavior of the female golden hamster (*Mesocricetus auratus* Waterhouse), *Anim. Behav.* 8:687–691.

Payne, A. P., and Swanson, H. H., 1972b, Neonatal androgenization and aggression in the male golden hamster, *Nature* 239:282–283.

Ohno, S., 1979, *Major Sex Determining Genes,* Springer-Verlag, Berlin, Heidelberg, New York.

Ohno, S., and Weiler, C., 1961, Sex chromosome behavior pattern in germ and somatic cells of *Mesocricetus auratus, Chromosoma* 12:362–372.

Ohno, S., Begak, U. W., and Begak, M. L., 1964, X-autosome ratio and the behavior pattern of individual x-chromosomes in placental mammals, *Chromosoma* 15:14–30.

Ortiz, E., 1945, The embryological development of the wolffian and müllerian ducts and the accessory reproductive organs of the golden hamster (*Cricetus auratus*), *Anat. Rec.* 92:371–389.

Ortiz, E., 1947, The postnatal development of the reproductive system of the golden hamster (*Cricetus auratus*) and its reactivity to hormones, *Physiol. Zool.* 20:45–67.

Paup, D. C., Coniglio, L. P., and Clemens, L. G., 1972, Masculinization of the female golden hamster by neonatal treatment with androgen or estrogen, *Horm. Behav.* 3:123–131.

Paup, D. C., Coniglio, L. P., and Clemens, L. G., 1974, Hormonal determinants in the development of masculine and feminine behavior in the female hamster, *Behav. Biol.* 10: 353–363.

Payne, A. P., 1974a, A comparison of the effects of androstenedione, dihydrotestosterone and testosterone propionate on aggression in the castrated male golden hamster, *Physiol. Behav.* 13:21–26.

Payne, A. P., 1974b, Neonatal androgen administration and aggression in the female golden hamster during interactions with males, *J. Endocrinol.* 63:497–506.

Payne, A. P., 1976. A comparison of the effects of neonatally administered testosterone propionate and dihydrotestosterone on aggressive and sexual behaviour in the female golden hamster, *J. Endocrinol.* 69:23–31.

Payne, A. P., 1977, Changes in aggressive and sexual responsiveness of male golden hamsters after neonatal androgen administration, *J. Endocrinol.* 73:331–337.

Payne, A. P., 1979, Neonatal androgen administration and sexual behaviour: Behavioural responses and hormonal responsiveness of female golden hamsters, *Anim. Behav.* 27:242–250.

Payne, A. P., and Swanson, H. H., 1970, Agonistic behaviour between pairs of hamsters of the same and opposite sex in a neutral observation area, *Behaviour* 36:259–269.

Payne, A. P., and Swanson, H. H., 1973, The effects of neonatal androgen administration on the aggression and related behaviour of male golden hamsters during interactions with females, *J. Endocrinol.* 58:627–636.

Phoenix, C. H., Goy, R. W., Gerall, A. A., and Young, W. C., 1959, Organizing action of prenatally administered testosterone propionate on the tissues mediating mating behavior in the female guinea pig, *Endocrinology* 65:369–382.

Richards, M. P. M., 1969, Effects of oestrogen and progesterone on nest building in the golden hamster, *Anim. Behav.* 17:356–361.

Rowell, T. E., 1961, Maternal behaviour in non-maternal golden hamsters (*Mesocricetus auratus*), *Anim. Behav.* 9:11–15.

Ruppert, P. H., and Clemens, L. G., 1981, The role of aromatization in the development of sexual behavior of the female hamster (*Mesocricetus auratus*), *Horm. Behav.* 15:68–76.

Saksela, E., and Moorhead, P. S., 1962, Enhancement of secondary constrictions and the hetero chromatic x in human cells, *Cytogenetics* 1:225–244.

Stern, J. M., and Strait, T., 1983, Reproductive success, postpartum maternal behavior, and masculine sexual behavior of neonatally androgenized female hamsters, *Horm. Behav.* 17:208–224.

Sundell, G., 1962, The sex ratio before uterine implantation in the golden hamster, *J. Embryol. Exp. Morphol.* 10:58–63.

Swanson, H. H., 1966, Modification of the reproductive tract of hamsters of both sexes by neonatal administration of androgen or oestrogen, *J. Endocrinol.* 36:327–328.

Swanson, H. H., 1970, Effects of castration at birth in hamsters of both sexes on luteinization of ovarian implants, oestrous cycles and sexual behaviour, *J. Reprod. Fertil.* 21:183–186.

Swanson, H. H., and Crosley, D. A., 1971, Sexual behaviour in the golden hamster and its modification by neonatal administration of testosterone propionate, in: *Hormones and Development* (M. Hamburgh and E. J. W. Barrington, eds.), Appleton-Century Croft, New York, pp. 677–687.

Swanson, L. J., and Campbell, C. S., 1979, Induction of maternal behavior in nulliparous golden hamsters (*Mesocricetus auratus*), *Behav. Neural Biol.* 26:364–371.

Swanson, L. J., Desjardins, C., and Turek, F. W., 1982, Aging of the reproductive system in the male hamster: Behavioral and endocrine patterns, *Biol. Reprod.* 26:791–799.

Tiefer, L., 1970, Gonadal hormones and mating behavior in the adult golden hamster, *Horm. Behav.* 1:189–202.

Tiefer, L., and Johnson, W. A., 1971, Female sexual behaviour in male golden hamsters, *J. Endocrinol.* 51:615–620.

Vandenberg, J. G., 1971, The effects of gonadal hormones on the aggressive behaviour of adult golden hamsters (*Mesocricetus auratus*), *Anim. Behav.* 19:589–594.

Vomachka, A. J., and Greenwald, G. S., 1979, The development of gonadotropin and steroid hormone patterns in male and female hamsters from birth to puberty, *Endocrinol.* 105:960–966.

Wachtel, S., 1979, Primary sex determination, *Arthritis Rheum.* 22:1200–1210.

Wachtel, S. S., Ohno, S., Koo, G. C., and Boyse, E., 1975, Possible role of H-Y antigen in primary sex determination, *Nature (London)* 257:235–236.

Whalen, R. E., and Etgen, A. M., 1978, Masculinization and defeminization induced in female hamsters by neonatal treatment with estradiol benzoate and RU-2858, *Horm. Behav.* 10:170–177.

White, M. R., 1947, Effects of hormones on embryonic sex differentiation in the golden hamster I. Androgenic effects in treated males, *Anat. Rec.* 99:397–426.

Whitsett, J. M., 1975, The development of aggressive and marking behavior in intact and castrated male hamsters, *Horm. Behav.* 6:47–57.

8

Female Sexual Behavior

C. SUE CARTER

In a review of reproduction in the golden hamster published in 1968, Kent devoted less than four pages to a relatively exhaustive survey of the existing literature on female sexual behavior. The current review will summarize findings that have appeared in the intervening 15 years. In that period, research on the female hamster increased exponentially. The hamster has taken over an important position, probably second only to the domestic rat, in its popularity as a laboratory subject for studies of female sexual behavior.

The present review will include (1) a description of the sexual behavior of the female golden hamster, including studies of stimulus factors regulating precopulatory behaviors, the induction of receptivity, and the duration of receptivity, (2) experimental evidence regarding the hormonal and neurochemical correlates of these behaviors, and (3) a brief speculation on the evolutionary significance of certain unique features of the female sexual behavior of the golden hamster. Many aspects of hamster reproduction including the characteristics of the estrous cycle, chemical communication, and neuroanatomical substrates of reproduction are detailed elsewhere in this volume, and are summarized here only as necessary for continuity.

1. THE DESCRIPTION OF FEMALE SEXUAL BEHAVIOR

The estrous cycle in the female hamster is typically four days in length and is highly regular (Orsini, 1961). Sexual behavior or receptivity, usually indexed by the lordosis posture, first appears about 1 to 2 hr before dark on the day of estrus. Lordosis can be elicited for about 12 to 20 hr in unmated females (Frank and Fraps, 1945; Kent, 1968). Receptivity typically precedes ovulation by 10 to 12 hr (Orsini, 1961). (In the

---

C. SUE CARTER • Departments of Psychology and Ecology, Ethology, and Evolution, University of Illinois, Champaign, Illinois 61820.

current review, the 24-hr period from lights-off, which encompasses most of the period of sexual receptivity and ovulation, will be termed the day of estrus.)

Sexual behavior in the female hamster, including the lordosis posture characteristic of estrus, was first described in detail by Reed and Reed (1946, p. 8):

> The female often runs a few steps forward before assuming the copulatory pose. Once she has assumed the pose, she maintains it without change for minutes at a time. She extends her fore legs but flexes them slightly at the elbow, extends the hind legs and elevates the pelvis, elongates the body and straightens the back, spreads the hind feet wide apart, and raises her tail. The eyes are glazed, fixed, and may be half-closed. Copulations may continue for more than half an hour, during most of which time the female remains rigidly in the copulatory position, although towards the end of the period she usually becomes restless and may move about between the less frequent copulations.

Direct observations of the sexual behavior of the golden hamster in nature are not available and little is known of their natural history (Murphy, 1971). Seminatural observations by Ciaccio and Lisk (1979), in conjunction with standard laboratory studies, have suggested that the female hamster is relatively asocial on the two days following ovulation. However, the female actively solicits the male on the day before behavioral estrus (Johnston, 1975). On this day the female exhibits high levels of vaginal marking and holds an immobile "prelordosis" posture in response to male attention. Males tended to follow such females and in the observations of Ciaccio and Lisk (1979) were seen to enter an underground burrow that had been prepared by the female. Under these conditions, the female often sealed the burrow and the male and female remained within through one sleep cycle. Before dark on the night of estrus, the female arose first, approached the sleeping male, and activated him by physical contact. These estrous females assumed the lordosis posture and mating began in the burrow system. Sexual behavior continued for about 1 hr. As mating ceased the female began to attack the male and forced him from the burrow.

2. PRECOPULATORY BEHAVIORS

Proceptivity in the female hamster has been defined by Beach et al. (1976) as male-directed behaviors on the part of a female that appear to increase the male's "responsiveness to the female and raise the probability that he will initiate copulatory activity." In the laboratory when estrous female hamsters are placed with an experienced male, lordosis is often seen within a few seconds. Under these conditions the expression of proceptive behaviors is minimal. However, some components of proceptivity in the hamster may appear a day or so before estrus (Ciaccio et al., 1979; Johnston, 1979; Steel, 1979, 1980).

If the estrous female hamster encounters a partner that does not immediately attend to her or mount her, she will typically approach and assume the lordosis posture in front of the partner. If that fails to elicit mounting the estrous female may nuzzle, lick, and bite and in some cases will mount the partner (Dieterlan, 1959; Hsü and Carter, 1982). She may also emit ultrasonic vocalizations (Floody et al., 1977).

Proceptivity has been operationally defined in hamsters by measures such as the tendency of estrous females (1) "to spend reliably more time" next to a caged male versus a diestrous female, (2) "to exhibit tonic immobility as distinguished from lordosis" while near a male's cage, and (3) "to exhibit full lordosis" in proximity to a male (Beach *et al.*, 1976). These authors found that diestrous female hamsters did not show a male–female preference and spent relatively little time near caged animals of either sex. About 2 hr after the first appearance of lordosis, unmated estrous females showed a dramatic increase in the time near the male. This apparent interest in the male began to disappear a few hours before behavioral receptivity (lordosis) was terminated.

Male–female interactions in the hamster vary as a function of the female's estrous cycle. Steel (1979b, 1980) found that females showed their highest levels of approach behavior and tended to maintain proximity with the male on the day of estrus. Females were most likely to avoid the male on the day after estrus. Approach and proximity measures were intermediate on the other days of the cycle. Ciaccio *et al.* (1979) had similar findings, noting that fighting occurred most often on the days following estrus. They also recorded an immobilization posture (which they termed "prelordosis") that became prominent about 8 hr prior to the onset of lordosis. Steel (1981) found that estrogen alone produces levels of proceptive behaviors similar to those seen during peak receptivity (estrus) and found little evidence that progesterone promotes these proceptive behaviors.

Estrous females, in natural or hormone-induced estrus, may direct mounts toward other animals. Mounting under these conditions may be considered a proceptive behavior (Hsü and Carter, 1982).

As in many species, the attractiveness of the female hamster may vary as a function of the female's estrous condition. However, attempts to demonstrate, when the behavior of the female is controlled, that male hamsters prefer estrous versus diestrous females have generally not been successful (Darby *et al.*, 1975; Landauer *et al.*, 1978; Johnston, 1974, 1980). As noted before, in laboratory and seminatural studies, vaginal marking increased prior to the day of estrus (Johnston, 1977, 1979; Ciaccio and Lisk, 1979). The quantity or quality of chemical signals (for example, deposits of vaginal secretions) that presumably attract the male may change as a function of the behavior of the female. It has been reported that sexually experienced male hamsters are able to discriminate and show a preference for vaginal secretions (Carmichael, 1980) or soiled bedding (Johnston, 1980) from estrous versus diestrous females. This discrimination appeared to rely on vaginal secretions since the preference for bedding disappeared after surgical vaginectomy (Kwan and Johnston, 1980; Johnston, 1980).

In social interactions, female hamsters emit 34 to 42 kHz vocalizations (Sales, 1972; Floody and Pfaff, 1977b,c). Both Sales and Floody and his associates (Floody *et al.*, 1977) have concluded that female ultrasounds function primarily as "sexual advertisement" or proceptive signals capable of communicating the female's location and state of sexual receptivity. The presence of a male or male-soiled bedding increased the rate of ultrasonic vocalizations and estrous female hamsters emitted higher rates of ultrasound production than did nonestrous females. Maximal rates of ultrasound production were elicited following estrogen plus progesterone treatment. Estrogen alone produced intermediate levels and progesterone alone did not have a significant effect on call rates (Floody et al., 1979).

Payne and Swanson (1970) have suggested that the female communicates her estrous condition to the male primarily by her behavior. Overt female-initiated aggression typically has been seen during the nonreceptive phase and ceases when the female becomes receptive (Beach et al., 1976; Floody and Pfaff, 1977a).

Proceptive behaviors in estrous females were depressed following auditory damage (Beach et al., 1976). In addition, following olfactory bulb removal, ultrasound production and scent marking were reduced (Kairys et al., 1980). Lordosis was not affected by these manipulations supporting the hypothesis that the neurosensory factors regulating proceptivity differ from those controlling receptivity (Murphy, 1976). In addition, the findings of Steel (1981) that estrogen, but not progesterone, promotes proceptivity supports the contention that proceptivity and receptivity have somewhat independent mechanisms of control.

3. LORDOSIS

Lordosis in the estrous female hamster is elicited in the presence of a male hamster that is either investigating or mounting her. Experiments have shown that pressure on the flanks, rump or perineum reliably facilitates lordosis. Lordosis in female hamsters also could be elicited by lightly brushing the hair or by air puffs. Touch to the upper body was also sometimes effective (Kow et al., 1976). The most sensitive areas are innervated by the dorsal roots of spinal nerves L1, L2, L5, L6, and S1 (Kow and Pfaff, 1977).

Somatosensory stimuli were maximally effective when presented in the presence of a male (Kow et al., 1976). Males that were confined in a transparent perforated container or that were anesthetized were much less effective in lordosis elicitation (Murphy, 1974).

Olfaction was not necessary for the induction of female sexual behavior in this species (Carter, 1973b). Audition, vision, or hearing also were not essential for lordosis elicitation in the female hamster. However, stimuli from a confined or anesthetized male summated to enhance the effectiveness of somatosensory stimuli (Noble, 1973; Murphy, 1974). Females that had had brief male contact (even without intromissions) assumed the lordosis posture more readily and held the posture for a longer duration than unprimed females (Noble, 1973).

Cycling female hamsters placed with a male on the day of estrus typically assume a lordosis posture within a few seconds. Although the female's legs remain stationary she continues to play an active role in the copulatory interaction through pelvic adjustments. The male usually stimulates the perineal region of the female through licking and pelvic contact. Such contact elicits pelvic movements in the direction of the stimulus and tail deviation away from the stimulus (Noble, 1979). These adjustments apparently facilitate the orientation of the male. When males that had not been previously aroused (by female exposure) were placed with genitally anesthetized females, copulatory behavior was inhibited and variable. Sexually aroused males, however, mated normally with genitally anesthetized females (Noble, 1980).

4. HORMONAL FACTORS INDUCING ESTRUS

Ovarian hormones are essential for the induction of estrus (as measured by prolonged lordosis) in the female golden hamster. Female sexual behavior generally disappears in this species following ovariectomy, although an occasional ovariectomized female will show a brief immobile posture (Tiefer, 1970).

During the estrous cycle in intact females, estrogen levels increase for approximately two days, followed by a preovulatory surge of progesterone (Chapter 2). Simulations of this hormone pattern have been successful in eliciting lordosis in ovariectomized female hamsters. As originally demonstrated by Frank and Fraps (1945), lordosis resembling that in natural estrus is readily induced after ovariectomy if estrogen pretreatment is followed by progesterone (Kent, 1968; Tiefer, 1970). Although female hamsters are capable of showing lordosis postures as early as 18 days of age (Diamond et al., 1974), natural ovarian cycles with estrous behavior typically appear about 35 days of age (Diamond and Yanagimachi, 1970).

A number of treatment regimens have elicited lordosis in the ovariectomized female. In most recent studies, hormones have been injected subcutaneously in oil suspension. Steroids can also be administered in Silastic implants. Estradiol benzoate (EB) is most often used for injections because the benzoate slows the release properties of the hormone and allows a more sustained hormone exposure, thought to mimic the hormone production pattern of the ovary. Pulsatile hormonal stimulation is also effective, since separated pulse injections of unesterified estradiol will induce receptivity (Clark and Roy, unpublished).

Lordosis induction in this species has been shown to be dose dependent (Kent, 1968; Meyerson, 1970). Injections of as little as 0.3 μg of EB (administered twice, 48 and 24 hr before testing) and 0.05 mg of progesterone (given about 4 hr prior to testing) induced lordosis in about one third of the females tested (Carter and Porges, 1974). Full receptivity in recently ovariectomized females is routinely elicited following injections of 1.0 μg or more of EB (usually given on two consecutive days) and 0.25 mg or more of progesterone. Larger doses of EB (up to 333 μg/day) did not produce measurably different lordosis responses than did 1 μg. However, large doses of progesterone (5.0 mg) slightly increased the duration of the period of estrus (Carter and Porges, 1974).

Removal of the uterus inhibited lordosis in hormone-treated female hamsters. Opposite effects have been reported in female rats. The mechanisms of these effects remain to be explained (Siegel et al., 1979).

Implants of very small amounts of estradiol in the ventromedial hypothalamus were effective in facilitating lordosis when subsequent systemic progesterone was given (DeBold et al., 1982b). In that study implants in other brain regions were relatively ineffective, except in areas such as the anterior hypothalamus where there were indications that the hormone had leaked into the peripheral circulation. An earlier study by Ciaccio and Lisk (1973/74) reported that anterior hypothalamic implants of estradiol were behaviorally effective. (More posterior sites were not thoroughly explored.) De-Bold and associates (1982b) suggest that the implants of Ciaccio and Lisk may have been primarily acting through leakage from the implants into other regions.

Lordosis in ovariectomized females was facilitated by estradiol-17, β, but not by estradiol-17, α (even at doses of 100 μg) (Whitsett et al., 1978). Receptivity was blocked by synthetic antiestrogens, such as CI-628 (Morin et al., 1976).

Various metabolites of progesterone have been tested in estrogen-primed female hamsters. 5-Alpha-pregnan-3,20-dione was much less potent in lordosis induction than was progesterone, and other metabolites were even less effective (Johnson et al., 1976; Humphrys et al., 1977).

Attempts to localize brain sites for progesterone's lordosis facilitative effects have not been successful. Progesterone implants in various hypothalamic and mesencephalic areas have not been found to consistently induce receptivity in estrogen-primed female hamsters (DeBold et al., 1976).

As in many other mammals, estrogen alone has been found to be adequate to elicit female sexual responses (Feder, 1981). However, in the hamster a minimum of about six days of daily EB treatment is required for the reliable induction of lordosis and even after 14 days or more of EB treatment lordosis durations were not as long as those observed in natural estrus or after EB plus progesterone (Carter et al., 1973). Estrogen-alone-injected females required more stimulation to elicit lordosis and frequently walk-ed out of the lordosis posture.

Progesterone alone did not induce female sexual behavior in ovariectomized females, even when administered over a period of weeks. Testosterone alone (even after 32 days of injection) was only weakly effective in the elicitation of lordosis (Tiefer, 1970). Those responses that were seen after long-term testosterone may reflect the aromatization of testosterone to estrogen.

In this species, female sexual behavior continued following hypophysectomy (when appropriate, steroid replacement was given) (Carter and Steele, unpublished). These results indicate that pituitary hormones are not essential for the induction of lordosis.

Luteinizing hormone-releasing hormone (LHRH), which can regulate pituitary function, is capable of facilitating lordosis in rats (Moss and McCann, 1973; Pfaff, 1973; Dudley and Moss, 1976). Attempts in my laboratory to use LHRH to elicit or facilitate lordosis in estrogen-primed hamsters have not been successful (Carter, Bohnsak, Summerfelt, and Ramirez, unpublished observations). Systemic injections of LHRH in doses ranging between 100 ng and 10 μg were given to ovariectomized EB-primed females. In one experiment, a 6-day EB-priming treatment was used to examine the possibility that LHRH might be facilitatory in female hamsters that were partially receptive or at least heavily primed with EB. The effects of LHRH were also examined in hypophysectomized females. Luteinizing hormone-releasing hormone has also been directly implanted into the third ventricle (J. F. DeBold, personal communication). Under all of these circumstances, LHRH treatments were without significant behavioral effect. These experiments do not exclude a role for LHRH in the sexual behavior of the female hamster, but do indicate that LHRH treatments that are similar to those that were effective in the rat were not effective in the hamster.

Prostaglandin E_2 (PGE_2) has been shown to facilitate lordosis in estrogen-primed female hamsters (Buntin and Lisk, 1979). The facilitatory effects of PGE_2 were observed in ovariectomized and adrenalectomized females, indicating that progesterone does not play a role in the lordosis-promoting action of PGE_2. Further evidence for the independence of the action of PGE_2 comes from the observation that treatment with a prostaglandin inhibitor, indomethacin, did not significantly reduce estrogen plus progesterone-induced lordosis.

In rats, either prostaglandins or LHRH may facilitate lordosis. There is also evidence that LHRH release may be potentiated by the prostaglandins and it has been proposed that the facilitatory effects of the prostaglandins may be mediated through LHRH (Dudley and Moss, 1976). The present lack of evidence for a facilitatory role for LHRH in the hamster leaves open the question of the cross-species generality of this hypothesis.

Another level of control over female sexual responses may come from neurotransmitters and in particular the biogenic amines, such as serotonin, the catecholamines, and acetylcholine (reviewed by Carter and Davis, 1977; Crowley and Zemlan, 1981). Meyerson (1970) demonstrated that drugs which increased monoamine levels (pargyline and nialamide) inhibited lordosis, whereas drugs that depleted monoamines (reserpine and tetrabenazine) did not disrupt female sexual behavior. In rats, research has pointed to a particularly important inhibitory role for serotonin.

Serotonin is also a precursor in the production of melatonin. There is some evidence that melatonin (produced in highest concentrations in the pineal gland) can inhibit reproductive behavior. We also hypothesized that melatonin, or other pineal products, might have a behavioral role in the hamster (Carter, Docimo, and Zack, unpublished). To examine this possibility, we observed female sexual responses in ovariectomized animals treated with estradiol benzoate and progesterone. The lordosis responses of pinealectomized females were normal and did not differ from operated or unoperated controls. In a separate experiment, we also looked at the effects of pinealectomy on behavioral responses to daily injections of EB (3.3 μg/day). In this case pinealectomized females were slightly more responsive than operated (sham pinealectomized) controls. However, anesthetic controls (receiving injections of sodium pentobarbital but not surgery) showed an intermediate level of lordosis and did not differ from the other two groups. When compared with other studies of long-term EB-treated females from our laboratory, the results for the sham-operated control group appeared depressed, and there was no marked enhancement in the pinealectomized females. The responses of individual hamsters to long-term EB are variable and we have not attempted to further explore this phenomenon. At present we can only speculate that injury to the pineal during surgery (perhaps due to hemmorhage) might have altered the secretory patterns of the organ or subsequent tissue sensitivity to pineal products.

In hamsters (Carter et al., 1978), parachlorophenylalanine (PCPA, 360 mg/kg), which can inhibit serotonin synthesis, rapidly facilitated lordosis in estrogen-primed females. Parachlorophenylalanine was even more effective when administered after six days of EB priming. Parachlorophenylalanine also facilitated lordosis in EB-treated females after both ovariectomy and adrenalectomy, indicating that this drug can induce or potentiate lordosis in the absence of adrenal secretions such as progesterone. Methysergide (a serotonin receptor agonist) also facilitated lordosis, offering further support for a role for serotonin in the regulation of female sexual behavior.

Alpha-methyl-para-tyrosine (aMPT, which inhibits catecholamine synthesis) likewise induced lordosis after six days of EB priming. However, aMPT also produced an elevation in serum progesterone (in ovariectomized, adrenally intact females). Other drugs that affect the catecholamine systems, including pimozide (Carter et al., 1978), amphetamine, and chlorpromazine (Carter et al., 1977), had little effect on female sexual behavior in the hamster.

Recent rat research has suggested a complex role for the cholinergic system in female sexual behavior. Lindström (1972) found that the cholinergic agonist pilocarpine rapidly inhibited lordosis in estrogen-plus-progesterone-activated female hamsters. Attempts to block the inhibitory effects of pilocarpine with reserpine or PCPA (monoaminergic inhibitors) were not successful in the hamster (in contrast to the rat). In ovariectomized estrogen-primed female hamsters, pilocarpine treatment facilitated lordosis and methyl-atropine (which does not cross the blood–brain barrier) prevented this facilitation suggesting that pilocarpine's facilitatory effects were due to actions outside of the central nervous system. In my laboratory (Carter, Bohnsak, and Bahr, unpublished), we have replicated Lindström's pilocarpine experiment and confirmed his observation that lordosis was induced in ovariectomized estrogen-primed female hamsters. We subsequently measured serum progesterone and found that progesterone levels were approximately 2.5 times greater in pilocarpine-treated females than in saline controls. Our further attempts to examine the possible role of the adrenal as a mediator of the behavioral effects of pilocarpine were unsuccessful because the drug was debilitating and often lethal when given following adrenal surgery in the hamster. The current evidence for this species suggests, at least for one cholinergic agonist (pilocarpine), that the observed lordosis induction may in part reflect the effects of adrenal progesterone that can be released by cholinergic stimulation. However, these results do not exclude a central role for acetylcholine.

Opiates also have been implicated in female sexual behavior in the hamster. Relatively small doses of morphine that were without other obvious behavioral effects interfered with the pelvic adjustment movements of the female hamster. These effects of morphine were reversed by treatment with the opiate receptor agonist, naloxone. Naloxone alone had little effect on the sexual behavior of female hamsters (Ostrowski et al., 1979). Intracerebroventricular (ICV) injections of morphine did not affect lordosis durations or pelvic adjustments. These results may suggest either that the observed behavioral effects of morphine are peripheral or that the ICV injections were not effective for some other reason.

The neurochemical correlates of receptivity induction in this species remain open to further investigation. In addition, neurochemical factors regulating proceptivity and attractivity have not been segregated from factors influencing receptivity for the hamster.

5. FACTORS REGULATING THE DURATION OF FEMALE SEXUAL RECEPTIVITY

Natural estrus has been reported to last for approximately 12 to 20 hr in the female hamster. The onset of estrus is clearly influenced by ovarian secretions including estrogen and progesterone. Ovariectomy approximately 4 hr before the expected onset of behavioral receptivity prevented the appearance of lordosis. Progesterone, but not estrogen, injections restored lordosis in these females. However, if females were allowed to become behaviorally receptive, subsequent progesterone injections were not effective in eliciting lordosis (Ciaccio and Lisk, 1971a). Ovariectomized females that were estrogen-primed and then received progesterone also became refractory to subsequent progesterone treatments. The duration of the refractory period was a function of the mode of progesterone treatment. Long-lasting preparations of progesterone (suspended in sesame oil) blocked lordosis for a longer period of time than did a presumably shorter-acting treatment of progesterone in propylene glycol (Ciaccio and Lisk, 1971b). (This capacity of one progesterone exposure to reduce or block the response to a subsequent treatment has been termed the biphasic effect of progesterone.)

The inhibitory effects of progesterone are not absolute. When females that had shown the biphasic effect of progesterone were given a subsequent, large dose of progesterone (1.5 mg, 7.5 times the inducing dose), they responded at least briefly with a second period of receptivity (Lisk and Reuter, 1980). In the hamster, the inhibitory effects of progesterone were most potent when the progesterone was first given 48 hr after an initial estrogen priming (termed a sequential treatment). When estrogen and progesterone were given concurrently, a much higher dose of progesterone was necessary to produce subsequent reductions in lordosis.

The mechanisms of these inhibitory effects remain the subject of controversy. It has been proposed that treatments that inhibit receptivity act by (1) interfering with the priming actions of estrogen, either before or after the cellular uptake and binding of estrogen, (2) block the actions of progesterone, and/or (3) inhibit the release or action of LHRH.

To test the possibility that progesterone might act on behavior like an antiestrogen, the inhibitory effects of the antiestrogenic compound CI-628 have been compared with those of progesterone. CI-628 blocked lordosis in estrogen-treated hamsters whether given concurrently or sequentially. However, the effective inhibitory doses of CI-628 were about ten times larger than those of progesterone. Dihydrotestosterone (DHT) did not inhibit subsequent responses to progesterone when given either concurrently with estrogen or in the sequential paradigm. However, if DHT was given 10 hr before estrogen priming, then DHT was inhibitory (DeBold et al., 1978b).

Attempts to demonstrate that progesterone inhibits the cellular uptake of estrogen have been generally unsuccessful. Progesterone, given at the same time or 1 hr after tritiated estradiol, did not significantly reduce the accumulation of radioactivity in hypothalamus, cortex, pituitary, or uterus. Progesterone treatment also did not affect the depletion or replenishment of cytoplasmic estrogen receptors in the hypothalamus (DeBold *et al.*, 1976).

Studies in hamster uterus have shown that progesterone increases enzyme activity (phosphatase) that inactivates estrogen receptors in cell nuclei (MacDonald *et al.*, 1982); analogous studies have not been reported for brain tissue.

Although the hypothalamus has been implicated in the estrogen-related induction of lordosis (DeBold *et al.*, 1982b), progesterone implanted in the hypothalamus did not facilitate lordosis. Implants in another area, the mesencephalon, did, however, produce reductions in receptivity (DeBold *et al.*, 1976). These results support the argument that the actions of estrogen and progesterone are separable. Further evidence for the separability of the effects of estrogen and progesterone comes from a report indicating that the dose of progesterone that facilitates lordosis is lower than the dose that inhibits sexual behavior. In addition, 5-alpha-dihydroprogesterone, which is only weakly facilitatory, produced a marked inhibition of later receptivity (DeBold *et al.*, 1982a).

In the male hamster, estrogen or androgen may be used to maintain or induce either male or female sexual response patterns. (It is widely assumed that at least some of androgen's actions occur after intracellular conversion to an estrogen.) Progesterone has been shown to be a potent inhibitor of androgen- or estrogen-induced lordosis in both sexes. However, progesterone did not interfere with male copulatory behavior maintained (after castration) with either testosterone or estradiol. These results were interpreted as additional support for the hypothesis that progesterone's excitatory and inhibitory effects have separate mechanisms and/or sites of action (DeBold *et al.*, 1978a).

6. THE EFFECTS OF COPULATORY STIMULI ON FEMALE BEHAVIOR

In virtually all of the species that have been studied, female sexual behavior may be influenced by the presence of copulatory stimuli from the male (see, for example, Goldfoot and Goy, 1970; McDermott and Carter, 1980; Huck *et al.*, 1982). Reductions in the duration of estrus are readily apparent in the female hamster following mating (Carter and Schein, 1971). Depressions in lordosis durations, often accompanied by increases in female aggression, have been observed within the first hour after the onset of copulation. Mating also decreases the proceptive behaviors of the female hamster and diminishes the tendency of the male to approach her (Steel, 1979a).

The rate of decline in female receptivity is a function of the amount and spacing of copulatory stimulation. Even 10 min of mounting with intromissions produced significant depressions in lordosis durations when a period of hours passed between exposures (Carter and Schein, 1971; Buntin *et al.*, 1981). (Well-rested male hamsters typically provide 10 to 25 intromissions and several ejaculations in a 10-min period.)

Receptivity was clearly depressed following approximately 36 or more intromissions. Male mounting alone had only a slight depressive effect. Intromissions with mounts but without ejaculations (provided by introducing a series of males that were removed before they ejaculated), were highly effective in reducing the female's receptivity. Simulated intromissions, using a soft-tipped probe, also slightly reduced subsequent receptivity. When simulated intromissions were combined with mounting (from a diapered male), the reductions in receptivity approached those that followed normal intromissions with or without ejaculation (Carter, 1973a). The inhibitory effects of mating on female receptivity have been replicated by Buntin *et al.* (1981), who in addition found that the introduction into the vagina of a piece of Silastic® tubing, intended to mimic a seminal plug, also hastened the decline in receptivity.

We have attempted to prevent the inhibitory effects of copulatory stimuli by blocking genital sensory input. Circumvaginal injections of a local anesthetic interfered somewhat with locomotion, but did not prevent postcopulatory reductions in receptivity (Carter, 1973a). Likewise, severing the pelvic and pudendal nerves did not produce significant changes in the postcopulatory behavior of females in natural estrus (Diakow and Carter, unpublished). (The only obvious behavioral effect of these denervations was an inability of the females to perform pelvic adjustments during copulation, making it necessary for the experienced males used in that experiment to compensate more than usual to achieve intromission.)

Depressions in receptivity following mating occurred whether the female hamster was in (1) natural estrus, (2) estrus induced by exogenous estradiol followed by progesterone, or (3) estrus induced by estradiol only (Carter *et al.,* 1976). It was not necessary for the ovaries, adrenals, (Carter, 1972), pituitary gland (Carter *et al.,* 1976), pineal gland (Carter, Docimo, and Zack, unpublished), or olfactory bulbs (Carter, 1973b) to be present in order to observe postcopulatory changes in receptivity. However, hormones can modulate the inhibitory effects of mating. In one experiment, females were brought into estrus with Silastic® pellets of estrogen and progesterone. The postcopulatory depressions in these females lasted for nine or more days. Females in estrus induced by estrogen plus progesterone injections recovered rapidly if progesterone injections were spaced at two-day rather than one-day intervals. Mating in conjunction with prolonged progesterone exposure produced more long-lasting depressions in receptivity than those seen in unmated females exposed to progesterone for several days. Females in estrogen-only-induced estrus showed rapid, short-term reductions in receptivity; however, behavioral recovery was apparent within one day (Carter *et al.,* 1976). These results suggest that short-term (approximately one to two days) inhibitory reductions in receptivity can occur in the absence of progesterone. Longer-lasting postcopulatory depressions apparently depend in some manner on the continued presence of progesterone (Carter *et al.,* 1976; Lisk and Reuter, 1980).

The mechanisms responsible for postcopulatory abbreviations in receptivity remain obscure. Lordosis durations continue to show expected reductions following mating in hormone-treated females that have been ovariectomized, adrenalectomized, hypophysectomized, or pinealectomized, indicating that the observed postcopulatory depression in receptivity does not depend on secretory events in these organs.

Attempts to block the inhibitory effects of mating with pharmacological treatments have been generally unsuccessful. Since drugs that inhibit monoaminergic function are capable of facilitating lordosis in ovariectomized, estrogen-primed female hamsters (Carter and Davis, 1977; Carter et al., 1978), we hypothesized that these neurotransmitter systems might play a role in the rapid depressions that followed mating. For example, injections of serotonin antagonists, including PCPA and methysergide, facilitated the onset of lordosis (Carter et al., 1978). These drugs did not, however, significantly reduce the inhibitory effects of mating (Carter, unpublished). We likewise hypothesized that LHRH, which plays a dynamic role in the induction of pregnancy and which can influence female sexual behavior in rats (Dudley and Moss, 1976), could be involved in the postcopulatory depression in receptivity. (LHRH injection does not inhibit receptivity in hamsters.) The inhibitory effects of mating were not blocked by systemic LHRH injections (Carter, Bohnsak, Summerfelt, and Ramirez, unpublished). Attempts to demonstrate a critical role for endogenous opiates as mediators of the inhibitory effects of mating have also been generally unsuccessful. Neither naloxone (4 mg/kg) nor naltrexone (100 mg/kg), which are potent morphine (opiate) antagonists, reliably blocked mating-induced decreases in female sexual behavior (Ostrowski et al., 1981).

Short-term mating-timed abbreviations in the duration of behavioral estrus could reduce the predator vulnerability of the female. Since hormone levels in pregnancy can be nearly as high as those measured in estrus (Baranczuk and Greenwald, 1973), mechanisms for producing long-term reductions in the behavioral sensitivity to hormones might function to reduce infertile mating in pregnancy. Lordosis has been observed during pregnancy in the hamster (Krehbiel, 1952). In my laboratory we observed that lordosis responses during pregnancy were rare, generally brief, and limited to days 11 to 14 of pregnancy (when endogenous hormones are particularly high). When exogenous estrogen and progesterone injections were given during pregnancy, females treated on days 5 to 7 and tested on day 7 were less responsive than females given identical treatments on days 10 to 12 and tested on day 12 (Carter, unpublished). Lordosis in the latter group was slightly depressed, but approached the levels seen in nonpregnant ovariectomized animals.

7. SUMMARY AND DISCUSSION

Female sexual behavior in the golden hamster is regulated by ovarian hormones. Estrogen facilitates proceptivity, including the willingness of the female to spend time with the male. Estrogen priming followed by progesterone elicits within a few hours the onset of sexual receptivity, which is characterized by lordosis. After longer-term (greater than 24 hr) exposure to progesterone, receptivity declines. Exposure to copulatory stimuli, and in particular intromissions, also produces declines in receptivity. The effects of mating are rapid, usually resulting within a few hours in a total loss of receptivity. Increases in female agonistic responses are typical following mating. Mating-related depressions in receptivity can occur following ovariectomy, adrenalectomy, pinealectomy, or hypophysectomy. The effects of copulation are long-lasting when progesterone is present, suggesting a possible mechanism for the inhibition of sexual responses during pregnancy.

Pharmacological studies of female sexual responses in this species support the hypothesis that monoamines and particularly serotonin may play a direct role in the regulation of the onset of lordosis. Other hormones and neurotransmitters are also probably involved in this behavior. However, pharmacological manipulations can trigger the release of adrenal hormones, such as progesterone. Because of problems associated with studying the behavior of adrenalectomized hamsters, it has been difficult to partition the adrenally and nonadrenally mediated effects of pharmacological manipulations. Hormone and drug treatments, thus far, have been relatively ineffective in preventing either the inhibitory effects of prolonged progesterone or copulation. Additional research on these problems, possibly focusing on receptor phenomena, is needed.

The golden hamster has become a very popular subject in recent studies of behavior endocrinology. In part this is because of the relatively regular nature of the four-day estrous cycle and because the female shows a uniquely well-defined and prolonged sexual posture. Work to date indicates that the endocrine control of the induction of female sexual behavior in the hamster shares many basic features with other short-cycling spontaneous ovulators including the rat, mouse, and guinea pig (Feder, 1981). Although housing female hamsters in large groups tends to suppress the expression of lordosis (Brown and Lisk, 1978; Lisk et al., 1980), the regularity of the estrous cycle in singly housed females is notable in comparison to other species. We know little about the social organization of this species in nature. However, laboratory studies strongly suggest that the female hamster is solitary, except during the proestrus-estrous period (Ciaccio et al., 1979; Ciaccio and Lisk, 1979). One adaptation that could be correlated with this solitary life style might be the development of an estrous cycle that relies heavily on photoperiod or other geophysical factors, rather than social determinants. In more social species, such as rats, mice, and gerbils, cycle irregularity is common and may provide reproductive flexibility that is lacking in the hamster. The restricted geographic distribution of the golden hamster (found in a limited area in the Middle East) may reflect in part this reproductive intransience.

In my view, studies of the reproductive behaviors of the female hamster, when examined in a cross-species perspective, provide an excellent opportunity to gain insights into the possible evolutionary origins of patterns of sexual behavior. For example, in contrast to the golden hamster, the highly social prairie vole shows no estrous cycle, has a socially induced estrus, and is an induced ovulator (Richmond and Conaway, 1969; Getz and Carter, 1980; Carter et al., 1980). The Norway rat, which is a relatively social species with a global zoogeography, appears capable of exploiting virtually any photoperiodic or environmental conditions. Rats can show photoperiodically timed estrus (Feder, 1981) or under adverse conditions may rely on socially determined estrus and ovulation (Johns et al., 1978).

Along with many aficionados of this species, I feel that the major drawback to the study of the golden hamster is the continuing absence of data on its natural history. [See

Murphy (1971) for one heroic attempt to gain this information.] In spite of this problem, the golden hamster has emerged as one of the most valuable subjects available for laboratory studies of female sexuality.

ACKNOWLEDGMENTS. I would like to acknowledge the useful comments on this manuscript by Dr. Edward Roy. Published and unpublished research in this paper has been supported by grants from the National Institute of Mental Health, National Institutes of Health (HD 16679), the National Science Foundation (BNS 79-25713), and a Biomedical Research Support Grant (BRSG RR-07030 to the University of Illinois).

REFERENCES

Baranczuk, R., and Greenwald, G. S., 1973, Peripheral levels of estrogen in the cyclic hamster, *Endocrinology* 92:805–812.

Beach, F. A., Stern, B., Carmichael, M., and Ransom, E., 1976, Comparisons of sexual receptivity and proceptivity in female hamsters, *Behav. Biol.* 18:473–487.

Brown, S. M., and Lisk, R. D., 1978, Blocked sexual receptivity in grouped female hamsters, the result of contact induced inhibition, *Biol. Reprod.* 18:829–833.

Buntin, J. D., and Lisk, R. D., 1979, Prostaglandin E2-induced lordosis in estrogen-primed female hamsters: Relationship to progesterone action, *Physiol. Behav.* 23:569–575.

Buntin, J. D., Ciaccio, L. A., and Lisk, R. D., 1981, Temporal aspects of mating-induced inhibition of sexual receptivity and its recovery in the female golden hamster, *Behav. Neural Biol.* 31:443–456.

Carmichael, M. S., 1980, Sexual discrimination by golden hamsters (*Mesocricetus auratus*), *Behav. Neural Biol.* 29:73–90.

Carter, C. S., 1972, Postcopulatory sexual receptivity in the female hamster: The role of the ovary and adrenal, *Horm. Behav.* 3:261–265.

Carter, C. S., 1973a, Stimuli contributing to the decrement in sexual receptivity of female golden hamsters, *Anim. Behav.* 21:827–834.

Carter, C. S., 1973b, Olfaction and sexual receptivity in the female golden hamster, *Physiol. Behav.* 10:47–51.

Carter, C. S., and Davis, J. M., 1977, Biogenic amines, reproductive hormones and female sexual behavior: A review, *Biobehav. Rev.* 1:213–224.

Carter, C. S., and Porges, S. W., 1974, Ovarian hormones and the duration of sexual receptivity in the female golden hamster, *Horm. Behav.* 5:303–315.

Carter, C. S., and Schein, M. W., 1971, Sexual receptivity and exhaustion in the female golden hamster, *Horm. Behav.* 2:191–200.

Carter, C. S., Michael, S. J., and Morris, A. H., 1973, Hormonal induction of female sexual behavior in male and female hamsters, *Horm. Behav.* 4:129–141.

Carter, C. S., Landauer, M. R., Tierney, B. M., and Jones, T., 1976, Regulation of female sexual behavior in the golden hamster: Behavioral effects of mating and ovarian hormones, *J. Comp. Physiol. Psychol.* 90:839–950.

Carter, C. S., Daily, R. F., and Leaf, R., 1977, Effects of *d*-amphetamine, chlorpromazine, chloradiazepoxide and oxazepam on sexual responses in male and female hamsters, *Psychopharmacology* 55:195–201.

Carter, C. S., Bahr, J. M., and Ramirez, V. D., 1978, Monoamines, estrogen and female sexual behavior in the golden hamster, *Brain Res.* 144:109–121.

Carter, C. S., Getz, L. L., Gavish, L., McDermott, J. L., and Arnold, P., 1980, Male-related pheromones and the activation of female reproduction in the prairie vole (*Microtus ochrogaster*), *Biol. Reprod.* 23:1038–1045.

Ciaccio, L. A., and Lisk, R. D., 1971a, Hormonal control of cyclic estrus in the female hamster, *Am. J. Physiol.* 221:936–942.

Ciaccio, L. A., and Lisk, R. D., 1971b, The role of progesterone in regulating the period of sexual receptivity in the female hamster, *J. Endocrinol.* 50:201–207.

Ciaccio, L. A., and Lisk, R. D., 1973/74, Central control of estrous behavior in the female golden hamster, *Neuroendocrinology* 13:21–28.

Ciaccio, L. A., and Lisk, R. D., 1979, Organization of reproductive behavior relative to the tunnel of the female golden hamster, in: *Eastern Conference on Reproductive Behavior*, New Orleans, Louisiana,

Ciaccio, L. A., Lisk, R. D., and Reuter, L. A., 1979, Pre-lordotic behavior in the hamster: A hormonally modulated transition from aggression to sexual receptivity, *J. Comp. Physiol. Psychol.* 93:771–780.

Crowley, W. R., and Zemlan, F. P., 1981, The neurochemical control of mating behavior, in *Neuroendocrinology of Reproduction, Physiology and Behavior* (N. T. Adler, ed.), Plenum Press, New York, pp. 65–85.

Darby, E. M., Devor, M., and Chorover, S. L., 1975, A presumptive sex pheromone in the hamster: Some behavioral effects, *J. Comp. Physiol. Psychol.* 88:496–502.

DeBold, J. F., Martin, J. V., and Whalen, R. E., 1976, The excitation and inhibition of sexual receptivity in female hamsters by progesterone: Time and dose relationships, neural localization and mechanisms of action, *Endocrinology* 99:1519–1527.

DeBold, J. F., Morris, J. L., and Clemens, L. G., 1978a, The inhibitory actions of progesterone: Effects on male and female sexual behavior of the hamster, *Horm Behav* 11:28–41.

DeBold, J. F., Ruppert, P. H., and Clemens, L. G., 1978b, Inhibition of estrogen-induced sexual receptivity of female hamsters: Comparative effects of progesterone, dihydrotestosterone and an estrogen antagonist, *Pharmacol. Biochem. Behav.* 9:81–86.

DeBold, J. F., Pleim, E. T., and Sarokhan, N. E., 1982a, Relative effectiveness of progestins for facilitation and inhibition of sexual receptivity in female hamsters, in: *Conference on Reproductive Behavior*, East Lansing, Michigan,

DeBold, J. F., Malsbury, C. W., Harris, V. S., and Malenka, R., 1982b, Sexual receptivity: Brain sites of estrogen action in female hamsters, *Physiol. Behav.* 29:589–593.

Diamond, M., and Yanagimachi, R., 1970, Reproductive development in the female golden hamster in relation to spontaneous estrus, *Biol. Reprod.* 2:223–229.

Diamond, M., Mast, M., and Yanagimachi, R., 1974, Reproductive development and induced estrus in the prepubertal hamster, *Horm. Behav.* 5:129–133.

Dieterlan, F. von., 1959, Das verhalten des syrischen goldhamsters (*Mesocricetus auratus* waterhouse), *Z. Tierpsychol.* 16:47–103.

Dudley, C. A., and Moss, R. L., 1976, Prostaglandin E2. Facilitative action on the lordotic response, *J. Endocrinol.* 71:457–458.

Feder, H. H., 1981, Estrous cyclicity in mammals, in: *Neuroendocrinology of Reproduction, Physiology and Behavior* (N. T. Adler, ed.), Plenum Press, New York, pp. 279–348.

Floody, O. R., and Pfaff, D. W., 1977a, Aggressive behavior in female hamsters: The hormonal basis for fluctuations in female aggressiveness correlated with estrous state, *J Comp Physiol Psychol* 91:443–464.

Floody, O. R., and Pfaff, D. W., 1977b, Communication among hamsters by high-frequency acoustic signals: I. Physical characteristics of hamster calls, *J. Comp. Physiol. Psychol.* 91:794–806.

Floody, O. R., and Pfaff, D. W., 1977c, Communication among hamsters by high-frequency acoustic signals: III. Responses evoked by natural and synthetic ultrasounds, *J Comp Physiol Psychol* 91:820–829.

Floody, O. R., Pfaff, D. W., and Lewis, C. D., 1977, Communication among hamsters by high-frequency acoustic signals: II. Determinants of calling by females and males, *J. Comp. Physiol. Psychol.* 91:807–819.

Floody, O. R., Merkle, D. A., Cahill, T. J., and Shopp, G. M., Jr., 1979, Gonadal hormones stimulate ultrasound production by female hamsters, *Horm. Behav.* 12:172–184.

Frank, A. H., and Fraps, R. M., 1945, Induction of estrus in the ovariectomized golden hamster, *Endocrinology* 37:357–361.

Getz, L. L., and Carter, C. S., 1980, Social organization in *Microtus ochrogaster* populations, *Biologist* 62:56–69.

Goldfoot, D. A., and Goy, R. W., 1970, Abbreviation of behavioral estrus in guinea pigs by coital and vagino-cervical stimulation, *J. Comp. Physiol. Psychol.* 72:426–434.

Hsü, C.-H., and Carter, C. S., 1982, Factors influencing male sexual behavior in female hamsters, in: *Abstracts of Conference on Reproductive Behavior,* East Lansing, Michigan.

Huck, U. W., Carter, C. S., and Banks, E. M., 1982, Natural or hormone induced sexual and social behaviors in the female brown lemming, *Lemmus trimucronatus, Horm. Behav.* 16:199–207.

Humphrys, R. R., Vomachka, A. J., and Clemens, L. G., 1977, Induction of lordosis in the female hamster utilizing metabolites of progesterone, *Horm. Behav.* 9:358–361.

Johns, M. A., Feder, H. H., Komisaruk, B. R., and Mayer, A. D., 1978, Urine-induced reflex ovulation in anovulatory rats may be a vomeronasal effect, *Nature (London)* 272:446–447.

Johnson, W. A., Billiar, R. B., and Little, B., 1976, Progesterone and 5α-reduced metabolites: Facilitation of lordosis behavior and brain uptake in female hamsters, *Behav. Biol.* 18:489–497.

Johnston, R. E., 1974, Sexual attraction function of golden hamster vaginal secretion, *Behav. Biol.* 12:111–117.

Johnston, R. E., 1975, Sexual excitation function of hamster vaginal secretion, *Anim. Learn. Behav.* 3:161–166.

Johnston, R. E., 1977, The causation of two scent-marking behaviour patterns in female hamsters (*Mesocricetus auratus*), *Anim. Behav.* 25:317–327.

Johnston, R. E., 1979, Olfactory preferences, scent marking, and "proceptivity" in female hamsters, *Horm. Behav.* 13:21–39.

Johnston, R. E., 1980, Responses of male hamsters to odors of females in different reproductive states, *J. Comp. Physiol. Psychol.* 94:894–904.

Kairys, D. J., Magalhaes, H., and Floody, O. R., 1980, Olfactory bulbectomy depresses ultrasound production and scent marking by female hamsters, *Physiol. Behav.* 25:143–146.

Kent, G. C., Jr., 1968, Physiology of reproduction, in: *The Golden Hamster* (R. A. Hoffman, P. F. Robinson, and H. Magalhaes, eds.), Iowa State University, Ames.

Kow, L.-M., and Pfaff, D. W., 1977, Sensory control of reproductive behavior in female rodents, *Ann. N. Y. Acad. Sci.* 290:72–97.

Kow, L.-M., Malsbury, C. W., and Pfaff, D. W., 1976, Lordosis in the male golden hamster elicited by manual stimulation: Characteristics and hormonal sensitivity, *J. Comp. Physiol. Psychol.* 90:26–40.

Krehbiel, R. H., 1952, Mating of the golden hamster during pregnancy, *Anat. Rec.* 113:117–121.

Kwan, M., and Johnston, R. E., 1980, The role of vaginal secretion in hamster sexual behavior: Males' responses to normal and vaginectomized females and their odors, *J. Comp. Physiol. Psychol.* 94:905–913.

Landauer, M. R., Banks, E. M., and Carter, C. S., 1978, Sexual and olfactory preferences of naive and experienced male hamsters, *Anim. Behav.* 26:611–621.

Lindström, L. H., 1971, The effect of pilocarpine and oxotremorine on oestrous behaviour in female rats after treatment with monoamine depletors or monoamine synthesis inhibitors, *Eur. J. Pharmacol.* 15:60–65.

Lindström, L. H., 1972, The effect of pilocarpine and oxotremorine on hormone-activated copulatory behavior in the ovariectomized hamster, *Arch. Pharmacol.* 275:233–241.

Lisk, R. D., and Reuter, L. A., 1980, Relative contributions of oestradiol and progesterone to the maintenance of sexual receptivity in mated female hamsters, *J. Endocrinol.* 87:175–183.

Lisk, R. D., Langenberg, K. K., and Buntin, J. D., 1980, Blocked sexual receptivity in grouped female golden hamsters: Independence from ovarian function and continuous group maintenance, *Biol. Reprod.* 22:237–242.

MacDonald, R. G., Okulicz, W. C., and Leavitt, W. W., 1982, Progesterone-induced inactivation of nuclear estrogen receptor in the hamster uterus is mediated by acid phosphatase, *Biochem. Biophys. Res. Commun.* 104:570–576.

McDermott, J. L., and Carter, C. S., 1980, Ovarian hormones, copulatory stimuli and female sexual behavior in the Mongolian gerbil, *Horm Behav* 14:211–223.

Meyerson, B. J., 1970, Monoamines and hormone activated oestrous behaviour in the ovariectomized hamster, *Psychopharmacologia* 18:50–57.

Morin, L. P., Powers, J. B., and White, M., 1976, Effects of the antiestrogens, MER-25 and CI-628, on rat and hamster lordosis, *Horm. Behav.* 7:283–291.

Moss, R. L., and McCann, S. M., 1973, Induction of mating behavior in rats by luteinizing hormone-releasing factor, *Science* 181:177–179.

Murphy, M. R., 1971, Natural history of the Syrian golden hamster—A reconnaissance expedition, *Am. Zool.* 11:632.

Murphy, M. R., 1974, Relative importance of tactual and nontactual stimuli in eliciting lordosis in the female golden hamster, *Behav. Biol.* 11:115–119.

Murphy, M. R., 1976, Olfactory impairment, olfactory bulb removal and mammalian reproduction, in: *Mammalian Olfaction, Reproductive Processes, and Behavior* (R. L. Doty, ed.), Academic Press, New York, pp. 95–117.

Noble, R. G., 1973, Facilitation of the lordosis response of the female hamster (*Mesocricetus auratus*), *Physiol. Behav.* 10:663–666.

Noble, R. G., 1979, The sexual responses of the female hamster: A descriptive analysis, *Physiol. Behav.* 23:1001–1005.

Noble, R. G., 1980, Sex responses of the female hamster: Effects of male performance, *Physiol. Behav.* 24:237–242.

Orsini, M. W., 1961, The external vaginal phenomena characterizing the stages of the estrous cycle, pregnancy, pseudopregnancy, lactation, and the anestrous hamster *Mesocricetus auratus* Waterhouse, *Proc. Anim. Care Panel* 11:193–206.

Ostrowski, N. L., Stapleton, J. M., Noble, R. G., and Reid, L. D., 1979, Morphine and naloxone's effect on sexual behavior of the female golden hamster, *Pharmacol. Biochem. Behav.* 11:673–681.

Ostrowski, N. W., Noble, R. G., and Reid, L. D., 1981, Opiate antagonists and sexual behavior in female hamsters, *Pharmacol. Biochem. Behav.* 10:881–888.

Payne, A. P., and Swanson, H. H., 1970, Agonistic behaviour between pairs of hamsters of the same and opposite sex in a neutral observation area, *Behaviour* 36:259–267.

Pfaff, D. W., 1973, Luteinizing hormone-releasing factor potentiates lordosis behavior in hypophysectomized ovariectomized female rats, *Science* 182:1148–1149.

Reed, C. A., and Reed, R., 1946, The copulatory behavior of the golden hamster, *J. Comp. Psychol.* 39:7–12.

Richmond, M. E., and Conaway, C. H., 1969, Induced ovulation and oestrus in *Microtus ochrogaster*, *J. Reprod. Fertil.* 6:357–376.

Sales, G. D., 1972, Ultrasounds and mating behaviour in rodents with some observations on other behavioural situations, *J. Zool.* 168:149–164.

Siegel, H. I., Cohen, P., and Rosenblatt, J. S., 1979, The effect of hysterectomy on hormone-induced lordosis behavior in hamsters, *Physiol. Behav.* 23:851–853.

Steel, E., 1979a, Short-term, postcopulatory changes in receptive and proceptive behavior in the female Syrian hamster (*Mesocricetus auratus*), *Horm. Behav.* 12:280–292.

Steel, E., 1979b, Male-female interaction throughout the oestrous cycle of the Syrian hamster (*Mesocricetus auratus*), *Anim Behav* 27:919–929.

Steel, E., 1980, Changes in female attractivity and proceptivity throughout the oestrous cycle of the Syrian hamster (*Mesocricetus auratus*), *Anim. Behav.* 28:256–265.

Steel, E., 1981, Control of proceptive and receptive behavior by ovarian hormones in the Syrian hamster (*Mesocricetus auratus*), *Horm. Behav.* 15:141–156.

Tiefer, L., 1970, Gonadal hormones and mating behavior in the adult golden hamster, *Horm. Behav.* 1:189–202.

Whitsett, J. M., Gray, L. E., and Bediz, G. M., 1978, Differential influence of stereoisomers of estradiol on sexual behavior of female hamsters, *J. Comp. Physiol. Psychol.* 92:7–12.

9

Male Sexual Behavior

HAROLD I. SIEGEL

1. INTRODUCTION

This chapter deals primarily with a detailed description of the components and pattern-ing of male copulatory behavior, the effects of castration and replacement therapy with a variety of steroids, and the hormonal induction of female sexual behavior in males. Other topics have not been studied extensively and therefore can only be discussed briefly. These include the ontogeny of male sexual behavior, the effects of aging, and the effects of pharmacological agents. Additional material relevant to male copulatory behavior can be found in this volume; for example, see Chapter 7 on neonatal hormone manipulations, Chapter 14 on circadian factors, Chapter 11 on the involvement of central nervous system structures, Chapter 6 on chemical and auditory factors, and Chapter 4 on endocrine processes.

2. ONTOGENY

In an early report (Bond, 1945), males were observed copulating by 30 days of age, but were not fertile until almost two weeks later. Using the penile smear method in which a moistened toothpick is rotated between the prepuce and the glans penis, Vandenbergh (1971) found sperm in the smears at about 42 days.

In a more extensive analysis of the physiological and behavioral changes associated with sexual maturity, Miller *et al.* (1977) were able to identify two separate processes. The first included anatomical growth (body weight, seminal vesicle weight, and flank gland size), testosterone levels, and increases in the components of mating behavior (mounts with and without intromissions) and this development was linear over the interval between 3 to 8 weeks of age (Fig. 1). The second process was related to

---

HAROLD I. SIEGEL • Institute of Animal Behavior and Department of Psychology, University College, Rutgers University, Newark, New Jersey 07102.

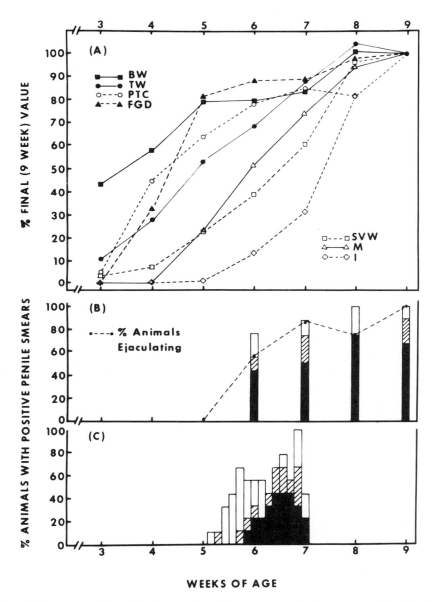

Figure 1. Measurements of sexual development in male hamsters expressed as the percentage of week 9 values. (A) BW = body weight, TW = testis weight, PTC = plasma testosterone concentration, FGD = flank gland diameter, SVW = seminal vesicle weight, M = mounts, I = intromissions. (B) and (C) Sperm concentration in penile smears. Open bars = few sperm, hatched bars = moderate, solid bars = dense. Reprinted with permission of the publisher by Miller *et al.* (1977).

increasing competence in mating behavior and was curvilinear over the same age range; at 5 weeks of age, head and side mounts were three times greater than rear mounts, whereas at 9 weeks of age, rear mounts were five times as frequent as the head/side mounts. Interestingly, this increasing competence in appropriate mounting behavior occurred in the absence of sexual experience because each male was tested only once during the study. Castration at 21 days of age resulted in reductions in seminal vesicle weight, flank gland diameter, and copulatory behavior although one third of these males continued to show some mounting behavior (see also Warren and Aronson, 1956). Relative to the intact males, males castrated at 21 days of age showed less rear mounting behavior, further suggesting that the increasing copulatory competence depends on androgen.

3. DESCRIPTION OF MALE COPULATORY BEHAVIOR

Copulatory behavior in the hamster has been described by Reed and Reed (1946), Beach and Rabedeau (1959), and most comprehensively by Bunnell et al. (1977). Bunnell et al. have classified the hamster copulatory pattern as consisting of multiple intromissions and multiple ejaculations. Thrusting during intromissions, labeled "long intromissions," is an unusual pattern among rodents and has been observed under two conditions. If they are tested until sexual exhaustion, males begin showing long intromissions as they approach satiety. Both normal and long intromissions are sometimes combined in a given ejaculatory sequence, but long intromissions never culminate in an ejaculation. Males also show this pattern during their first copulatory series if 60-sec delays are imposed between intromissions (Miller and Sessions, 1972).

The following discussion summarizes the major findings of Bunnell et al. (1977). The pattern of copulatory behavior is perhaps most variable when the male and female are first placed together. Some females immediately adopt the lordotic posture enabling some males to achieve intromissions within seconds (also reported by Reed and Reed, 1946). More commonly, the initial mount(s) does not involve an intromission and follows an interval of sniffing and licking of the female.

Clearly the mount and intromission latencies depend on each partner's motivation and on their appropriate postural orientations. It is not uncommon for a male to mount inappropriate parts of the female's body and an additional problem is that the male may often be smaller than the female. "The tail of the female is also a hindrance, as it pokes into the abdomen of the male if he stands on both hind feet, but by raising one foot the male can divert the tail into the inguinal region beneath the elevated leg" (Reed and Reed, 1946, p. 9). The female also facilitates orientation by making adjustment responses during copulation. Males tested with females whose genital regions were anesthetized showed longer latencies to mount and achieve intromission and fewer of these males displayed intromissions and ejaculations (Noble, 1980a).

By checking for sperm in vaginal smears, Bunnell et al. (1977) were able to distinguish intromissions with and without ejaculations. Intromissions are characterized by a constant rate of pelvic thrusting just prior to penile insertion. Intromissions with ejaculations differ in that they are accompanied by an increase in thrusting just prior to full intromission. This change in thrusting rate (and observable vibration of the testes) proved more reliable in distinguishing an ejaculation than previous criteria of increased thrusting amplitude, spasmodic flexion of the elevated rear leg, and increased duration of vaginal penetration (Beach and Rabedeau, 1959).

Both Beach and Rabedeau (1959) and Bunnell et al. (1977) tested their animals until sexual satiety and recorded essentially similar results. The first ejaculatory series required a larger number of intromissions and a longer latency than subsequent series. As expected, the median postejaculatory interval increased as the test progressed. Interestingly, Miller and Sessions (1972) showed that removing the female for 60 sec after each intromission leading to the first ejaculation resulted in a higher intromission frequency (29 compared with 17 on the pretest), whereas imposing a 60-sec delay only after the first intromission resulted in a lower frequency (8). Further, both the single and multiple delay groups had shorter postejaculatory intervals, the period from the first ejaculation to the next intromission.

Compared with comparable data from rats (Beach and Jordan, 1956), Beach and Rabedeau (1959) pointed out that hamsters show shorter interintromission intervals, shorter postejaculatory intervals, more ejaculations per satiety test, and quicker recovery from prior exhaustion tests.

Two additional findings from the studies of Bunnell et al. (1977) should be mentioned. The first is that the Coolidge effect was obtained in 18 of 23 tests among five males. The Coolidge effect refers to the renewed display of copulatory behavior in sexually satiated males (no intromissions for 15 min) who are presented with a second female. These tests always involved the introduction of a new female and therefore precluded the possibility that copulation would have resumed if the original female was removed and then replaced. Miller (1971; see Bunnell et al., 1977) also used a satiety criterion of 15 min without an intromission, removed the female and then introduced either the same, a recently mated, or an unmated female. The second female elicited copulations in at least one half of the tests regardless of condition. However, the unmated female condition was associated with the shortest mount and intromission latencies and the greatest number of mounts, intromissions, and ejaculations.

The final point here concerns the complete behavioral repertoire of the male and female during copulation tests. Relative to other species studied, Bunnell et al. (1977) reported that during the interval between the first intromission and ejaculation, the male hamster is "exceptionally channelized toward copulatory activity." These activities primarily consist of "pursuit-mount" and "genital grooming" and occupy more than 80% of this ejaculation latency period compared with the 20% shown by male rats. One difference between rats and hamsters that probably accounts for this channelization is that female rats spend less than 10% of this interval in the lordotic posture, whereas female hamsters adopt the posture for virtually the entire period.

The maintenance of the lordotic posture also probably accounts for the behavior of the male when presented with two sexually receptive females at the same time (Dewsbury et al., 1979). Although there was little change in the pattern of copulatory behavior, the males tended to stay with the same female until an ejaculation and then shifted their attention to the other female. Both females received sexual stimulation and more pups were delivered in the two-female condition.

4. EFFECTS OF CASTRATION

Castration of the hamster results in an increase in body weight, a decrease in adrenal gland weight (Swanson, 1967), the absence of spontaneous seminal emissions (Beach and Eaton, 1969), and a decrease in the size and pigmentation of the flank glands (Vandenbergh, 1973).

Castration also results in a gradual decline in copulatory behavior. Pauker (1948) reported that males achieved an average of 1.8 copulations (mounts with intromissions) per minute one week after castration and less than 0.2 copulations four weeks later (removal of the prostate and seminal vesicles had no additional effect). Several studies have examined the decline in sexual activity following castration and they will be discussed before presenting the effects of hormone replacement.

Bunnell and Flesher (1965) studied the display of mounts and intromissions as a function of time since androgen removal. Hormone withdrawal was accomplished by either castrating animals or stopping the administration of androgen in previously castrated animals. Three groups, each including both types of males, were first tested 45, 60, or 75 days after androgen removal and were then tested three more times at five-day intervals. There were no differences among the three groups and by the fourth test slightly more than one half of the males were still mounting and about one third of the subjects were still achieving intromissions.

The effects of experience were studied (Bunnell and Kimmel, 1965) in groups of males initially matched on a composite sexual activity score (mounts plus intromissions). In the high-experience condition, males were given nine additional 5-min tests with sexually receptive females, whereas males in the low-experience condition were exposed to nonestrous females on nine occasions. Males in both experience conditions were then tested intact or after castration. The sexual activity scores of the low-experience intact males were initially lower, but soon increased to the scores of the high-experience intact males. Sexual behavior continued for almost six weeks after castration, but during this period the high-experience males had higher sexual activity scores than the low-experience males.

Both sexual experience and the frequency of testing were found to affect the postcastration decline in copulatory behavior (Lisk and Heimann, 1980). Four groups were given sexual experience in the form of three weekly tests. One group remained intact and was tested at twelve weekly intervals and three of these groups were castrated and tested weekly, every third week or every sixth week. A fifth group was castrated without prior sexual experience and tested six weeks later.

The effect of castration was shown by comparing the intact and castrated groups tested weekly. Differences in mount and intromission frequencies and latencies were seen four weeks after castration. Intromissions were rarely seen six weeks after castration. Males permitted precastration sexual experience showed higher levels of mounts and intromissions and shorter mount and intromission latencies during the first three to four weeks of testing than the inexperienced castrated animals. Over the first six weeks of testing, the more frequently tested males had lower mount and intromission frequencies and shorter mount latencies than the groups tested every third or sixth week.

5. HORMONE REPLACEMENT

This section summarizes the results of a large number of studies on the ability of testosterone, other androgens, estrogen, and progesterone to restore copulatory behavior in castrated animals. It should be briefly noted that androgen has also been shown to have related effects and these include the resumption of spontaneous seminal emissions (Beach and Eaton, 1969), an increase in the size and pigmentation of the flank glands (Vandenbergh, 1973), and the return of the male's interest in diestrous females (measured by proximity to her and olfactory investigation) (Steel, 1982).

5.1. Testosterone

The evidence is quite clear that testosterone is capable of restoring male copulatory behavior. Daily injections of 0.075 or 0.1 mg beginning three months after castration significantly increased the display of intromissions, whereas 0.001 mg was ineffective (Beach and Pauker, 1949). Testosterone propionate (TP) at daily doses of 0.2 mg (Tiefer, 1970) or 0.3 mg (Noble and Alsum, 1975) resulted in levels of copulatory behavior that were comparable to those of intact males. Tiefer and Johnson (1973) compared the effectiveness of testosterone with that of TP when given at 0.1 mg/day for 54 days beginning one month after castration. Testing at nine intervals during the period of hormone administration showed that the TP-treated males displayed more intromissions and ejaculations than the testosterone-treated animals.

The issue is not whether testosterone can restore sexual behavior, but whether it needs to be aromatized. Aromatase activity has been found in the preoptic hypothalamic area and the anterior limbic cortex (Callard et al., 1979). The effects of another androgen, androstenedione, and testosterone metabolites are discussed next.

5.2. Androstenedione

Christensen et al. (1973) found that androstenedione (0.5 mg/day), an aromatizable androgen, was equal to testosterone and significantly more effective than two nonaromatizable androgens, dihydrotestosterone (DHT) and androsterone, on the display of mount and intromission frequencies. At lower daily doses (0.1 mg), Tiefer and Johnson (1973) also found comparable effects of androstenedione and testosterone on the frequency of mounts, intromissions, and ejaculations, but both of these hormones resulted in lower levels of copulatory behavior than TP.

At identical doses ranging from 0.05 to 1 mg/day, Whalen and DeBold (1974) reported that androstenedione was at least as effective as testosterone on the percentage

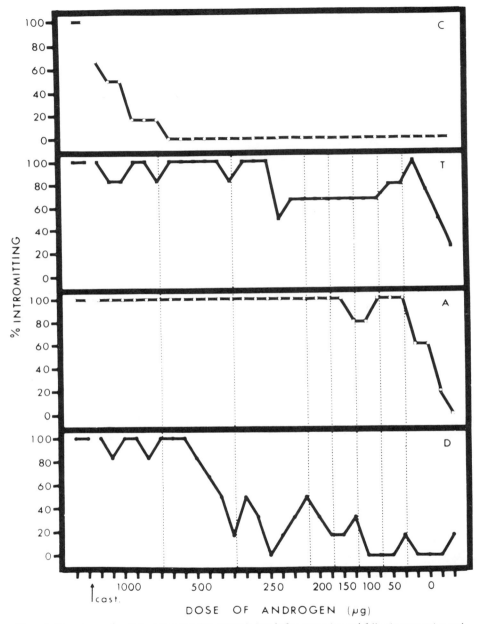

Figure 2. Percentages of male hamsters showing intromissions before castration and following castration and treatment with vehicle (C) or progressively decreasing doses of testosterone (T), androstenedione (A), or dihydrotestosterone (D). The percentages of the same males showing mounts and ejaculations follow patterns similar to those for intromissions. Reprinted with permission of the publisher by Whalen and DeBold (1974).

of their males showing mounts, intromissions, and ejaculations (Fig. 2). However, Payne and Bennett (1976) showed that although androstenedione maintained seminal vesicle weight, it was less effective than testosterone or DHT on the components of copulatory behavior.

5.3. Dihydrotestosterone (DHT)

The effects of DHT, a nonaromatizable androgen, on male sexual behavior have been examined in several studies. Christensen *et al.* (1973) used castrated, testosterone-treated males who displayed at least ten intromissions on a pretest. These males were then treated for ten weeks with 0.5 mg DHT and tested every two weeks. The DHT-treated males showed fewer mounts and intromissions than a testosterone-treated group, and by the last week of testing their mount and intromission scores were comparable to those of castrated controls receiving oil.

Sexually experienced castrated males were given DHT daily at 1 mg for the first six weeks, 0.5 mg for the next six weeks, 0.25 mg for another six weeks, and then the dose was reduced by 0.05 mg at two-week intervals (Whalen and DeBold, 1974). The effectiveness of DHT on the percentage of males displaying mounts, intromissions, and ejaculations was related to the dose (see Fig. 2). At 1 mg/day, DHT was equal to the same dose of testosterone. From 0.5 to 0.05 mg/day, the number of males showing all components of sexual behavior declined to levels lower than those of the group treated with identical doses of testosterone. Thus, at the same dose of DHT, the results of this study are comparable to those of Christensen *et al.* (1973). However, at twice the dose, 1 mg/day, DHT completely restored the full copulatory pattern.

Noble and Alsum (1975) administered 0.3 mg/day of the propionate, longer-acting, form of DHT. Compared with males given the same dose of TP, the DHT males showed somewhat lower frequencies of mounts and intromissions and significantly lower frequencies of ejaculations although both groups had equivalent glans penis weights. The partial restoration by DHT on copulatory behavior in this case using 0.3 mg compared with the ineffectiveness of 0.5 mg used by Christensen *et al.* (1973) may have been due to the use of the propionate form of the hormone.

Dihydrotestosterone at a dose of 0.5 mg three times per week for nine weeks resulted in levels of copulatory behavior that were essentially comparable to those shown by sham-castrated males (Payne and Bennett, 1976). Castrated males receiving oil showed a 36% increase in body weight compared with a 4% increase for testosterone-treated and a 19% increase for DHT-treated males. Dihydrotestosterone was equal to testosterone in maintaining seminal vesicle weights. Finally, DeBold and Clemens (1978) found that a relatively low dose of DHT (0.15 mg/day for ten weeks) was capable of restoring mounts, intromissions, and ejaculations but only in a minority of the males (Fig. 3).

5.4. Estrogen, Progesterone, and Hormone Combinations

The results of aromatizable and nonaromatizable androgens on copulatory behavior in the hamster are not conclusive. Depending on the dosage, DHT has been shown to

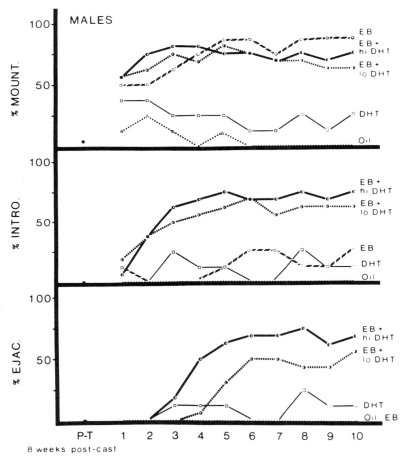

Figure 3. Percentages of male hamsters showing mounts, intromissions, and ejaculations at an 8-week postcastration pretest (P-T) and following treatment with dihydrotestosterone (DHT; high dose = 0.15 mg, low dose = 0.03 mg) and/or estradiol benzoate (EB, 1- and 5-μg groups have been combined). Animals were tested once/week for weeks. Reprinted with permission of the publisher by DeBold and Clemens (1978).

be either completely or only partially effective. Further, full restoration of sexual behavior by a nonaromatizable androgen does not preclude a role for aromatization. This section summarizes the effects of estrogen with and without progesterone and the combination of these hormones with androgens on male sexual behavior.

Males treated with estradiol benzoate (EB) mounted as frequently as those treated with TP, but failed to intromit or ejaculate (Tiefer, 1970; Noble and Alsum, 1975; DeBold and Clemens, 1978) (see Fig 3). However, Johnson (1975) found that even mount frequencies were quite low in castrated males treated with EB alone or in combination with progesterone. Compared with males receiving TP only, males given TP plus high daily doses of progesterone showed a delay of only several days in regaining intromissions (DeBold *et al.*, 1978). The addition of progesterone given to EB-treated males had no effect.

Estradiol benzoate increased the frequency of mounts and DHT restored all components of copulatory behavior in a small percentage of males (DeBold and Clemens, 1978). The combination of estradiol and DHT, both metabolites of testosterone, resulted in the complete pattern of sexual behavior in the majority of males (Fig. 3). Females treated with androgen during the early postnatal period also responded to the combined estradiol and DHT treatment, whereas nonandrogenized females responded poorly. Taken together, these results suggest that both estradiol and DHT may play a role in normal male sexual behavior and that early androgen increases the animals' sensitivity to one or both hormones.

This dual component model of testosterone action (DeBold and Clemens, 1978) in which estradiol is responsible for "inciting sexual activity" by increasing mounting behavior and DHT is responsible for "organizing that activity into the complete copulatory sequence" has received recent experimental attention.

In the first of these studies, Lisk and Bezier (1980) implanted TP, estradiol, or DHT into the anterior hypothalamus. TP was also implanted into the posterior hypothalamus. The percentage of the castrated groups showing mounting behavior during the period of hypothalamic implantation ranged from 70% to 88% for TP and estradiol and only 27% for DHT. Only TP (posterior hypothalamic implants) facilitated intromissions but in only 55% of the group.

In a follow-up study, Lisk and Greenwald (1983) tested the effects of hypothalamically implanted TP and EB with or without subcutaneous Silastic® implants of DHT. Both TP and EB resulted in precastration levels of mounting but not intromissions. The addition of DHT facilitated intromission behavior equally in both TP and EB implanted animals. Thus, the central nervous system administration of testosterone and estrogen required the peripheral administration of DHT for the display of high levels of the complete copulatory pattern. Although hypothalamic implants of DHT did not affect mounts or intromissions, it is possible that other implant sites may have been effective.

In summary, the normal hormonal control of male sexual behavior remains unclear. Systemically administered testosterone, but not always androstenedione, completely restored sexual activity. Hypothalamically administered testosterone restored mounting behavior, but only partially facilitated intromission behavior. The effects of DHT on the copulatory pattern depend on the particular conditions of the studies. The effectiveness of DHT has been shown to be extremely limited or to be comparable to that of testosterone. The only demonstrated effect of estrogen, given peripherally or centrally, has been to increase mounting behavior. The combination of relatively low doses of estrogen and DHT or of hypothalamic estrogen or testosterone and systemic DHT is effective on the display of the complete pattern of sexual behavior and therefore supports the dual testosterone metabolite model and suggests a role for aromatization.

6. INDUCTION OF FEMALE SEXUAL BEHAVIOR

Before discussing the hormonal induction of female sexual behavior in males, it is important to briefly describe the typical female copulatory pattern. In a number of other rodents, the receptive female assumes the lordotic posture in response to individual mounts. The female hamster may adopt the lordotic posture when first placed with the male even in the absence of tactile stimulation (Bunnell *et al.*, 1977). More commonly, she first shows lordosis after an interval of sniffing and licking of her head, body, and pudendum by the male. Although the female may move away from the male after the initial attempts at copulation, she soon shows lordosis again and often maintains the posture for the duration of the ejaculatory interval and even the postejaculatory interval. The most common measures of female receptivity are latency to the first display of the lordotic posture and the total lordosis duration during the test (typically 5 or 10 min).

6.1. Estrogen

When tested with intact males, 40% of castrated males showed lordosis and this percentage increased to 90% after they received 6 μg EB (Tiefer, 1970). Total lordosis durations were short in both cases and the display of lordosis did not extend beyond the time during which the males mounted. Groups treated with 6 or 100 μg EB both showed lordosis, again of short duration, and the higher dose resulted in a greater percentage of the more rigid, female type, of lordosis (Tiefer and Johnson, 1971).

Carter *et al.* (1973) injected 6.6 μg EB daily for 15 days and tested their males on days 1, 2, 3, 6, 9, 12, and 15. Total lordosis durations increased significantly on day 6 and were in the range of 150 to 300 sec (of a 600-sec test) during the remaining tests. Males receiving a total of 6.6 or 666 μg EB split into two injections one day apart showed few lordoses, very short total lordosis durations, and long latencies to the first lordosis. Ten days of 6 μg EB resulted in 100% of castrated males showing lordosis compared with 20% of the oil-treated group (Noble and Alsum, 1975). Lordosis durations increased as a function of a 32-day injection schedule of 6 μg EB for days 1 to 16 and 200 μg EB for days 17 to 32, but these durations were still less than those recorded in females (Johnson, 1975).

In contrast to the previous studies showing some degree of receptivity in castrated males treated with estrogen, DeBold *et al.* (1978) found their animals completely unreceptive despite two weeks of daily injections of 5 or 50 μg EB. It is possible, as these authors point out, that the repeated testing employed in the previously mentioned studies contributed to the display of female sexual behavior.

6.2. Progesterone

The addition of progesterone in estrogen-treated castrated males has been shown to greatly facilitate the display of the female copulatory pattern. Tiefer (1970) and Tiefer and Johnson (1971) reported that compared with estrogen alone, the combination of estrogen and progesterone resulted in more lordoses, more rigid lordoses, more lordoses that outlasted the interval during which the males were mounted, and less aggression.

Males given a single injection of progesterone after 15 days of EB treatment had shorter latencies to show lordosis, greater total lordosis durations, and longer single lordosis durations that approached the limit of the 10-min test than males tested after EB only (Carter *et al.*, 1973). Similar findings were obtained when progesterone was

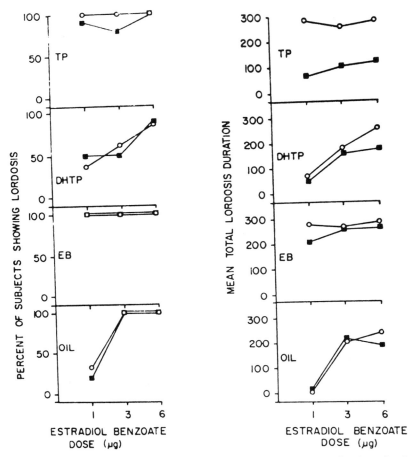

Figure 4. Percentages of male hamsters showing lordosis and their mean total lordosis durations after chronic treatment with either testosterone propionate (TP), dihydrotestosterone propionate (DHTP), estradiol benzoate (EB), or oil plus sequential injections of various doses of EB plus progesterone. Reprinted with permission of the publisher by Noble and Alsum (1975).

given to males injected only twice with EB at either 3.3 or 333 μg per injection. When given progesterone, there were no differences between the high- and low-dose EB groups. Further, chronically EB-treated males given progesterone did not differ from similarly treated females; all animals showed lordosis and total lordosis durations approached the entire length of the test period (Noble and Alsum, 1975) (Fig. 4).

Although a single injection of progesterone in EB-treated animals significantly elevated total lordosis durations, chronic progesterone administered along with EB resulted in an inhibition in both males and females (DeBold *et al.,* 1978). Thus, although progesterone can inhibit hormone-induced receptivity in males, as mentioned earlier, it had only a small effect on the inhibition of testosterone-induced male behavior.

6.3. Androgens

Although quite effective in stimulating male sexual behavior, TP also resulted in the display of lordosis in 67% of the males (compared with 11% of the females). Lordosis durations were short, did not exceed mount durations, and were generally not of the more rigid, female type of posture (Tiefer, 1970; Johnson, 1975).

Chronic TP treatment followed by various doses of EB and progesterone increased the percentage of males showing lordosis (about 75% to 100%) compared with males given TP only (10%) (Noble and Alsum, 1975) (see Fig. 4). At 3 and 6 μg EB (but not 1 μg) plus progesterone, chronically treated TP males had shorter total lordosis durations than oil-treated males given identical doses of EB and progesterone. Long-term DHT treatment resulted in a minority of the males showing lordosis and this increased to 90% after the administration of the highest dose of EB (6 μg) plus progesterone (see Fig. 4). Lordosis durations after the lowest and highest doses of EB did not differ between the DHT- and oil-treated groups.

Finally, DeBold et al. (1978) reported a lack of receptivity in castrated males receiving 0.15 or 0.5 mg TP for two weeks. Two additional weeks of TP injections followed by an injection of progesterone 4 hr prior to the test for receptivity resulted in a small facilitation of lordosis durations in both males and females given the higher dose of TP. The addition of progesterone for the entire four weeks inhibited the facilitation of lordosis durations by TP only in both sexes.

The results taken together show that males are capable of showing lordosis behavior and this capacity is enhanced with repeated treatment with estrogen. The addition of progesterone greatly facilitates the display of lordosis, lordosis durations, the rigidity of the response, and eliminates the need for higher doses of estrogen. Testosterone generally resulted in inferior lordosis responses relative to those of estrogen, and although DHT has not been studied as extensively its effects were comparable to those of oil. The facilitation of TP-induced lordosis behavior was achieved with lower doses of EB than that of DHT-induced female behavior. This is at least consistent with the aromatization hypothesis in which DHT would be expected to require larger amounts of estrogen than TP to facilitate lordosis behavior.

7. EFFECTS OF MATING

Murphy et al. (1979) reported that beta endorphin levels were much higher in males during the 30 sec after their fifth ejaculation than in unmated animals and were slightly but significantly higher after a nonejaculatory intromission following their fifth ejaculation. Males tested after five ejaculations also showed a transient analgesia in response to a nociceptive stimulus (Noble, 1980b; Cruz et al., 1980). Latencies to respond to the stimulus increased only slightly after five intromissions, when endorphin levels would not be expected to be as high.

8. PHARMACOLOGICAL EFFECTS

Both methadone and naltrexone treatment inhibited sexual activity (Murphy et al., 1979). The ability to achieve erections and intromissions was reduced by methadone, whereas naltrexone pretreatment blocked the effect. In certain doses, morphine totally abolished male sexual behavior and opiate antagonists reduced the frequency of mounts without intromissions and decreased the proportion of intromissions to ejaculation (Noble, 1980b).

Chlordiazepoxide and oxazepam, both benzodiazepines, and chlorpromazine, a neuroleptic, reduced mount, intromission, and ejaculation frequencies (Carter *et al.*, 1977). Dextroamphetamine had no significant effect on male copulatory behavior.

9. AGING

Bond (1945) found that two-year-old males were still capable of ejaculating. The effect of age has been studied more systematically by Swanson *et al.* (1982). Males were tested between 12 to 24 months of age. There were no effects of the ages studied on intromission frequency, intromission latency, or postejaculatory latency. Although the ejaculatory latency increased gradually and significantly after 20 months of age, 80% of the males at all ages up to 24 months were still displaying the entire copulatory sequence.

ACKNOWLEDGMENTS. Supported by Biomedical Support Grant (FR 07059-18) and Rutgers Research Council Grant (URF-G-84-830-NK-62) to H.I.S. and USPHS Grant MH-08604 to Jay S. Rosenblatt. I wish to thank Ms. Marilyn Dotegowski for assistance and Ms. Cynthia Banas for preparation of the illustrations. Institute of Animal Behavior Publication No. 399.

REFERENCES

Beach, F. A., and Eaton, G., 1969, Androgenic control of spontaneous seminal emissions in hamsters, *Physiol. Behav.* 4:155–156.

Beach, F. A., and Jordan, L., 1956, Sexual exhaustion and recovery in the male rat, *Quant. J. Exp. Psychol.* 8:121–133.

Beach, F. A., and Pauker, R. S., 1949, Effect of castration and subsequent androgen administration upon mating behavior in the male hamster (*Cricetus auratus*), *Endocrinology* 45:211–221.

Beach, F. A., and Rabedeau, R. G., 1959, Sexual exhaustion and recovery in the male hamster, *J. Comp. Physiol. Psychol.* 52:56–61.

Bond, C. R., 1945, The golden hamster (*Cricetus auratus*): Care, breeding, and growth, *Physiol. Zool.* 18:52–59.

Bunnell, B. N., and Flesher, C. K., 1965, Copulatory behavior of male hamsters as a function of time since androgen withdrawal, *Psychon. Sci.* 3:181–182.

Bunnell, B. N., and Kimmel, M. E., 1965, Some effects of copulatory experience on postcastration mating behavior in the male hamster, *Psychon. Sci.* 3:179–180.

Bunnel, B. N., Boland, B. D., and Dewsbury, D. A., 1977, Copulatory behavior of golden hamsters (*Mesocricetus auratus*), *Behaviour* 61:180–206.

Callard, G., Hoffman, R., Petro, Z., and Ryan, K., 1979, In vitro aromatization and other androgen transformations in the brain of the hamster, *Biol. Reprod.* 21:33–38.

Carter, C. S., Michael, S. J., and Morris, A. H., 1973, Hormonal induction of female sexual behavior in male and female hamsters, *Horm. Behav.* 4:129–141.

Carter, C. S., Daily, R. F., and Leaf, R., 1977, Effects of chlordiazepoxide, oxazepam, chlorpromazine, and d-amphetamine on sexual responses in male and female hamsters, *Psychopharmacology* 55:195–201.

Christensen, L. W., Coniglio, L. P., Paup, D. C., and Clemens, L. G., 1973, Sexual behavior of male golden hamsters receiving diverse androgen treatments, *Horm. Behav.* 4:223–230.

Cruz, S. E., Ostrowski, N. L., and Noble, R. G., 1980, Mating and responsiveness to a nociceptive stimulus, *Bull. Psychon. Sci.* 16:55–56.

DeBold, J. F., and Clemens, L. G., 1978, Aromatization and the induction of male sexual behavior in male, female, and androgenized female hamsters, *Horm. Behav.* 11:401–413.

DeBold, J. F., Morris, J. L., and Clemens, L. G., 1978, The inhibitory actions of progesterone: Effects on male and female sexual behavior of the hamster, *Horm. Behav.* 11:28–41.

Dewsbury, D. A., Lanier, D. L., and Oglesby, J. M., 1979, Copulatory behavior of Syrian golden hamsters in a one-male two-female test situation, *Anim. Learn. Behav.* 7:543–548.

Johnson, W. A., 1975, Neonatal androgenic stimulation and adult sexual behavior in male and female golden hamsters, *J. Comp. Physiol. Psychol.* 89:433–441.

Lisk, R. D., and Bezier, J. L., 1980, Intrahypothalamic hormone implantation and activation of sexual behavior in the male hamster, *Neuroendocrinology* 30:220–227.

Lisk, R. D., and Greenwald, D. P., 1983, Central plus peripheral stimulation by androgen is necessary for complete restoration of copulatory behavior in the male hamster, *Neuroendocrinology* 36:211–217.

Lisk, R. D., and Heimann, J., 1980, The effects of sexual experience and frequency of testing on retention of copulatory behavior following castration in the male hamster, *Behav. Neural Biol.* 28:156–171.

Miller, C. R., and Sessions, G. R., 1972, Effects of prolonged interintromission delays on the copulatory behavior of the male golden hamster, *Psychon. Sci.* 29:288–290.

Miller, L. L., Whitsett, J. M., Vandenbergh, J. G., and Colby, D. R., 1977, Physical and behavioral aspects of sexual maturation in male golden hamsters, *J. Comp. Physiol. Psychol.* 91:245–259.

Murphy, M. R., Bowie, D. L., and Pert, C. B., 1979, Copulation elevates plasma β-endorphin in the male hamster, *Soc. Neurosci. Abstr.* 5:470.

Noble, R. G., 1980a, Sex responses of the female hamster: Effects on male performance, *Physiol.* 24:237–242.

Noble, R. G., 1980b, Hamster sexual functioning: Modulation by opiate agonists, opiate antagonists, and exercise, in: *Eastern Conference on Reproductive Behavior,* New York, New York, p. 18.

Noble, R. G., and Alsum, P. B., 1975, Hormone dependent sex dimorphisms in the golden hamster (Mesocricetus auratus), *Physiol. Behav.* 14:567–574.

Pauker, R. S., 1948, The effect of removing seminal vesicles, prostate, and testes on the mating behavior of the golden hamster *Cricetus auratus, J. Comp. Phsyiol. Psychol.* 41:252–257.

Payne, A. P., and Bennett, N. K., 1976, Effect of androgens on sexual behaviour and somatic variables in the male golden hamster, *J. Reprod. Fertil.* 47:239–244.

Reed, C. A., and Reed, R., 1946, The copulatory behavior of the golden hamster, *J. Comp. Psychol.* 39:7–12.

Steel, E., 1982, Testosterone-dependent non-copulatory behaviour in male hamsters (Mesocricetus auratus), *J. Endocrinol.* 95:387–396.

Swanson, H. H., 1967, Effects of pre- and post-pubertal gonadectomy on sex differences in growth, adrenal and pituitary weights of hamsters, *J. Endocrinol.* 39:555–564.

Swanson, L. J., Desjardins, C., and Turek, F. W., 1982, Aging of the reproductive system in the male hamster: Behavioral and endocrine patterns, *Biol. Reprod.* 26:791–799.

Tiefer, L., 1970, Gonadal hormones and mating behavior in the adult golden hamster, *Horm. Behav.* 1:189–202.

Tiefer, L., and Johnson, W. A., 1971, Female sexual behavior in the male golden hamster, *J. Endocrinol.* 51:615–620.

Tiefer, L., and Johnson, W. A., 1973, Restorative effect of various on copulatory behavior of the male golden hamster, *Horm. Behav.* 4:359–364.

Vandenbergh, J. G., 1971, The penile smear: An index of sexual maturity in male golden hamsters, *Biol. Reprod.* 4:234–237.

Vandenbergh, J. G., 1973, Effects of gonadal hormones on the flank gland of the golden hamster, *Horm. Res.*
    4:28–33.
Warren, R., and Aronson, L., 1956, Sexual behavior in castrated-adrenalectomized hamsters maintained on
    DCA, *Endocrinology* 58:293–304.
Whalen, R. E.,and DeBold, J. F., 1974, Comparative effectiveness of testosterone, androstenedione and
    dihydrotestosterone in maintaining mating behavior in the castrated male hamster, *Endocrinology*
    95:1674–1679.

10

Parental Behavior

HAROLD I. SIEGEL

1. INTRODUCTION

At the time of this writing, only a few laboratories are actively engaged in studying parental behavior in the hamster. This is not unusual and reflects the history of this topic in this species since the initial experimental studies in the 1960s. Several reasons account for this steady but slow pursuit of the biological and psychological determinants of parental care. Parental behavior even in the well-studied laboratory rat has had an early beginning but a rather short history. Although studies on parental care in the rat began at the end of the last century (Small, 1899), with the notable exception of the research of Wiesner and Sheard (1933), it has been during the last two decades that research has been the most prolific and has produced major scientific advances.

Relatively speaking, and almost regardless of the behavior in question, studies in the hamster have tended to lag behind comparable studies in the rat and other more common laboratory species. The more recent utilization of the hamster as a research animal and the related issues of unfamiliarity with the species and the lack of a substantial data base have probably contributed to fewer active research programs. In hamsters, the absence of solid field studies in general and on parental behavior in particular can be considered handicaps in the initiation of appropriate research. Finally, the little that is known about this species, from both anecdotal and scientific sources, includes the phenomenon of cannibalism, a topic that perhaps few investigators wish to think about and even fewer wish to study.

One should not be discouraged by the scarcity of information or by the reasons presented above. As mentioned, major advances have often only recently been made in the more commonly studied species and the framework in which these advances were made may now be applied, although cautiously, to the hamster. Clearly, this volume provides evidence that the hamster is gaining in popularity as a research animal and also

HAROLD I. SIEGEL • Institute of Animal Behavior and Department of Psychology, University College, Rutgers University, Newark, New Jersey 07102.

provides the foundation of a strong data base. Field data are accumulating (see Chapter 1, this volume) and the continued emphasis on ecological, ethological, sociobiological, as well as physiological studies may serve to focus attention on behaviors such as infanticide.

2. CHAPTER OBJECTIVES

There are four aims in this chapter. The first is to briefly summarize the major theoretical framework of parental care that was developed for the rat and has since been applied to other species (Rosenblatt et al., 1979; Rosenblatt and Siegel, 1981). Second, descriptive and experimental studies in the hamster will be examined to determine if this organizing framework is applicable to this species. Third, when appropriate, results of studies in the hamster will be compared with those of other species. Fourth, methodological considerations will be treated as they arise and particular gaps in existing information will be noted.

It is also important to mention topics that will be presented only briefly or not at all. The effects of hormonal differentiation as it relates to parental behavior is covered elsewhere in this volume (see Chapter 7). The discussion of the neural bases of parental care is found in Chapter 11. Studies on maternal aggression are discussed in both this chapter and Chapter 12 on aggression. Finally, other topics such as placentophagia, studied extensively in rats (Kristal, 1980), have not been investigated systematically in the hamster.

3. THEORETICAL FRAMEWORK

The framework developed to account for the organization of maternal behavior in the rat (Rosenblatt et al., 1979; Rosenblatt and Siegel, 1981) is straightforward and easily summarized. There are two main phases; an *onset* phase associated with the endocrine events of pregnancy and a *maintenance* phase related to stimuli from the litter during lactation. By providing young foster pups, the onset of maternal behavior in the rat can be shown to develop during the final 24 to 36 hr prepartum. The specific hormonal regulation for the initiation of maternal care involves primarily rising levels of estrogen prepartum. The effectiveness of the estrogen is determined at least in part by the prior period of high progesterone levels that then decline in association with the rise in estradiol. Other factors such as oxytocin may also be involved (Pedersen and Prange, 1979), but its role is less well understood at this time. The more important point here is not which hormone is involved, but that some endogenous factor changes during pregnancy and in turn this leads to an increase in the female's responsiveness to pups at parturition.

The postpartum female faces a problem in that she must remain highly maternal for 2 to 3 weeks until weaning, yet the hormonal profile that initiated her respon-siveness at delivery has changed drastically. Although the female could rely on the endocrine conditions of lactation for the maintenance of her behavior, the results of studies involving organ removal and exogenous hormone injections have demonstrated that hormonal manipulations have little if any effect. Instead, the parturient animal's maternal care is synchronized with the developmental demands of her litter. During the early lactation period when the pups are most dependent on her care, the female displays high levels of licking, nest building, nursing, and retrieving. As lactation advances, the degree of her involvement is inversely related to the degree of the developmental independence of her litter.

Experimental evidence further supports the importance of these litter stimuli. Replacing relatively young, constant-age pups for the female's own growing litter results in the maintenance of high levels of maternal care and the delay (for a number of months) of weaning-directed behaviors. Alternatively, sustituting older pups for the female's own young pups hastens the decline in maternal responsiveness.

The two phases, a hormonal onset and a nonhormonal maintenance, suggest that a shift must occur. This shift in the regulation of the mother's behavior occurs during the *transition* phase that begins at parturition and is completed within an as yet undetermined interval, but one that probably lasts from several hours to two days postpartum. The transition occurs at the time when the endocrine stimuli responsible for the intitation of maternal care is at its peak or beginning to decline and at the same time that the litter stimuli are most potent. Removing the litter during the transition period prevents the shift from occurring and hence results in deficiencies in maternal responsiveness when pups are replaced at a later time.

One last phenomenon is pertinent to the framework. Sensitization is a process whereby maternally naive females and males can be induced to show components of maternal care by daily exposure to young pups. In rats, sensitization requires 4 to 7 days and results in the display of licking, retrieving, nest building, and adopting the lactating posture (crouching) even though milk secretion does not occur. Although sensitization also results in changes in estrous cyclicity, sensitization latencies are similar in intact, gonadectomized, or hypophysectomized rats. This nonhormonal basis of maternal behavior is probably important throughout the parental cycle. During the onset phase, pregnancy-related hormonal events may act on this maternal capacity to essentially reduce latencies to zero so that the parturient female is able to respond immediately to pups at parturition. Also, stimuli from the litter may act on this same capacity to maintain the female's responsiveness. In rats, there is evidence for a common neural site (the medial preoptic area) (Numan, 1974) for both hormonal and nonhormonal contributions to maternal care.

The framework described has been most useful in organizing the hundreds of studies on maternal behavior in rats and other species as well as providing guidelines for subsequent research. Applying this organization to the hamster will add to the comparative psychobiology of parental care and will suggest directions for future studies.

4. PREGNANCY

Counting the day after mating as day 1, gestation lasts for 16 days. When exposed to a 14-hr light–10-hr dark photoperiod, hamsters deliver an average of 9 to 12 pups

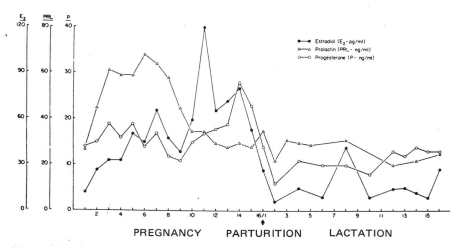

Figure 1. Circulating levels of estradiol, progesterone, and prolactin during pregnancy and lactation in the hamster. [Data from Baranczuk and Greenwald (1974) and Bast and Greenwald (1974). Reprinted with permission of the publisher from Siegel and Rosenblatt (1980).]

several hours after lights on. Parturition requires approximately 2 hr and occurs within a narrow range of time among individually housed females. For reference, the hormonal profiles of estradiol, progesterone, and prolactin during pregnancy and early lactation are shown in Fig. 1.

A number of behavioral changes occur during gestation. Although wheel running steadily declines during pregnancy, time spent in the nest and in sleeping decreases between days 4 and 5 and between days 9 and 10 (Richards, 1966a). An extensive description of the behavior of the pregnant hamster was reported by Daly (1972). Animals lived in either regular, wire-mesh cages or in burrow boxes constructed to allow the observer to view the females below ground level. As pregnancy advanced, animals in both conditions spent more time in the nest and eating, showed a modest increase in sleeping time, and gained weight in a linear fashion. Animals in the cage condition showed an initial increase in time spent gnawing soon after mating that then declined over pregnancy to premating levels near term. The pattern of digging in animals in burrow boxes showed the same pattern; an initial increase followed by a decline to premating levels as parturition approached. Daly argued that the underlying motivation of digging and gnawing is essentially identical and suggested that in nature, pregnancy signals the construction of a more elaborate burrow for litter-related activities.

4.1. Nest Building

Daly (1972) distinguished changes in burrowing from changes in nest building. Over the last four days prepartum, the females showed their largest decrease in digging and their largest increase in nest building. Richards (1966a) reported that nest building begins to increase shortly after mating and rises steadily until parturition in females whose nests were removed daily. In ovariectomized animals, relatively large implants of both estrogen and progesterone for 18 days resulted in an increase in nest building comparable to that observed in pregnant hamsters (Richards, 1969). Similarly treated castrated males showed only a small increase in nest building and neonatal hormone manipulations, castration in males and testosterone administration in females, failed to affect the sex differences after hormone implantation. Swanson and Campbell (1979a) were unable to observe an increase in nest building during pregnancy measured by the incorporation of cotton into the nests.

The occurrence and timing of nest building during gestation remains unclear. If nest building is temporally related to other maternal activities (for example, aggression and maternal responsiveness during pregnancy), then one could search for common physiological mechanisms. At this time, due to the different methods used to measure nest building and to the varying results, the relationship between this behavior and other pup-related behaviors cannot be determined.

4.2. Sexual Receptivity

The hamster is somewhat unusual in that she displays lordosis behavior during pregnancy. Lordosis responding was observed on days 10 and 14 of gestation although not on days 2 or 6 (Wise, 1974). As measured by the shorter durations of the lordotic posture during gestation, the display of "receptivity" appears incomplete, although it is interesting that the female would show this behavior at all.

4.3. Aggression

Gestation enhances the normally high levels of aggression found in nonpregnant females (Wise, 1974). Aggression during pregnancy was higher than during the estrous cycle or pseudopregnancy. In comparison to females tested on days 2 and 6 of pregnancy, a greater percentage of animals on day 10 showed fights, pounces, and chases and their latency to pounce, but not to fight, was shorter. Levels of aggression were even higher during lactation, especially within the first five days postpartum. This will be discussed in Section 6.4.

4.4. Maternal Responsiveness during Pregnancy

The fact that parturient hamsters are maternally responsive whereas virgins typically attack and cannibalize young pups suggests the importance of some factor related to pregnancy or parturition. Although the birth process may serve to focus the mother's attention on her litter, several studies have clearly shown that maternal responsiveness occurs prepartum and therefore parturition is not necessary for the onset of maternal care.

In the first of these studies, Richards (1966b) tested virgins, lactating animals within 24 hr after parturition, and pregnant females within 24 hr prepartum. The tests consisted of placing three pups, less than 24 hr of age, directly into the females' nests. In the case of the parturient animals, their own litters were removed prior to the tests.

There were no differences in the number of animals in each condition who sniffed and licked the pups and engaged in nest building. However, relative to the lactating groups, the virgins spent less time sniffing, licking, and nest building. Seven of eight of the lactating females displayed nursing behavior during the 15-min test and only one female attacked any of the pups. In contrast, none of the virgins showed nursing behavior and instead, all nulliparous animals killed all of the test pups.

The pregnant animals' responses were intermediate between those of the lactating and virgin groups. The pregnant animals were similar to the parturient animals in terms of the number of each group observed sniffing, licking, nest building, and nursing as well as the durations of each of these behaviors. The pregnant animals differed from the lactating group in that the majority of the former group attacked and ate pups. Whereas the virgins each killed all three pups, the pregnant animals killed an average of 1.4 pups. As Richards (1966b) explained, it was common for the pregnant animals to kill the first and possibly the second pup they encountered, but then to behave maternally to the remainder. Richards concluded that maternal responsiveness increased sometime between mating and the end of gestation and that a further increase occurred at parturition.

Siegel and Greenwald (1975) looked for an increase in maternal responsiveness during the final 24 hr prepartum. They used a single pup, less than 24 hr of age, and tested the same females repeatedly beginning one day prepartum. The single pup was placed not in the nest, but in the opposite corner of the cage. Seventy-five percent of the nulliparous pregnant animals showed an abrupt prepartum change from cannibalism to maternal responsiveness (retrieving and crouching over the pup). The mean interval between the first display of maternal care and the onset of parturition was 2.5 hr. All of their primiparous pregnant animals became maternal prepartum, an average of 6.7 hr before parturition. The primiparous animals became maternal slightly earlier on the morning of delivery and gave birth somewhat later.

Siegel and Greenwald (1975) next attempted to alter the prepartum onset of maternal responsiveness by administering either 1 or 10 μg estradiol and/or 0.1 mg progesterone approximately 24 hr prepartum, a time when serum levels of both steroids are declining. The hormone-treated groups became maternal at about the same time on the morning of day 16 as the oil-treated group. Estradiol alone had no effect on the timing of parturition, but progesterone delayed delivery by 8 hr. Therefore, the groups treated with progesterone alone or in combination with estradiol were maternal for 8.5 to 9.5 hr prepartum. These experiments show that parturition per se is not essential for the display of maternal care. Further, although the changes from cannibalism to maternal care occurred abruptly in each female, the cumulative percent curves for the groups showed a more gradual change whereby some animals began retrieving and crouching as much as 15 hr prepartum.

The two most recent studies on the prepartum onset of maternal care sought to determine whether maternal responsiveness could be observed earlier in gestation and

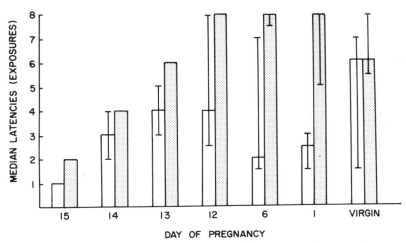

Figure 2. Median number of pup exposures to show maternal care in virgin and pregnant hamsters tested initially on days 1, 6, 12, 13, 14, and 15 of gestation. The 25th and 75th percentiles are shown except in those cases when the percentiles are equal to the median. Groups were tested twice daily by exposing them to either one pup placed outside of the nest or three pups placed directly into the nest. ☐, Three pups in nest; ▦, one pup outside nest. [Reprinted with permission of the publisher from Siegel *et al.* (1983a).]

both are quite similar in terms of design, results, and conclusions. Buntin *et al.* (unpublished data) tested hamsters on days 0 (day of mating), 5, 9, 13, and 15 by placing three young pups once daily into the corner of the cage that was opposite from the females' nests. Siegel *et al.* (1983a) tested hamsters twice daily on days 1, 6, 12, 13, 10, and 15 by placing either three newborn pups directly into the females' nests or one pup into the opposite corner of the cage (Fig. 2). In both studies, testing continued for several days or until parturition.

Results from these two studies confirm the earlier findings of Richards (1966b) and Siegel and Greenwald (1975) that maternal responsiveness develops abruptly and only during the latter stages of gestation. In both studies, the first increase in maternal responsiveness occurred in the animals initially tested on day 13, but this increase was not apparent until 24 to 48 hr later, i.e., days 14 to 15. The level of responsiveness after two days of testing in these day 13 animals is determined by both endogenous factors as well as by the daily exposure to pups (sensitization).

A clearer picture of the effect of the endogenous changes can be seen in the groups tested initially on day 15. Approximately 50% of the day 15 animals in the Buntin *et al.* (unpublished) study and a similar percentage of day 15 animals exposed to one pup outside of the nest in the Siegel *et al.* (1983a) study were maternal. Fully 100% of the day 15 animals exposed to three pups in the nest (Siegel *et al.*) were maternal on their first test. The results of both studies point to an important endogenous change occurring sometime between days 14 to 15 that enables females to display maternal care to young test pups.

The obvious next question pertains to the identity of these endogenous factors. As already discussed, injections of estrogen and progesterone on day 15 of gestation when levels of these hormones are declining did not affect the prepartum onset of maternal care. Siegel and Rosenblatt (unpublished data) attempted to simulate steroid patterns during pregnancy by injecting estradiol and progesterone for 10 days in ovariectomized virgin hamsters. The animals were exposed to pups two days after the injections were stopped to mimic the normal decline of these hormones (see Fig. 1). Latencies to the onset of retrieving and crouching were not shorter than those of an oil-treated control group.

Buntin (unpublished data) has employed pregnancy termination, a method used successfully to identify the relevant hormonal factors associated with the onset of maternal behavior in the rat (see Rosenblatt *et al.*, 1979). Pregnancy termination in the hamster might be expected to produce hormonal changes similar to those related to natural termination at parturition. However, the latencies of pregnancy-terminated hamsters on day 13 were not different from those of sensitized virgin controls.

More recently, we have injected late pregnant hamsters with ergocryptine, a prolactin blocker (Siegel and Ahdieh, unpublished data). Injected animals delivered normally on the morning of day 16, but cannibalized their entire litters within 8 hr after parturition. Conclusions on the role of prolactin are premature because the specificity of the effect has not yet been established.

From the discussion thus far, it is clear that some unidentified factor associated with pregnancy contributes to the change from cannibalism to maternal care. In the untreated pregnant rat, maternal behavior also appears during the final 1 to 2 days prepartum. Under appropriate conditions, e.g., pregnancy termination, an increase in maternal responsiveness can be observed around midpregnancy and the steroids and not prolactin are most critical. Perhaps due to the shorter gestation period, the hamster becomes responsive to pups only very late during pregnancy. The steroids have not been implicated in maternal care in the hamster. Whether the important endocrine factors are more similar to those in the rabbit in which prolactin is apparently involved (see Rosenblatt and Siegel, 1981) remains to be determined.

5. PARTURITION

The studies in the previous section suggest that parturition is not essential for the onset of maternal care. However, the pregnant hamster normally first interacts with her litter at parturition and this process may act to focus her attention on her offspring and in turn to allow her to make the transition from the endogenous to the exogenous regulation of her maternal behavior.

The most extensive description of parturition and the immediate postpartum period was recorded by Rowell (1961a). She reported that a 50% increase in the respiration rate was reliably associated with delivery during the next several hours. The females appear restless for some time prepartum and may show a sudden movement that Rowell compares with "that of a drowsing person bitten by an insect. . ."

The first appearance of the first pup is preceded by either overt straining or continuous licking of the genital region. The pup is cleaned and the birth fluids and umbilical cord are consumed within several minutes. The placenta is either consumed at this time or it is pouched or added to the food pile. The mother cleans herself and then sleeps until the next birth, 10 to 30 min later.

Rowell (1961a) also classified certain births as abnormal because they included some degree of cannibalism. Importantly, abnormal births composed 25% of all observed births. The most common category of abnormal deliveries were those of "ticklish" females. Within several hours after parturition, "the mother suddenly becomes very restless, jerking away from the pups as if they had become red hot." These females make several trips out of the nest, each one preceded by the pups' attempts at nursing, and then kill and eat some or all of the litter.

The second type of abnormal birth involves females who deliver their pups outside of the nest. This situation increases the likelihood that the pups will be ignored or cannibalized. It is also of interest that pups that are born in the nest but are later displaced accidentally by the mother are also less likely to survive.

Rowell points out that it is not uncommon for a mother to eat one of her pups while nursing the rest of her litter. This occurs more often after rather than during parturition. This is also observed in the case of stillborn pups that may temporarily be placed on the food pile.

The significance of Rowell's description is that atypical births occurred in such a sizeable number of deliveries. Further, the next section discusses the behavior of the postpartum hamster and will show that cannibalism often continues beyond the day of parturition and into the early lactation period.

6. THE POSTPARTUM PERIOD

The parturient hamster shows high levels of a number of maternal behaviors including nest building, sniffing and licking of the pups, nursing, retrieving, and maternal aggression. As lactation proceeds, the mother shows a gradual decline in these behaviors and a concomitant increase in nonmaternal behaviors including gnawing, rearing, climbing, environmental sniffing, grooming, and moving about the cage (Daly, 1972; Swanson and Campbell, 1979a). Retrieval tests show a marked increase in latency beginning on day 15 in primiparous females and on day 18 in multiparous females.

Daly (1972) compared the behavior of mothers rearing their litters in standard lab cages with those rearing their litters in subterranean burrow boxes. One difference was that the cage situation may have accelerated the weaning process; these mothers stopped nursing and began avoiding their nests several days earlier than the burrow box mothers. The other difference was that cage mothers were "forced" to retrieve pups more often. Daly suggested that the extra retrievals represent an "extreme unnatural over-elicitation in the usual laboratory setting."

6.1. Postpartum Cannibalism

It is tempting to speculate that the extra maternal burden of added retrievals in the lab cage situation plays some role in the most striking characteristic of maternal care in the hamster, namely, postpartum cannibalism. It would not account for the cannibalism described by Rowell (1961a) that occurs soon after parturition, but pup destruction continues into the first week of lactation. Swanson and Campbell (1979a) reported that all eight of their primiparous mothers reared their first litters but four of them failed with their second litters; two entire litters were neglected or eaten soon after parturition and two were destroyed between days 10 and 14 of lactation.

These reports of pup destruction are confirmed by anecdotal as well as experimental observations. Depending on the particular group of animals studied, as high as two thirds of the pups of untreated lactating hamsters do not survive to weaning. Although it is more common for a mother to kill a portion of her litter, it is not all that uncommon for a mother to cannibalize her entire litter. Although a satisfactory explanation for this cannibalism has not been provided, the phenomenon has been subjected to experimental analysis.

Day and Galef (1977) have performed the most extensive studies of cannibalism. In the first experiment, they examined the effects of disturbance of the mother (by daily confinement to a holding cage for 5 min/day for the first ten days of lactation), maternal age (mating occurred at the first estrus or four or eight weeks later), parity, litter size at parturition, pup condition (birth weight and the presence of subcutaneous hemorrhages or color abnormalities), and the relationship between the degree of cannibalism on three successive litters. The investigators were unable to demonstrate an effect of any of these factors. There were nonsignificant trends for a relationship between the degree of cannibalism on the second and third litters and for more cannibalism to be associated with larger litters.

Overall, Day and Galef reported that cannibalism occurred in approximately 80% of their 88 observed litters. The mothers killed an average of almost two pups per litter

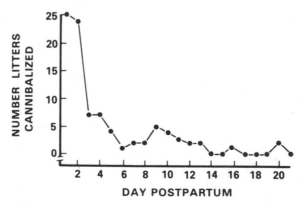

Figure 3. Number of litters in which cannibalism occurred as a function of time postpartum (n = 63 litters). [Reprinted with permission of the publisher from Day and Galef (1977).]

and most of the cannibalism was restricted to the first three days postpartum (Fig. 3). The authors stated that all cannibalism was completed by the fifth day postpartum, but the results cited above (Swanson and Campbell, 1979a) demonstrate that cannibalism can occur after the pups have begun to eat solid food and after their eyes have opened.

According to Day and Galef (1977), litters culled to four pups suffered little cannibalism, whereas litters maintained at their parturient sizes, by daily replacement of lost pups, were cannibalized more often. Although the replacement procedure continued for the first ten days postpartum, the increase in cannibalism stopped by day 5.

Day and Galef next addressed the question of why their mothers stopped cannibalizing by day 5 of lactation. One reason could be that there are particular cues from 1- to 5-day-old pups that elicit cannibalism and these cues change with age. To test this possibility, litters were maintained at their parturient sizes for seven days postpartum by the daily replacement with newborn pups. For the most part, these mothers stopped cannibalizing by day 4 suggesting that developmental changes in the pups are not responsible for the cessation of pup destruction.

Perhaps the duration of cannibalism is related to an endogenous maternal factor, time since parturition. Different groups of mothers were each given four additional pups, of the same age as their own litters, on either day 0, 2, 5, or 9 postpartum. The results were clear; only those mothers presented with extra pups on days 0 and 2 showed an increase in cannibalism.

In their final study, Day and Galef provided mothers on the day of parturition with either 2, 4, or 6 newborn foster pups. An additional experimental group was given two newborn foster pups at the same time that four of their own pups were removed. Control mothers were given four foster pups in exchange for four of their own pups. The degree of cannibalism was almost directly proportional to the number of pups added or removed. Control mothers whose litter sizes were not changed killed about two pups, whereas mothers given two or four pups killed an average of almost four and six pups, respectively. Mothers given six pups cannibalized an average of almost seven pups and mothers whose litters were reduced by two killed a mean of less than one pup.

Day and Galef concluded that postpartum cannibalism is not a deficit in parental care, but instead an organized behavior through which the mother can regulate the number of pups in relation to possible environmental demands. The authors further concluded that the postpartum regulation of litter size would be especially adaptive in view of irregular environmental conditions of food and rainfall that are associated with the hamsters' natural habitat.

Although Day and Galef did demonstrate that hamsters adjust their litter sizes through cannibalism, they provided no empirical support for their "environmental hypothesis." All of their animals were housed under normal laboratory conditions of *ad libitum* food and water. It could be argued that these lab cages in which food is placed on the tops of wire lids are sufficiently different from normal field conditions in which the hamster may accumulate a food hoard in a burrow.

Miceli and Malsbury (unpublished data) have successfully related the extent of postpartum cannibalism with an environmental factor, the opportunity to hoard food pellets. Beginning before conception, hamsters were provided with food jars containing about 500 g of pellets. These jars were placed into the females' cages and were replenished twice weekly allowing the animals to actively hoard the food. Control animals were placed on a fixed-interval schedule of food delivery. These animals had considerably less food than the hoard group, but their daily food supplies were greater than the amounts eaten by other animals in comparable reproductive states. By day 5 of lactation, litters in the hoard group were twice as large as those in the control group and this difference persisted until the end of the study on day 15.

Although these results are interesting and support the "environmental hypothesis," this explanation for cannibalism is still not entirely satisfactory. Clearly, hamsters cannibalize pups despite sufficient supplies of food and water. Even when given the opportunity to hoard relatively large amounts of food, hamsters still cannibalize; the hoard group killed an average of one third of their pups. Further, why do some animals destroy their entire litters?

One other observation may be relevant to this issue. As already mentioned, cannibalism occurs at times other than during the early lactation period. Hamsters occasionally kill pups during the second week of lactation and pregnant animals more than 1 to 2 days prepartum kill pups. Also, virgin animals typically cannibalize young pups. Therefore, cannibalism in the hamster appears to be more pervasive than to function only as a means to regulate litter size. The relationship between cannibalism and the more aggressive solitary nature of this species needs to be determined as well as the relationship between cannibalism in the nonparturient state with its display during the early postpartum period.

One serious problem is that it is not known whether hamsters in nature also cannibalize portions of their litters. In other words, cannibalism may be a laboratory artifact. The opportunity to hoard food may be only one laboratory factor related to cannibalism. Lab conditions of space, temperature, photoperiod, and the opportunity to burrow are only a few variables that need to be examined. Further studies of these and other factors under more natural conditions are required.

6.2. Maintenance of Mother–Litter Bond

Another important aspect of the early postpartum period pertains to the formation of the mother–litter bond that will sustain parental care until weaning. The establishment of this bond is probably as essential as it is in the rat. In both species, changes during pregnancy initiate maternal care. In the rat, these physiological events are no longer present postpartum and there is now sufficient evidence that the mother regulates her maternal care in terms of litter-related stimuli (see Rosenblatt et al., 1979). This section addresses the formation of the mother–litter bond in the hamster.

Among mammals, mother–litter separation during the early postpartum period results in a decline in subsequent maternal responsiveness (see Rosenblatt and Siegel, 1981). In hamsters, Siegel and Greenwald (1978) permitted females either 1, 24, or 48 hr of contact with their litters beginning immediately postpartum. Different groups of animals in each of these three conditions were offered young pups at daily intervals. Females allowed 1 or 24 hr of contact were maternal for only three and four days, respectively, whereas animals permitted 48 hr were still maternal after a two-week interval without pups.

In a similar study, Buntin *et al.* (1979) showed that 36 to 48 hr of postpartum litter contact were sufficient for one third of these animals to be immediately maternal three and six weeks later. The remaining animals became maternal within 72 hr after the reintroduction of pups.

6.3. Mother–Litter Synchrony

The mother–litter relationship can also be viewed in terms of behavioral and physiological synchrony. Rowell (1960) tested lactating females with foster pups between days 1 to 30 of age. The mothers' responses were categorized as either acceptance, rejection in which the pups were treated as a conspecific intruder or food, or ambivalence. Ambivalent behaviors included aspects of both acceptance and rejection; for example, the mothers might retrieve the strange pups to the nest and then kill them.

Primiparous and multiparous animals showed similar frequencies of acceptance, rejection, and ambivalence. Pups between 7 to 10 days of age were more readily accepted by mothers at all stages of lactation. Test pups less than 14 days of age were never treated as a conspecific rival and pups older than 14 days were never treated as food. Both age of the strange pups relative to the ages of the mothers' own litters and time since parturition, to some extent regardless of the age of the test pups, were factors that determined the types of responses displayed.

In a follow-up study, mothers were maintained with pups that were either 2 to 6, 6 to 10, 10 to 14, or 14 to 18 days of age (Rowell, 1960). Again, some of the mothers' behaviors depended on pup age, whereas other responses were related to the time since parturition. Weight gain of their foster pups, especially the younger ones, was poor and was related to time since parturition rather than the needs of the particular litter. Similar results were found when the mothers' time spent out of the nest was measured. In contrast, the extent of pup licking was related to the ages of the litters rather than time since parturition.

More recently, Swanson and Campbell (1980, 1981) removed the litters of females on days 12 to 13 of lactation and replaced them with 3-day-old litters. The animals were observed until the younger litters were weaned (defined as the absence of nursing during observation periods on two consecutive days). Pup weights, nursing durations, and retrieval latencies were measured and compared with control females rearing their own litters.

The results point to somewhat separate mechanisms controlling lactation and maternal care (Figs. 4 and 5). Although the foster litters gained weight more slowly than the control litters, they were capable of prolonging the mothers' parental behavior. Relative to the control mothers, the nursing behavior of the experimental females continued for an additional 14 days. The decline in nursing behavior as well as the physiological events leading to the resumption of estrous cyclicity were related to the developing pups and not time since parturition. Nursing declined when the pups of both groups of mothers were similar in weight. Also in both groups, retrieval latencies

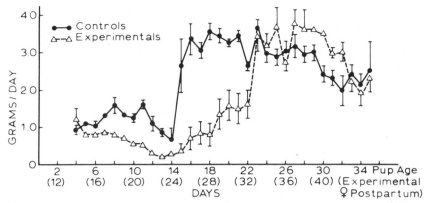

Figure 4. Mean (± SE) change in pup weight in grams/day over days of age and days postpartum in control and experimental litters. Control females reared their own pups, whereas the litters of the experimental females on day 12 of lactation were replaced with 3-day-old pups. [Reprinted with permission of the publisher from Swanson and Campbell (1980).]

were maximal several days after nursing began to decline: day 18 in controls and day 33 in the experimental group. In contrast to behavioral weaning and the temporally associated initiation of an estrous cycle, milk production and prolactin release are more closely related to time since parturition and not age of the pups.

The behavior of the postpartum hamster is similar to that of the rat and is

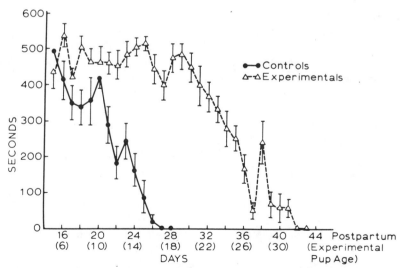

Figure 5. Mean (± SE) number of seconds of nursing behavior observed per 10-min period over days postpartum. Control females reared their own pups, whereas the litters of the experimental females on day 12 of lactation were replaced with 3-day-old pups. [Reprinted with permission of the publisher from Swanson and Campbell (1980).]

synchronized with the growth and increased independence and mobility of the off-spring. Hamsters, however, differ from rats in their inability to prolong lactation when maintained with younger pups. Nicoll and Meites (1959) and Bruce (1961) have demonstrated an extensive prolongation of lactation of up to one year in rats whose litters were exchanged every ten days with younger foster pups. Thus, it appears that the factors regulating the duration of both parental care and lactation can be more easily dissociated in the hamster than in the rat.

6.4. Maternal Aggression

Another important postpartum behavior is maternal aggression. Testing females with males in neutral arenas, Wise (1974) showed that lactating animals, especially those tested during the first five days postpartum, were more aggressive than pregnant, cycling, and pseudopregnant animals.

One obvious hormonal correlate of the high levels of postpartum aggression is prolactin. Wise and Pryor (1977) tested the effect of ergocornine, a blocker of prolactin release, on the aggression of postpartum hamsters. Three groups were injected on days 4 to 7 of lactation with either ergocornine, ergocornine plus prolactin, or the vehicles. The females were ovariectomized to eliminate any effect of ovarian hormones and were tested with males in neutral arenas on day 7 of lactation.

Ergocornine virtually eliminated aggression, whereas the control group displayed high levels of pounces, fights, chases, and rolls. The aggressive behavior of the ergocornine plus prolactin group was intermediate between that of the control and drug only groups. Perhaps higher doses or more frequent injections of the prolactin would have more fully restored levels of aggression.

A most interesting finding in this study was that 70% of the ergocornine only and 42% of the ergocornine plus prolactin groups killed their entire litters during the four-day injection schedule. It should be recalled that we also found destruction of entire litters within several hours after parturition in females treated with ergocryptine beginning on day 15 of gestation (Siegel and Ahdieh, unpublished data). Although more research is needed on the role of prolactin, these studies are the only ones to implicate a role for any endogenous factor on maternal behavior.

Siegel et al. (1983b) investigated maternal aggression by testing lactating animals on days 5 and 15 in their home cages with female intruders. Different groups of females were tested in the presence of their litters or at the end of a 6-hr separation. Relative to cycling controls, the lactating animals showed higher levels of fights, attacks, chases, and opponent retreats. Unlike the results of Wise (1974), we found no decline in aggression between days 5 and 15 of lactation and this may have been due to our particular experimental conditions of home cage tests with female intruders.

Six-hour mother—litter separations had no effect on the typical high levels of aggression, but a more recent study (Giordano et al., unpublished data) showed that aggression declines dramatically after a 24-hr interval of separation. Further, a 0.5- or 2-hr mother—litter reunion following a 24-hr period of separation restored high levels of aggression.

6.5. Weaning

Although weaning can be thought of as a continuous process, Rowell (1961a) observed three major structural changes in the family unit correlated with particular ages of pup development. At 23 to 25 days into the postpartum period, the mother is no longer bound to the nest in which she raised her litter and builds a new but smaller nest in a different location. The building of a new nest continues at weekly intervals until the litter leaves.

Between 29 to 34 days of age, the pups no longer attempt to nurse and instead must now begin to avoid their mother's occasional attacks. At sexual maturity, play fighting among the littermates stops and the mother is no longer aggressive. It should be pointed out that these observations were made on animals living in a 6 ft × 1 ft enclosure and as Rowell (1961a) suggested, the family unit in nature may separate before the onset of sexual maturity. The aggression shown by the mother one month postpartum may represent either the normal culmination of the weaning process or an artifact of the lab situation in which neither the mother nor her litter is permitted to disperse.

7. THE TRANSITION PHASE

The transition period is hypothesized to connect the endogenously mediated onset of maternal behavior with the exogenously mediated maintenance of parental care during lactation. The transition phase is thought to begin sometime after parturition when the physiological factors responsible for the onset phase have waned. Some aspects of postpartum maternal care appear dependent on internal maternal factors. These include the duration of lactation, a relatively small effect on the acceptance of strange pups (Rowell, 1960; Swanson and Campbell, 1980), and the restriction of cannibalism to the first few postpartum days (Day and Galef, 1977).

Despite the existence of these endogenous maternal factors associated more absolutely with time since parturition, it is clear that stimuli associated with the litter affect the course of parental care. Evidence for the importance of these external stimuli includes the prolongation through litter replacement of the quality and intensity of the maternal behaviors normally associated with the early stage of lactation (Swanson and Campbell, 1980).

Ovarian hormones can probably be excluded as a major factor in the maintenance phase of maternal care. Ovariectomy has no apparent effect during the first week postpartum (Wise and Pryor, 1977). Although prolactin is currently a more likely candidate for contributing to both the onset and the maintenance phases, a definitive role in either awaits further data.

The research on the necessity of litter contact on later maternal responsiveness may provide a clue as to the duration of the transition period. These studies show the importance of the first 24 to 48 hr of postpartum mother–litter interaction (Siegel and Greenwald, 1978; Buntin *et al.*, 1979). Once primed by endogenous factors to respond immediately and maternally at birth, the female requires this initial period of contact with her litter to enable her to behave appropriately to the subsequently changing needs of her litter during the period of lactation and weaning.

It is interesting that the transition period occurs during the period of greatest pup cannibalism. Perhaps this cannibalism reflects a somewhat unstable transition. Richards (1966c) speculated that the evolutionary pressures that may have condensed gestation to 16 days have not been entirely matched to the display of parental care. As Richards pointed out, golden hamsters have a gestation period that is four to six days shorter than it is in other hamster species and in the rat, and as he demonstrated (Richards, 1966c), hamsters are much more likely to respond maternally to pups that are at least six days of age.

8. RESPONSES OF MATERNALLY NAIVE HAMSTERS TO PUPS

This section summarizes the effects of presenting pups to otherwise nonmaternal hamsters. Studies on females and males as well as juveniles are covered along with the variety of methodologies that have been employed. Finally, a role for sensitization in the regulation of maternal care is described.

The results from the initial studies in which adult animals were tested with pups were quite clear (Rowell, 1961a; Richards, 1966b; Noirot and Richards, 1966). The most typical response toward young pups was cannibalism and this occurred in nulliparous as well as in nonlactating but experienced females. Slightly less cannibalism was observed when the pups were placed into the females' nests or when animals were first exposed to either 1- or 9-day-old pups and then two days later to 5-day-old pups. The least amount of cannibalism by virgin animals was displayed toward pups between 6 to 10 days of age (Richards, 1966c) and this is essentially identical to the situation in females at all stages of lactation (Rowell, 1960).

Marques and Valenstein (1976) tested intact females and males in neutral pens with 2- to 9-day-old pups. Although the test lasted for 5 hr, approximately equal halves of the female group either cannibalized or carried pups during the first 10 min. All of the males carried pups, most within the first 10 min, and no male ever attacked any pup. It should be noted that carrying was scored even if the pups were not carried to the nest.

In other studies, the percentage of females cannibalizing on their first exposure to pups is typically much higher than 50%. The pups used in the Marques and Valenstein study (1976) included both those that more often elicit cannibalism (less than 6 days of age) and those that more often are treated maternally (between 6 and 9 days of age). Further, testing in a neutral cage may have also contributed to the higher percentage of maternal responsiveness.

8.1. Sensitization

Until recently, there had been no attempt to sensitize virgin hamsters by repeated exposure to pups. By presenting 6- to 12-day-old pups to adult nulliparous hamsters twice per day in their home cage, Swanson and Campbell (1979b) reported a mean latency of 18 hr measured from the first test until the animals retrieved and crouched over the pups. Females sensitized 11 weeks earlier required a mean of 28.5 hr during their second exposure to pups. Almost all of the females in these two cases killed pups on their first exposure.

Adult male hamsters were also sensitized, but with a mean latency of 72 hr that was significantly longer than it was in females. The males showed more incomplete retrievals and never adopted the lactating position characteristic of lactating females and often observed in sensitized females. Unlike the reports of Marques and Valenstein (1976), Swanson and Campbell (1979b) reported that males more frequently injured pups than females.

A sensitization latency of 18 hr is somewhat surprising considering the hamster's tendency to cannibalize. Perhaps the 6- to 12-day-old pups, the most readily accepted by both lactating and virgin females, accounted for the relatively short latencies. Can hamsters be sensitized with very young pups that are typically cannibalized during a single test and how would these latencies compare with those of animals exposed to older pups?

Siegel and Rosenblatt (1978) tested virgin females twice daily with three pups between 1 and 2 days of age. The animals were tested in their home cages and the pups were placed on the floor of the cage diagonally opposite from each female's nest. The mean latency for the onset of retrieving and crouching was 1.8 days or 43 hr. Additionally, this study demonstrated that the ovaries are not essential for sensitization; ovariectomized females had virtually identical latencies to those of the intact animals.

Further studies (Siegel, unpublished data) have focused on particular test conditions that affect sensitization latencies. Generally speaking, the least effective method consists of presenting one young pup on the floor of the cage outside of the nest, once per day. The test conditions that result in the shortest latencies consist of presenting three pups in the females's nests, twice per day. Studies are currently underway to test the effectiveness of these individual factors (number of pups, placement of pups in the cage, and number of tests per day).

In a recent study, Miceli and Malsbury (1982) tested the effect of the availability of a food hoard previously shown to suppress postpartum cannibalism on sensitization latencies. One group, the hoard group, had access to up to 350 g of food pellets located in a jar at the end of a wire mesh tunnel attached to the cage. Sham-hoard females could see and smell the food through a perforated lid, but were permitted access to 30 to 40 g of pellets, twice the average daily intake. Hoard females displayed maternal behavior with significantly shorter latencies than those of the sham group.

Two studies have examined the responses of juveniles to pups. In the first, Rowell (1961b) placed 1- to 8-day-old pups into the cages of juveniles whose mothers had been removed. Licking and mouthing of the pups were the first behaviors shown by some litter members as early as 15 days of age. There were no apparent differences between male and female juveniles and some animals in each litter displayed maternal responses by 19 days of age. Cannibalism was observed at a mean of 37 days and this cannibalism was delayed in litters living in cages without their mothers.

In the second study (Siegel et al., unpublished data), males and females were tested in individual cages at 20, 33, or 55 days of age. There were no differences in the latencies to retrieve the 1- to 2-day-old test pups. Median latencies to the onset of maternal care were approximately two days for both sexes at all three ages tested. Prior to showing maternal responses, many of the animals cannibalized one or more pups.

8.2. Role of Sensitization

How does the study of sensitization add to our understanding of the regulation of parental care? First, these studies show a powerful effect of pup stimulation on the display of parental behavior. This supports the hypothesis that postpartum maternal care may depend primarily on external cues from the litter. This capacity to respond maternally to pups is observed in juvenile and adult males and estrous cycling as well as in lactating and virgin females that had been ovariectomized. Thus, sensitization does not depend upon any particular endocrine profile.

Second, it would be possible to identify the critical physiological factors responsible for the onset of maternal care by treating nulliparous animals with one or more suspected agents. The most appropriate factors might result in a shortening of the latency to respond maternally to pups compared with those of vehicle-treated animals. Despite the already relatively short sensitization latencies (two days compared with four to six days in rats), it may still be possible to find a more immediate effect especially considering that the prepartum change from cannibalism to maternal care is abrupt.

The third reason to study sensitization is to determine those maternal characteristics that are related to pregnancy, parturition, and lactation. As discussed earlier, maternal aggression can be compared in lactating and sensitized groups that differ in terms of their reproductive conditions. Other factors such as the age of the pups that more often elicit maternal care and the importance of the nest site can also be studied in both types of animals. The data thus far show similarities for both lactating and sensitized females. Both groups respond maternally more often to 6- to 10-day-old pups and to pups in their nest. These factors are therefore not strictly related to the postpartum environment or condition of the female.

There are also parallels in terms of cannibalism. Pregnant, lactating, and virgin females all show some degree of pup destruction. This suggests that cannibalism is a behavior normally found in the behavioral repetoire of the hamster. Postpartum cannibalism has been discussed in terms of ecological considerations in relation to resource availability, as an artifact of laboratory housing conditions, and a relatively fragile transition from the onset to the maintenance phase of maternal behavior, and may also be considered a response of "solitary" animals to "intruders." The fact that nonparturient hamsters also cannibalize pups does not necessarily preclude any of these hypotheses, but it does illustrate that pup destruction is not a characteristic associated solely with the early postpartum period.

9. CONCLUSIONS AND FUTURE DIRECTIONS

Although it is clear that the onset of maternal behavior is related to an endogenous and perhaps short-latency physiological change during late pregnancy, the identity of the factor(s) is unknown. The only hormone implicated is prolactin, but only a few factors have yet been studied. Do hormones play a role in nest building? When does the pregnant female begin to construct its maternal nest? These questions have received contradictory answers depending on how nest building was measured.

There is also ample experimental support for a maintenance phase that is mediated by stimuli from the developing pups. The delay in the weaning process after the substitution of younger litters, the optimal stimuli provided by 6- to 10-day-old pups, and relatively short sensitization latencies all point to a major contribution of the offspring. In this regard the important sensory factors are not known. On the other hand, there are effects related more to the absolute time interval following parturition that include the incidence of cannibalism and the decline in lactation despite replacement with younger litters. The mechanisms underlying these effects need to be examined as well as the possible involvement of prolactin and other hormones.

Some evidence for a transition period has been presented. Exactly when the transition is completed is unclear. Can the transition phase be accelerated or delayed? Does parturition contribute to the mother's transition?

It is expected that the study of parental care in the hamster in the field, in the lab, and in seminatural conditions will attract further investigation, and hopefully the framework presented here will help guide this research. Although the framework may be generally applicable to a variety of mammals, it is clear that the precise mechanisms vary from species to species.

ACKNOWLEDGEMENTS. Supported by Biomedical Support Grant (FR 07059-18) and Rutgers Research Council Grant (URF-G-84-830-NK-62) to H.I.S. and USPHS Grant MH-08604 to Jay S. Rosenblatt. I wish to thank Ms. Marilyn Dotegowski for assistance and Ms. Cynthia Banas for preparation of the illustrations. Institute of Animal Behavior Publication No. 400.

REFERENCES

Baranczuk, R., and Greenwald, G. S., 1974, Plasma levels of oestrogen and progesterone in pregnant and lactating hamsters, *J. Endocrinol.* 63:125–135.

Bast, J. D., and Greenwald, G. S., 1974, Daily concentrations of gonadotropins and prolactin in the serum of pregnant or lactating hamsters, *J. Endocrinol.* 63:527–532.

Bruce, H. M., 1961, Observations on the suckling stimulus and lactation in the rat, *J. Reprod. Fertil.* 2:17–34.

Buntin, J. D., Jaffe, S., and Lisk, R. D., 1979, Physiological and experiential influences on pup-induced maternal behavior in female hamsters, Presented at *Eastern Conference on Reproductive Behavior*, New Orleans, Louisiana, p. 67.

Daly, M., 1972, The maternal behaviour cycle in golden hamsters, *Z. Tierpsychol.* 31:289–299.

Day, C. S. D., and Galef, B. D., 1977, Pup cannibalism: One aspect of maternal behavior in golden hamsters, *J. Comp. Physiol. Psychol.* 91:1179–1189.

Kristal, M. B., 1980, Placentophagia: A biobehavioral enigma (or *de gustibus non disputandem est*), *Neurosci. Biobehav. Rev.* 4:141–150.

Marques, D. M., and Valenstein, E. S., 1976, Another hamster paradox: More males carry pups and fewer kill and cannibalize young than do females, *J. Comp. Physiol. Psychol.* 90:653–657.

Miceli, M. O., and Malsbury, C. W., 1982, Availability of a food hoard facilitates maternal behaviour in virgin female hamsters, *Physiol. Behav.* 28:855–856.

Nicoll, C. S., and Meites, J., 1959, Prolongation of lactation in the rat by litter replacement, *Proc. Soc. Exp. Biol. Med.* 101:81–82.

Noirot, E., and Richards, M. P. M., 1966, Maternal behavior in virgin female golden hamsters: Changes consequent upon initial contact with pups, *Anim. Behav.* 14:7–10.

Numan, M., 1974, Medial preoptic area and maternal behavior in the female rat, *J. Comp. Physiol. Psychol.* 87:746–759.

Pedersen, C. A., and Prange, A. J., Jr., 1979, Induction of maternal behavior in virgin rats after intra-cerebroventricular administration of oxytocin, *Proc. Natl. Acad. Sci. U.S.A.* 76:6661–6665.

Richards, M. P. M., 1966a, Activity measured by running wheels and observation during the estrous cycle, pregnancy, and pseudopregnancy in the golden hamster, *Anim. Behav.* 14:450–458.

Richards, M. P. M., 1966b, Maternal behaviour in the golden hamster: Responsiveness to young in virgin, pregnant, and lactating females, *Anim. Behav.* 14:310–313.

Richards, M. P. M., 1966c, Maternal behaviour in virgin female golden hamsters (*Mesocricetus auratus* Waterhouse): The role of the age of the test pup, *Anim. Behav.* 10:303–309.

Richards, M. P. M., 1969, Effects of oestrogen and progesterone on nest building in the golden hamster, *Anim. Behav.* 17:356–361.

Rosenblatt, J. S., and Siegel, H. I., 1981, Factors governing the onset and maintenance of maternal behavior among nonprimate mammals, in: *Parental Care in Mammals* (D. J. Gubernick and P. H. Klopfer, eds.), Plenum Press, New York, pp. 13–76.

Rosenblatt, J. S., Siegel, H. I., and Mayer, A. D., 1979, Progress in the study of maternal behavior in the rat: Hormonal, nonhormonal, sensory, and developmental aspects, in: *Advances in the Study of Behavior*, Volume 10 (J. S. Rosenblatt, R. A. Hinde, C. Beer, and M.-C. Busnel, eds.), Academic Press, New York, pp. 225–311.

Rowell, T. E., 1960, On the retrieving of young and other behaviour in lactating golden hamsters, *Proc. Zool. Soc. London* 135:264–282.

Rowell, T. E., 1961a, The family group in golden hamsters: Its formation and break-up, *Behaviour* 17:81–93.

Rowell, T. E., 1961b, Maternal behaviour in non-maternal golden hamsters (*Mesocricetus auratus*), *Anim. Behav.* 9:11–15.

Siegel, H. I., and Greenwald, G. S., 1975, Prepartum onset of maternal behavior in hamsters and the effects of estrogen and progesterone, *Horm. Behav.* 6:237–245.

Siegel, H. I., and Greenwald, G. S., 1978, Effects of mother-litter separation on later maternal responsiveness in the hamster, *Physiol. Behav.* 21:147–149.

Siegel, H. I., and Rosenblatt, J. S., 1978, Short-latency induction of maternal behavior in nulliparous hamsters, presented at *Eastern Conference on Reproductive Behavior*, Madison, Wisconsin, p. 52.

Siegel, H. I., and Rosenblatt, J. S., 1980, Hormonal and behavioral aspects of maternal care in the hamster: A review, *Neurosci. Biobehav. Rev.* 4:17–26.

Siegel, H. I., Clark, M. C., and Rosenblatt, J. S., 1983a, Maternal responsiveness during pregnancy in the hamster (*Mesocricetus auratus*), *Anim. Behav.* 31:497–502.

Siegel, H. I., Giordano, A. L., Mallafre, C. M., and Rosenblatt, J. S., 1983b, Maternal aggression in hamsters: Effects of stage of lactation, presence of pups, and repeated testing, *Horm. Behav.* 17:86–93.

Small, W. S., 1899, Notes on the psychic development of the young white rat, *Am. J. Psychol.* 11:80–100.

Swanson, L. J., and Campbell, C. S., 1979a, Maternal behavior in the primiparous and multiparous golden hamster, *Z. Tierpsychol.* 50:96–104.

Swanson, L. J., and Campbell, C. S., 1979b, Induction of maternal behavior in nulliparous golden hamsters (*Mesocricetus auratus*), *Behav. Neural Biol.* 26:364–371.

Swanson, L. J., and Campbell, C. S., 1980, Weaning in the female hamster: Effect of pup age and days postpartum, *Behav. Neural Biol.* 28:172–182.

Swanson, L. J., and Campbell, C. S., 1981, The role of the young in the control of the hormonal events during lactation and behavioral weaning in the golden hamster, *Horm. Behav.* 15:1–15.

Wiesner, B. P., and Sheard, N. M., 1933, *Maternal Behavior in the Rat*, Oliver & Boyd, London.

Wise, D. A., 1974, Aggression in the female golden hamster: Effects of reproductive state and social isolation, *Horm. Behav.* 5:235–250.

Wise, D. A., and Pryor, T. L., 1977, Effects of ergocornine and prolactin on aggression in the postpartum golden hamster, *Horm. Behav.* 8:30–39.

11

Neural Basis of Reproductive Behavior

CHARLES W. MALSBURY, MARIO O. MICELI, and CHARLES W. SCOUTEN

1. INTRODUCTION

We believe that the study of the organization and function of the brain is the most fascinating scientific enterprise possible (molecular genetics notwithstanding). However, because of the complexity of the mammalian brain, in order to have any chance of success at this enterprise, one must pick one's questions with great care. Why study the neural basis of reproductive behaviors?

First, there are a variety of important questions waiting to be answered. There are also certain strategic advantages to studying the neural mechanisms underlying reproductive behaviors in rodents. One is the relative ease with which these behaviors can be studied under laboratory conditions. Another is their hormone dependence (at least in the case of sexual behaviors). That is, these behaviors can be effectively "turned" on or off by manipulations of the appropriate gonadal steroid hormones. Identification of the critical hormonal stimuli for a particular behavior provides a great advantage in studying its neural basis, since the hormone(s) can be used to probe the brain in an effort to solve the old, but still important, question of localization of function. This is illustrated by the experiments described below (Section 3.1.) that have led to the identification of the ventromedial hypothalamus as the primary site of action of estradiol in inducing sexual receptivity. Radioactively labeled steroids can be used to determine the sites of hormone uptake in the brain, and intracerebral implants can be used to determine which of these sites are relevant for sexual behaviors. As Bard (1940) has said, "The important point from an experimental point of view is that if you can localize a process, you can then study it; but if you cannot localize it, attempts to study it are apt to be abortive."

In our own research on the lordosis response, we have chosen to study the hamster.

CHARLES W. MALSBURY, MARIO O. MICELI, and CHARLES W. SCOUTEN • Department of Psychology, Memorial University of Newfoundland, St. John's, Newfoundland, Canada A1B 3X9.

This is not because we believe that *comparing* rat and hamster data will necessarily lead to new insights, but because the hamster has certain advantages over the rat for these studies (discussed in Section 3.1.). If one hopes to understand the neural mediation of behavior, we believe it is important to choose the most suitable behavioral model. The sensory and hormonal determinants of lordosis are relatively well-understood (see Chapter 8, this volume), and this also aids the search for CNS determinants. The same advantages apply to the study of CNS determinants of male sexual behavior in the hamster, but may not apply, at least to the same degree, in the study of hamster maternal behaviors (see Section 4).

2. MALE SEXUAL BEHAVIOR

2.1. Olfactory Systems

Recent advances in neuroanatomical tracing techniques have enabled significant progress in describing the limbic connections of the main and accessory olfactory systems in the hamster. The demonstration that male hamster sexual arousal is uniquely dependent on chemosensory input (Murphy and Schneider, 1970) has perhaps been equally important in driving this progress. A number of investigators realized that the dependence of male hamster copulatory behavior on olfaction could provide a significant advantage in understanding the neural basis of this behavior. This area of research provides an example of how neuroanatomical and neurobehavioral findings can interact in a very productive way.

Although olfactory information probably plays a role in male mating behavior in most mammalian species, olfactory input is not essential for copulation in male rats, rabbits, cats, dogs, or monkeys. Whether olfaction is essential for copulation in mice remains unresolved (Murphy, 1976). In contrast, the hamster is clearly dependent on olfactory input. Murphy and Schneider (1970) first demonstrated that bilateral olfactory bulbectomy completely eliminated interest in receptive females. The bulbectomy effect has since been replicated in several laboratories. This demonstration is important, because it allows the search for the neural determinants of sexual arousal to focus on the olfactory pathways.

During the 1970s it became clear that bulbectomy produces behavioral changes that cannot be attributed to anosmia alone. These were termed "nonsensory" deficits (Cain, 1974; Edwards, 1974). One of the first questions that arose following the demonstration of the bulbectomy effect in hamsters was whether the males stopped copulating because they were anosmic or because of additional effects caused by bulbectomy. This deceptively simple question was addressed by attempting to produce anosmia at the periphery, instead of by destroying the bulbs, and then comparing the results. One method of doing this is to infuse a zinc sulphate solution through the nasal cavities and thereby destroy the olfactory receptors in the nasal epithelium. Two studies using zinc sulphate (Devor and Murphy, 1973; List et al., 1972) and one using procaine to anesthetize the receptors (Doty and Anisko, 1973) found a temporary loss of mating,

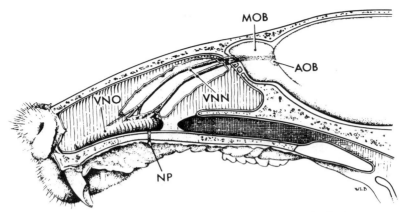

Figure 1. Parasagittal section through hamster nasal cavity. Exposure shows left side of nasal septum and left lateral surface of brain. Mucosa covering the septum has been dissected away in part to expose the vomeronasal nerves (VNN) leading from the vomeronasal organ (VNO), along the medial surface of the main olfactory bulb (MOB), to terminate in the accessory olfactory bulb (AOB). NP, nasopalatine canal. [This figure is taken from Winans and Powers (1977) and is reprinted with permission from Elsevier Biomedical Press B.V.]

suggesting that anosmia did eliminate sexual arousal. However, Powers and Winans (1973) found that although zinc sulphate treatment produced anosmia as measured by the lack of ability to locate odors in a Y maze, these males continued to mate normally. These authors suggested that the apparently conflicting outcomes might be a result of treatments applied in different studies having more or less effect on the vomeronasal (VN) receptors within the VN organ (reviewed by Scalia and Winans, 1976). Because of the structure of the organ (see below) and differences in the exact ways in which zinc sulphate was infused, there could be great variability between studies in the amount of damage to the VN receptors. Only one study has directly examined the VN epithelium following zinc sulphate treatment (Winans and Powers, 1977).

The VN organ of the hamster is a bilaterally paired tubular structure with a pore at the anterior end (see Fig. 1) (also see Wysocki, 1979). Thus gaseous or liquid odor solutions can only gain access to the receptors lining the inside of the tube by entering through this pore. In fact, stimuli may normally only enter the organ when they are "pumped" in (Meredith, 1980). The organ is located on the anterior floor of the nasal cavity, whereas the olfactory epithelium is located dorsally and more caudally, covering the walls of the nasal cavity.

The importance of the VN receptors for male hamster copulatory behavior was demonstrated by studying the effect of cutting the VN nerves. This can be done without doing great damage to the main olfactory bulbs by cutting at the point where the nerves course along the medial surfaces of the main olfactory bulbs toward their terminations in the accessory olfactory bulbs (see Fig. 1). Vomeronasal nerve cuts produced a complete loss or severe deficit in sexual arousal in approximately 40% of males (Powers and Winans, 1975; Winans and Powers, 1977). The remaining males stopped copulating when they were given the additional treatment of a nasal infusion of zinc sulphate. The interpretation favored by Powers and Winans is that both the main olfactory system (receptors in nasal epithelium) and accessory olfactory system (receptors in VN organ) contribute to sexual arousal. Winans and Powers (1977) demonstrated that in their hands zinc sulphate infusions selectively affected the olfactory epithelium, destroying 90% of this surface in most males, while leaving the VN epithelium intact. The fact that 40% of their males stopped copulating following a relatively selective VN deafferentation (VN nerve cuts), whereas no males stopped following zinc sulphate treatment alone, indicates that VN receptors provide the major input to forebrain circuits mediating sexual arousal.

The significance of the VN organ has recently been demonstrated using a different method of disrupting its function. Meredith and O'Connell (1979) have demonstrated that electrical stimulation of the nasopalatine nerve, which carries autonomic efferents to the VN organ, activates a vasomotor pumping mechanism that sucks fluids into the organ. Behavioral studies were then done in which efferent control of the pump was eliminated by cauterizing the nasopalatine nerves. When this was done in combination with nasal infusions of zinc sulphate, copulatory behaviors were severely reduced (Meredith, 1980; Meredith *et al.*, 1980). As in previous studies (Powers and Winans, 1975; Winans and Powers, 1977), zinc sulphate treatment alone had no effect. The changes produced by nasopalatine nerve cautery plus zinc sulphate were nearly as severe as those produced by VN nerve cuts plus zinc sulphate. These data suggest that pump activation is necessary for VN stimulation in this species and that VN stimulation is important for normal copulatory behavior.

These behavioral findings are particularly exciting because of anatomical studies in the rat, rabbit, and opossum that showed that the main and accessory olfactory bulbs have nearly completely independent projections to the temporal lobe (Scalia and Winans, 1976). This segregation of the terminal fields of axons from the main and accessory bulbs led to the concept of dual olfactory systems that might then have relatively separate functional roles. In 1978, Davis *et al.* published results showing the same type of separation of the main olfactory and VN projections in the hamster (see Fig. 2).

The dramatic effects of olfactory bulbectomy are then probably a result of destroying both the main and accessory olfactory input to the temporal lobe. The projection neurons of the main and accessory bulbs send their axons to the temporal lobe via separate parts of the lateral olfactory tract (LOT), sometimes distinguished by the terms LOT (from main bulb) and accessory olfactory tract (from accessory bulb). There are three reports of the effects of cutting the LOT (axons from both systems). Although the results differ in interesting ways, all three agree that, as predicted, such cuts produce a loss or severe deficits in copulation (Devor, 1973; Macrides *et al.*, 1976; Marques *et al.*, 1982).

Devor (1973) developed the technique of selectively cutting the LOT without damaging overlying neural structures. All males with bilateral LOT cuts stopped copulating, but the nature of the behavioral changes produced depended on the rostral-caudal level of the cuts. Males with cuts rostral to the olfactory tubercle behaved similar to bulbectomized males in that they showed little interest in the female, no mounts or

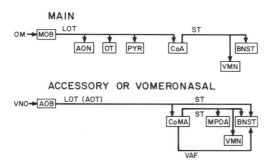

Figure 2. A schematic diagram of some of the olfactory and amygdaloid projections that may be important for reproductive behaviors in the hamster. The connections are based mainly on the work of Davis *et al.* (1978), Kevetter and Winans (1981a,b) and Lehman and Winans (1983). Abbreviations: AOB, accessory olfactory bulb; AON, anterior olfactory nucleus; AOT, accessory olfactory tract; BNST, bed nucleus of the stria terminalis; CoA, cortical nucleus of the amygdala (C1 and C2); CoMA, cortical nucleus (C3) and medial nucleus of the amygdala; LOT, lateral olfactory tract; MOB, main olfactory bulb; MPOA, medial preoptic area; OM, olfactory mucosa; OT, olfactory tubercle; PYR, pyriform cortex; ST, stria terminalis; VAF, ventral amygdalofugal pathway; VMN, ventromedial nucleus of the hypothalamus; VNO, vomeronasal organ.

intromissions, and were unable to locate a buried food pellet. In contrast, animals with LOT cuts at more caudal levels were evidently aroused in that they spent a great deal of time sniffing and licking the female, but they did not copulate. Most of these animals were not anosmic, since they could locate buried food. These interesting dissociations suggested that chemosensory input to structures at the level of the olfactory tubercle is sufficient for investigatory responses to odors from the female or food, but that the more caudal projections of the LOT (possibly to corticomedial amygdala) are necessary for copulation. Marques *et al.* (1982) confirmed that cutting both main and accessory olfactory bulb efferents in the LOT eliminated copulation. However, they ware unable to replicate Devor's finding that more caudal cuts (caudal to the olfactory tubercle) spared the precopulatory responses of sniffing and licking the female. The reason for this difference is not clear, but Marques *et al.* have suggested that the persistence of precopulatory behaviors in Devor's animals may have been a result of more extensive preoperative experience in that study. Marques *et al.* provide further evidence that the olfactory projections to the more rostral terminal regions are not sufficient to support copulation and that the projections to the corticomedial amygdala may be most important. At the completion of the behavioral observations, they used injections of tritiated amino acids into the olfactory bulbs and the autoradiographic method to directly visualize any remaining bulb efferents traveling in the LOT to their terminal fields. With this method they were able to demonstrate that most animals with partial sparing of accessory olfactory bulb projections to the amygdala did continue to exhibit investigatory and copulatory behaviors postoperatively. Interestingly, in some of these males with partial cuts, copulatory activity progressively declined over a month of postoperative testing. The cause of this unusual delayed deficit is not yet clear.

2.2. Forebrain-Hypothalamus

In the rat, olfactory information gains access to the mediodorsal (MD) thalamic nucleus and frontal neocortex via neurons of the endopiriform nucleus and adjacent olfactory tubercle (Price and Slotnick, 1983). This area is rostral to the corticomedial nucleus of the amygdala, and if the anatomy of the system is similar in the hamster, olfactory projections to these neurons were probably spared by Devor's more caudally placed cuts. Thus the sparing of investigatory behaviors in that study (Section 2.1) may have been due to the sparing of olfactory input to thalamocortical pathways.

Reep and Winans (1982) have demonstrated in the hamster that the MD nucleus has reciprocal connections with an area of frontal cortex adjacent to the rhinal sulcus. Sapolsky and Eichenbaum (1980) have studied the effects of lesions in this thalamocortical pathway on measures of olfactory function and male sexual behavior in hamsters. They found that lesions of the MD nucleus or frontal cortex did not produce anosmia, but did reduce or eliminate preferences for the odor of an estrous female and did affect mating behavior. Males copulated after either of these lesions with normal latencies to mount and intromit. However, their behavior was inefficient. Males in both groups showed a greater number of incorrectly oriented mounts and a greater amount of time investigating the female's head and flanks as opposed to the genital region. These changes in mating were interpreted as resulting from a lack of ability to discriminate odors, since the odor of the female's vaginal secretions probably normally serves to orient the male prior to mounting. Thus although thalamocortical pathways do not play a critical role in mating, they are important for the male's preference for the odor of an estrous female. This function could be quite important for mating under more natural conditions in which odors might guide the male's approach to a female from a distance. Thus the importance of these pathways for mating may be underestimated in the usual testing situation where mating is observed in small spaces.

More dramatic changes in male copulatory behavior have been seen following damage to the piriform cortex and amygdala in monkeys, cats, and rats (Malsbury and Pfaff, 1974). Rat studies have shown that lesions that include the corticomedial nuclei produce decrements in this behavior. For example, Harris and Sachs (1975) found that lesions of the corticomedial region, but not the basolateral region, increased the latency to copulate and the latency to ejaculate once mating had begun. This was due primarily to an abnormally high number of intromissions preceding ejaculation. In the hamster, lesions of the rostral part of the medial nucleus produced greater decrements in that copulation was eliminated and investigation of the female's anogenital region was severely reduced (Lehman et al., 1980; Lehman and Winans, 1982). As in the rat, lesions of the basolateral region were without effect. Since the effects of medial nucleus lesions were identical to those produced by eliminating olfactory and VN input to the amygdala (Section 2.1), the authors concluded that medial nucleus lesions produced their behavioral effects by interrupting the relay of chemosensory information to the bed nucleus of the stria terminalis (BNST) and medial preoptic area—medial anterior hypothalamus (MPO–MAH). Recently Lehman and Winans (1983) have demonstrated anatomically that information from medial nucleus neurons can reach the BNST via a ventral amygdalofugal (VAF) pathway in addition to the well-known stria terminalis projection (see Fig. 2). They have also studied the behavioral effects of cutting these two amygdalofugal pathways separately or in combination. When the stria terminalis alone was cut, most males did mate on at least one of the weekly postoperative mating tests (Lehman et al., 1983). In those males that did mate, changes in the temporal pattern of copulation were seen. These changes are similar to those reported following such cuts in male rats, i.e., deficits in the initiation of mounting behavior and an increased number of intromissions preceding ejaculation (Paxinos, 1976). Thus stria terminalis cuts did not produce the same effects as lesions of the medial nucleus, i.e., an elimination of mating behavior. Selective parasagittal cuts of the VAF pathway made with a retractable wire knife at the level of the MPO–MAH also did not eliminate copulation. In fact, these VAF pathway cuts had very little effect on copulation by themselves. However, when cuts were made with a razor blade that damaged portions of the BNST and destroyed both stria terminalis and VAF afferents to the BNST, severe deficits in the initiation of copulation were seen (Lehman and Winans, unpublished). None of the males with this type of combined cut ejaculated postoperatively. This suggests that information from the medial nucleus that is important for copulation is carried in both the stria terminalis and VAF. Apparently the integrity of either pathway is sufficient to support copulation. Since olfactory-VN information is essential, this implies that chemosensory-related information is carried in both pathways.

The MPO–MAH is essential for male copulatory behavior in that lesions of this area have eliminated copulation in every species studied to date (Hansen et al., 1982) including the golden hamster (Bergondy et al., 1982). Does chemosensory information trigger copulation in male hamsters by acting on MPO neurons? The stria terminalis does project from the medial amygdala to the MPO (Kevetter and Winans, 1982a,b). However, Kendrick (1982) recently reported the surprising finding that in male rats corticomedial amygdala neurons that project to the MPO do not respond to electrical stimulation of the olfactory bulb or accessory olfactory bulb. If such is the case in the hamster, this would raise the possibility that the arousing properties of female odors may be mediated by a separate population of corticomedial amygdala neurons that respond to these odors and terminate selectively in the BNST. This chemosensory-related information could reach the BNST via either the VAF or the stria terminalis. The MPO might receive chemosensory-related information indirectly via short axon connections from the BNST (Paxinos et al., 1978). Alternatively, projections from the BNST to areas other than the MPO could mediate the effects of female odors on male hamster sexual arousal.

A study of the effect of selective BNST lesions in male hamsters has not yet been conducted. Potegal et al. (1981) have found that lesions of the anteroventral septal area, which included parts of the BNST, disrupt copulation in male hamsters. However, these males showed increased attacks on restrained, nonaggressive intruder males, as well as on receptive females in the mating tests. Thus the reduced copulation may have been secondary to an increase in aggression. Lesions of the BNST in rats have produced effects similar to, but perhaps somewhat more severe, than the effects produced by stria terminalis lesions (Emery and Sachs, 1976). Males with BNST lesions sometimes show deficits in initiating copulation, and when they do copulate they may have prolonged interintromission intervals suggesting deficits in sexual arousal. The most consistent finding, as in the case of lesions of the corticomedial amygdala (Harris and Sachs, 1975) and of the stria terminalis (Paxinos, 1976) is an increased number of intromissions preceding ejaculation. The finding that damage to the corticomedial amygdala or its connections with the MPO–BNST region increases the number of intromissions to ejaculation in the rat is interesting because this type of change is *not* seen following olfactory bulbectomy in this species (Larsson, 1979). This indicates that, at least in the rat, the corticomedial amygdala and stria terminalis do more than simply transmit chemosensory information to other forebrain areas.

The physiology of the stria terminalis projections to the MPO is influenced by testosterone in the rat (Kendrick and Drewett, 1979; Kendrick, 1982). Similar electrophysiological studies have not been conducted in the hamster, but the presence of androgen target neurons in the corticomedial amygdala, BNST, and MPO (Doherty and Sheridan, 1981) suggest androgen action on these pathways in the hamster as well.

2.3. Summary

The demonstration of the critical importance of the olfactory bulbs has led to detailed studies of the role of the olfactory pathways in male mating behavior. Discoveries include the demonstration of the importance of the vomeronasal organ and its central projections, and the demonstration of a vomeronasal pumping mechanism which may provide for central control of stimulus access to vomeronasal receptors. One of the terminal regions for efferents of the accessory olfactory bulb (vomeronasal system) is the medial nucleus of the amygdala. Lesions of this nucleus produce a complete loss of mating behavior. Recently a ventral amygdalofugal pathway has been demonstrated that may relay some aspect of olfactory information from the medial nucleus to the BNST. Knife-cut studies indicate that this pathway is important for male sexual arousal.

3. FEMALE SEXUAL BEHAVIOR

3.1. Lordosis

3.1.1. The Ventromedial Hypothalamus

Research on the neural control of sexual behavior has been conducted for the past 40 years, and much of this work has focused on the role of the hypothalamus (Bard, 1940; Brookhart *et al.*, 1940). However, until recently the localization of hypothalamic regions important for female copulatory behavior was not clear. Although it had been known since the 1940s that damage to the medial hypothalamus could eliminate sexual receptivity in rodents, the exact region critical for this effect was not known (reviewed in Malsbury *et al.*, 1977). Now, the question of the major hypothalamic site of estrogen's action in inducing receptivity appears to have been answered.

A number of studies published since 1977 demonstrate the critical importance of the region of the ventromedial nucleus (VMN) of the hypothalamus. Studies using the autoradiographic method have demonstrated estrogen-concentrating neurons in and around the VMN in hamster (Krieger et al., 1976) and rat (Pfaff and Keiner, 1973; Stumpf et al., 1975). Estrogen implants there can induce receptivity in hamster (see Fig. 3) and rat (Davis et al., 1979; Rubin and Barfield, 1980), whereas lesions of the region can eliminate receptivity in hamster (Malsbury et al., 1977) and rat (Mathews and Edwards, 1977; Pfaff and Sakuma, 1979). Recently, we have found that cyclohexi-mide, a protein synthesis inhibitor, can reduce receptivity in hamsters when injected into the VMN region (DeBold and Malsbury, 1983).

The role of the VMN region in mediating this behavior requires that information pass out of the hypothalamus to influence extrahypothalamic brain regions and eventually the spinal motor neurons directly controlling the behavior. During the past few years, our laboratory has made significant progress in identifying the neural connec-tions of the VMN region that are necessary for lordosis in the hamster. A series of reports of the effects of hypothalamic knife cuts have been published since 1978. In the first experiments a Halasz-type, rotating wire, bayonet-shaped knife blade was used. The most significant finding was that 180 degree half-cylinder cuts placed anterolateral to the VMN can eliminate receptivity, but similar cuts placed posterolateral to the VMN have little or no effect. We concluded that neural pathways critical for receptivity pass in or out of the VMN region in an anterolateral and not a posterior, or post-erolateral direction (Malsbury et al., 1978). Similar studies in the rat later confirmed the importance of these anterolateral connections (Phelps and Nance, 1979; Pelps et al., 1980; Yamanouchi and Arai, 1979).

The effectiveness of the half-cylinder cuts rostral to the VMN began to define the critical connections. To further define the critical pathway or pathways, a different type of knife was used to make cuts confined to the sagittal plane (Malsbury and Daood, 1978). The most important finding was that receptivity was reduced or eliminated following bilateral sagittal-plane cuts placed lateral to the medial anterior hypoth-alamus–VMN (MAH–VMN). In contrast, sagittal-plane cuts placed lateral to the medial preoptic area–MAH (MPO–MAH) never disrupted lordosis. Similar MPO–MAH cuts in female rats also failed to disrupt lordosis (Numan, 1974). Because of the ease of measuring lordosis in hamsters, we were able to define the most critical rostral-caudal region for disrupting lordosis quite precisely. This region extends from the level of the suprachiasmatic nuclei back to the rostral one third of the VMN (see Fig. 4). The effectiveness of sagittal-plane cuts was later confirmed in the rat (Manogue et al., 1980; Pfeifle et al., 1980), and is completely consistent with the results following Halasz-type cuts discussed above, as the lateralmost sweep of those cuts interrupts the same connections.

At about this time neuroanatomical studies revealed several long-axon projection pathways out of the VMN region (Conrad and Pfaff, 1976b; Krieger et al., 1979; Saper et al., 1976). The location of our most effective sagittal-plane cuts indicated that they interrupted a subset of these VMN efferents that project laterally out of the hypoth-alamus. Many of these axons travel close to the base of the brain in the region of the supraoptic commissures (SOC). This particular projection pathway sends axons to the amygdala and the midbrain and pontine tegmentum. We proposed that cutting VMN

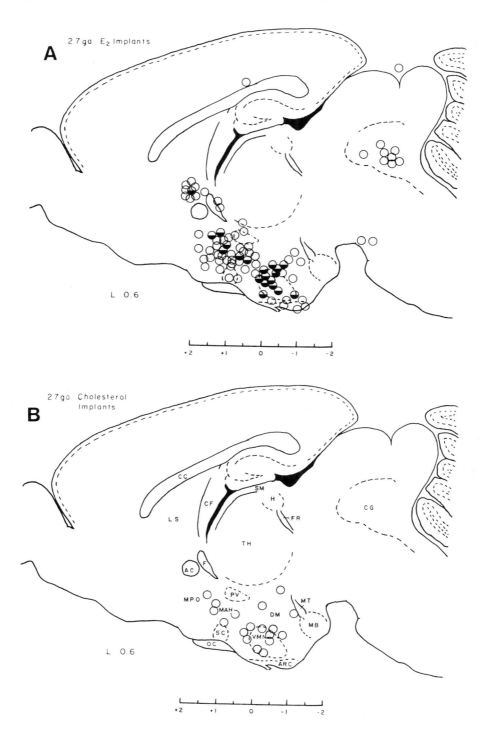

efferents in the SOC might disrupt sexual receptivity by removing an estrogen-sensitive input to one or more of these terminal regions.

However, our first sagittal cuts were placed at the medial border of the medial forebrain bundle (MFB) and thus also severed medial hypothalamic connections that travel in the MFB. The first major test of whether the axons in the SOC are critical was made by simply varying the lateral position of the same type of bilateral sagittal-plane cut (Malsbury et al., 1979). If the hypothesis were correct, then both medially and laterally placed cuts should reduce receptivity, since the axons of the SOC pass through the lateral hypothalamus. In this experiment near-lateral (NL) cuts were placed at or just lateral to the fornix, whereas far-lateral (FL) cuts were placed at the lateral edge of the MFB. Both NL and FL cuts reduced lordosis in response to both sexually active males and manual stimulation. Near-lateral but not FL cuts also increased agonistic behavior and produced obesity. Since both NL and FL cuts severed axons traveling in the region of the SOC, these data supported our hypothesis that the SOC connections are critical for sexual receptivity. The increased agonistic behavior and obesity that follow NL cuts are attributed to interruption of medial hypothalamic connections that travel in the MFB.

The difference between the effects produced by NL and FL cuts is important for understanding the neural systems controlling receptivity. The lack of agonistic behavior in the FL group demonstrates that the reduction in receptivity can be produced independent of a release of agonistic behavior or an increase in irritability that might interfere with the elicitation of lordosis. The lack of obesity in the FL group demonstrates that the reduction in receptivity can be produced independent of metabolic changes associated with the hyperinsulinemia that might result from VMN region lesions and NL cuts. Rather, these results support the idea that the disruption of the lordosis response is a primary effect of cuts, either NL or FL, that interrupt the SOC connections of the ventromedial hypothalamus. See Table 1 for a comparison of the similarities and differences between the effects of VMN lesions, Halasz-type cuts, and sagittal-plane cuts.

In order to further characterize the effects of sagittal-plane hypothalamic cuts, we examined the effects of unilateral NL or FL cuts. The hamster has an advantage over the rat here, since lordosis can be elicited and maintained by stroking the fur on only one side of the body (Kow et al., 1976). This enables one to look for lateralized effects of unilateral brain damage. We found that both NL and FL unilateral cuts caused marked lordosis deficits in response to stroking the flank contralateral, but not ipsilateral, to

---

Figure 3. (A) Sagittal section (0.6 mm off midline) showing the locations of implants of estradiol-17B in female hamsters. Half-filled circles indicate implant sites associated with sexual receptivity. Open circles indicate sites with no effects on sexual behavior. (B) Sagittal section showing the locations of cholesterol implants in female hamsters. None of the cholesterol implants resulted in sexual receptivity. Abbreviations: AC, anterior commissure; ARC, arcuate nucleus; CC, corpus callosum; CF, commissure of the fornix; CG, central gray; DM, dorsomedial hypothalamus; F, fornix; LS, lateral septal area; MAH, medial anterior hypothalamus; MB, mamillary body; MPO, medial preoptic area; MT, mamillothalamic tract; OC, optic chiasm; SC, suprachiasmatic nucleus; VMN, ventromedial nucleus. [This figure is taken from DeBold et al. (1982) and is reprinted with permission from Pergamon Press Ltd.]

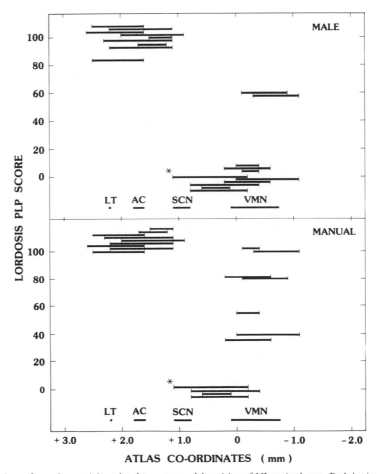

Figure 4. Loss of sexual receptivity related to rostrocaudal position of NL sagittal cuts. Each horizontal line represents the position and length of the bilateral cuts in each female. The position of each line on the x axis shows the rostrocaudal extent of the hypothalamus through which the cuts reached the base of the brain bilaterally. Thus, although the same knife blade was used to make all cuts, slight asymmetries in the rostrocaudal positions of the cuts reduced the length of the area that was cut bilaterally in any particular female. Anatomical landmarks are indicated by the labeled horizontal lines just above the x axis. The position of each line on the y axis shows the lordosis postlesion performance score (PLP). A low score indicates poor performance. Lines that represent scores of 100 or 0 are spaced above 100 or below 0 on the y axis only so that their positions on the x axis are not obscured. The lordosis scores in tests in the male's cage are shown in the top section of the figure and the scores in tests with manual stimulation are shown in the bottom section. The nine females in the MPO–MAH group are represented by the cluster of lines in the upper left corner of each section of the figure and the 11 females in the MAH–VMN group are represented by the lines scattered in the right half of each section. Abbreviations: AC, anterior commissure; LT, lamina terminalis; SCN, suprachiasmatic nucleus; VMN, ventromedial nucleus. [This figure is taken from Malsbury and Daood (1978) and is reprinted with permission from Elsevier Biomedical Press B.V.]

Table 1. Effects of Hypothalamic Surgery on Measures of Social Behavior and Weight Regulation in Female Hamsters[a]

Hypothalamic surgery	Lordosis response	Agonistic behavior	Body weight
VMN[a] lesions	▼	▲	▲
Halasz cuts			
360° around VMN	▼	▲	▲
180° AL to VMN	▼	▲	▲
180° PL to VMN	—	—	—
Sagittal cuts			
MPO–MAH, NL	—	—	▼
MAH–VMN, NL	▼	▲	▲
MAH–VMN, FL	▼	—	—

[a]A summary of the effects of bilateral hypothalamic surgeries on measures of social behavior and weight regulation in ovariectomized, hormone-primed female hamsters. Postoperative decreases in these measures are indicated by downward-pointing arrowheads and upward-pointing arrowheads indicate increases. When group data are considered, all of the types of lesions that produced decreases in lordosis also produced increases in agonistic behavior and body weight, *except* FL cuts at the level of the MAH-VMN, see bottom row (Malsbury *et al.*, 1979). These dissociations are important, because they show that the lesion-produced lordosis deficits are not necessarily secondary to other changes that produce increases in agonistic behavior and body weight (see text).

[b]Abbreviations: AL, antero-lateral; FL, far lateral; MAH, medial anterior hypothalamus; MPO, medial preoptic area; NL, near lateral; PL, posterolateral; VMN, ventromedial nucleus of the hypothalamus.

the cut (Ostrowski *et al.*, 1981). In contrast, these cuts failed to impair lateral displacement, a pelvic adjustment that occurs during lordosis in response to unilateral perineal stimulation (Noble, 1979). Several additional tests of sensory responsiveness yielded no evidence of a unilateral sensory neglect. These results allow us to rule out illness, purely motor impairments, or other "whole body" problems as a cause of the lordosis deficits, since the affected females could display normal, bilaterally symmetrical lordosis in response to stroking the ipsilateral flank. Because both NL and FL cuts produced similar lateralized deficits, these results again support the hypothesis that both types of cuts are effective because of damage to the same lateral connections.

Although all of our data up to this point are consistent with the idea that hypothalamic knife cuts disrupt lordosis by cutting VMN region efferents projecting anterolaterally, such cuts destroy hypothalamic *afferents* as well. Such cuts may produce their behavioral effect by cutting both types of connections. Two of the most important questions we wish to answer are: (1) What are the identities of the cell groups whose axons are severed by these sagittal cuts? and (2) Which of these cell groups are important for lordosis? We have developed a technique that we have called "labeling knife cuts" that will enable us to answer the first question (Scouten *et al.*, 1982). We have found that by drying a solution of horseradish peroxidase (HRP) tracer on the retractable blade used to make the cuts, HRP can be applied to the tissue as the cut is being made. The HRP is then taken up by cut axons and transported back to their perikarya of origin. This method allows us to directly visualize the cells whose axons are severed by knife cuts that reduce receptivity. In addition, since lordosis can be tested within the postoperative survival period, it is possible to correlate the degree of behavioral deficit

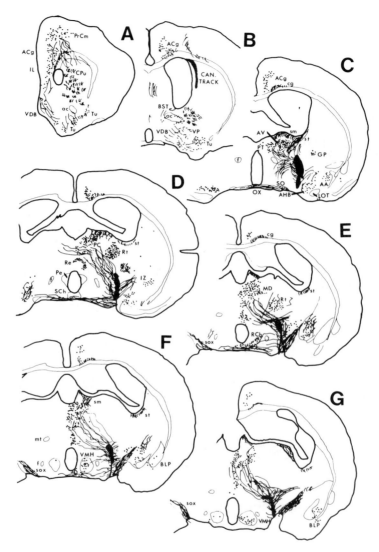

Figure 5. The locations of cell bodies containing HRP reaction product (labeled cells) following an FL labeling cut at the level of the SOC-ventromedial hypothalamus that produced a deficit in lordosis. The cut is shown in sections C to H. Labeled axons are also indicated. Abbreviations: AC, anterior commissure; BC, brachium conjunctivum; BLA, basolateral amygdala; BNST, bed nucleus of the stria terminalis; CA, central amygdala; CIN, cingulate cortex; CST, corticospinal tract; CU, cuneate nucleus; DB, diagonal band; DBN, nucleus of the diagonal band; DP, decussation of the pyramids; DR, dorsal raphe; F, fornix; GP, globus pallidus; HC, habenular commissure; IC, internal capsule; IL, infralimbic cortex; MET, medial and midline thalamus; MT, mamillothalamic tract; MR, medial raphe; NT, spinal nucleus of the trigeminal; NTS, nucleus of the solitary tract (A2); OC, optic chiasm; OT, optic tract; PBG, parabigeminal nucleus; PCM, precentral medial cortex;

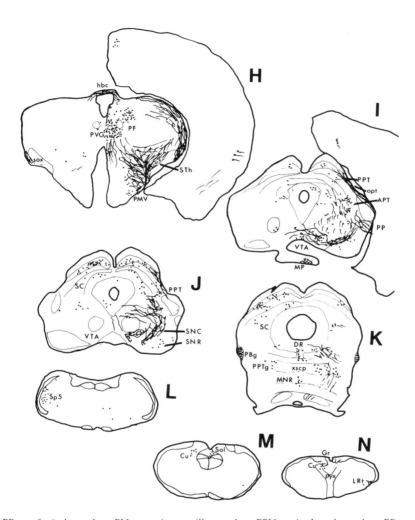

PF, parafascicular nucleus; PM, posterior mamillary nucleus; PPN, peripeduncular nucleus; PR, paramedian reticular nucleus; PVG, periventricular gray (A11); R, red nucleus; RC, retrochiasmatic area; RL, lateral reticular nucleus (A1); RS, retrosplenial cortex; RT, reticular nucleus of the thalamus; SC, superior colliculus; SCN, suprachiasmatic nucleus; SM, stria medullaris; SNC, substantia nigra pars compacta; SNR, substantia nigra pars reticulata; SOC, supraoptic commissure; ST, stria terminalis; SUT, subthalamic nucleus; TAV, anteroventral thalamus; TPC, tegmental pedunculopontine nucleus; TPO, posterior thalamus; TU, olfactory tubercle; VM, ventromedial hypothalamus; VT, ventral tegmental area of Tsai (A10).

and the presence of labeled cells in particular brain areas within individual animals. This may provide an answer to the second question.

We are currently analyzing the results of such labeling cut experiments. Hamsters were given unilateral hypothalamic sagittal-plane cuts, either NL or FL, using a wire knife whose blade had been precoated with HRP. These animals were tested for lateralized lordosis deficits after five days, killed immediately after testing, and the brains processed for HRP histochemistry (Mesulam, 1978). To determine whether there was a lateralized lordosis deficit, flank stroking was applied either contralateral or ipsilateral to the hypothalamic cut. The difference between mean latency to elicit lordosis with contralateral compared with ipsilateral stroking was divided by 60 sec (the maximum test duration) to obtain a percentage indicating the degree of lateralized deficit. A score of 100% would indicate we were unable to elicit lordosis by stroking the contralateral flank, but obtained an immediate response to touching the ipsilateral flank. Figure 5 depicts the labeling following an FL cut that produced a 60% lateralized deficit.

The large number of structures containing labeled cells in Fig. 5, extending from the frontal poles to the decussation of the pyramids, is striking. Labeled cells can be seen in several areas that have been proposed to play a role in the lordosis behavior of rodents, including the VMN, brainstem noradrenergic areas (Hansen *et al.*, 1981), the peripeduncular nucleus of the lateral midbrain (Carrer, 1978), the deep layers of the superior colliculus (Rose 1982), and the midbrain central gray (Pfaff, 1980). However, similar numbers of labeled cells were found in many of these areas following similar cuts that did not impair lordosis. If cutting axons from the same area in another animal does not cause a lordosis deficit, their damage in this animal is unlikely to be important for the observed deficit. In addition, several of these areas are not labeled in the brains of animals that had severe lordosis deficits and sensitive HRP reactions (as indicated by the presence of labeled cells in other regions). Thus, it is hoped that comparisons of several cases for which we have both behavioral and anatomical data will reduce to one or a very few the number of areas in which the presence of labeled cells correlates with the degree of behavioral deficit. The behavioral deficits resulting from our cuts must then be due to the damage to axons from a subset of these areas.

3.1.2. Forebrain Inhibitory Influences

We have shown that electrical stimulation of the lateral septal area and adjacent BNST (Zasorin *et al.*, 1975) and the MPO–MAH (Malsbury *et al.*, 1980) can disrupt ongoing lordosis and interfere with its elicitation in tests with sexually active males. Conversely, Vomachka *et al.* (1982) have reported that septal lesions can facilitate lordosis in hamsters, an effect that has been reported previously in rats (Nance *et al.*, 1974). These regions contain estrogen-concentrating neurons (Krieger *et al.*, 1976). It is possible that receptivity is partly dependent on estrogenic inhibition of a lordosis-inhibitory action of such neurons. However, studies using intracerebral implants of estrogens tend to indicate that estrogenic stimulation restricted to any one of these areas is neither necessary nor sufficient to induce receptivity in rat or hamster (Davis *et al.*, 1979; DeBold *et al.*, 1982; Rubin and Barfield, 1980).

3.1.3. Hypothalamic Connections with the Dorsal Midbrain and Pontine Tegmentum

Pfaff (1980) has presented a model for the neural control of lordosis in which samotosensory inputs from the spinal cord and estrogen-sensitive inputs from the hypothalamus interact in the midbrain to enable lordosis to occur. Pfaff has emphasized the importance of the central gray in this circuitry, and he has found that in the rat electrical stimulation there can facilitate lordosis, whereas lesions restricted to the central gray can disrupt lordosis. Lesions of the laterally adjacent tegmentum that did not encroach on the central gray had no effect on rat lordosis (Fig. 6 in Sakuma and Pfaff, 1979). In the hamster, however, alhough lesions restricted to the central gray can eliminate lordosis, deep tectal lesions that produced little or no central gray damage had equally severe effects (Muntz et al., 1980). Floody and O'Donohue (1980) also examined the effects of central gray lesions in hamsters and found that lesions restricted to the central gray prevented lordosis on some but not all postoperative tests. On tests in which lordosis was elicited, its duration was reduced.

In summary, lesion studies indicate that neurons in the dorsal midbrain and pontine tegmentum are important for lordosis in rats and hamsters. In the hamster, these neurons may be scattered from the central gray into the laterally adjacent dorsal tegmentum, whereas in the rat, they may be more confined to the central gray.

Muntz et al. (1980) have discussed these species differences and have suggested that they may be related to differences in the somatosensory determinants of lordosis, as well as the tonic (hamster) versus phasic (rat) nature of the response in the two species. In the hamster, brushing of the fur at any one point over a wide region of the rear half of the body is sufficient to elicit and maintain lordosis for minutes at a time (Kow et al., 1976). In the rat, brushing of the fur is ineffective. Cutaneous stimulation of the flanks, followed by pressure on the rump, tailbase, and perineum is required to elicit the brief response (Pfaff, 1980). Single-unit recordings from the dorsal midbrain and pons in anesthetized hamsters have shown that neurons that respond to light tactile stimulation of the type that will elicit lordosis in awake females are most common at the level of the posterior superior colliculus and intercollicular zone (reviewed in Muntz et al., 1980). Such lordosis-relevant neurons are found within and lateral to the central gray and in the deep layers of the superior colliculus (Rose, 1978, 1982). It is likely that the deep tectal-dorsal tegmental lesions that eliminate lordosis in the hamster (Muntz et al., 1980) interfere with the normal function of these neurons either by destroying their perikarya, destroying their outputs, or destroying their inputs from the hypothalamus. These possibilities are not mutually exclusive.

The hypothalamic connections with the dorsal midbrain and pons that are important for lordosis are beginning to be described, but important questions remain. Precisely which neural populations are connected monosynaptically and what routes do these connections follow? Which of the demonstrated anatomical connections carry information relevant to lordosis? It is known that neurons in and around the ventromedial nucleus do project to the dorsal midbrain. Injections of HRP either within or lateral to the central gray produce labeled perikarya in the ventromedial nucleus in rat (Morrell et al., 1981), and hamster (Malsbury, personal observations). Recently Morrell and Pfaff (1982) have directly demonstrated that some of these VMN region projection neurons do contain estrogen receptors. Direct projections from the VMN region to midbrain and pontine levels have also been mapped using the autoradiographic method in rats (Krieger *et al.*, 1979; Saper *et al.*, 1976). These studies demonstrate that VMN region neurons send axons to the midbrain via two main routes: (1) via a direct descending projection in the periventricular gray and (2) via a more circuitous route first projecting anterolaterally in the area of the SOC and then descending to the midbrain. Our studies of the effects of hypothalamic knife cuts demonstrate that this latter projection via the SOC is most important for lordosis. Sagittal-plane cuts that sever the lateral pathway can eliminate lordosis in hamsters (Malsbury and Daood, 1978), whereas Halasz-type cuts posterolateral to the VMN region that sever the periventricular, medial projections have no effect (Malsbury *et al.*, 1978). Some of these laterally projecting axons end in the lateral midbrain in the peripeduncular nucleus (PPN), whereas others may continue dorsomedially to terminate in the dorsal midbrain in the deep tectum and intercollicular region.

In the rat, lesions of the peripeduncular nucleus can eliminate lordosis (Carrer, 1978), but it is possible that such lesions are effective because of the interruption of VMN axons passing through on their way to more caudal and medial areas of termination (Edwards and Pfeifle, 1981).

The direct VMN projection to the dorsal midbrain via the lateral pathway suggests a basis for the similarity of the behavioral effects of unilateral dorsal midbrain lesions (Muntz *et al.*, 1980) and unilateral hypothalamic cuts through the lateral pathway in hamsters (Ostrowski *et al.*, 1981). Both types of lesions produced a lateralized deficit in which unilateral flank stimulation was much less effective when applied to the side of the body contralateral to the lesion.

3.1.4. Summary

Studies in rat and hamster strongly suggest that the ventromedial hypothalamus is the primary site of estrogen's action in inducing receptivity. This role requires that information pass out of the hypothalamus to influence extrahypothalamic brain regions. In a series of studies using hypothalamic knife cuts, we were the first to determine that the lateral connections of the VMN region are critical for sexual receptivity. Some lateral projections of the VMN descend to the dorsal midbrain and pontine tegmentum where hormonal information may intereact with sensory input to enable lordosis to occur. However, our sagittal-plane cuts also sever hypothalamic afferents (Fig. 5) and fibers of passage, and it remains to be determined whether the loss of any of these afferents is important for the behavioral effects of such cuts.

3.2. Ultrasonic Vocalizations

High-frequency vocalizations (ultrasounds) are a type of courtship behavior common to male and female rodents (Barfield *et al.*, 1979; Floody, 1979). Hamster ultrasounds are thought to facilitate copulatory behavior by communicating location and hormonal status (in females) and possibly by arousing the partner of the opposite sex (Floody, 1979). Recent studies similar in design to those of the neural control of lordosis and male copulatory behavior have been fruitful in identifying neural structures important for ultrasound production in hamsters and other rodents.

3.2.1. Olfactory Systems

As in the case of male hamster copulatory behavior, but unlike lordosis (Carter, 1973), chemosensory stimuli and the olfactory systems are important for stimulating ultrasound emissions. Exposure to bedding soiled by a hamster of the opposite sex increases the rate of ultrasonic vocalizations (Floody, 1979). Olfactory bulbectomy in females markedly reduced vocalization rates during exposure to soiled bedding (Kairys et al., 1980). Bulbectomized females also showed reduced vocalization rates following a brief interaction with a male. We have some evidence that the decrease in vocalization rates in bulbectomized hamsters may not be a nonsensory deficit (Cain, 1974), since bulbectomized females show normal elevations in vocalization rates during exposure to synthetically produced ultrasounds, but make fewer vocalizations than control females following a 5-min exposure to a test male (Miceli, unpublished). Presumably, the bulbectomized females failed to respond to chemosensory cues desposited by the test male while it was in the female's cage. It appears then that the function of the olfactory system(s) with regard to ultrasonic vocalizations is the processing and relaying of chemosensory information. The relative importance of the main and accessory olfactory systems for ultrasound production in the hamster has not yet been determined; but in male mice the accessory, but not the main, olfactory system is critically important for olfactory-elicited vocalizations (Bean, 1982).

3.2.2. Forebrain-Hypothalamus

Studies have also focused on hypothalamic control of ultrasonic vocalizations in hamsters. Floody and his associates have presented evidence of opposing hypothalamic control mechanisms of ultrasound production. Merkle and Floody (1979) reported that lesions of the VMN area, which abolished or reduced lordosis, had no effect on female hamster ultrasounds. More recently, Floody (1983) reported that similar VMN area lesions can actually increase vocalizations in both males and females. In contrast, lesions of the MPO did not reduce lordosis, but did decrease rates of ultrasonic calling (Merkle and Floody, 1979). We have recently evaluated the effects of sagittal-plane hypothalamic knife cuts on ultrasound production in female hamsters (Miceli and Malsbury, 1982b). Near-lateral and FL type cuts were placed as in the study of lordosis (see above), but along the MPO–MAH rather than the MAH–VMN (see Fig. 6). The more lateral (FL) cuts were placed so as to spare MPO–MAH outflow through the MFB, whereas NL cuts were placed between the MPO–MAH and the MFB. We found that NL and FL cuts were equally effective in reducing vocalization rates. Moreover, the cuts were selective in that they did not reduce spontaneously occurring vocalization (i.e., baseline vocalizations recorded prior to the introduction of a male), but blocked the increase in call rates following exposure to a male. This selective reduction of vocalizations was quite similar to that seen in bulbectomized females (Miceli, unpublished), and is consistent with the idea that preoptic-hypothalamic knife cuts reduce ultrasound pro-

MALE MATING,
ULTRASONIC CALLS   MATERNAL   LORDOSIS

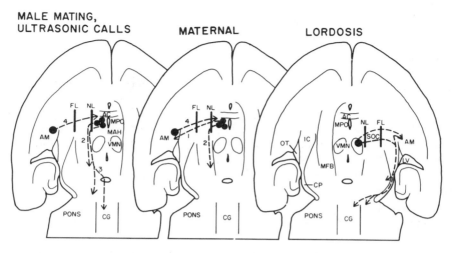

Figure 6. Three schematic horizontal sections of the brain summarize our current knowledge and hypotheses concerning some of the neural pathways important for hamster reproductive behaviors. Three regions of the forebrain are emphasized: (1) the MPO–MAH, (2) the region of the VMN, and (3) the medial amygdala (AM). The origins of the projections are indicated by large dots and axons by dashed lines. Hypothalamic sagittal-plane cuts are indicated by the lines labeled NL for near-lateral and FL for far-lateral. Four pathways are numbered in the two diagrams on the left. Pathways numbered 1 to 3 represent projection pathways of the MPO–MAH that have been described in the rat (Conrad and Pfaff, 1976a) and are presumed to exist in the hamster as well. Number 1 represents ventral efferents that cross the lateral hypothalamus at the level of the optic chiasm. Number 2 represents efferents that descend in the medial part of the MFB. Number 3 represents descending efferents that first turn laterally and then course medially in the periventricular gray of the posterior hypothalamus, midbrain, and pons. Number 4 represents the VAF pathway recently demonstrated in the hamster by Lehman and Winans (1983) that projects to the BNST. The lateral projections of the VMN region that travel in the region of the SOC are indicated in the diagram on the right. These axons are believed to terminate in the medial amygdala (AM), in the midbrain PPN (2nd arrowhead), and in the deep tectum and dorsal tegmentum of the caudal midbrain and rostral pons (last two arrowheads). (See text for explanation of the following summary.) Pathways important for: male sexual behaviors = 2 and 4; female sexual behaviors, lordosis = SOC, ultrasounds = 3 and 4; maternal behaviors, regulation of pup cannibalism = 1 and 4, nest building = 2.

duction by severing fibers important for chemosensory stimulation of ultrasounds. This idea is given further consideration below.

3.2.3. *Brainstem*

Preoptic-hypothalamic connections with the midbrain central gray may also be important for ultrasound production in hamsters and other rodents, just as VMN–midbrain connections are important for lordosis. Lesions of the central gray with maximal damage at the level of the decussation of the brachium conjunctivum greatly reduced ultrasounds in female hamsters (Floody and O'Donohue, 1980). In contrast to the effects of bulbectomy (Miceli, unpublished) and preoptic-hypothalamic knife cuts (Miceli and Malsbury, 1982b), central gray lesions produced a global reduction in ultrasonic call rates. Both spontaneously occurring and male-exposure-induced vocalizations were reduced in lesioned females. Lesioned females also showed reduced vocalizations during and after exposure to synthetically produced ultrasounds. These results from animals with central gray lesions are complimented by those of a brain stimulation study of anesthetized male rats in which Yajima *et al.* (1980) recorded ultrasonic and audible vocalizations during electrical stimulation of various sites in the basal forebrain and dorsal midbrain. The most sensitive sites yielding ultrasonic vocalizations in their animals were in the midline posterior hypothalamus and dorsomedial central gray. Stimulation sites that elicit ultrasonic vocalizations in anesthetized male rats form a pattern corresponding to the periventricular fiber system through which the MPO–MAH sends descending projections to the central gray (Conrad and Pfaff, 1976a,b). Our NL, but not FL, preoptic level knife cuts apparently transected MPO–MAH efferents heading toward the periventricular system through the MFB; yet, the two types of knife cut were equally effective in reducing ultrasonic calling. Evidently, NL and FL cuts reduce ultrasonic vocalizations by interrupting different axonal systems. We suspect that reduced ultrasound rate in animals with NL cuts is primarily due to damage to MPO–MAH efferents descending first through the MFB and then coursing medially to join the periventricular system, heading toward the central gray; and that FL cuts have similar effects on vocalizations by severing *afferents* to the MPO–MAH and neighboring tissue. For example, our FL (and possibly NL) cuts may have severed the VAF pathway from the medial amygdala to BNST (Lehman and Winans, 1983). Although we have no direct evidence, we propose that an olfactory–forebrain circuit as described for control of male hamster copulatory behavior may underly mating-related ultrasonic vocalizations in both male and female hamsters, and possibly other species. This hypothesis would require evidence that the amygdala and BNST participate in the control of rodent ultrasonic vocalizations. With this in mind, it is interesting to note that the amygdala and amygdalofugal fibers are important for (audible) social and affective vocalizations in cats and monkeys (Hilton and Zbrozyna, 1963; Jurgens, 1982).

3.2.4. Summary

The MPO–MAH may be one site at which estrogens and androgens exert their effect on ultrasounds (Floody, 1979). The MPO–MAH may also be involved in the processing of chemosensory information that stimulates ultrasound emissions, through a direct projection from the medial amygdala via the ST (compare Kendrick, 1982) or more indirectly via the VAF and BNST (see Fig. 2). Thus, the MPO–MAH could mediate hormonal and chemosensory control of ultrasound production by modulating neural mechanisms in the central gray. The role of the lower brainstem in the control of ultrasonic vocalizations in hamsters has not yet been examined. Rat studies indicate that the central gray projections to the lateral and dorsomedial medulla and ultimately to motor neurons in nucleus ambiguus (which control ultrasound-production musculature) form the final link in a central pathway underlying ultrasounds (Wetzel *et al.*, 1980; Yajima *et al.*, 1981, 1982). These studies should serve to guide future work on lower brainstem control of ultrasonic vocalization in hamsters.

4. MATERNAL BEHAVIOR

Maternal behavior in hamsters does not easily lend itself to the analysis of its neural control. One obstacle to progress here is the difficulty that has been encountered in determining whether hormones play an important role in controlling these behaviors (Siegel and Rosenblatt, 1980). Particularly troublesome is the fact that hamsters are highly cannibalistic toward their young (see Miceli and Malsbury, 1982a; Chapter 10, this volume) and that very little is known about what causes cannibalism or its biological significance. Consequently, it is difficult to assess what constitutes normal maternal behavior patterns in this species, and therefore a well-defined baseline of maternal behavior against which the effects of CNS manipulations can be compared is not yet available. Another related problem is that experimental manipulations (endocrinological, neural) appear not to affect maternal behavior directly, but instead alter the animal's tendency to cannibalize pups. For example, a lesion of a given area might abolish any interest a rat has in her litter, but increase cannibalistic behavior in the hamster. Pup cannibalism has a behavioral topography similar to predatory-attack behavior, especially in nonlactating females. Experiments in which cannibalistic behavior rather than maternal behavior per se is altered are therefore difficult to interpret because it may be impossible to distinguish whether a neural system underlying maternal or agonistic/predatory behaviors has been altered.

4.1. Olfactory Systems

Despite these pitfalls, some progress has been made in the study of neural control of maternal behavior in hamsters. In an experiment on virgin females, Marques (1979) examined the roles of the main and accessory olfactory systems in pup-directed behaviors. Groups of virgin females were either bulbectomized, given VN cuts, treated with zinc sulphate nasal infusions, or given combined VN cuts and zinc sulphate treatment. Bulbectomy produced a syndrome that may be best described as a loss of response consistency normally seen in intact nonlactating animals over repeated tests with foster pups. That is, normal animals usually either retrieve or attack and cannibalize foster pups on a given test, and are generally consistent with their response over temporally spaced repeated testing. Marques (1979) found that most of his bulbectomized animals that either cannibalized or retrieved foster pups during preoperative tests showed no interest in or response toward pups during postoperative tests. In addition, some bulbectomized-preoperative cannibals retrieved pups postoperatively, whereas some preoperative pup retrievers cannibalized postoperatively. In contrast, peripheral sensory deafferentation appeared to suppress pup cannibalism and, possibly as a consequence, facilitated maternal behaviors. This facilitation was seen in animals with VN cuts and not in zinc-sulphate-treated animals. However, zinc sulphate did potentiate the facilitation produced by VN cuts when the treatments were combined. Nest building, a component of maternal behavior, has also been studied in hamsters with central olfactory damage. In an early study, Goodman and Firestone (1973) found that bulbectomized female virgins built poor nests and showed poor operant responding for nesting material. Similarly, Marques (1979) found that virgin and postparturient bulbectomized hamsters built poor or no nests. This nest-building deficit was not observed in peripherally deafferented animals.

As in virgins, bulbectomy in postparturient females produces severe alterations of normal pup-directed behaviors. Marques (1979) found that over 70% of his bulbectomized females (animals bulbectomized and tested as virgins) cannibalized their entire litter within the first two days postpartum. He also described a pattern of disorganized maternal behavior (pups and paper toweling randomly scattered about the cage) in animals with surviving litters and in the cannibals before they had devoured their litter. We found that more conservative bulbectomy that spares the frontal pole and posterior portions of the bulb (including the accessory olfactory bulb) produced a similar disruption in postpartum behavior (Miceli, unpublished). Approximately 80% of bulbectomized animals cannibalized their entire litter early in the postpartum period. Of these animals, more than half cannibalized older (day 10) foster pups in tests later in the postpartum period, whereas most of the remaining animals showed no interest in test pups and only two of 16 of these animals showed pup retrieval and other maternal behaviors. In contrast to earlier reports (Goodman and Firestone, 1973; Marques, 1979), changes in pup-directed behavior in our bulbectomized dams were not accompanied by nest-building or food-hoarding deficits. Neonatal bulbectomy apparently has less effect on subsequent postpartum maternal behavior, since only 20% of neonatally bulbectomized females cannibalized their entire litter and 13% culled their litters down to one pup (Leonard, 1972).

4.2. Forebrain-Hypothalamus

More recently we have focused on preoptic-hypothalamic pathways important for hamster maternal behavior. In one study, we compared the effects of anterior cuts placed immediately lateral to the MPO–MAH and more posterior cuts placed lateral to the MAH–VMN on lordosis and maternal behaviors (Marques et al., 1979). Animals were tested for responses to foster pups both preoperatively and postoperatively as virgins, their lordosis responses were measured during mating, and they were finally tested for maternal responsiveness during the postpartum period. The anterior cuts, although having no effect on lordosis, did have dramatic effects on maternal responsiveness. Unlike in rats, in which this type of knife cut abolishes maternal behavior (Numan, 1974), the primary effect of anterior cuts was on cannibalistic behavior. All virgin animals with anterior cuts cannibalized pups on postoperative tests. On postpartum tests, approximately half of anterior cut animals retrieved pups and displayed other maternal behaviors. However, all animals cannibalized their entire litter by the first five days postpartum. In contrast, more posterior knife cuts that reduced lordosis behavior had no effect on maternal or cannibalistic behaviors in virgin animals tested with foster pups. During lactation, the posterior cuts appeared to have reduced the otherwise normal level of pup cannibalism, since these animals weaned larger litters than controls.

On the basis of rather indirect data from the rat, it has been proposed that the MPO–MAH connections important for maternal behavior travel in the MFB (Numan,

1974). To test this idea in the hamster, we have evaluated the effects of NL (identical to the anterior cuts above) and FL cuts along the MPO–MAH (Miceli and Malsbury, 1982a). By placing FL cuts along the lateral border of the MFB, we were able to selectively cut MPO–MAH efferents that cross the MFB and lateral preoptic area/hypothalamus (Conrad and Pfaff, 1976a,b) without interrupting MPO–MAH efferent projections through the MFB. We had earlier observed that surgically intact animals allowed access to food pellets inside their cages and thus able to construct and maintain a food hoard cannibalize fewer of their pups than conventionally maintained animals that normally do not have the opportunity to hoard food (Miceli and Malsbury, unpublished). Therefore, we compared the effects of NL and FL cuts in animals with or without food hoards. This additional factor enabled us to determine whether knife cuts that increase pup cannibalism also disrupt food hoarding, and whether knife cuts abolish an apparent regulation of litter size in accordance with food supply. With the exception of nest building, NL and FL rostral cuts were equally effective in altering pup-directed behavior. In postoperative tests, nonlactating knife-cut animals were more likely to cannibalize foster young than controls. This effect was less consistent (significantly so) if the knife-cut animals were allowed access to a food hoard. During postpartum testing, knife-cut animals were not reliably different from sham operates in 8-min tests of maternal behavior on days 1 and 2 of lactation. However, the majority of food-rationed, knife-cut animals cannibalized their entire litter during the remainder of lactation. On the other hand, knife-cut animals with access to a food hoard cannibalized fewer pups than their rationed counterparts, but reared smaller litters than corresponding surgical controls. These findings suggest that preoptic-hypothalamic knife cuts in hamsters interrupt a fiber pathway(s) that is directly related to the suppression of pup cannibalism and not maternal behavior per se. In addition, the factorial design enabled us to conclude that knife cuts do not abolish an apparent regulation of litter size as a function of feeding condition, since access to a hoard did partially ameliorate the effects of both NL and FL cuts. Lastly, that FL cuts, which selectively sever MPO–MAH efferents through the FL preoptic area/hypothalamus, are as effective as NL cuts in disrupting pup directed behavior (nest building being the only exception) suggests that at least one subset of MPO–MAH efferents important for this behavior in this species travels via a far lateral trajectory (see Fig. 6). The importance of this FL pathway for maternal behavior is also suggested by the results of our recent study of the rat (Miceli *et al.*, 1983).

4.3. Summary

Chemosensory pup stimuli, processed primarily in the accessory olfactory system, may normally facilitate cannibalistic behavior and perhaps as an indirect result of this inhibit the expression of maternal responsiveness (Marques, 1979). However, olfactory bulbectomy results in highly cannibalistic behavior (Marques, 1979) even under optimal feeding and testing conditions (Miceli, unpublished), an effect unlike that produced by peripheral sensory deafferentation. This provides a clear example of a nonsensory deficit produced by bulbectomy in the hamster (Cain, 1974; Edwards, 1974). Nest building requires no obvious chemosensory stimulation and the bulbectomy-produced deficits in this behavior again suggest that bulbectomy produces a general disruption of limbic system function (Cain, 1974). Thus, the olfactory bulbs function in a manner more complex than just processing pup chemosensory stimuli in the control of pup-directed behavior. The integrity of the lateral connections of the MPO–MAH appear to be critical for adequate suppression of pup cannibalism, especially during the postpartum period. Our recent data indicate that the lateral connections of the MPO–MAH that cross the MFB play an important, and previously unsuspected, role in hamster and rat maternal behavior. Hopefully, further work will determine the neuroanatomical identity and function of MPO–MAH axons whose integrity is necessary for normal maternal behavior.

5. CONCLUSIONS

Figure 6 summarizes our current hypotheses concerning some of the neural pathways important for hamster reproductive behaviors. Three schematic horizontal sections of the brain depict connections that may be important for male mating and ultrasonic calls (on the left), maternal behaviors (in the middle section), and lordosis (on the right). The positions of NL and FL sagittal-plane cuts that affect ultrasounds, maternal behaviors, and lordosis, as demonstrated in our laboratory, are also indicated.

5.1. Male Mating (Fig. 6)

Lehman and Winans (unpublished) have shown that combined cuts of the stria terminalis (not shown) and VAF (4) can disrupt male hamster sexual behavior. Knife-cut studies done in rats indicate the critical importance of MPO–MAH connections with the MFB (2). However the effects of such cuts (NL) on male hamster mating have not yet been described. These data suggest that pathways 2 and 4 are important for male mating behavior in the hamster.

5.2. Female Sexual Behaviors (Fig. 6)

Studies from our laboratory have demonstrated the importance of the lateral projections of the VMN region for lordosis. These projections leave the hypothalamus in the region of the SOC as indicated in the diagram on the right. Both NL and FL cuts at the level of the MAH–VMN interrupt this pathway and produce similar lordosis deficits (Malsbury et al., 1979; Ostrowski et al., 1981). More rostrally placed NL and FL cuts, as shown in the diagram on the left, reduce the number of ultrasonic vocalizations emitted by receptive females following exposure to a male (Miceli and Malsbury, 1982b). This may be due to interruption of the VAF (4) that may normally carry chemosensory information to the BNST that could then influence the MPO–MAH. Medial preoptic area–MAH lesions reduce ultrasonic vocalizations in receptive female hamsters (Merkle and Floody, 1979). In the rat, electrical stimulation along the periventricular pathway (3) and central gray elicits ultrasounds (Yajima *et al.*, 1980). These data suggest that pathways 3 and 4 may be important for ultrasonic calls.

5.3. *Maternal Behaviors (Fig. 6)*

Both NL and FL cuts at the level of the MPO–MAH, as shown in the middle diagram, can disrupt maternal behavior in hamsters in that they increase pup cannibalism (Miceli and Malsbury, 1982a). We believe this may be due to interrupting the MPO–MAH projections that travel laterally just above the optic chiasm (1). The VAF (4) is probably also interrupted by these cuts. However, it is not clear how interruption of the VAF might contribute to this effect. Near-lateral cuts produced more severe deficits in nest building in our study, and this may be due to interruption of MPO–MAH connections with the MFB (2). Whether these MFB connections are afferents to or efferents of the MPO–MAH is not known.

Figure 6 provides a rough visualization of some of the pathways through which steroid-sensitive neurons of the forebrain might modulate reproductive behaviors, since the origins of all of the pathways shown contain neurons that concentrate gonadal steroids as demonstrated in autoradiographic studies (Krieger *et al.*, 1976; Doherty and Sheridan, 1981). Whether any of the long-axon projection neurons in these areas are themselves hormone sensitive is not yet known (compare Morrell and Pfaff, 1982). If not, hormone-sensitive interneurons could serve to influence the function of the projection pathways indicated.

ACKNOWLEDGMENTS. Most of the research from Dr. Malsbury's laboratory has been supported by grants from the National Institute of Mental Health since 1976. Recent work has also been supported by Memorial University of Newfoundland and the Natural Sciences and Engineering Research Council of Canada (NSERC), grant no. A7907. Mario Miceli is the holder of an NSERC postgraduate fellowship. The authors would like to thank Michael Lehman, Sarah Winans and J. B. Powers for supplying us with unpublished manuscripts describing their recent studies of male hamster sexual behavior, and Genevieve Bouzane for her excellent help in preparing this ms. Special thanks go to Dr. Winans for providing us with Fig. 1, and Dr. Powers for his comments on an earlier version of the ms.

REFERENCES

Bard, P., 1940, The hypothalamus and sexual behavior, *Res. Publ. Assoc. Res. Nerv. Ment. Dis.* 20:551–579.

Barfield, R. J., Auerbach, P., Geyer, L., and McIntosh, T. K., 1979, Ultrasonic vocalizations in rat sexual behavior, *Am. Zool.* 19:469–480.

Bean, N. J., 1982, Olfactory and vomeronasal mediation of ultrasonic vocalizations in male mice, *Physiol. Behav.* 28:31–37.

Bergondy, M. L., Winans, S. S., and Powers, J. B., 1982, Copulatory and chemoinvestigatory deficits following medial preoptic area lesions in male golden hamsters, *Soc. Neurosci. Abstr.* 8:972.

Brookhart, J. M., Dey, F. L., and Ranson, S. W., 1940, Failure of ovarian hormones to cause mating reactions in spayed guinea pigs with hypothalamic lesions, *Proc. Soc. Exp. Biol. Med.* 44:61–64.

Cain, D. P., 1974, The role of olfactory bulb in limbic mechanisms, *Psychol. Bull.* 81:654–671.

Carrer, H. F., 1978, Mesencephalic participation in the control of sexual behavior in the female rat, *J. Comp. Physiol. Psychol.* 92:877–887.

Carter, C. S., 1973, Olfaction and sexual receptivity in the female golden hamster, *Physiol. Behav.* 10:47–52.

Conrad, L. C. A., and Pfaff, D. W., 1976a, Efferents from the medial basal forebrain and hypothalamus in the rat. I. An autoradiographic study of the medial preoptic area, *J. Comp. Neurol.* 169:185–220.

Conrad, L. C. A., and Pfaff, D. W., 1976b, Efferents from the medial basal forebrain and hypothalamus in the rat. II. An autoradiographic study of the anterior hypothalamus, *J. Comp. Neurol.* 169:221–262.

Davis, B. J., Macrides, F., Youngs, W. M., Schneider, S. P., and Rosene, D. L., 1978, Efferents and centrifugal afferents of the main and accessory olfactory bulbs in the hamster, *Brain Res. Bull.* 3(1):59–72.

Davis, P. G., McEwen, B. S., and Pfaff, D. W., 1979, Localized behavioral effects of tritiated estradiol implants in the ventromedial hypothalamus of female rats, *Endocrinology* 104:898–903.

DeBold, J. F., and Malsbury, C. W., 1983, Inhibition of sexual receptivity after intracranial cycloheximide infusions in female hamsters, *Brain Res. Bull.* 11:633–636.

DeBold, J. F., Malsbury, C. W., Harris, V. S., and Malenka, R., 1982, Sexual receptivity: Brain sites of estrogen action in female hamsters, *Physiol. Behav.* 29:589–593.

Devor, M., 1973, Components of mating dissociated by lateral olfactory tract transection in male hamsters, *Brain Res.* 64:437–441.

Devor, M., and Murphy, M. R., 1973, The effect of peripheral olfactory blockade on the social behavior of the male golden hamster, *Behav. Biol.* 9:31–42.

Doherty, P. C., and Sheridan, P. J., 1981, Uptake and retention of androgen in neurons of the brain of the golden hamster, *Brain Res.* 219:327–334.

Doty, R. L., and Anisko, J. J., 1973, Procaine hydrochloride olfactory block eliminates mounting in the male golden hamster, *Physiol. Behav.* 10:395–398.

Edwards, D. A., 1974, Non-sensory involvement of the olfactory bulbs in the mediation of social behaviors, *Behav. Biol.* 11:287–302.

Edwards, D. A., and Pfeifle, J. K., 1981, Hypothalamic and midbrain control of sexual receptivity in the female rat, *Physiol. Behav.* 26:1061–1067.

Emery, D. E., and Sachs, B. D., 1976, Copulatory behavior in male rats with lesions in the bed nucleus of the stria terminalis, *Physiol. Behav.* 17:803–806.

Floody, O. R., 1979, Behavioral and physiological analyses of ultrasound production by female hamsters (Mesocricetus auratus), *Am. Zool.* 19:443–456.

Floody, O. R., 1981, The hormonal control of ultrasonic communication in rodents, *Am. Zool.* 21:129–142.

Floody, O. R., 1983, Lesions of the ventromedial hypothalamus increase rates of ultrasonic vocalization in male and female hamsters, *Soc. Neurosci. Abstr.* 9:1079.

Floody, O. R., and O'Donohue, T. L., 1980, Lesions of the mesencephalic central gray depress ultrasound production and lordosis by female hamsters, *Physiol. Behav.* 24:79–85.

Goodman, E. D., and Firestone, M. I., 1973, Olfactory bulb lesions: Nest reinforcement and handling reactivity in hamsters, *Physiol. Behav.* 10:1–8.

Hansen, S., Stanfield, E. J., and Everitt, B. J., 1981, The effects of lesions of lateral tegmental noradrenergic neurons on components of sexual behaviour and pseudopregnancy in female rats, *Neuroscience* 6:1105–1118.

Hansen, S., Kohler, Ch., Goldstein, M., and Steinbusch, H. V. M., 1982, Effects of ibotenic acid-induced neuronal degeneration in the medial preoptic area and the lateral hypothalamic area on sexual behavior in the male rat, *Brain Res.* 239:213–232.

Harris, V. S., and Sachs, B. D., 1975, Copulatory behavior in male rats following amygdaloid lesions, *Brain Res.* 86:514–518.

Hilton, S. M., and Zbrozyna, A. W., 1963, Amygdaloid region for defense reactions and its efferent pathway to the brainstem, *J. Physiol.* 165:160–173.

Jurgens, U., 1982, Amygdalar vocalization pathways in the squirrel monkey, *Brain Res.* 241:189–197.

Kairys, D. J., Magalhaes, H., and Floody, O. R., 1980, Olfactory bulbectomy depresses ultrasound production and scent marking by female hamsters, *Physiol. Behav.* 25:143–147.

Kendrick, K. M., 1982, Inputs to testosterone-sensitive stria terminalis neurones in the rat brain and the effects of castration, *J. Physiol.* 323:437–449.

Kendrick, K. M., and Drewett, R. F., 1979, Testosterone reduces refractory period of stria terminalis neurons in the rat brain, *Science* 204:877–878.

Kevetter, G. A., and Winans, S. S., 1981a, Connections of the corticomedial amygdala in the golden hamster: I. Efferents of the "vomeronasal amygdala," *J. Comp. Neurol.* 197:81–98.

Kevetter, G. A., and Winans, S. S., 1981b, Connections of the corticomedial amygdala in the golden hamster: II. Efferents of the "olfactory amygdala," *J. Comp. Neurol.* 197:99–112.

Kow, L.-M., Malsbury, C. W., and Pfaff, D. W., 1976, Lordosis in the male golden hamster elicited by manual stimulation: characteristics and hormonal sensitivity, *J. Comp. Physiol. Psychol.* 90:26–40.

Krettek, J. E., and Price, J. L., 1977, Projections from the amygdaloid complex to the cerebral cortex and thalamus in the rat and cat, *J. Comp. Neurol.* 172:687–722.

Krieger, M. S., Morrell, J. I., and Pfaff, D. W., 1976, Autoradiographic localization of estradiol-concentrating cells in the female hamster brain, *Neuroendocrinology* 22:193–205.

Krieger, M. S., Conrad, L. C. A., and Pfaff, D. W., 1979, An autoradiographic study of the efferent connections of the ventromedial nucleus of the hypothalamus, *J. Comp. Neurol.* 183:785–816.

Larsson, K., 1979, Features of the neuroendocrine regulation of masculine sexual behavior, in: *Endocrine Control of Sexual Behavior* (C. Beyer, ed.), Raven Press, New York, pp. 77–163.

Lehman, M. N., and Winans, S. S., 1982, Vomeronasal and olfactory pathways to the amygdala controlling male hamster sexual behavior: Autoradiographic and behavioral analyses, *Brain Res.* 240:27–41.

Lehman, M. N., and Winans, S. S., 1983, Evidence for a ventral non-strial pathway from the amygdala to the bed nucleus of the stria terminalis in the male golden hamster, *Brain Res.* 268:139–146.

Lehman, M. N., Winans, S. S., and Powers, J. B., 1980, Medial nucleus of the amygdala mediates chemosensory control of male hamster sexual behavior, *Science* 210:557–560.

Lehman, M. N., Powers, J. B., and Winans, S. S., 1983, Stria terminalis lesions alter the temporal pattern of copulatory behavior in the male golden hamster, *Behav. Brain Res.* 8:109–128.

Leonard, C. M., 1972, Effects of neonatal (day 10) olfactory bulbectomy on social behavior in the female golden hamster (Mesocricetus auratus), *J. Comp. Physiol. Psychol.* 80:208–215.

Lisk, R. D., Zeiss, J., and Ciaccio, L. A., 1972, The influence of olfaction on sexual behavior in the male golden hamster (Mesocricetus auratus), *J. Exp. Zool.* 181:69–78.

Macrides, F., Firl, A. C., Jr., Schneider, S. P., Bartke, A., and Stein, D. G., 1976, Effects of one-stage or serial transections of the lateral olfactory tracts on behavior and plasma testosterone levels in male hamsters, *Brain Res.* 109(1):97–110.

Malsbury, C. W., and Daood, J. T., 1978, Sexual receptivity: Critical importance of supraoptic connections of the ventromedial hypothalamus, *Brain Res.* 159:451–457.

Malsbury, C. W., and Pfaff, D. W., 1974, Neural and hormonal determinants of mating behavior in adult male rats. A review, in: *Limbic and Autonomic Nervous Systems Research* (L. V. DiCara, ed.), Plenum Press, New York, pp. 85–136.

Malsbury, C. W., Kow, L.-M., and Pfaff, D. W., 1977, Effects of medial hypothalamic lesions on the lordosis response and other behaviors in female golden hamsters, *Physiol. Behav.* 19:223–237.

Malsbury, C. W., Strull, D., and Daood, J., 1978, Half-cylinder cuts antero-lateral to the ventromedial nucleus reduce sexual receptivity in female golden hamsters, *Physiol. Behav.* 21:79–87.

Malsbury, C. W., Marques, D. M., and Daood, J. T., 1979, Sagittal knife cuts in the far-lateral hypothalamus reduce sexual receptivity in female hamsters, *Brain Res. Bull.* 4:833–842.

Malsbury, C. W., Pfaff, D. W., and Malsbury, A. M., 1980, Suppression of sexual receptivity in the female hamster: Neuroanatomical projections from medial preoptic and anterior hypothalamic electrode sites, *Brain Res.* 181:267–284.

Manogue, K. R., Kow, L.-M., and Pfaff, D. W., 1980, Selective brainstem transections affecting reproductive behavior of female rats: The role of hypothalamic output to the midbrain, *Horm. Behav.* 14:277–302.

Marques, D. M., 1979, Roles of the main olfactory and vomeronasal systems in the response of the female hamster to young, *Behav. Neural Biol.* 26:311–329.

Marques, D. M., Malsbury, C. W., and Daood, J. T., 1979, Hypothalamic knife cuts dissociate maternal behaviors, sexual receptivity and estrous cyclicity in female hamsters, *Physiol. Behav.* 23:347–355.

Marques, D. M., O'Connell, R. J., Benimoff, N., and Macrides, F., 1982, Delayed deficits in behavior after transection of the olfactory tracts in hamsters, *Physiol. Behav.* 28:353–365.

Matthews, D., and Edwards, D. A., 1977, Involvement of the ventromedial and anterior hypothalamic induction of receptivity in the female rat, *Physiol. Behav.* 19:319–326.

Meredith, M., 1980, The vomeronasal organ and accessory olfactory system in the hamster, in: *Chemical Signals, Vertebrates, and Aquatic Invertebrates* (D. Muller-Schwarze and R. M. Silverstein, eds.), Plenum Press, New York, pp. 303–326.

Meredith, R. M., and O'Connell, R. J., 1979, Efferent control of stimulus access to the hamster vomeronasal organ, *J. Physiol.* 286:301–316.

Meredith, M., Marques, D. M., O'Connell, R. J., and Stern, F. L., 1980, Vomeronasal pump: Significance for male hamster sexual behavior, *Science* 207:1224–1226.

Merkle, D. A., and Floody, O. R., 1979, Preoptic and VMH lesions separate proceptive and receptive components of female sex behaviors, *Soc. Neurosci. Abstr.* 5:470.

Mesulam, M.-M., 1978, Tetramethyl benzidene for horseradish peroxidase neurohistochemistry: A non-carcinogenic blue reaction-product with superior sensitivity for visualizing neural afferents and efferents, *J. Histochem. Cytochem.* 26:106–117.

Miceli, M. O., and Malsbury, C. W., 1982a, Sagittal knife cuts in the near and far lateral preoptic area-hypothalamus disrupt maternal behaviour in female hamsters, *Physiol. Behav.* 28:857–867.

Miceli, M. O., and Malsbury, C. W., 1982b, Sagittal knife cuts in the near and far lateral preoptic area-hypothalamus reduce ultrasonic vocalizations in female hamsters, *Physiol. Behav.* 29:953–956.

Micéli, M. O., Fleming, A. S., and Malsbury, C. W., 1983, Disruption of maternal behavior in virgin and postparturient rats following sagittal plane knife cuts in the preoptic area-hypothalamus, *Behav. Brain Res.* 9:337–360.

Morrell, J. I., and Pfaff, D. W., 1982, Characterization of estrogen-concentrating hypothalamic neurons by their axonal projections, *Science* 217:1273–1276.

Morrell, J. I., Greenberger, L. M., and Pfaff, D. W., 1981, Hypothalamic, other diencephalic, and telencephalic neurons that project to the dorsal midbrain, *J. Comp. Neurol.* 201:589–620.

Muntz, J. A., Rose, J. D., and Shults, R. C., 1980, Disruption of lordosis by dorsal midbrain lesions in the golden hamster, *Brain Res. Bull.* 5:359–364.

Murphy, M. R., 1976, Olfactory impairment, olfactory bulb removal, and mammalian reproduction, in: *Mammalian Olfaction, Reproductive Processes and Behavior* (R. L. Doty, ed.), Academic Press, New York, pp. 96–118.

Murphy, M. R., and Schneider, G. E., 1970, Olfactory bulb removal eliminates mating behavior in the male golden hamster, *Science* 167:302–304.

Nance, D. M., Shryne, J., and Gorski, R. A., 1974, Septal lesions: Effects on lordosis behavior and pattern of gonadotropin release, *Horm. Behav.* 5:73–81.

Noble, R. G., 1979, The sexual responses of the female hamster: A descriptive analysis, *Physiol. Behav.* 23:1001–1006.

Numan, M., 1974, Medial preoptic area and maternal behavior in the female rat. *J. Comp. Physiol. Psychol.* 87:746–759.

Ostrowski, N. L., Scouten, C. W., and Malsbury, C. W., 1981, Reduced lordosis response following unilateral hypothalamic knife cuts, *Physiol. Behav.* 27:323–329.

Paxinos, G., 1976, Interruption of septal connections: Effects on drinking, irritability and copulation, *Physiol. Behav.* 17:81–88.

Paxinos, G., Emson, P. C., and Cuello, A. C., 1978, Substance P projections to the entopeduncular nucleus, the medial preoptic area and the lateral septum, *Neurosci. Lett.* 7:133–136.

Pfaff, D. W., 1980, *Estrogens and Brain Function,* Springer-Verlag, New York.

Pfaff, D. W., and Keiner, M., 1973, Atlas of estradiol-concentrating cells in the central nervous system of the female rat, *J. Comp. Neurol.* 151:121–158.

Pfaff, D., and Sakuma, Y., 1979, Deficit in the lordosis reflex of female rats caused by lesions in the ventromedial nucleus of the hypothalamus, *J. Physiol.* 288:203–210.

Pfeifle, J. K., Shivers, M., and Edwards, D. A., 1980, Parasagittal hypothalamic knife cuts and sexual receptivity in the female rat, *Physiol. Behav.* 24:145–150.

Phelps, C. P., and Nance, D. M., 1979, Sexual behavior and neural degeneration following hypothalamic knife cuts, *Brain Res. Bull.* 4(3):423–430.

Phelps, C. P., Nance, D. M., and Saporta, S., 1980, Reduced ovarian function and lordosis behavior following hypothalamic knife cuts, *Brain Res. Bull.* 5:531–537.

Potegal, M., Blau, A., and Glusman, M., 1981, Effects of anteroventral septal lesions on intraspecific aggression in male hamsters, *Physiol. Behav.* 26:407–412.

Powers, J. B., and Winans, S. S., 1973, Sexual behavior in peripherally anosmic male hamsters, *Physiol. Behav.* 10:361–368.

Powers, J. B., and Winans, S. S., 1975, Vomeronasal organ: critical role in mediating sexual behavior of the male hamster, *Science* 187:961–963.

Price, J. L., and Slotnick, B. M., 1983, Dual olfactory representation in the rat thalamus: An anatomical and electrophysiological study, *J. Comp. Neurol.* 215:63–77.

Reep, R. L., and Winans, S., 1982, Efferent connections of dorsal and ventral agranular insular cortex in the hamster, *Mesocricetus auratus, Neuroscience* 7:2609–2636.

Rose, J. D., 1978, Midbrain and pontine unit responses to lordosis-controlling forms of somatosensory stimuli in the female golden hamster, *Exp. Neurol.* 60(3):499–508.

Rose, J. D., 1982, Midbrain distribution of neurons with strong, sustained responses to lordosis trigger stimuli in the female golden hamster, *Brain Res.* 240:364–367.

Rubin, B. S., and Barfield, R. J., 1980, Priming of estrous responsiveness by implants of 17-beta-estradiol in the ventromedial hypothalamic nucleus of female rats, *Endocrinology* 106:504–509.

Sakuma, Y., and Pfaff, D. W., 1979, Mesencephalic mechanisms for integration of female reproductive behavior in the rat, *Am. J. Physiol.* 237:R285–R290.

Saper, C. B., Swanson, L. W., and Cowan, W. M., 1976, The efferent connections of the ventromedial nucleus of the hypothalamus of the rat, *J. Comp. Neurol.* 169(4):409–442.

Sapolsky, R. M., and Eichenbaum, H., 1980, Thalamocortical mechanisms in odor-guided behavior. II. Effects of lesions of the mediodorsal thalamic nucleus and frontal cortex on odor preferences and sexual behavior in the hamster, *Brain Behav. Evol.* 17:276–290.

Scalia, F., and Winans, S. S., 1976, New perspectives on the morphology of the olfactory system: Olfactory and vomeronasal pathways in mammals, in: *Mammalian Olfaction, Reproductive Processes and Behavior* (R. L. Doty, ed.), Academic Press, New York, pp. 8–28.

Scouten, C. W., and Malsbury, C. W., 1981, Labeling knife cuts used to trace a pathway necessary for lordosis in the hamster, *Soc. Neurosci. Abstr.* 7:753.

Scouten, C. W., Harley, C. W., and Malsbury, C. W., 1982, Labeling knife cuts: A new method for revealing the functional anatomy of the CNS demonstrated on the noradrenergic dorsal bundle, *Brain Res. Bull.* 8:229–232.

Siegel, H. I., and Rosenblatt, J. S., 1980, Hormonal and behavioral aspects of maternal care in the hamster: A review, *Neurosci. Biobehav. Rev.* 4:17–26.

Stumpf, W. E., Sar, M., and Keefer, D. A., 1975, Atlas of estrogen target cells in rat brain, in: *Anatomical Neuroendocrinology* (W. E. Stumpf and L. D. Grant, eds.), S. Karger AG, Basel, Switzerland, pp. 104–119.

Vomachka, A. J., Richards, N. R., II, and Lisk, R. D., 1982, Effects of septal lesions on lordosis in female hamsters, *Physiol. Behav.* 29:1131–1135.

Wetzel, D. M., Kelley, D. B., and Campbell, B. A., 1980, Central control of ultrasonic vocalizations in neonatal rats: I. Brainstem motor nuclei, *J. Comp. Physiol. Psychol.* 94:596–605.

Winans, S. S., and Powers, J. B., 1977, Olfactory and vomeronasal deafferentation of male hamsters: Histlogical and behavioral analyses, *Brain Res.* 126:325–344.

Wysocki, C. J., 1979, Neurobehavioral evidence for the involvement of the vomeronasal system in mammalian reproduction, *Neurosci. Biobehav. Rev.* 3:301–341.

Yajima, Y., Hayashi, Y., and Yoshii, N., 1980, The midbrain central gray substance as a highly sensitive neural structure for the production of the ultrasonic vocalization in the rat, *Brain Res.* 198:446–452.

Yajima, Y., Hayashi, Y., and Yoshii, N., 1981, Identification of ultrasonic vocalization substrates determined by electrical stimulation applied to the medulla oblongata in the rat, *Brain Res.* 229:353–363.

Yajima, Y., Hayashi, Y., and Yoshii, N., 1982, Ambiguus motoneurons discharging closely associated with ultrasonic vocalizations in rats, *Brain Res.* 238:445–451.

Yamanouchi, K., and Arai, Y., 1979, Effects of hypothalamic deafferentation on hormonal facilitation of lordosis in ovariectomized rats, *Endocrinol. Jpn.* 26:307–312.

Zasorin, N. L., Malsbury, C. W., and Pfaff, D. W., 1975, Suppression of lordosis in the hormone-primed female hamster by electrical stimulation of the septal area, *Physiol. Behav.* 14:595–600.

12

Aggressive Behavior

HAROLD I. SIEGEL

1. INTRODUCTION

The study of aggressive behavior in hamsters has focused on five major areas and this chapter is organized accordingly. The first topic involves a description of the aggressive and submissive behaviors associated with intraspecific encounters. As will be seen, the effectiveness of a variety of experimental treatments is in part determined by which behaviors are chosen for measurement. The second section discusses the different testing conditions that have been used because these too represent potential sources of discrepancy among investigators. The third section analyzes the role of hormones, those administered exogenously and those secreted normally during the reproductive cycle. The effect of damage to central nervous system structures is the fourth topic and the final section includes peripheral influences (e.g., role of the flank glands, harderian glands, and so forth).

The effect of neonatal hormone manipulations on aggressive behavior in adulthood is covered in Chapter 7 (this volume). Also not covered in this chapter is interspecific aggression and the reader is referred to the experiments by Van Hemel (1977) and the extensive series of studies by Polsky (1974, 1976, 1977a,b, 1978a,b,c).

2. BEHAVIORAL DESCRIPTIONS

There have been several detailed descriptions of agonistic behavior in hamsters (Dieterlin, 1959; Grant and Mackintosh, 1963; Lerwill and Makings, 1971; Floody and Pfaff, 1977). Figure 1 is adapted from Grant and Mackintosh and illustrates many of the behaviors that have been quantified in a large number of studies of aggressive behavior.

HAROLD I. SIEGEL • Institute of Animal Behavior and Department of Psychology, University College, Rutgers University, Newark, New Jersey 07102

Figure 1. Social postures shown by hamsters. Clockwise from top left: left, nose—right, nose; left, submissive posture—right, offensive upright posture; left, offensive upright posture—right, defensive sideways posture; left, defensive upright posture—right, offensive sideways posture; right, defensive upright posture. [Reprinted with permission of the publisher from Grant and Mackintosh (1963).]

Behaviors classified as Introductory Acts include Attend, Approach, Investigate, and Nosing and are restricted to whisker contact. There are Upright and Sideways postures, each with offensive and defensive subdivisions. The Offensive Upright and Sideways postures are characterized by the animal holding its forepaws apart displaying its dark ventral markings. In the Defensive postures, the position of the forepaws precludes the display of the chest patches (but see final section on Chest Patches). The behaviors labeled Flag and Evade are head and forebody movements away from the opponent, respectively, whereas Threat and Thrust are similar movements made toward the other animal. Finally, in the Full Submissive posture, the animal lies on its back and may remain in this position after the opponent has moved away.

The behaviors described above have been used by Lerwill and Makings (1971) in their analysis of agonistic encounters in male hamsters. A quantitative flow chart was developed based on 28 observations of seven pairs consisting of 239 "resolved encounters" defined when the dominant animal no longer chased the retreating opponent (Fig. 2). The observations were made in the home cage of one member of each pair and except for some minor differences, the behavioral sequences were quite similar for residents and intruders.

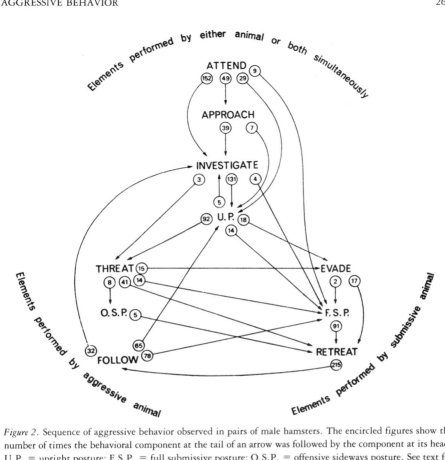

Figure 2. Sequence of aggressive behavior observed in pairs of male hamsters. The encircled figures show the number of times the behavioral component at the tail of an arrow was followed by the component at its head. U.P. = upright posture; F.S.P. = full submissive posture; O.S.P. = offensive sideways posture. See text for further details. [Reprinted with permission of the publisher from Lerwill and Makings (1971).]

Several other aspects of Fig. 2 should be noted. Investigate includes Nosing, Follow includes Chase, and the Upright posture is a combination of both offensive and defensive elements shown by the dominant and submissive animals, respectively.

The authors reported that dominance and submission were established quickly in the "sparring phase" associated with the display of the various postures. Interestingly, there was no home cage advantage in that the intruders became dominant in 140 of the 239 recorded encounters. Actual attacks with biting were rarely seen and instead there were many instances of "premature retreats" that occurred at almost any point in the sequence, particularly after investigation and the adoption of the upright posture of the dominant male. Because these premature retreats were not included, the behavioral sequences are actually more variable than Fig. 2 suggests. The other main source of variability consists of repetitive sequences, especially in the Follow–Chase phase.

The results of more recent studies (Floody and Pfaff, 1977) are presented here to permit a comparison with the methodology and results of Lerwill and Makings (1971).

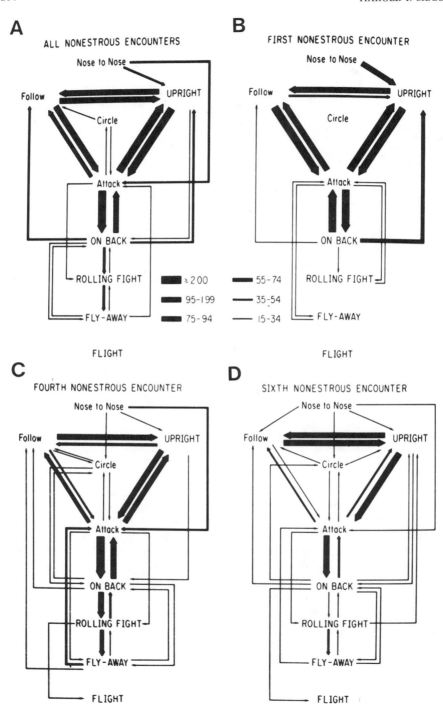

A ALL NONESTROUS ENCOUNTERS

B FIRST NONESTROUS ENCOUNTER

C FOURTH NONESTROUS ENCOUNTER

D SIXTH NONESTROUS ENCOUNTER

In the Floody and Pfaff studies, the subjects were older females (110 to 160 days of age) and were tested in neutral cages. In addition, the pairs were matched in terms of body weight and stage of the estrous cycle and the same pairs were observed during six nonestrous tests. (Differences over the estrous cycle and on the day of behavioral estrus are discussed in Section 4.)

Figure 3, based on the analysis of filmed observations, shows the results of the first, fourth, and sixth encounters and also includes a summary of all six pairings. Clearly, several of the behavioral labels in Fig. 3 differ from those in Fig. 2. The sequences begin with Nosing and move on to either Upright, Follow, or Circle. Upright postures are not classified offensively or defensively, but instead in terms of the position of the upright female's head, i.e., close to the opponent's head or trunk. Follow refers to the slow pursuit of one female by the other whereby the nose of the following animal is in close proximity to the anogenital region of the second female. Circle can be considered mutual following.

Attacks were initiated from the Upright and Sideways postures and resulted in Sideways and Upright postures, respectively, on the part of the opponent. As shown in Fig. 3, attacks often lead to several other behaviors including On Back, Rolling Fight, and Fly Away. On Back refers to an animal lying on its back with its opponent generally in close proximity in the Upright posture. Rolling Fight, also called locked fighting, occurs when the two opponents are wrapped perpendicularly around each other and also includes bitings and vocalizations. Fly Away refers to the rapid moving away of one member of the pair and according to Floody and Pfaff is characteristic of extremely aggressive interactions.

Whereas Attack and On Back were fairly common in the initial encounters, the transition from On Back to more violent behavior (Rolling Fight and Fly Away) was relatively uncommon. In other words, initial elements "recycled" back to milder forms of interaction. The transition from On Back to Rolling Fight and Fly Away was more frequent in the fourth encounter for the pairs. By the sixth pairing, the total number of transitions was greatly reduced; for example, there was an average of 18.8, 8.5, and 3.0 attacks/interaction in the first, fourth, and sixth encounters, respectively.

The authors make the point that the establishment of dominance should be associated with a reduction in the more violent forms of agonistic behavior. If this is correct, then it would appear that dominance was established not on the first encounter but instead during one of the later encounters. The results of Lerwill and Makings (1971) as well as those of many other investigators suggest that dominance is established rapidly during the first encounter of a particular pair.

Thus, although an extended series of encounters may be necessary to evaluate dominance within pairs, Floody and Pfaff go on to argue that the ecology and "social intolerance" of the hamster may preclude such extended interactions in the natural

Figure 3. Mean transitional frequencies (mean number of transitions per pair per encounter) among female hamsters' agonistic postures summarized for (A) all six nonestrous encounters, (B) the first encounter, (C) the fourth encounter, and (D) the sixth encounter. See text for further details. [Reprinted with permission of the publisher from Floody and Pfaff (1977).]

environment. Therefore, the study of the physiological and behavioral determinants involved in the initial encounter may be more ecologically valid.

These two analyses of agonistic behavior illustrate interesting similarities and differences. Both show quantifiable sequences of interacting behavior with plenty of room for flexibility (premature retreats and recycling back to initial behaviors). Generally, the postures shown in Fig. 1 have their place in both of these descriptions as well as in many other studies discussed later. There are, however, several differences, the most striking being the presence or absence of Attack and the more violent On Back, Rolling Fight, and Fly Away. Although it is difficult to specify precisely the reason for the discrepancy, several likely possibilities pertain to the different methodologies.

Lerwill and Makings (1971) used relatively young adult males observed for 3 min in home cage tests, whereas Floody and Pfaff (1977) used older females matched for body weight and stage of the estrous cycle and tested for 10 min in neutral cages. Subjects in both studies were housed individually from weaning or soon thereafter, but because of the age at testing the animals in the Floody and Pfaff study had been caged separately for a longer interval. The different variables incorporated in just these two studies include sex of the subject, sex of the opponent, type of test cage, age, length of social isolation, test duration, single or multiple pairings of the same animals, and matched or nonmatched pairs (body weight). Each of these variables has been shown to affect the course of agonistic encounters. In addition, there are other factors that can contribute to the results of studies on aggression; for example, Lerwill and Makings (1971) tested their animals at various times during the light–dark cycle and reported higher levels of interaction during the dark phase. Before discussing the role of hormones and neural and peripheral structures, the next section analyzes methodological variables.

3. METHODOLOGICAL CONSIDERATIONS

The aim of this section is not to exhaustively discuss each variable that has been shown to affect the results of agonistic encounters. There are, however, a number of variables that have relatively clear effects. Further, the direction of the effect of one variable may be affected in different ways by the incorporation of other test conditions. These considerations are important to keep in mind when generalizing from one study to another, attempting to resolve interlaboratory discrepancies, and designing future experiments.

3.1. Sex of the Opponents

Although the effects of organ removal and exogenous hormone treatments are discussed in the next section, it is generally true that females are more aggressive than males. However, this statement has two parts. The first issue is how do males and females compare when paired together and the second issue is whether female–female pairs display more or less aggression than male–male pairs. Most often, nonestrous females show more aggression and are dominant over males when placed in mixed sex pairs (Payne and Swanson, 1970). These same investigators compared same-sex pairs and showed that there were no differences in the number of tests with sparring only, number of tests with sparring and fights, number of decisive interactions per test, or the number of tests in which the same opponent won all of the interactions.

Marques and Valenstein (1977) showed that when matched for body weight, females were generally dominant over males although the incidence of aggression was relatively low. It was concluded that the female determines the level of aggression and the level is heavily influenced by the sex and reproductive condition of the opponent. Tiefer (1970) reported that although females were generally more aggressive, both males and females initiated similar numbers of fights when each was paired with a male opponent.

In a somewhat unusual laboratory situation, Goldman and Swanson (1975) placed a pregnant animal in each of four enclosures measuring 80 × 90 cm. After eight weeks the mothers were removed, and one week later the connecting doors between the enclosures were opened. At this point, females accounted for more than 90% of all attacks and of these attacks, 65% were directed at males. At 120 days of age, the pattern of aggression shifted in that males initiated 90% of the attacks and 80% of these were still directed at other males. Although this situation is quite unusual, the results demonstrate that either sex can become highly aggressive depending on the experimental conditions.

Taken together, these studies illustrate that although intact females are more aggressive than males, the particular conditions of each study and of the interacting animals must be considered. Types of subjects and their opponents have included intact, gonadectomized, and bulbectomized males and females, nonestrous and estrous females, as well as animals that have been muzzled and tethered. The results of studies involving these types of animals will be discussed later.

3.2. Housing Conditions

The effects of housing animals individually for some period of time prior to testing has typically been shown to increase submissive distance and aggressiveness although not as consistently in males. Brain (1972) specifically compared males and females that had been isolated or grouped (same sex, ten animals per cage) for more than 60 days. Greater percentages of isolated males and females displayed aggressive behavior and these animals also showed a greater number of attacks, an increase in total attack time, and a decreased latency to attack. Isolated females were more aggressive than isolated males and there were no differences between group-housed males and females.

Wise (1974) showed that isolation for more than 29 days was associated with increased pouncing and fighting among estrous cycling animals and Huang and Hazlett (1974) showed that the distance at which five-week isolate males showed submissive behavior was greater than that in animals housed four per cage. Grelk et al. (1974) found that isolated, ovariectomized females showed more aggression toward ovariectomized, previously grouped females than nonisolated ovariectomized females.

No differences were found in the number of fights between grouped and isolated castrated males when paired with group-housed castrated males. Interestingly, estradiol benzoate (EB) and testosterone increased the amount of fighting in grouped females but not in isolated females. These same hormones increased fighting in both grouped and isolated males.

3.3. Body Weight

Body weight is often not a factor in agonistic encounters because many investigators closely match the opponents' weights. When examined experimentally, a number of studies have demonstrated that heavier animals maintain an advantage in establishing dominance and displaying higher levels of aggression relative to their lighter opponents (Boice et al., 1969; Drickamer et al., 1973; Drickamer and Vandenbergh, 1973; Vandenbergh, 1971). More specifically, Payne and Swanson (1970) found correlation coefficients of 0.76 for males and 0.91 for females between dominant interactions and body weights. Marques and Valenstein (1977) showed that when matched for body weight, females were dominant over males in 5 of 7 cases (two were neutral). When heavier, the typically less aggressive male became dominant in 4 of 10 cases (the lighter females were dominant in 3 of 10 cases and the remaining ones were classified as neutral).

Not all studies have demonstrated an advantage for the heavier animal. Other variables may possibly interact with body weight as a determinant of dominance and levels of aggression. Garrett and Campbell (1980) found that lighter males exposed to 6-hr light–18-hr dark photoperiods and undergoing testicular regression became dominant over males who had previously defeated them under 14-hr light–10-hr dark conditions.

3.4. Age

In general, young animals are not especially aggressive and some investigators have used these animals as opponents specifically for that reason (Brain, 1972). Dieterlin (1959) reported that 30- to 40-day-old males were observed defending their living quarters. In the most systematic study, Whitsett (1975) began testing males at 30 days of age following either castration or sham surgery at 24 days. These animals were tested with the same opponents every five days until they reached 50 days of age and then with different opponents every ten days until the age of 80 days. Dominance was established by 35 days of age and remained stable, whereas the mean number of attacks per pair per test increased more than fourfold over the course of the study.

3.5. Other Subject Variables

A number of other factors related to the opponents can affect the results of aggressive encounters. The animal's prior history in aggression tests can influence its subsequent level of aggression. Females' prior displays of dominance or neutrality remained constant when these animals were later tested with intact males, although not with females or castrated males (Marques and Valenstein, 1977).

On the basis of previous tests, Murphy (1976a) selected males that showed either high or low levels of aggressive behavior. The less aggressive males were then bilaterally bulbectomized to further reduce their aggression. This approach was further modified by Potegal et al. (1980a,b) to separate the aggression of one opponent from that influenced by the behavior of the other animal. Males were initially chosen on the basis of their levels of aggression. The more aggressive males were then individually housed to increase their aggressiveness and the less aggressive males were then group housed, bilaterally bulbectomized, and tethered in the testing cage. Among other findings to be discussed later, the use of these standardized opponents allowed the investigators to distinguish between the effects of androgen on agonistic postures from those on overt attacks.

3.6. Test Cage

As mentioned in the previous section, Lerwill and Makings (1971) found that intruders were most successful in establishing dominance than their home cage opponents. More commonly, however, the evidence suggests a relatively strong home cage advantage.

Murphy (1976a) used males who had already displayed higher levels of aggression and they showed longer attack durations when tested in their home cages than when tested in neutral cages. Castrated males are generally less aggressive than intact males, but show more aggression when tested in their home cages with intact intruders (Payne, 1973, 1974a,b). Similarly, intact males are less aggressive than females in neutral cages but increase their aggressiveness in their home cages when presented with female intruders. Only one of a total of more than 1200 attacks recorded in encounters between lactating animals tested in their home cages with cycling females was initiated by the intruder (Siegel et al., 1983). Finally, Grant et al. (1970) showed that the effect of coloring the chest patches and forearms with black dye depended on the test cage. The dyed animal rarely initiated aggression when tested as the intruder, but was more likely to display aggressive behavior when tested as the home cage resident.

3.7. Other Test Condition Variables

Lerwill and Makings (1971) reported that the frequency and duration of aggressive behavior were higher during the dark phase of the photoperiod. The results of Landau (1975a,b) will be discussed later, but he also demonstrated that aggression levels were greater during the early part of the dark phase.

Other variables to be considered in future studies include the size of the test cage (Lawlor, 1963) and repeated pairings of the same opponents that may lead to stable dominance relationships (Whitsett, 1975). A related issue is the repeated use of the same intruders; Buntin et al. (1981) used the same males to test more aggressive females and presented these males with receptive females between aggression tests to insure that the males would continue to approach females.

3.8. Measures of Aggression

This section is meant to point out how aggression has been measured. Most investigators record the frequency or duration of one or, more commonly, several of the behaviors described earlier. Latencies to the first display of these behaviors are also often measured. In addition, researchers have sometimes measured the outcome of an encounter or the dominance of one animal over the other. Dominance has been defined in a number of ways, for example: (1) a particular frequency of chasing and biting of an opponent per observation period (Marques and Valenstein, 1977); (2) the display of these behaviors followed by their cessation as an indication that dominance was established (Lerwill and Makings, 1971); (3) the appearance of offensive behavior in one animal accompanied by defensive behavior in the other (Landau, 1975a); or (4) apparatus-related measures such as water or tunnel dominance (Boice et al., 1969).

Investigators have occasionally made certain distinctions in their measures of aggression. As already mentioned, Potegal et al. (1980a,b) have distinguished between overt attacks and the display of aggressive postures. Hammond and Rowe (1976) separately measured low-intensity behaviors (wrestling, sparring, and offensive postures without biting) and high-intensity behaviors (fights, rolling fights with biting, and attack-chase sequences). Particular lesions affected these categories differently.

In studies on the effects of androgen administered in adulthood or during the postnatal period, Payne (1974a, 1976) recorded composite aggressive and defensive scores. Aggressive behavior included upright and sideways offensive postures, fights, bites, chases, and attacks and defensive behavior included upright and sideways postures and behaviors such as evade, flee, and the adoption of the full submissive posture. Whereas composite scores simplify results, Wise (1974) points out that not all measures correlate well with each other and therefore composites may be misleading.

4. HORMONE EFFECTS: FEMALE

4.1. Estrous Cycle

Several studies have examined changes in aggression as a function of the stage of the estrous cycle. These data have also formed the basis of further studies aimed at identifying the hormones and hormonal combinations that contribute to aggressiveness. Counting day 4 of the cycle as the day of behavioral estrus, there is general agreement that this period is associated with less aggression. Under certain conditions, however, sexual receptivity and female aggression are not mutually exclusive. Wise (1974) reported that females can show both aggression and receptivity on particular days of pregnancy and lactation, although the receptivity measured as the duration of the lordotic posture is relatively weak compared with that shown during the estrous cycle. Murphy (1978) presented males, either Syrian or Turkish, to sexually receptive Turkish females. While Turkish males elicited lordosis in 100% of the tests and attacks in 0% of the tests, the Syrian males elicited lordosis and attacks in 20% and 80% of the tests, respectively.

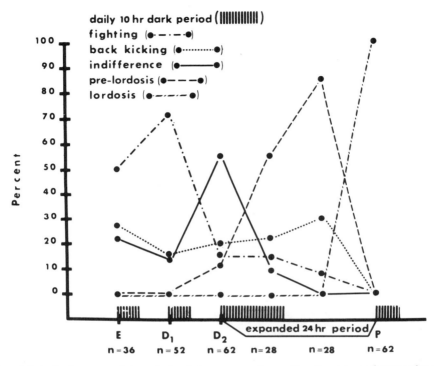

Figure 4. Behavioral responses of the cyclic female hamster toward a sexually active male as a function of stage of the estrous cycle. Animals were tested for mating for 10 min during the dark phase of the photoperiod on each day of the cycle and also 8 and 16 hr prior to the onset of receptivity. [Reprinted with permission of the publisher from Ciaccio *et al.* (1979).]

Although there is a decrease of aggression on the day of receptivity, the relative amount of aggressiveness shown on the other days of the cycle is not clear. Ciaccio *et al.* (1979) and Payne and Swanson (1970) report that the highest levels of aggression are observed on day 1 and 2, i.e., the day of vaginal estrus and the first day of diestrus (Fig. 4). Floody and Pfaff (1977) recorded the longest latencies and the lowest frequencies of aggressive behavior on the day of behavioral estrus, but did not detect quantitative differences over the next three days. Wise (1974) reported an increase in the number of tests in which pounces and fights occurred as the day of receptivity approached and explained these results as a consequence of the males showing more and more interest in the females.

The effect of receptivity is dramatic and investigators generally do not use receptive females as experimental animals or as opponents unless they are specifically interested in this phenomenon. Drickamer and Vandenbergh (1973) reported less fighting among groups of females if the dominant female was receptive. Ciaccio *et al.* (1979) observed the "prelordotic posture" in which the back is arched, the tail is raised, and the animal remains immobile for brief periods of time. This posture is shown more often as the time of receptivity nears and interestingly, it is also shown by submissive females in female–female pairs and by males defeated by females. This is therefore a submissive posture that is influenced by aggressive encounters and in females it also signals the onset of sexual receptivity.

4.2. Ovariectomy and Estrogen and Progesterone

From the previous discussion of aggression during the estrous cycle, it might be concluded that estrogen alone increases aggressive and progesterone either alone or when combined with estrogen decreases aggressiveness. Unfortunately, only the effects of the combination of estrogen followed by progesterone are clear at this time and even this statement requires a qualification. The administration of progesterone in estrogen-primed hamsters results in a significant decline in aggression (Kislak and Beach, 1955; Tiefer, 1970; Floody and Pfaff, 1977; Ciaccio et al., 1979), but despite the continued presence of these hormones, levels of aggression begin to rise in less than 24 hr (Ciaccio et al., 1979).

The effects of ovariectomy with and without either estrogen or progesterone are contradictory. Ovariectomy without hormone replacement has been shown to increase aggressiveness (Ciaccio et al., 1979), increase aggressiveness in hamsters individually housed but not those housed in groups (Grelk et al., 1974), decrease aggressiveness (Kislak and Beach, 1955; Payne and Swanson, 1971a, 1972a), or produce no change (Vandenbergh, 1971; group-housed only, Grelk et al., 1974).

Treatment of ovariectomized hamsters with estrogen has been reported to decrease aggression (Ciaccio et al., 1979), increase aggression (Kislak and Beach, 1955), increase aggression in grouped but not isolated animals (Grelk et al., 1974), or result in no appreciable change (Vandenbergh, 1971; Payne and Swanson, 1971a, 1972a; Floody and Pfaff, 1977). Progesterone alone has been shown to have no effect (Kislak and Beach, 1955; Floody and Pfaff, 1977), regardless of the animals' housing conditions (Grelk et al., 1974).

However, Payne and Swanson (1971a, 1972a) reported that progesterone increased aggressive success (number of wins/number of tests with a winner) by either direct or indirect means depending on the sex of the opponent. When placed with female opponents, progesterone-treated ovariectomized hamsters showed an increase in aggressive behavior. Male opponents tested with progesterone-treated females showed a decline in their aggressiveness and therefore the females displayed relatively more aggression and were more successful in winning the encounters.

There is no doubt that some of these discrepancies are a result of some of the methodological considerations discussed in the previous section. These studies also differ in terms of the dosages of hormone administered, the duration of the periods of hormone stimulation, possible carryover effects of one hormone treatment to the next, and the interval between ovariectomy and the initiation of treatment.

4.3. Testosterone

Generally, testosterone propionate (TP) in doses from 0.025 to 1.0 mg/day has been shown to affect levels of aggression in female hamsters (Vandenbergh, 1971; Payne and Swanson, 1971a, 1972a; Hammond and Rowe, 1976; Floody and Pfaff, 1977). It should be noted that aggression in TP-treated ovariectomized females is compared with that in ovariectomized animals and the effect of ovariectomy is not clear.

Grelk et al. (1974) found that TP increased aggression in group-housed females but not in isolates. Drickamer and Vandenbergh (1973) reported that ovariectomized females receiving either 0.5 or 1 mg TP every other day for three weeks consistently ranked first or second among groups of four females. Females receiving 0.1 mg always ranked third and oil-treated animals were ranked last. Additionally, increasing doses of the hormone corresponded to increasing values of the flank gland index. One main difference among these studies was that Drickamer and Vandenbergh observed their animals for extended observation periods instead of relatively brief 5- to 10-min tests.

4.4. Other Hormones

Several other hormones have been tested for their effects on aggression in female hamsters. Floody and Pfaff (1977) found no difference in the levels of aggression between hypophysectomized animals treated with either prolactin or saline and their levels were similar to those found in intact, nonestrous females. Wise and Pryor (1977) treated females previously shown to be highly aggressive with ergocornine, a blocker of prolactin release. There was no effect of the reduced prolactin levels on the aggression shown by these females toward male intruders. It should be pointed out that these same investigators found a reduction in aggression after ergocornine administration in lactating animals (see Section 4.5).

Using a third approach to the issue of the involvement of prolactin in female aggression, Buntin et al. (1981) transplanted pituitaries to the kidney capsules of ovariectomized females, a procedure resulting in the continuous release of prolactin. Three weeks later, these females were paired with both males and sham-transplanted females in a neutral arena. The pituitary-transplanted females were clearly more aggressive than either of their opponents. The major differences between this study and those of Floody and Pfaff (1977) and Wise and Pryor (1977) pertain to the duration of treatment (minimum of three weeks of elevated prolactin release versus three days of prolactin or ergocornine injections) and the type of prolactin (endogenous versus exogenous ovine prolactin).

No effect of adrenocorticotropic hormone (ACTH) has been demonstrated (Brain and Evans, 1974; Floody and Pfaff, 1977). Follicle-stimulating hormone and luteinizing hormone in hypophysectomized females resulted in levels of aggression that were not different from those in either saline-treated controls or intact nonestrous females (Floody and Pfaff, 1977). It should be noted that Floody and Pfaff also showed that the aggressiveness of intact nonestrous females was not different from that in ovariectomized and adrenalectomized females suggesting that adrenal secretions play little if any role.

4.5. Pregnancy and Lactation

The only systematic comparison of aggression during the estrous cycle, pseudopregnancy, pregnancy, and lactation was performed by Wise (1974). Females were individually housed and tested once with weight-matched, individually housed, intact males. Although the pseudopregnant animals were slightly more aggressive than the cycling females, the pregnant and especially the lactating animals were the most aggressive. Pregnant animals were more aggressive on day 10 than they were on days 2 and 6, and females tested during the first five days of lactation were more aggressive than those tested between days 15 and 20.

The possible hormonal basis for the increased aggression shown by lactating

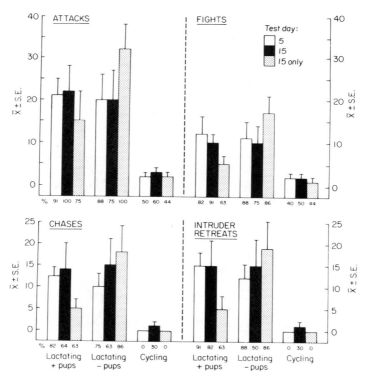

Figure 5. Mean (± SE) number of attacks, fights, chases, and intruder retreats initiated by lactating and cycling hamsters. Percentages of the groups initiating at least one of the behaviors per test are shown below the bars. Animals were tested for 10 min in their home cages with a female intruder on both days 5 and 15 of lactation or on day 15 only. Cycling animlas were tested either twice, ten days apart, or once at the later test to equate periods of social isolation. [Reprinted with permission of the publisher from *Siegel et al.* (1983).]

animals was examined (Wise and Pryor, 1977). Females were ovariectomized on day 4 of lactation and were then treated over the next three days with either ergocornine, ergocornine plus prolactin, or the vehicle. Each animal was tested once with a male in a large neutral enclosure. The control females showed the highest levels of rolls, fights, pounces, and chases. The ergocornine animals showed a dramatic reduction in aggression, whereas that of the ergocornine plus prolactin group was intermediate. Although the exogenous prolactin might have been expected to fully restore aggressive behavior to that of the controls, there are several reasons that may explain why it did not: the use of ovine prolactin, the dose of prolactin relative to the dose of ergocornine, and perhaps other effects of the drug not related to the reduction in hormone levels.

The study of aggression during lactation is complicated by the presence of the litter. Are the high levels of aggression a function of the physiological conditions associated with lactation, the pups, or the combination of the two? Siegel et al. (1983) studied animals in their home cages on both days 5 and 15 of lactation in the presence of the pups or after a 6-hr mother–litter separation. In this case, lactating females displayed more aggression toward female intruders on both days compared with cycling controls isolated for comparable intervals (Fig. 5). Levels of aggression were essentially similar on both days of lactation and the 6-hr separation between mother and the litter did not result in any reduction in aggression.

More recently, Giordano et al. (1983) have shown that a 24-hr mother–litter separation results in levels of aggression that are similar to those observed in cycling controls. Separation intervals of 24 hr followed by 30 or 120 min mother–litter reunions significantly increased all measures of aggression recorded.

5. HORMONAL EFFECTS: MALE

5.1. Castration and Androgens

The effects of gonadectomy and hormone replacement on aggression in the male are more conclusive than they are in the female. Vandenbergh (1971) showed that castration in adulthood resulted in decreases in nosing behavior, flank marking, dislodging (a milder form of aggression) and elicited fewer attacks, chases, and fights (combined measure) from the intact opponents. The administration of 0.5 mg TP every other day for 14 days increased levels of marking, dislodging, and the combined measure of attacks, fights, and chases. Grelk et al. (1974) also showed that levels of aggressive behavior were reduced by castration and increased by TP in animals living in both grouped and isolated conditions.

Drickamer et al. (1973) observed six groups of castrated males and within each group the animals received either 0.1, 0.5, or 1.0 mg TP or oil every other day for three weeks. Rankings of dominance were similar in each group. Males receiving the two highest doses of TP were always ranked first or second, the males treated with the lowest dose were ranked third, and the oil-treated animals were ranked last. The dominance rankings corresponded to an index of flank gland size and its degree of pigmentation and to the frequency of flank gland marking. Rowe and Swanson (1977) confirmed the castration-induced decline in aggressive behavior and showed that the systemic injection or topical placement of TP onto the flank glands increased dominant behavior comparable to that observed in intact males.

In a series of studies (Payne, 1973, 1974a,b; Payne and Swanson, 1972b), the aggressive behavior of intact and castrated males toward intact, gonadectomized, and gonadectomized plus hormone-treated males and females was examined. Intact males showed more aggression and won more encounters than intact male intruders, whereas castrated males were shown to be either somewhat more aggressive than or comparable to these intruders. In neutral test cages, however, castrated males were less aggressive and were subordinate to both intact males and females. As intruders, intact males initiated and elicited more attacks than castrated males.

Testosterone propionate replacement resulted in increased aggression when the treated animals were provided with male intruders. Both androstenedione and dihydrotestosterone were also effective in restoring levels of aggression. In fact, androstenedione was the most effective androgen in raising aggression levels but was unable to maintain seminal vesicle weight or prevent the castration-induced elevation in body weight.

Finally, intact males show little aggression toward intact females. They display more aggression when paired with ovariectomized females and this level of aggression increased when provided with ovariectomized females previously treated with TP. Castration plus TP did not increase the aggressive behavior of males paired with intact females.

These studies demonstrate that the neutrality of the test cage and the sex and reproductive condition of the opponents must be considered when drawing conclusions on the role of hormones in aggression. Marques and Valenstein (1977) reported that females showed more aggressive behavior when paired with castrated males than when paired with intact males. In fact, females who did not previously show much aggression when tested with males became much more aggressive when tested with castrated males (or other females).

To separate the effects of castration and androgen replacement on the treated animal from those affected by the behavior of the opponent, Potegal et al. (1980b) pretested their subjects and then paired males that showed high levels of aggression with males that were not highly aggressive. The highly aggressive males had been individually housed and tested in their home cages to further increase their aggressiveness, whereas the nonaggressive animals had been group housed and bilaterally bulbectomized before becoming tethered intruders. The castrated and intact home-cage males were not different in terms of the duration and temporal distribution of their attack bouts, but the castrated animals showed an increase in the latency to the first attack and reductions in the displays of offensive postures. These effects of castration were not reversed with androgen replacement therapy suggesting that they are not androgen dependent.

Not all studies have shown that male aggression declines after castration and increases after androgen therapy. Although TP did increase levels of aggression, castration did not reduce their levels (Tiefer, 1970). Whitsett (1975) studied the ontogeny of aggression in intact—intact and castrate—castrate pairs. Initial testing began at 30 days of age and continued every five or ten days until the age of 80 days. In both types of pairs, aggressive behavior and even flank marking, despite the absence of the flank glands in the castrated animals, increased with age. In intact—castrate pairs tested at the older ages, with or without previous testing, there was no difference in the display of aggressive behavior.

In another developmental study, groups of four litters lived in separate compartments of a large enclosure. At about 9 weeks of age, the connecting doors separating the different compartments were opened. Mothers were first observed to attack their litters when the pups reached 36 days of age. The first attacks distinguishable from play fights were observed at 46 days of age. Females initiated 90% of these attacks, but this changed dramatically over the next several weeks when the connecting doors were opened. By the last two weeks of the total 120-day observation period, males were initiating the vast majority of attacks and of these, 80% were against other males. Although these results have no bearing on the effects of castration or hormone treatment, they are included to show that males can display high levels of aggression especially in male—male encounters.

The final study related to the role of androgens showed an inverse relationship between levels of testosterone and aggression (Garrett and Campbell, 1980). Males exposed to a photoperiod of 6-hr light—18-hr dark underwent testicular regression, had correspondingly low testosterone levels, but displayed shorter attack latencies and higher frequencies of attack when compared with males exposed to a normal photoperiod of 14-hr light—10-hr dark. Furthermore, there was an upward shift in dominance of these males over their heavier counterparts who had previously defeated them in the 14-hr light—10-hr dark situation. When the males exposed to short days were subsequently exposed to the typical amount of light, they showed increased levels of testosterone and decreased levels of aggression. To account for these data, Garrett and Campbell suggested that in nature males would need to be aggressive during short-daylength months to compete for their own resources. During the breeding season, high levels of aggressiveness would preclude the opportunity for mating and hence these levels should decline. Although these suggestions are interesting, the results on which they are based are difficult to reconcile with the laboratory studies on castration and androgen treatment. The lab studies demonstrate that males, intact and castrated and given TP, do not generally attack females but intermale aggression appears more androgen dependent.

5.2. *Estrogen and Progesterone*

Except for the lack of effect reported by Tiefer (1970), estrogen has been shown to increase the aggressiveness of castrated males. Using male opponents, Vandenbergh (1971) showed that 0.05 mg EB increased the frequencies of dislodging and attacks, chases, and fights to those shown by castrated males given TP. In doses ranging from 1 to 50 μg/day, EB increased aggression in castrated males housed individually and those housed in groups. Estradiol benzoate treatment of castrated males increased their level of aggression to that of intact males when paired with intact males, but had no effect when the animals were paired with intact females, and in this case the females were clearly more aggressive (Payne and Swanson, 1972b).

Progesterone increased the aggression of isolated, castrated males especially at the higher doses (0.5 and 1.0 mg/day) (Grelk *et al.*, 1974). Payne and Swanson (1971b, 1972b) treated castrated males with either progesterone or ovarian tissue implanted under the kidney capsule. Untreated castrated males showed less aggression when paired with either intact males or females. The addition of the ovarian tissue resulted in higher levels of aggression when compared with intact females. The effect of exogenous progesterone depended on the sex of the opponent. With female opponents, progesterone increased aggressiveness to that of the intact females. With male opponents, progesterone produced no direct change in the treated males, but they were relatively more aggressive because their intact male opponents showed a reduction in their own level of aggression.

5.3. Other Hormones

Brain and Evans (1974) reported that injections of ACTH every other day for 21 days had no effect on fighting behavior in male hamsters. Landau (1975a,b) showed a nocturnal rhythm in aggressive behavior that began shortly after lights off and then tested the effects of adrenalectomy with and without cortisol acetate on the levels and rhythmicity of aggression in males. Adrenalectomy abolished the nocturnal rhythm, but had no effect on levels of aggression. It should be noted that the adrenalectomized males were treated chronically with deoxycorticosterone acetate, a mineralocorticoid, to insure survival after surgery. Males given deoxycorticosterone plus cortisol, a glucocorticoid, also showed no nocturnal rhythm but higher levels of aggression than the sham-operated controls. The lack of rhythmicity may have been due to the use of cortisol implants that prevented cyclic hormonal fluctuations. This is supported by the data showing a relative increase in aggression during the dark and not the light phase when compared with that of the sham-operated animals.

6. NEURAL BASIS

6.1. Limbic System

Most of the studies on the involvement of neural structures on aggression have focused on the septum and amygdala. In general, animals with septal lesions have displayed increased levels of aggression, decreased hoarding and nest building, and no change in measures of emotionality. Bilateral septal lesions in males resulted in increased frequencies of aggressive encounters although of relatively short duration and no effect on open-field behavior or responsiveness to the presentation of a probe, air-blasting, handling, and capture (Johnson et al., 1972; Sodetz et al., 1967).

Sodetz and Bunnell (1970) showed that the level of aggression following damage to the medial and lateral septal nuclei depended on both the animals' preoperative aggressive encounters and the type of opponent. Although septal-lesioned males increased their levels of fights, threats, and attacks, those males that were submissive in preoperative tests showed no increase in aggression postoperatively. The level of aggression in septal-lesioned males was greater when they were paired with previously dominant males or other septal-lesioned animals. There were no postoperative changes in emotionality, and regardless of their preoperative histories septal-lesioned animals showed less hoarding, less nest building, but more social investigation of other males.

There were no sex or age differences among males and females receiving septal lesions at 25 or 45 days of age (Janzen and Bunnell, 1976). All septal-lesioned animals displayed increased aggressiveness toward intact opponents of the same sex and also increased levels of shock-induced aggression (Shipley and Kolb, 1977). In both studies these lesions produced deficits in nest building and hoarding.

Potegal et al. (1981a) paired males previously shown to be moderately aggressive and then given anteroventral or posterior septal–anterior thalamic lesions with nonaggressive, muzzled males. Lesions involving the posterior septum had no effect on aggression or activity. Both lesioned groups showed deficits in nesting, hoarding, and sexual behavior. The anteroventral septal-lesioned group showed a dramatic increase in attack rates and not only defeated sham-operated controls and animals with the other type of lesion, but attacked sexually receptive females, an unusual finding. Further, electrical stimulation of the septum inhibited the normally high level of attacks on nonaggressive (group-housed, muzzled, analgesic-treated) males (Potegal et al., 1981b). The lowest threshold for attack inhibition was found in the mid and ventral septum.

Bilateral amygdala lesions resulted in a decline in territorial and shock-induced aggression and nest building (Shipley and Kolb, 1977). Bunnell et al. (1970) reported that their males with amygdala lesions showed more indifference to other animals. Males shown to be highly aggressive before the lesions showed a decrease in aggression, whereas males previously shown to be submissive showed less submissive behavior postoperatively. Animals that were socially inexperienced preoperatively displayed intermediate levels of aggression and submissive behavior postoperatively. They were less aggressive and less submissive than the males who preoperatively were most aggressive and most submissive, respectively.

Medial preoptic area lesions resulted in decreased levels of both high-intensity (fights, attacks, and chases) and low-intensity (wrestling and sparring) aggressive responses in females (Hammond and Rowe, 1976). Subsequent testing after ovariectomy and androgen replacement showed no change. When compared with controls, females with anterior hypothalamic lesions showed similar levels of high-intensity but increased levels of low-intensity aggressive behavior. Ovariectomy reduced the frequency of low-intensity responses to that of controls and there was no additional effect of the androgen.

6.2. Olfaction

Bilateral olfactory bulbectomy, lateral olfactory tract transection, and peripherally induced anosmia (zinc sulfate infusions) resulted in large deficits in a variety of behaviors including sexual behavior, hoarding, nest building, scent marking, preference for vaginal odor, and several measures of aggression (Murphy and Schneider, 1970; Devor and Murphy, 1973; Hilger and Rowe, 1975; Macrides et al., 1976; Murphy, 1976a). Unilateral bulbectomy did not produce these behavioral changes unless it was accompanied by the reversible closing off of the contralateral nostril. The hamster has separate nasal passages and this procedures results in one intact olfactory bulb and one functional nostril but no connection between them; control animals were unilaterally bulbectomized and their ipsilateral nostrils were clipped (Devor and Murphy, 1973; Murphy, 1976a).

Bilaterally bulbectomized males rarely initiated an attack even when tested in their home cages, but did fight back when attacked in the home cage of a highly aggressive male (Murphy, 1976a). Additionally, these centrally or peripherally induced anosmic males spent less time investigating other animals. Because they fought back when attacked, the bulbectomized males were not considered to be generally debilitated. The effects of bulbectomy were also not mediated by alterations in androgen secretion because testes weights and flank gland sizes were not different from controls. Finally, eye enucleation was performed in additional males as another means to reduce sensory input (Murphy, 1976b). These males showed no reduction in aggression and instead had longer attack durations than controls.

The effects of bilateral bulbectomy in females were not as dramatic as those in males (Hilger and Rowe, 1975). Bulbectomized females showed a decrease in attacks and chases, but no change in their investigation of other animals or in wrestling or fighting.

7. PERIPHERAL INFLUENCES

7.1. Flank Gland

The flank glands, first seen in the 10-day-old fetus, are slightly raised organs located bilaterally in a costovertebral position. They consist of sebaceous tissue, hair follicles, and dermal melanocytes, all of which are androgen dependent (Algard et al., 1964; Giegel et al., 1971; Hamilton and Montagna, 1950).

The flank gland measures approximately 8 mm in diameter in intact males and 2 mm or less in intact females. Castration shrinks the glands to the size found in females and results in decreases of DNA and RNA, whereas androgen replacement increases thymidine and uridine uptake (Giegel et al., 1971). Testosterone propionate but not EB increases the size and degree of pigmentation in the flank glands of both gonadectomized males and females (Vandenbergh, 1973).

Drickamer et al. (1973) and Drickamer and Vandenbergh (1973) devised an index of flank gland development that is a function of the size and absorptive properties of the gland. They showed positive correlations between the index and dominance in both males and females. Castration reduced the gland index and the display of dominance and androgen replacement increased both measures. The most dominant males and females also showed the most marking behavior. As mentioned earlier, topical application of TP onto the glands of castrated males also increased aggression, possibly a result of odor or taste (Rowe and Swanson, 1977).

In contrast to these relationships between marking, androgen, and dominance, Whitsett (1975) found no correlation between dominance and flank gland index in intact—intact or castrate—castrate encounters. In mixed pairs, i.e., intact—castrate, the intact animals showed more marking and more attacking. Finally, Macrides *et al.* (1976) found an inverse relationship between testosterone levels and the frequency of scent marking among sham-operated and lateral olfactory tract-transected males.

7.2. Vaginal Discharge

The presence of vaginal discharge inhibits aggression in male—male pairs. Murphy (1973, 1976a) showed that males display fewer attacks and shorter attack durations if their male partners (active or anesthetized) are scented with vaginal discharge or if the test cages contain small amounts of the substance. These results have been confirmed in terms of agonistic posturing toward stimulus males smeared with vaginal secretions (Johnston, 1975).

7.3. Urine

Smearing urine from intact males or females onto stimulus animals had no effect on levels of aggression (Evans and Brain, 1974). Payne (1974c) reported increased aggression among males when the intruders were treated with intact male urine. The urine of castrated males and intact females were no different and had no effect on eliciting attack from home cage males.

7.4. Chest Patches

The role of the chest patches could be an important variable. It should be noted that the display of the chest patches has been used to distinguish between offensive and defensive postures (Grant and Mackintosh, 1963) (see Fig. 1).

Grant *et al.* (1970) dyed the chest and forearms of adult males with black hair color. The dyed males tested in their home cages against a nondyed male were much more likely to initiate aggressive behavior. According to Grant *et al.*, the accidental display of the darkened areas elicited flight reactions on the part of the opponent that led to increased aggression by the dyed animals. Using dyed males and females of younger ages, just past weaning, Payne and Swanson (1972c) reported that the dyed animals won more encounters by a 2 : 1 margin.

In an extensive series of studies on the effect of the chest patches, Johnston (1976) used males matched for age and body weight. When tested under dim illumination, males with darkened chests, two to three times the size of the normally darker patches, won significantly more fights and won more encounters than controls whose backs were dyed. Yet, Johnston noted that the stereotyped postures required to display the chest patches occurred only after the winners and losers were determined.

Johnston (1976) next repeated the experiment in complete darkness, turning the lights on to determine the winners and losers only after the sounds of fighting were heard. In this case, the dyed animals won only 50% of the encounters. The behavior of male pairs were videotaped and this analysis calls into question the exact nature of the postures shown by the dominant and subordinate animals.

Johnston argues that the postures cannot be considered threatening because the patches only occasionally were displayed in a dramatic fashion and then only after dominance was established. Further, the posture showing off the chest patches more clearly was adopted by the subordinate and not the dominant animal. The larger patches may have been more effective in dim illumination because they produced fear, thus giving the dyed male an advantage, and in fact the typical patch display may be more important to advertise submission. Some support for this idea comes from a study on the aggression of males whose eyes had been enucleated (Murphy, 1976b). These males showed longer attack durations possibly because they could not see the submissive display of their opponents' chest patches.

7.5. Harderian Gland

The harderian gland functions to lubricate the nictitating membrane. The prophyrin concentration of the glands of males is low, increase 100-fold after castration, and can be reduced by a variety of androgens (Hoffman, 1971; Payne et al., 1975). In females, the prophyrin levels are highest on the day of estrus and lowest the next day (Payne et al., 1977). Also, these levels are highest during periods of increased fertility.

The effect of harderian gland homogenates dabbed around the eyes and snout of intruder males was studied by Payne (1977). The aggression of the home-cage males decreased when paired with a male smeared with homogenate of females, whereas the effect of male homogenate was equal to that of water.

8. CONCLUSIONS

The hamster has proved to be a useful species for the study of aggression. Aggressive behavior is readily elicited and easily quantified. A great deal has been learned about the organization of the behavioral components, the effect of hormones, the role of central and peripheral structures, and the variables that affect the outcome of aggressive encounters. However, there are large areas of research that have not yet been explored. It is hoped that future research will continue on its present course and also employ seminatural studies of aggression on the one hand and the methods of neurochemistry and molecular endocrinology on the other.

ACKNOWLEDGMENTS. Supported by Biomedical Support Grant (FR 07059–18) and Rutgers Research Council Grant (URF-G-84-830-NK-62) to H.I.S. and USPHS Grant MH-08604 to Jay S. Rosenblatt. I wish to thank Ms. Marilyn Dotegowski for assistance and Ms. Cynthia Banas for preparation of the illustrations. Institute of Animal Behavior Publication No. 401.

REFERENCES

Algard, F. T., Dodge, A. H., and Kirkman, H., 1964, Development of the flank organ (scent gland) of the Syrian hamster. I. Embryology, *Am. J. Anat.* 114:435–455.

Boice, R., Hughes, D., and Cobb, C. J., 1969, Social dominance in gerbils and hamsters, *Psychon. Sci.* 16:127–128.

Brain, P. F., 1972, Effects of isolation/grouping on endocrine function and fighting behavior in male and female golden hamsters (*Mesocricetus auratus* Waterhouse), *Behav. Biol.* 7:349–357.

Brain, P., and Evans, C. M., 1974, Some recent studies on the effects of corticotrophin on agonistic behaviour in the house mouse and the golden hamster, *J. Endocrinol.* 61:xxxix–xl.

Bunnell, B. N., Sodetz, F. J., Jr., and Shalloway, D. I., 1970, Amygdaloid lesions and social behavior in the golden hamster, *Physiol. Behav.* 5:153–161.

Buntin, J. D., Catanzaro, C., and Lisk, R. D., 1981, Facilitatory effects of pituitary transplants on intraspecific aggression in female hamsters, *Horm. Behav.* 15:214–225.

Ciaccio, L. A., Lisk, R. D., and Reuter, L. A., 1979, Prelordotic behavior in the hamster: A hormonally modulated transition from aggression to sexual receptivity, *J. Comp. Physiol. Psychol.* 93:771–780.

Devor, M., and Murphy, M. R., 1973, The effect of peripheral olfactory blockade on the social behavior of the male golden hamster, *Behav. Biol.* 9:31–42.

Dieterlen, F., 1959, Das Verhalten des syrischen Goldhamster, *Zeit. Tierpsychol.* 16:47–103.

Drickamer, L. C., and Vandenbergh, J. G., 1973, Predictors of social dominance in the adult female golden hamster (*Mesocricetus auratus*), *Anim. Behav.* 21:564–570.

Drickamer, L. C., Vandenbergh, J. G., and Colby, D. R., 1973, Predictors of dominance in the male golden hamster (*Mesocricetus auratus*), *Anim. Behav.* 21:557–563.

Evans, C. M., and Brain, P. F., 1974, Some influences of sex steroids on the aggressiveness directed towards golden hamsters (*Mesocricetus auratus* Waterhouse) of both sexes by trained fighter individuals, *J. Endocrinol.* 61:xlvi–xlvii.

Floody, O. R., and Pfaff, D. W., 1977, Aggressive behavior in female hamsters: The hormonal basis for fluctuations in female aggressiveness correlated with estrous state, *J. Comp. Physiol. Psychol.* 91:443–464.

Garrett, J. W., and Campbell, C. S., 1980, Changes in social behavior of the male golden hamster accompanying photoperiodic changes in reproduction, *Horm. Behav.* 14:303–318.

Giegel, J. L., Stolfi, L. M., Weinstein, G. D., and Frost, P., 1971, Androgenic regulation of nucleic acid and protein synthesis in the hamster flank organ and other tissues, *Endocrinology* 89:904–909.

Giordano, A. L., Siegel, H. I., and Rosenblatt, J. S., 1983, Effects of mother-litter separation and reunion on maternal aggression and pup mortality in lactating hamsters, presented at *Conference on Reproductive Behavior*, Medford, Massachusetts.

Goldham, L., and Swanson, H. H., 1975, Developmental changes in pre-adult behavior in confined colonies of golden hamsters, *Dev. Psychobiol.* 8:137–150.

Grant, E. C., and Mackintosh, J. H., 1963, A comparison of the social postures of some common laboratory rodents, *Behaviour,* 21:246–259.

Grant, E. C., Mackintosh, J. H., and Lerwill, C. J., 1970, The effect of a visual stimulus on the agonistic behaviour of the golden hamster, *Z. Tierpsychol.* 27:73–77.

Grelk, D. F., Papson, B. A., Cole, J. E., and Rowe, F. A., 1974, The influence of caging conditions and hormone treatments on fighting in male and female hamsters, *Horm. Behav.* 5:355–366.

Hamilton, J. B., and Montagna, W., 1950, The sebaceous glands of the hamster. I. Morphological effects of androgens on integumentary structures. *Am. J. Anat.* 86:191–233.

Hammond, M. A., and Rowe, F. A., 1976, Medial preoptic and anterior hypothalamic lesions: Influences on aggressive behavior in female hamsters. *Physiol. Behav.* 17:507–513.

Hilger, W. N., Jr., and Rowe, F. A., 1975, Olfactory bulb ablation: Effects on handling reactivity, open-field behavior, and agonistic behavior in male and female hamsters, *Physiol. Psychol.* 3:162–168.

Hoffman, R. A., 1971, Influence of some endocrine glands, hormones and blinding on the histology and prophyrins of the Harderian glands of golden hamsters, *Am. J. Anat.* 132:463–478.

Huang, D., and Hazlett, B. A., 1974, Submissive distance in the golden hamster *Mesocricetus auratus, Anim. Behav.* 22:467–472.

Janzen, W. B. and Bunnell, 1976, Septal lesions and the recovery of function in the juvenile hamster, *Physiol. Behav.* 16: 445–452.

Johnson, D. A., Poplawsky, A., and Bieliauskas, L., 1972, Alterations of social behavior in rats and hamsters following lesions of the septal forebrain, *Psychon. Sci.* 26:19–20.

Johnston, R. E., 1975, Sexual excitation function of hamster vaginal secretion, *Anim. Learn. Behav.* 3:161–166.

Johnston, R. E., 1976, The role of dark chest patches and upright postures in the agonistic behavior of male hamsters, *Mesocricetus auratus, Behav. Biol.* 17:161–176.

Kislak, J. W., and Beach, F. A., 1955, Inhibition of aggressiveness by ovarian hormones, *Endocrinology* 56:684–692.

Landau, I. T., 1975a, Light-dark rhythms in aggressive behavior of the male golden hamster, *Physiol. Behav.* 14:767–774.

Landau, I. T., 1975b, Effects of adrenalectomy on rhythmic and non-rhythmic aggressive behavior in the male golden hamster, *Physiol. Behav.* 14:775–780.

Lawlor, M., 1963, Social dominance in the golden hamster, *Bull. Br. Psychol. Soc.* 16:1–14.

Lerwill, C. J., Makings, P., 1971, The agonistic behaviour of the golden hamster *Mesocricetus auratus* (Waterhouse), *Anim. Behav.* 19:714–721.

Macrides, F., Firl, A. C., Jr., Schneider, S. P., Bartke, A., and Stein, D. G., 1976, Effects of one-stage or serial transections of the lateral olfactory tracts on behavior and plasma testosterone levels in male hamsters, *Brain Res.* 109:97–109.

Marques, D. M., and Valenstein, E. S., 1977, Individual differences in aggressiveness of female hamsters: Response to intact and castrated males and to females, *Anim. Behav.* 25:131–139.

Murphy, M. R., 1973, Effects of female hamster vaginal discharge on the behavior of male hamsters, *Behav. Biol.* 9:367–375.

Murphy, M. R., 1976a, Olfactory stimulation and olfactory bulb removal: Effects on territorial aggression in male Syrian golden hamsters, *Brain Res.* 113:95–110.

Murphy, M. R., 1976b, Blinding increases territorial aggression in male Syrian golden hamsters, *Behav. Biol.* 17:139–141.

Murphy, M. R., 1978, Oestrous Turkish hamsters display lordosis toward conspecific males but attack heterospecific males, *Anim. Behav.* 26:311–312.

Murphy, M. R., and Schneider, G. E., 1970, Olfactory bulb removal eliminates mating behavior in the male golden hamster, *Science* 167:302–304.

Payne, A. P., 1973, A comparison of the aggressive behaviour of isolated intact and castrated male golden hamsters towards intruders introduced into the home cage, *Physiol. Behav.* 10:629–631.

Payne, A. P., 1974a, A comparison of the effects of androstenedione, dihydrotestosterone and testosterone propionate on aggression in the castrated male golden hamster, *Physiol. Behav.* 13:21–26.

Payne, A. P., 1974b, The aggressive response of the male golden hamster towards males and females of differing hormonal status, *Anim. Behav.* 22:829–835.

Payne, A. P., 1974c, The effects of urine on aggressive responses by male golden hamsters, *Aggress. Behav.* 1:71–79.

Payne, A. P., 1976, A comparison of the effects of neonatally administered testosterone, testosterone propionate and dihydrotestosterone on aggressive and sexual behaviour in the female golden hamster, *J. Endocrinol.* 69:23–31.

Payne, A. P., 1977, Pheromonal effects of Harderian gland homogenates on aggressive behaviour in the hamster, *J. Endocrinol.* 73:191–192.

Payne, A. P., and Swanson, H. H., 1970, Agonistic behaviour between pairs of hamsters of the same and opposite sex in a neutral observation area, *Behaviour* 36:259–269.

Payne, A. P., and Swanson, H. H., 1971a, Hormonal control of aggressive dominance in the female hamster, *Physiol. Behav.* 6:355–357.

Payne, A. P., and Swanson, H. H., 1971b, The effect of castration and ovarian implantation on aggressive behaviour of male hamsters, *J. Endocrinol.* 51:217–218.

Payne, A. P., and Swanson, H. H., 1972a, The effect of sex hormones on the aggressive behaviour of the female golden hamster (*Mesocricetus auratus* Waterhouse), *Anim. Behav.* 20:782–787.

Payne, A. P., and Swanson, H. H., 1972b, The effect of sex hormones on the agonistic behavior of the male golden hamster (*Mesocricetus auratus* Waterhouse), *Physiol. Behav.* 8:687–691.

Payne, A. P., and Swanson, H. H., 1972c, The effect of a supra-normal threat stimulus on the growth rates and dominance relationships of pairs of male and female golden hamsters, *Behaviour* 42:1–7.

Payne, A. P., McGadey, J., Moore, M. R., and Thompson, G., 1975, The effects of androgen manipulation on cell types and prophyrin content of the Harderian gland in the male golden hamster, *J. Anat.* 120:615–616.

Payne, A. P., McGadey, J., Moore, M. R., and Thompson, G., 1977, Cyclic and seasonal changes in Harderian gland activity in the female golden hamster, *J. Endocrinol.* 72:41P.

Polsky, R. H., 1974, Effects of novel environment on predatory behavior of golden hamsters, *Percept. Mot. Skills* 39:55–58.

Polsky, R. H., 1976, Conspecific defeat, isolation/grouping, and predatory behavior in golden hamsters, *Psychol. Rep.* 38:571–577.

Polsky, R. H., 1977a, The ontogeny of predatory behaviour in the golden hamster (*Mesocricetus a. auratus*). I. The influence of age and experience, *Behaviour* 61:26–57.

Polsky, R. H., 1977b, The ontogeny of predatory behaviour in the golden hamster (*Mesocricetus a. auratus*). II. The nature of the experience, *Behaviour* 61:58–81.

Polsky, R. H., 1978a, Influence of eating dead prey on subsequent capture of live prey in golden hamsters, *Physiol. Behav.* 20:677–680.

Polsky, R. H., 1978b, The ontogeny of predatory behaviour in the golden hamster (*Mesocricetus a. auratus*). III. Sensory preexposure, *Behaviour* 63:175–191.

Polsky, R. H., 1978c, The ontogeny of predatory behaviour in the golden hamster (*Mesocricetus a. auratus*). IV. Effects of prolonged exposure, ITI, size of prey and selective breeding, *Behaviour* 65:27–42.

Potegal, M., Blau, A., Black, M., and Glusman, M., 1980a, A technique for the study of intraspecific aggression in the golden hamster under conditions of reduced target variability, *Psychol. Rec.* 30:191–200.

Potegal, M., Blau, A. D., Black, M., and Glusman, M., 1980b, Effects of castration of male golden hamsters on their aggression toward a restrained target, *Behav. Neural Biol.* 29:315–330.

Potegal, M., Blau, A., and Glusman, M., 1981a, Effects of anteroventral septal lesions in intraspecific aggression in male hamsters, *Physiol. Behav.* 26:407–412.

Potegal, M., Blau, A., and Glusman, M., 1981b, Inhibition of intraspecific aggression in male hamsters by septal stimulation, *Physiol. Psychol.* 9:213–218.

Rowe, E. A., and Swanson, H. H., 1977, A comparison of central and peripheral effects of testosterone propionate on social interactions in the male golden hamster, *J. Endocrinol.* 72:39P–40P.

Shipley, J. E., and Kolb, B., 1977, Neural correlates of species-typical behavior in the Syrian golden hamster, *J. Comp. Physiol. Psychol.* 91:1056–1073.

Siegel, H. I., Giordano, A. L., Mallafre, C. M., and Rosenblatt, J. S., 1983, Maternal aggression in hamsters: Effects of stage of lactation, presence of pups, and repeated testing, *Horm. Behav.* 17:86–93.

Sodetz, F. J., and Bunnell, B. N., 1970, Septal ablation and the social behavior of the golden hamster, *Physiol. Behav.* 5:79–88.

Sodetz, F. J., Matalka, E. S., and Bunnell, B. N., 1967, Septal ablation and effective behavior in the golden hamster, *Psychon. Sci.* 7:189–190.

Tiefer, L., 1970, Gonadal hormones and mating behavior in the adult golden hamster, *Horm. Behav.* 1:189–202.

Vandenbergh, J. G., 1971, The effects of gonadal hormones on the aggressive behavior of adult golden hamsters (*Mesocricetus auratus*), *Anim. Behav.* 19:589–594.

Vandenbergh, J. G., 1973, Effects of gonadal hormones on the flank gland of the golden hamster, *Horm. Res.* 4:28–33.

Van Hemel, P. E., 1977, Interspecific attack on mice and frogs by golden hamsters (*Mesocricetus auratus*), *Bull. Psychon. Soc.* 9:186–188.

Whitsett, J. M., 1975, The development of aggressive and marking behavior in intact and castrated male hamsters, *Horm. Behav.* 6:47–57.

Wise, D. A., 1974, Aggression in the female golden hamster: Effects of reproductive state and social isolation, *Horm. Behav.* 5:235–250.

Wise, D. A., and Pryor, T. L., 1977, Effects of ergocornine and prolactin on aggression in the postpartum golden hamster, *Horm. Behav.* 8:30–39.

IV

Development and Individual Function

13

Behavioral Development in the Syrian Golden Hamster

THOMAS A. SCHOENFELD and CHRISTIANA M. LEONARD

1. INTRODUCTION

Study of behavioral development in hamsters had its formal beginnings in German ethology in the work of Dieterlen, Eibl-Eibesfeldt, and others. These studies offered some of the first descriptions of hamster pup behavior and established an important precedent for experimental inquiries into developmental mechanisms by the use of hamsters reared in isolation—the so-called Kaspar-Hauser animals (Dieterlen, 1959)—to test the importance of environmental stimuli on behavioral development.

In recent years the hamster has been an attractive subject for developmental study because investigators have recognized that its short gestation relative to other altricial animals may provide an important analytical tool. One question we wish to address in this review is whether a shorter period of fetal development thrusts the newborn hamster pup into its postnatal world in a less mature state than pups of other altricial rodents with longer gestations.

We will confine most of our remarks to the development of behavior in the individual Syrian golden hamster (*Mesocricetus auratus*), certainly the most widely studied of hamster species. Details of the development of adult social behaviors, such as sexual and aggressive behaviors, will be provided in the chapters of Section II in the present volume, although we will comment on several occasions about the social interactions of pups when they are still tied to the nest.

THOMAS A. SCHOENFELD • Worcester Foundation for Experimental Biology, Shrewsbury, Massachusetts 01545.   CHRISTIANA M. LEONARD • Department of Neuroscience, University of Florida College of Medicine, Gainesville, Florida 32610.

2. SENSORY AND MOTOR BEHAVIORS

A good starting place for a discussion of behavioral development is basic sensory and motor mechanisms, those functions that form the foundation for more complex behaviors. We first consider sensory development, focusing in particular on olfaction and vision, and then examine developmental changes in movement and activity.

2.1. Olfaction

The development of olfaction has been better studied than the development of most other sensory and motor functions in hamsters. Even so, it has not received the attention accorded olfactory development in rats (e.g., Alberts and Brunjes, 1978; Teicher and Blass, 1977; Rudy and Cheatle, 1977; Johanson and Hall, 1979; Pedersen and Blass, 1982; Tobach, 1977). Most studies in hamsters have been concerned with odor preferences and have not attempted to determine how the detection and discrimination of motivationally neutral odors changes as a function of age (cf. Alberts and May, 1980b; Johanson and Hall, 1979). Moreover, specific response patterns such as sniffing or head orienting have not been studied in any detail in the young hamster as they have in the rat (e.g., Welker, 1964; Alberts and May, 1980a). Consequently, the discussion in this section will be concerned less with olfactory processing per se and more with olfactory control of behavior in the developing hamster.

2.1.1. Normal Development

Preference and aversion studies have demonstrated that hamster pups can detect odors as early as P3–4* (Devor and Schneider, 1974; Cornwell, 1975; Crandall and Leonard, 1979). In such studies, pups indicate a preference by spending significantly more time over shavings of one odor than that of another (see Fig. 1). Pups younger than P3 have not been tested formally, although they reputedly respond differentially to lemon scent at birth (see Cromwell, 1975, p. 183). Hamsters probably can process odors at birth because (1) nipple attachment, which in the rat is under the control of olfactory stimuli (Teicher and Blass, 1977), occurs with nearly identical latencies in newborn hamsters and rats (Hall and Rosenblatt, 1979; see Section 3.2.1) and (2) a number of olfactory circuits in the brain are established in at least rudimentary form at birth as they are in the rat (Leonard, 1975; Grafe and Leonard, 1982; cf. Schwob and Price, 1978; Westrum, 1975).

At P4, pups readily avoid certain novel odors such as cedar (Devor and Schneider, 1974) and lemon (Cornwell, 1975). On the other hand, odor preferences in these neonates are more difficult to demonstrate. In two studies (Devor and Schneider, 1974; Gregory and Bishop, 1975), a preference for the odor of nest shavings over that of novel shavings emerged only gradually by P7–8 and then began to decline at P12–14

---

*Postnatal days 3–4. Where possible, we have adjusted age data according to PO = day of birth. Otherwise, a quoted age may be an overestimate by one day.

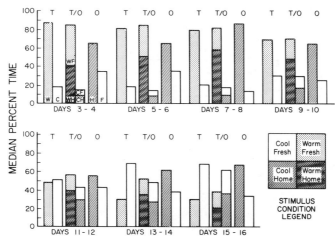

Figure 1. Comparison of changes in hamster pup preference behavior in three tests, T (Thermal-only), O (Odor-only), and T/O (Thermal/Odor), as a function of age. Data are median percent time spent over shavings of a particular odor and/or temperature observed in three 1-min trials for each test. For test T, shavings were warm (W) or cool (C). For test T/O, shavings were warm-fresh (WF), warm-home (WH), cool-fresh (CF), or cool-home (CH). For test O, shavings were home (H) or fresh (F). [From Crandall and Leonard (1979). Reprinted by permission of Academic Press, publisher, and the authors.]

(cf. Gregory and Pfaff, 1971, for rats). Under appropriate conditions, earlier preferences can be demonstrated, however. We have found, for example, that the use of multiple short tests reveals a preference for nest shavings as early as P3 (Crandall and Leonard, 1979) (Fig. 1). Three 1-min tests pitting nest shavings against fresh shavings may be more sensitive than a single test of 3 to 10 min (cf. Devor and Schneider, 1974; Gregory and Bishop, 1976). A preference for nest shavings demonstrated under these conditions is still less robust at P3 than at P8. In a bimodal test, in which a pup is confronted with both an olfactory and a thermal gradient, arranged orthogonally, 3-day-old hamster pups seem to ignore the olfactory gradient while attending to the thermal gradient, rushing into the heat whether or not it is paired with nest shavings (Fig. 1). It is not until P7–8 that a preference for nest odors clearly emerges independent of preference for warmth (Crandall and Leonard, 1979).

Interestingly, in the period P8–12, thermotaxis wanes and physiological thermoregulation improves (Leonard, 1974b; see Section 3.1). However, damage to the olfactory system (olfactory bulbectomy, lateral olfactory tract section) postpones by 4 to 7 days the normal decline in thermotaxis, whereas subsequent neural recovery (when it occurs, reinnervation of the olfactory tubercle) is coincident with a nearly normal decline (Leonard, 1978; Small and Leonard, 1983). Maturational changes in the olfactory system between P7 and P9, particularly changes in bulbar projections to the olfactory tubercle (Leonard, 1975; Small and Leonard, 1983; Grafe and Leonard, 1982; Schoenfeld *et al.,* 1979), may contribute to the emerging role of nest odors in guiding orientation and preference behaviors. Nest odors probably function in a way similar to nest warmth by keeping the pups tied to the nest (an olfactory "tether": Devor and Schneider, 1974). For the four odd days (P8–12) after the "thermal tether" begins to fray, pups are still incapable of actively searching for and consuming their own solid food (see Section 3.2.2) and must be kept close to the dam for suckling. An olfactory tether would therefore be adaptive to function as a protection against inadequate maternal retrieval during this period. Clearly, P8 does not mark the *onset* of olfaction in the hamster, but more specifically the beginning of a period in development when odors assume a more predominant and evolving role in the control of behavior.

At P12, hamster pups typically leave the nest for the first time as they begin to search for their own supply of solid food (Dieterlen, 1959; Daly, 1976; see Section 2.4.2). Two recent studies have documented changes in the development of olfactory-guided exploratory behavior that coincide with this first nest egression. Johnston and Coplin (1979) found that pups at P14 will avidly sniff and lick at glass slides containing a drop of a test substance such as vaginal discharge or carrot juice while ignoring a clean slide. Pups at P7 and P10, however, will spend little or no time at either the test slide or the clean slide, even when they are positioned only a centimeter away. Pilot studies conducted by these authors showed that the surge in sniffing frequently begins at P12 but peaks at P14. Sniffing of the test odorants was longer on the average at P14 than at either P17 or P20, but Johnston and Coplin also found that pups at P14 were more likely to sniff longer if their eyes had not yet opened. Thus, the decline in sniffing after P14 may reflect the introduction of competing stimuli and elicitation of behaviors incompatible with sniffing at the slides, rather than a decline in the saliency of odors for eliciting sniffing. The fact that all stimuli were more effective on sniffing at P14 than at other ages may reflect something more about the development of odor-induced sniffing than about olfactory processing per se, since hamster pups can in fact detect and discriminate both conspecific and novel test odorants well before P14 (see above). Interestingly, long and frequent bouts of sniffing do not become prominent in rats until about P11 (Alberts and May, 1980a). Perhaps in the hamster, the olfactomotor mechanisms that underlie sniffing are not fully operable until about this same time.

We have recently found that hamsters P12 and older show a surge in locomotor excursions in the presence of novel odors (Schoenfeld and Corwin, 1982). In our test paradigm, the movements of an entire litter (7 to 8 pups) were observed from the time that they were placed into a test cage until the time that they began to huddle together. The amount of prehuddling locomotion proved to be a sensitive indicator of differences in age and odor conditions. When the pups were introduced into a test cage containing partially soiled pine shavings, it took them about 2 to 3 min to reaggregate into one or two huddles, regardless of their age (P3–18). Up to the age of P12, pups took a comparable period of time to establish huddles regardless of the odor or the substrate (fresh, lemon-scented, or soiled shavings). In each case, the hamsters locomoted briefly about the cage before finding one another and settling into a huddle. However, at P12 and thereafter, pups took significantly longer to reaggregate into a huddle in the presence of novel odors, engaging in extensive prehuddling locomotor exploration of the test cage (Fig. 2). We interpret this increased responsiveness as reflecting a greater

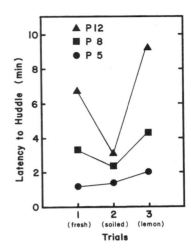

Figure 2. Duration of prehuddling exploration by litters of seven to eight hamster pups as a function of age and the odor condition of pine shavings in successive trials.

salience of novel odors at P12. Novel odors elicit exploration at P12 in spite of the pups' ultimate preference for conspecific odors as manifested in their eventually huddling (Leonard, 1982; Alberts and Brunjes, 1978; see Section 4.2.2). Indeed, we find that during prehuddling exploration P12 pups frequently come together as if they were settling into a huddle, only to disperse again and continue exploring for several more minutes. Moreover, the pups seem to already have the motor capability for exploratory locomotion at about P8. The style and pace of locomotion at this age is not much different than that seen at P12, and the duration of locomotion preceding the formation of a single huddle is nearly identical at P8 and P12 in the presence of partially soiled shavings (Fig. 2). An exploratory response to novel odors at P12 also does not reflect a newly acquired discrimination per se because pups will move away from such odors as early as P3 (see above).

Thus, nest egression at P12 may reflect an increased responsiveness to nonnest odors that may depend in part on maturing mechanisms for orientation to and exploration of all odors. Although the neural mechanisms underlying olfactory orientation and exploration remain largely unstudied (cf. Welker, 1964; Macrides, 1975), we are encouraged by recent correlative evidence that some primary olfactory and associational afferents of the olfactory tubercle and lateral entorhinal cortex establish connections at the same time (P9–12) that olfactory locomotor exploration emerges (Leonard, 1975; Schwob and Price, 1978; Schoenfeld and Corwin, 1982). Both the tubercle and the nucleus accumbens, to which the lateral entorhinal cortex projects, have direct access to systems for elicitation of locomotion through their projections to the ventral pallidum, lateral hypothalamus, and substantia nigra (Heimer and Wilson, 1975; Newman and Winans, 1980; Fink and Smith, 1980; Swanson *et al.,* 1982). Thus, it may be significant that a critical refinement of their connections with the olfactory system coincides with the maturation of odor-elicited locomotion.

2.1.2. Effect of Rearing Conditions

Dieterlen (1959) used the procedure of sensory isolation to determine whether the normal development of golden hamsters is dependent on early experience with odors. He removed pups from their mothers at P4–10 and raised them himself for several months before testing. These were called "Kaspar–Hauser" or "isolation-reared" animals. Elaborate measures were used to prevent the Kaspar–Hauser pups from being exposed to odors other than their own and Dieterlen's. Of 38 attempts, only 10 animals survived this artificial rearing. Nonetheless, these hamsters behaved quite like normally reared animals in their responses to conspecifics in agonistic and sexual encounters and in their ability to survive independently by building nests, acquiring food, and so forth. Although Dieterlen concluded from these results that a hamster's responses to odors are not based on early olfactory experience, his failure to completely isolate pups from birth makes this conclusion untenable. As we will discuss next, a more profitable procedure for examining the role of odor experience on olfactory-guided behavior has been to manipulate rather than eliminate the odors of the rearing environment.

In an extensive series of studies, Cornwell-Jones has investigated the influence of the postnatal olfactory environment on the expression of odor preferences and aversions by infant and juvenile hamsters. This work has provided evidence that exposure to certain odors at certain times during postnatal development can modify the effect that those and other odors have in controlling the behavior of developing hamsters. Initial observations indicated that the odor preferences of infant hamsters are extremely modifiable during the first seven to ten days postpartum (Cornwell, 1975, 1976). Rearing pups from birth in novel odors such as cedar or lemon neutralized, within four to five days, the normal aversion to these odors (Cornwell, 1975). In fact, only 1 hr of exposure to cedar was necessary to make pine-reared pups more tolerant at P6–7 of the once-aversive odor (Cornwell, 1976). However, on P8, 1 hr of exposure was no longer sufficient to modify olfactory preferences, implying that a critical period had passed.

Eventually, Cornwell-Jones discovered that long-term modifications in preference are only possible with extended exposure to a novel odor. In the P8 pups, for example, preference for cedar was demonstrable if they had been reared in cedar from birth rather than exposed to cedar for just an hour before testing (Cornwell, 1976). Extended exposure need not involve rearing in the odor right from birth, however. Exposure for three days before testing was found to be sufficient to make both natural cedar and cedar-nest odors preferable to infants (P10) and at least tolerable to juveniles (P30) (Cornwell-Jones, 1979). Similar results in altricial rats (see Cornwell-Jones, 1979) and dissimilar results in precocial species (spiny mice and guinea pigs: Porter and Etscorn, 1976; Carter and Marr, 1970) led to the conclusion that altricial rodents have a longer period of sensitivity to early olfactory experience than precocial rodents, one that extends beyond infancy and up to puberty (Cornwell-Jones, 1979). It should be noted, however, that altricial hamsters and rats are not completely alike in their responsiveness to odor experience. The odor preferences of even adult hamsters are modifiable by acute odor experience (Cornwell-Jones, 1979; Cornwell-Jones and Kovanic, 1981).

The adaptive value of olfactory modifiability in altricial pups is self-evident:

adapting to initially aversive odors in or near the nest helps to keep pups in the nest, to maintain the "olfactory tether." The adaptive value in older animals is less clear. For juvenile and adult hamsters, adaptability to novel botanical odors may make it easier for solitary animals to establish and inhabit independent nests, without the benefit of the presence of conspecific odors that colonial species such as rats may have. However, other factors may be involved. For example, one would expect that adult female hamsters would have difficulty in adapting to new nest odors after successive shuttles between isolation and maturity unless their odor preferences were extremely flexible. Although recent evidence from studies of juvenile hamsters suggests that the odor preferences of female hamsters may be more modifiable by experience than are the preferences of male hamsters or, for that matter, rats of either sex (Cornwell-Jones and Holder, 1979), such evidence may not pertain to the behavior of adult female hamsters because juvenile and adult responses to odor exposure have been found to differ (Cornwell-Jones, 1979). Unfortunately, only adult *males* have been studied directly, and they have been found to be *spontaneously* tolerant of novel botanical odors (Cornwell-Jones and Kovanic, 1981). Such tolerance may help males to adapt to and ignore unfamiliar odors while searching for a mate, and since it depends on testosterone, it may be peculiar to males. On the other hand, "extreme flexibility" in the face of new odors, expected of females, may not be all that different from "spontaneous tolerance" of new odors, as actually observed in males, and it may be that a hormone dependency in males does not reflect a real sex difference as such (e.g., female odors can induce a testosterone surge in males that might only *phasically* mediate odor tolerance). Certainly, the responsiveness of adult female hamsters to odor exposure should be examined, especially as a function of hormonal status (estrus cycle, pregnancy, lactation).

Two additional experimental paradigms for studying the effects of the olfactory rearing environment on olfactory development have recently been employed with rats. They should be mentioned here because they constitute conceptually new directions for the study of olfactory development and are thus relevant to such study in any mammalian species (see Leonard, 1981).

Cornwell-Jones has extended her work on odor experience in rats to test the role of central catecholamines in the neurological process of olfactory adaptation. She and her colleagues have found that juvenile and adult rats do not show expected changes in odor preferences following differential odor exposure when they are treated with the catecholaminergic neurotoxin 6-hydroxydopamine (6-OHDA) at birth (Cornwell-Jones et al., 1982). This result is similar to the finding that 6-OHDA interferes with the physiological modification of the kitten's visual cortex following alterations in visual stimulation (monocular deprivation) during a critical period in development (Kasamatsu and Pettigrew, 1979), suggesting that catecholamines, norepinephrine in particular, may be generally important to stimulus imprinting and sensory plasticity. It is interesting that drug treatment within several hours of birth did not interfere with the initial establishment of an odor preference after differential odor exposure that began immediately after drug administration and extended for the first ten days postpartum (Cornwell-Jones et al., 1982). Since odor preferences during the first week of life can be modified by exposure to a new odor for as little as 1 hr (Cornwell, 1976), perhaps olfactory imprinting in the Cornwell *et al.* study occurred before the 6-OHDA was able to interfere with catecholamine function in the olfactory system. An analysis of the timecourse of the drug's effects on catecholamine levels in the brain would help to test this possibility. Waiting one to two days between drug treatment and the beginning of differential odor exposure and alternating the order of treatment and exposure onset would provide additional tests of the role of catecholamines in the initial process of odor imprinting.

Several studies in rats have shown that manipulation of the prenatal olfactory environment can influence the control of behavior by odors after birth. Rat pups will attach to the nipples of an anesthetized dam that are lavaged with citral (a tasteless substance of lemon scent) if citral is injected into the amniotic fluid two days before the pups are born, whereas pups will reject a citral-treated nipple without prenatal exposure to citral (Pederson and Blass, 1982). Moreover, pairing of a novel substance introduced into the amniotic fluid with LiCl-induced toxicosis two days before birth can successfully condition rat pups to avoid the odor of the novel substance after birth (Smotherman, 1982). These demonstrations force us to expand the conceptualization of "rearing environment" to include the prenatal environment. Whether such manipulations would be equally successful in the hamster or other species remains to be investigated.

2.2. Vision

The development of visuomotor behavior in hamsters has been studied recently by Finlay *et al.* (1980) (see also Chapter 16, this volume). Visual guidance of behavior was not observed until after eye opening, contrary to findings in rats (Gottlieb, 1971; Routtenberg *et al.*, 1978). Preference for the dark (photophobia) did not exceed chance levels until the day of eye opening (P15). On the same day, hamster pups, even those with clipped whiskers, were able to detect the edge of a real cliff, as seen in increased latencies to fall or jump from the edge. On the other hand, a preference for dropping off the shallow rather than the deep side of the cliff, a rough measure of ability to discriminate a difference in depth of 10 cm, did not emerge until ten days after eye opening. Moreover, visual approach and orienting, using sunflower seeds as stimuli, did not develop in most pups until P20, five days after eye opening. Thus, by the end of the fourth week of life, when the maternal nest is abandoned for good and pups establish independent nest sites (Rowell, 1961; see Section 4.3), visual functioning has reached relative maturity as indicated by a variety of measures (Finlay *et al.*, 1980).

Most experimental tests of mechanisms in hamster visual development, using neurological and/or environmental manipulations, have for the most part ignored changes that might have emerged during the process of development in favor of focusing on changes manifested in adult behavior (cf. Schneider, 1970; Chalupa *et al.*, 1978; see Chapter 16, this volume). Finlay *et al.* (1980), however, have recently made observations in developing infant and juvenile hamsters that have revealed some of the early consequences of experimental perturbations to the developing visual system. In their study, lesions of the superior colliculus, an area important to visuomotor behavior in the adult (Schneider, 1969), were made in newborn hamster pups. Some of these animals were then reared in total darkness from P14 (a day before normal eye opening) until P28 when they were tested following 6 hr of exposure to light. Collicular lesions had no effect on the development of photophobia, but did alter the development of the other visuomotor behaviors mentioned earlier. The development of visual sensitivities to the presence of a cliff edge and to the height of a cliff were both slightly retarded by neonatal lesions, although both functions rose to normal levels within a few days. Visual approach and orienting were more severely disrupted and did not recover to normal levels. Dark rearing had no effect on sham-operated animals with intact superior colliculi. However, dark rearing exacerbated lesion-induced deficits in visual orienting and induced deficits in edge and depth perception that would normally have recovered from neonatal colliculectomy.

The range of these results may reflect the complex interaction of tectal and nontectal mechanisms in the visuomotor behavior of the neonate, but they may also depend on the degree of success of neural reorganization following an early lesion. Our concern here is with mechanisms of visual development (see Chapter 16, this volume, for discussion of neuroplasticity of the visual system), and at present there are few conclusions that can be made. The superior colliculus seems to be as important to peripheral visual orienting in the infant and juvenile as it is in the adult. It is less critical to other visuomotor behaviors in all age groups. Although dark rearing was ineffective in intact animals, visual experience may still play a role in visual development, according to reports that *patterned* deprivation of light (monocular deprivation, stroboscopic rearing) does produce deficits in visually guided behavior of adult hamsters (see Chalupa, 1981). The early impact of such restriction on infants and juveniles is unknown, however.

2.3. Other Senses

Vestibular reflexes appear soon after birth. Daly (1976) reports that pups do not self-right until P4–5, whereas Finlay et al. (1980) report slow self-righting at P3. In the rat, all pups will self-right at birth given sufficient time (Altman and Sudarshan, 1975), but the latency declines sharply between P0 and P4 (Almli and Fisher, 1977). Comparing righting latencies at P3 in the hamster (Finlay et al., 1980) and the rat (Almli and Fisher, 1977) reveals a similar stage of maturation at this age for the two periods. Moreover, hamster pups at P2 reveal functional vestibular sense in attempting to turn upward when facing down on an inclined ramp, although they are unsuccessful at completing the full 180° turn without falling (Corwin and Schoenfeld, unpublished data). Rat pups behave similarly at this age (Almli and Fisher, 1977).

Hamsters respond to somatosensory stimuli as soon as they are born. This can be seen in their responses to thermal and tactile cues and in their rooting behavior. The exothermic neonate hamster can readily return to the warm nest when displaced a short distance and is seen to struggle in a huddle with littermates for the most central and warmest part of the pile (Leonard, 1982; see Sections 3.1, 4.2.2). Pups also move quickly into the warm end of an artificial thermal gradient (Leonard, 1974b; see Section 3.1). In the first six days postnatal, hamster pups will show a whole body response, frequently withdrawal, to poking with a blunt probe (Finlay *et al.,* 1980; Corwin and Schoenfeld, unpublished data). After P6, pups are capable of a more localized response such as limb withdrawal (Finaly *et al.,* 1980). Not surprisingly, this localized tactile reflex predates the first appearance of scratching and grooming movements (Daly, 1976; see Section 3.4). Rooting behavior is seen at birth as a pup uses its snout to dive under and push past a crowd of other pups towards the dam (Dieterlen, 1959; see also Finlay *et al.,* 1980). It is also an important component of the active huddling behavior of pups during the first week after birth, as each pup in turn dives toward the warmest part of a huddle (Leonard, 1982; see Section 4.2.2).

Auditory development in the hamster is not well documented. Dieterlen (1959) reported that hamster pups first respond to sounds at P16–18, but the details of testing were omitted. Rats respond to sounds by P10, but only show a mature response at P15 (Almli and Fisher, 1977). Since the inner ear of hamsters is nearly complete histologically but immature in growth at P10 (Pujol *et al.,* 1975), quantitative testing of auditory sensitivity in hamster pups might reveal the same progression of function from P10 to P15 that is seen in rat pups.

The development of taste has not been studied in the hamster. However, we do expect that hamsters should be capable of rudimentary taste discrimination by P10, when they first begin to sample and ingest solid food (see Section 3.2.2). It is conceivable that they can taste even earlier, since rats display taste responsiveness more than a week before they are weaned to solid food (Hall and Bryan, 1981). Theory awaits data.

2.4. Activity and Movement

In treating sensory development thus far, we have reviewed numerous studies that have focused on the developing sensory control of movements such as sniffing and locomotion. In this section, we wish to consider briefly how some of the movements themselves are changing with the advancing age of the hamster pup. Also, we want to mark an important milestone in the development of activity and movement in any altricial animal, the time of nest egression.

2.4.1. Postnatal Changes in Mode of Movement

A striking feature of locomotion in the newborn hamster is its effectiveness in the face of its inefficiency. As with most altricial rodent pups, hamsters are unable to support the weight of their own body with their limbs when born. As described by Dieterlen (1959) (see also Crandall and Leonard, 1979), newborn pups lie on their bellies with limbs splayed out laterally. To move forward, the forelimbs only are used in an alternating rhythm, similar to "rowing," while the hindlimbs, not used, are dragged along. Despite the inefficient manner, hamster pups move quite quickly this way, even over pine shavings. This is particularly true of hamsters when they are placed in the cool end of a thermal gradient. Pups less than a week old will literally propel themselves into the warm end from the cool end of the gradient . Yet, they are quiescent when placed directly over warmth (Leonard, 1974b; Crandall and Leonard, 1979; see Section 3.1). Neonatal hamster pups also quickly reaggregate and huddle together when scattered in an open field (Leonard, 1982; Schoenfeld and Corwin, 1982).

Thermal cues are far less effective in guiding the movement of rat pups, but the difference probably does not involve fundamentally less efficient locomotor capability in rats. When neonatal rat pups begin to move along a thermal gradient and into the heat, they move somewhat more slowly than hamster pups do (Kleitman and Satinoff, 1982). A more important difference about rat pups, however, is that they do not begin to respond to the gradient until 40 to 60 min have elapsed (Kleitman and Satinoff, 1982), whereas hamster pups respond immediately (Leonard, 1974b; see Section 3.1). Cold seems to immobilize rat pups at first, but when they do begin to move along a thermal gradient, they can reach the hot end within several minutes (Kleitman and Satinoff, 1982). Thus, rats and hamsters are not greatly dissimilar in their basic capacity for locomotion during the first week. Rather, they seem simply to process thermal cues by different mechanisms (see Kleitman and Satinoff, 1982).

There are other examples of similarities in the development of movement in the two species. Both hamsters and rats have limited use of the hindlimbs when born and so move by the rowing motion of the forelimbs (Dieterlen, 1959; Altman and Sudarshan, 1975). Varying degrees of either forward locomotion or pivoting about the hindquarters are seen in both species during the first week (cf. Bolles and Woods, 1964; Altman and Sudarshan, 1975; Crandall and Leonard, 1979). Causes for the appearance of one or the other are poorly documented but seem to depend on a combination of factors, including changes in ambient and core body temperature, presence of conspecifics and degree of tactile stimulation from floor shavings and other objects (Corwin and Schoenfeld, unpublished). Crawling with all four limbs appears at P6–8 but soon gives way to walking (ventrum off the ground) by P10–15 in both hamsters and rats (Dieterlen, 1959; Crandall and Leonard, 1979; Altman and Sudarshan, 1975; Almli and Fisher, 1979). Eventually, a surge of activity occurs in both species by P15–20 (Dieterlen, 1959; Campbell and Mabry, 1972; Bolles and Woods, 1964). Although this surge is correlated with eye opening, it is not necessarily caused by eye opening for it builds gradually over several days rather than overnight (cf. Campbell and Mabry, 1972; Bolles, and Woods, 1964). The surge involves in part the appearance of new movements such as galloping while running and hopping without direction when surprised (Dieterlen, 1959; Bolles and Woods, 1964), as well as more exploratory-type movements such as rearing and climbing (Dieterlen, 1959; Daly, 1976; Altman and Sudarshan, 1975; cf. Campbell and Mabry, 1972; Campbell and Raskin, 1978). In the mouse, jumping behavior that also develops by P15 has been shown to be a genetically dominant trait, indicating that the timing of the development of jumping behavior probably confers some adaptive advantage to the 15-day-old mouse (Henderson, 1981). Such behavior presumably functions in the mouse as a temporary defensive strategy while the pup adjusts to its newly acquired visual capacities and before it has had sufficient experience to defend itself by other means against intruders (see Henderson, 1981). Hopping behavior in hamster and rat pups may serve a similar function.

2.4.2. Nest Egression

According to several reports (Dieterlen, 1959; Daly, 1976; Goldman and Swanson, 1975), hamster pups emerge from the maternal nest for the first time at P11–12 on the average. This first "foray" (Alberts and Leimbach, 1980) involves the first excursion by the pup in the world outside of the nest and permits two specific activities: (1) establishing a site away from the nest for urination and (2) foraging for nonmaternal supplies of food (Dieterlen, 1959; see Sections 3.2, 3.3). These excursions become more frequent with age until P25–30 when pups begin to establish their own nest sites. However, as long as the maternal nest is maintained, pups seem to return to it (Rowell, 1961).

The factors controlling nest exits and returns by hamster pups have not been studied, but initially odors undoubtedly predominate. Pups are more likely to engage in locomotor explorations of novel odors at the same age (P12) that nest egression is first seen (Schoenfeld and Corwin, 1982; see Section 2.1.1). Similarly, preference for nest odors over novel odors persists until about P15 (Devor and Schneider, 1974) and an inclination to huddle with conspecifics following bouts of locomotor exploration continues to at least P18 (Schoenfeld and Corwin, 1982). Maternal and conspecific odors may together comprise an olfactory "tether" (Devor and Schneider, 1974) that places limits on the extent of extra-nest excursions and thereby keeps the hamster family together for at least three weeks (see Section 4.2).

Nest egression by rat pups, which does not begin until about P19, seems to be largely determined by socially derived visual and auditory stimuli (Alberts and Leimbach, 1980). Egression is facilitated by the presence of the dam in the nest and is largely a reaction to moving and vocalizing conspecifics outside of the nest. Soon after the first nest egression, rat pups are then guided by conspecifics to the first feeding site (Galef and Clark, 1971). The role of visual cues in guiding hamster egression and ingression once the eyes open at P15 is unknown. Although nest exit in hamsters also occurs at about the time that solid food is first consumed, hamster pups apparently forage on their own without maternal guidance (Dieterlen, 1959). Moreover, hamsters do not orient well to sunflower seeds until P20, five days after eye opening (Finlay et al., 1980). Thus, visual stimuli probably do not play a major role in controlling hamster pup activity before the beginning of the fourth week.

3. MAINTENANCE BEHAVIORS

The young of altricial rodent species depend on the dam for a number of life-sustaining activities. At birth, hamster pups are incapable of foraging for and consuming solid food and their bodies cannot produce sufficient heat to regulate core temperature. Newborn pups are also incapable of eliminating waste by themselves and cannot clean themselves. Nonetheless, even neonates display some behaviors by which they help to maintain their own fitness. Pups are able to attach to and suckle milk from the dam's nipples without assistance from the dam. Moreover, they orient to and remain near the heat of the nest without constant need for maternal retrieval. In describing the maintenance behaviors of the developing hamster pup, we will be distinguishing between two kinds of developmental processes (see Oppenheim, 1980): the relatively late emergence of adult forms of adaptive response patterns (waste elimination, grooming), contrasted with metamorphic transitions between infant and adult forms of adaptive behavior (thermoregulation, ingestion).

3.1. Thermoregulation

Hamsters, like other rodents, cannot thermoregulate at birth. In fact thermoregulation takes an unusually long time to develop. Hissa, in a thorough analysis (1968), determined that no deviation from passive heat loss occurred until after P9 (see Fig. 3). Mice, rats, and, in his study, lemmings all showed some signs of thermoregulation by P4. Hamster pups have little brown fat and do not show an O_2 increase to cold stress until P12 or to injected noradrenalin until P17.

In contrast to this slow development of "physiological" thermoregulatory mechanisms, hamster pups demonstrate a finely tuned behavioral thermoregulation. They orient to very fine temperature gradients and locomote rapidly toward warmth (Leonard, 1974b). Then they continually adjust their position, maintaining their surface temperature constant. Depending on how well the artificial thermal conditions mimic those in the nest, the pups' temperature either drops slowly, at a rate of 0.1°C/min, or rises to 41° to 42° at which point the pups begin leaping about in some apparent discomfort. Under the conditions where the pups' core temperature drops, their belly temperature remains constant, leading to the suggestion (Leonard, 1982) that surface temperature is the regulated variable during the first postnatal week. Rats show a much less rapid behavioral orientation to temperature. Over a period of an hour rat pups will move down a gradient to a zone of thermal stability (Kleitman and Satinoff, 1982). A more rapid thermotaxis response can be demonstrated in P3 rat pups deprived for 24 hr (Johanson and Hall, 1980), but the response is lost by P6.

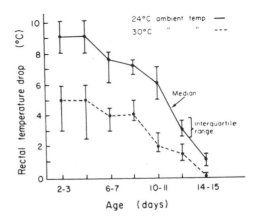

Figure 3. Effect of age on temperature regulation in the hamster: Rectal temperature drop after 15 min out of the nest at either 24°C or 30°C. The pups' temperature at the start of the test varied between 37 to 38°C. (N = 20 observations in each condition.) [From Leonard (1974b). Copyright 1974 by the American Psychological Association. Reprinted by permission of the publisher.]

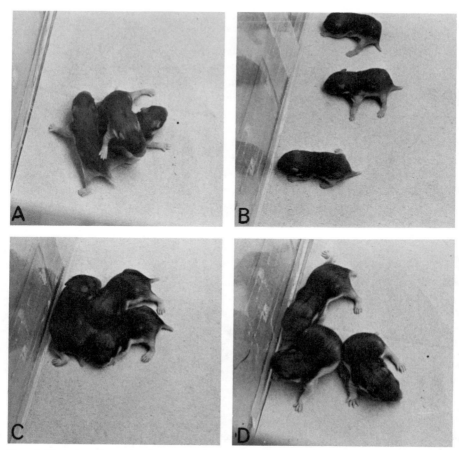

Figure 4. The variety of contact and noncontact behavior of hamster pups on a thermal gradient. (A) P5 pups rooting at 22°C (no gradient). (B) P5 pups spread out, away from the warm edge (34°C) on the strong gradient (warm edge at left). (C) P9 pups in passive contact at the warm edge (30°C) on the mild gradient. (D) P9 pups in a quiet huddle (the pups' ventra are off the substrate, and they are swinging their heads slowly back and forth). [From Leonard (1982). Copyright 1982 by the American Psychological Association. Reprinted by permission by the publisher.]

In the nest, hamster pups behaviorally thermoregulate by rooting, an active behavior in which each pup successively dives into the center of the pile, displacing its littermates (Fig. 4A). The amount of rooting shown is a function of the rate of core temperature drop and the age (Leonard, 1982; see Section 4.2.2). Pups whose temperature is dropping more gradually lie quietly together (Fig. 4C). If their core temperature starts to rise, pups younger than P8 separate (Fig. 4B), whereas older pups are more likely to maintain social contact (Fig. 4D) and allow their temperature to rise (Leonard, 1982).

Hamster pups are also known to vocalize, both audibly and ultrasonically, when core temperature is dropping (Okon, 1971). Rate of calling for both kinds of vocalizations is greatest in the first 10 min of exposure to cool ambient temperatures. After that time, only ultrasounds can be discerned. Hamster and rat pups differ in the development of vocalization. Hamsters vocalize from birth, whereas rats begin to vocalize at P2. Also, rats do not produce audible calls (Okon, 1971).

Many relatively sharp changes in the pup response to temperature occur at the end of the first postnatal week. The pups no longer rush headlong into the heat when placed on a thermal gradient (Leonard, 1974), they start to prefer areas with a combination of nest odor and warmth over areas in which each cue is available separately (Crandall and Leonard, 1979) (see Fig. 1), and their latency to form a huddle in a cool environment increases (Schoenfeld and Corwin, 1982) (see Fig. 2). They seem more tolerant of temperature changes in either direction, since they will wander around in the cold while their temperature drops (Leonard, 1974b; Small and Leonard, 1983) or will huddle together in the heat while their temperature rises (Leonard, 1982). Audible vocalizations become much less frequent and intense after the first week, concomitant with the decline in active rooting and emergence of homoiothermy. Thus, audible calls are closely associated with the early period of strong thermal orientation and exothermia in developing hamsters. Perhaps they constitute yet another strategy for behavioral thermoregulation. In providing an audible indication of pup and, indirectly, nest temperature, audible calls may help to entice a neglecting dam back to a nest that has begun to cool off in her absence (see Section 4.2.1).

In view of the wealth of evidence that olfactory cues increase in importance after P7 (Devor and Schneider, 1974; Gregory and Bishop, 1975; Crandall and Leonard, 1979; Schoenfeld and Corwin, 1982), we have hypothesized that the seeming neglect of temperature information after P7 reflects olfactory competition for the control of behavior (see Section 2.1.1). Although the pups have some rudimentary olfactory sensitivity at birth and prefer maternal shavings by at least P3 (Crandall and Leonard, 1979), thermal information is much more important in guiding their behavior (see Fig. 1). The pups probably perceive their mother and their littermates mainly as thermal stimuli. By the middle of the second postnatal week, however, the pups will maintain contact under a much greater variety of thermal situations.

Physiological regulation of core body temperature reaches maturity in the third week. Early in the third week, rectal temperature can be maintained to within a degree when hamster pups are left alone at moderately cool ambient temperatures (22 to 30°C: Leonard, 1974b, 1982; Okon, 1971) (Fig. 3), although more extreme conditions (e.g., 2 to 3°C for 1 hr) continue to induce hypothermia until P21 (Okon, 1971). By P15–16, hamsters no longer prefer warmth, even if paired with nest odors, and begin to show a preference for cool parts of the environment (Crandall and Leonard, 1979) (Fig. 1), probably to avoid heat stress. The achievement of homoiothermy coincides with the development of other capabilities and responses (sensorimotor function, other maintenance behaviors, nest building, hoarding) that give the hamster, at about P15, most of the tools needed for existence independent of the dam and its littermates (see Section 4.4).

3.2. Ingestion

In all mammals, the development of ingestion comprises two phases, one when pups suckle milk from the dam and a second when pups develop the ability to consume solid food. In the hamster, ingestion of solid food begins fairly early and these two phases overlap to a greater extent than in most other altricial mammals.

3.2.1. Suckling

Hamster pups make their first suckling attempts soon after birth. The first pups born are sometimes suckling from the teats by the time the next are delivered (Dieterlen, 1959; Rowell, 1961), although active suckling does not usually ensue until delivery of all pups is completed (Siegel and Rosenblatt, 1980). Hamsters are born with incisors with which they can pull out and secure the relatively small nipples of hamster dams (Hall and Rosenblatt, 1979; Lakars and Herring, 1980). Interestingly, there is a 1 mm^2 gap between the incisors of the newborn pup that is just big enough to hold a 1 to 2 mm diameter nipple (Lakars and Herring, 1980) (Fig. 5). Although the mouth and

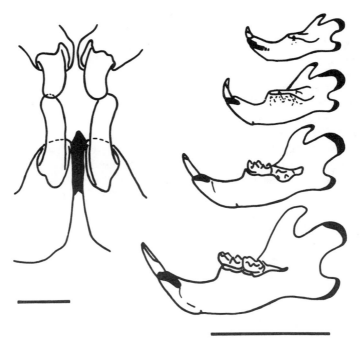

Figure 5. Dental and mandibular development in the hamster. On the left is a tracing of newborn dentoskeletal relations in anterior view, as seen in a stained and cleared skeleton. Scale = 1mm. On the right side are tracings of the lingual surfaces of mandibles showing dentoskeletal development at P0, P4, P9, P16 (top to bottom). Note the changes in incisal form after wear at P9 and the development of molars. Scale = 10 mm. [From Lakars and Herring (1980). Reprinted by permission of Alan R. Liss, Inc., publisher, and the authors.]

jaw can be opened by contraction of abductor muscles at birth, closure is accomplished passively by "elastic recoil of stretched oral tissues" (Lakars and Herring, 1980, p. 248). Attachment to the nipple may not involve biting per se but instead a more gentle grasping accomplished by compression of the diamond-shaped interincisal gap about the nipple.

The development of suckling behavior has been extensively studied in rat pups (see review by Blass et al., 1979). Using many of the same approaches employed to study rat suckling, Hall and Rosenblatt (1979) conducted a comparative study of suckling in hamster pups. Suckling by hamsters was found to go through several transitions during development that are similar to those seen in the development of suckling in rats. Right after birth, hamster pups rapidly attached to and suckled from the nipples of an anesthetized dam even if sated and even though the nipples produced no milk. This was followed by a period when the speed of attachment to a nipple became increasingly dependent on the degree of deprivation of suckling, food, and water. Eventually, hamster pups began to shift from nipple to nipple instead of remaining attached to one nipple as younger pups do (Fig. 6), an indication of their first response to the lack of milk let-down in the anesthetized dam. By P25, pups stopped attempts to suckle altogether, even under deprivation conditions.

Although the sequence of events in the development of suckling is similar in rats and hamsters, the timetable is not. Hall and Rosenblatt (1979) found that deprivation facilitates nipple attachment in hamsters for a relatively short period (P15–20), whereas in rats, deprivation persists as a suckling stimulus well beyond weaning. Deprived rat pups attach quickly even at P25–30; only at P35 did rat pups fail to attach even under deprivation (see Hall et al., 1977). Deprived hamster pups, however, cannot be induced to suckle at P25. In addition, nipple shifting begins sooner in hamsters, increasing sharply in incidence between P5 and P10 and peaking at P15 (Fig. 6). Only a few rats shift between nipples at P15; not until P21–22 do most rat pups display this behavior. The significance of this latter difference is not clear (see Hall and Rosenblatt, 1979). Hamster pups begin to eat solid food earlier and can be weaned earlier than rats (see

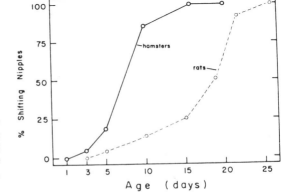

Figure 6. Percentage of deprived hamster and rat pups that shift nipples at least once during a 5-min test. [From Hall and Rosenblatt (1979). Reprinted by permission of John Wiley & Sons, Inc., publisher, and the first author.]

below), but these events are not directly caused by a failure of milk production and ejection because lactation, in the laboratory dam at least, usually extends beyond the time when nipple shifting reaches its peak (Siegel and Rosenblatt, 1980). On the other hand, there may be occasions in the wild when lactation declines earlier than usual because the dam's nipples are not regularly stimulated by suckling. This might happen if either she or the pups spend significant periods of time away from the nest, in an early dissolution of the family (see Section 4.3), or if the dam cannibalizes all but a few pups in the litter (see Siegel and Rosenblatt, 1980). Sensitivity to a decline in milk output, as displayed in nipple-shifting behavior, would provide the suckling pup with a means to regulate its milk intake.

3.2.2. Food Consumption

Suckling of liquid food—milk—is eventually supplemented by consumption of two other forms of food: semisolid digestive excreta, particularly maternal caecotrophe, and solid food comprising the maternal diet. "Caecotrophia" by hamster pups has been discussed extensively by Dieterlen (1959). Caecotrophe is a special form of fecal matter that originates in the caecum (Harder, 1949). It is typically softer and lighter in color than feces that are produced in the colon proper. It has important nutritive value for the pup because it contains vitamins and other nutrients such as protein that are essential to establish an adequate bacterial population in the gut of the pup (see Dieterlen, 1959; Leon, 1974). Dieterlen (1959) observed hamster pups sucking caecotrophe from their mother's anus as early as P5. A change in the appearance of the stool of pups coincided with the beginning of caecotrophia. Between P5 and P6, the stool goes from soft and yellowish-brown to a more solid and dark brown, a clear indication of a shift in pup diet occurring four to five days before solid food is first consumed. The onset of caecotrophia also coincides with the earliest appearance of active jaw closure (i.e., by muscle contraction) and a critical period in the development of the circuitry for proprioceptive control of jaw movement in hamsters (Lakars and Herring, 1980). Hamster pups continue to consume caecotrophe excreted by the dam for as long as they are suckling, up to P25. Neonatal guinea pigs are also reported to consume maternal caecotrophe, but they exhibit caecotrophia on the day of birth, when they also begin to both eat and suckle (Harder, 1949; cited in Dieterlen, 1959). In rats, on the other hand, maternal caecotrophe is only first produced in quantity and consumed by pups on P14 (Leon, 1974), much later than in hamsters, but still several days before solid food is first consumed by the pups. Thus, there is a clear relationship between the onset of caecotrophia and the onset of solid food consumption in rodents. It may be, as Leon (1974) has propsed, that maternal caecotrophe serves as "baby food" for the infant rodent, facilitating the transition from milk to solid food during the weaning process.

By P8 or so, hamster pups can be observed holding and mouthing pieces of the mother's food while lying in the nest. However, they do not actually begin to eat solid food until P10 (Dieterlen, 1959; Daly, 1976). Indeed, occlusion of the first molars, necessary for mastication of solid food, only occurs at about P10, whereas closure of the incisors, critical to biting but spread apart at birth, occurs a day or two earlier (Lakars and Herring, 1980) (Fig. 5). Until P12 the pups continue to lie in the nest to eat. After P12 or so, however, pups leave the nest (see Section 2.4.2) and procure their food from outside of the nest proper (Dieterlen, 1959). As noted above, suckling of milk continues after pups begin to consume solid food, but changes somewhat; that is, pups begin to nipple shift (see Section 3.2.1). Although adult hamsters may have difficulty compensating for an energy storage deficit (see Chapter 15, this volume), the coincidence of nipple shifting and eating during development may indicate that young hamsters, starting at about P12–15, are capable of responding to changes in food availability and of seeking and identifying alternate food sources. The extent to which pups at P10–25 actually regulate their consumption of various nutrients as taken from maternal milk, caecotrophe, and solid diet is not known.

3.3. Waste Elimination

Independent defecation appears before independent urination. The dam must help the newborn pup to eliminate waste, using gentle massage and licking of the ventrum and anus. Dieterlen (1959) claims that hamster pups can defecate without massage by about P6. By P7, he notes, the pups pull the boluses out of the rectum with their mouths and soon after, by P10, endeavor to place each bolus outside of the nest. Daly (1976) has reported more recently that golden hamsters do not defecate independently before P8 and usually not before P11. The reason for the discrepancy between these two reports is not apparent. By our own observations, golden hamsters defecate and urinate freely when tested alone at P10 under moderately stressful conditions (Corwin and Schoenfeld, unpublished). Younger animals did not excrete as freely in our test situation. One would expect that defecation can proceed more independently once solid fecal matter is produced in the intestinal tract, since rectal expulsion normally occurs reflexively from local stimulation by accumulating feces. Thus, the first production of a firm stool, at P6, consequent to the onset of caecotrophe ingestion at P5, should provide the appropriate stimulus for defecation independent of maternal massage, at about P6 as Dieterlen (1959) suggests.

Before P10, a pup's bladder is usually voided by maternal massage, although we always see one pup in six or seven that can urinate somewhat earlier. At P10, most pups release urine by themselves, but only outside of the nest (Dieterlen, 1959). It is not surprising that the ability of pups to urinate independently develops concomitantly with their inclination to leave the nest and to investigate odors outside of the nest (see Sections 2.1.1, and 2.4.2). Adult hamsters defecate anywhere but urinate only outside of the nest (Dieterlen, 1959). Moreover, urination is more ritualistic than defecation. Hamsters tend to seek a spot where urine has been deposited before, and then, when urinating, stand with legs apart, rear quarters lifted and backed against a wall (see Dieterlen, 1959). Until they establish their own nests, pups seek out the place chosen by the dam for urination. At first, they seem to lose the urine along the way to the dam's spot for urination, but then develop the ability to hold the urine until reaching the appropriate place (Dieterlen, 1959). This timetable corresponds roughly to Daly's (1976) observation that golden hamsters show independent urination as early as P11 but more usually at P14 and to our own observations, noted above, that hamsters can urinate outside of the nest at P10. Of course, in both urination and defecation, the development of the excretory muscles—in the bladder and rectum—may have a great deal to do with the observed timetables. For example, is maternal massage required because the excretory muscles are not operative in the first week? Although this issue may seem of minor importance, its clarification could aid the study of other developmental events to which the development of waste elimination may be related, such as the onset of feeding (see Section 3.2) and the onset of behaviors such as nest building that facilitate the breakdown of the family group (see Section 4.3).

3.4. Grooming

Golden hamsters show rudimentary groominglike movements as early as P2, usually simultaneous stroking by the two forepaws over the snout (Dieterlen, 1959). Although recognizing that these early movements are precursors of mature grooming, Dieterlen preferred to think of them as having a function more important to the neonate. He called these movements "Abwehrbewegung" (p. 66) or "movements of defense," since they are responses to irritative stimulation about the snout, as when a pup attempts to wipe about its snout immediately before and after sneezing (see Dieterlen, 1959). Since the forepaws do not actually touch the snout, it is doubtful that such movements serve this purpose.

More definitive grooming begins in the second week postpartum. Hamster pups begin to lick their forepaws by P8 (Dieterlen, 1959); a coordinated sequence of licking and brushing about the head and snout appears soon after at P10 (Dieterlen, 1959; Daly, 1976). Coordinated licking and brushing of the whole body first appears by P13 and is displayed by 50% of pups by P15 (Daly, 1976; Dieterlen, 1959). The age when all pups show complete body grooming has not been reported, however. Hindleg scratching also appears during this period (Dieterlen, 1959; Daly, 1976).

Rats seem to develop similarly (cf. Daly, 1976). They also begin to display rudimentary (contactless) grooming about the face by P2 (Bolles and Woods, 1964), more mature forms of snout and head grooming by P10, and both whole-body grooming and hindleg scratching by P13 (Bolles and Woods, 1964; Almli and Fisher, 1977). As with hamsters, half of all rat pups do not show mature forms of whole-body grooming until after P15 (Almli and Fisher, 1977).

These similarities mean that the development of grooming is not correlated with the development of other maintenance behaviors (ingestion, waste elimination), at least across species as divergent as rats and hamsters. Nevertheless, the emergence of grooming is suitably timed for hamsters. Head and snout grooming first appear in golden hamsters when they begin to consume solid food for the first time. Moreover, wholebody grooming is first displayed when hamster pups begin to eliminate urine and feces independently and spend considerable time outside of the nest away from the dam's careful licking. Interestingly, age difference among *hamster* species in the development of grooming *are* closely correlated with age differences in the development of ingestion and elimination (Daly, 1976).

4. THE FAMILY GROUP AS A SPECIAL FEATURE OF HAMSTER DEVELOPMENT

4.1. Affiliative Behavior in Social Isolates

Rowell (1961) was perhaps the first to observe that the establishment and maintenance of an affiliative family group by a hamster dam and her litter is an uncommon event for both because hamsters normally live in isolation from conspecifics. Undoubtedly, this intrusion into the dam's normally solitary existence requires special controls over her behavior (see Rowell, 1961; Siegel and Rosenblatt, 1980). However, what about the pups? How is their early affiliative behavior controlled? What behaviors characterize their affiliations within the maternal nest? What is the pups' role in maintaining the family unit during their early development? How does the transition from affiliative to agonistic existence occur? Since this particular issue has not received much attention since Rowell's observations of 20 years ago, we can only attempt a restatement of this important problem in light of the modicum of new evidence that exists.

The family cycle actually begins during late pregnancy when the dam starts to build a nest (Siegel and Rosenblatt, 1980; see Chapter 10, this volume). The family unit is then formally established with parturition and is maintained for about 25 to 30 days until the pups stop suckling and both the dam and the pups establish independent nest sites (Rowell, 1961). Two aspects of this cycle are particularly relevant to the issue of behavioral development. One concerns the behavioral relationship between members of the family during the first three weeks postpartum when the family unit is fairly tightly maintained. The other concerns the appearance in the third to fifth weeks of behaviors, asocial behaviors in particular, that mark the dissolution of the family group.

4.2. Early Family Relationships

4.2.1. Pup–Mother

The hamster pup has two kinds of affiliations in the nest: with its mother and with its littermates. The pup–mother relationship is usually the domain of discussions of maternal behavior (see Chapter 10, this volume) and so we have only a few things to say about it.

At birth, all of the pup's energies are geared toward securing warmth and nourishment, both of which are associated with the dam and secondarily with the nest and littermates. The pup responds to the tactile, thermal, and olfactory stimuli emanating from the dam and to the visceral—largely postingestive, preabsorptive—cues emanating from its digestive tract. It is really capable of little else, an indication of tremendous specialization.

The pup's vigorous rooting, suckling, and vocalizing in turn stimulate the dam. Although the onset of maternal behavior is probably a consequence of physiological changes within the dam associated with pregnancy and parturition, the *maintenance* of maternal behavior seems to depend in part on stimulation from the pups (see Siegel and Rosenblatt, 1980). The influence of the pups is greatest when the dam is exposed to them for at least 48 hr, especially when they are in the nest. Even then, some pups will be cannibalized in each litter. We hasten to add that, excepting the unusually high incidence of cannibalism, hamster dams are adequately maternal in most other respects (see Chapter 10, this volume), particularly when compared with a species such as the Mongolian gerbil (*Meriones unguiculatus*) that only rarely retrieves young pups displaced from the nest (see Cornwell-Jones and Azar, 1982). Although we have mentioned rooting, suckling, and vocalizing as "stimulating" behaviors, it is not clear exactly what the pups do that is most likely to help maintain maternal behavior. Both rooting and vocalizing are induced by changing thermal conditions and could be expected to elicit maternal crouching and retrieving, although the incidence of vocalizing while pups are in the nest and its effect on the dam have not been studied (see Section 3.1). The effect of suckling by hamster pups is likely to be enhanced by the presence, from birth, of incisors that are spaced apart just enough to enable a tight grasp on the dam's teats (see Section 3.2.1).

4.2.2. Pup–Pup

The second sort of affiliative relationship, that between the pups themselves, is most clearly seen in two forms: huddling and playfighting.

Huddling by hamster pups was first described by Rowell (1961), who called it " 'contact behavior,' a tendency to stay with other pups, to move with them, or just to eat or groom together" (p. 90). She observed that huddling and suckling develop and decline concomitantly with age. Both behaviors are associated with the maternal nest, and when the dam leaves that nest to establish another nest in the fourth to fifth weeks postpartum, the pups stop attempts to suckle further and establish independent nest sites without further huddling. In contrast, rat pups continue to suckle, especially when food deprived, well beyond the fourth week (see Section 3.2.1) and also huddle together throughout their lifetimes (Alberts, 1978a).

Alberts and Brunjes performed an elegant series of experiments in the rat (Alberts, 1978a,b; Alberts and Brunjes, 1978) demonstrating that thermal factors control the degree of huddling up to the age of P10, but that after P10 olfactory cues also begin to influence huddling. They found that before P10 pups did not maintain contact with cold pups and the surface area of a huddle of pups varied as a function of ambient temperature, which served to hasten or retard heat loss in warm or cool environments, respectively. Pups older than P10 preferred to huddle with an anesthetized conspecific than with an anesthetized gerbil that had the same thermotactile qualities, whereas younger pups showed no preference. Zinc sulfate-induced anosmia eliminated the preference in the older pups without altering the amount of huddling. The sensory control of huddling in rats thus involves a developmental transition from thermal to olfactory "dominance" (Alberts and Brunjes, 1978).

We have come to similar conclusions in studies of huddling in hamster pups. Huddling initially is controlled by thermal stimuli and also forms an important part of

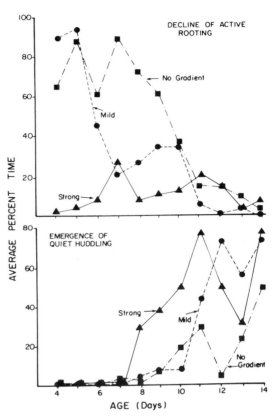

Figure 7. Effect of age and thermal gradient strength on two kinds of contact behavior in hamster pups. *Top:* Pups at room temperature (no gradient) root most of the time through P10. Pups on the strong gradient (22 to 34°C) never root. On the mild gradient (22 to 30°C), pups root most actively on P4 and P5. *Bottom:* Quiet huddling, characterized by grooming and sniffing in a crouched position, emerged at P8 on the strong graident. Pups tested at room temperature did not show an appreciable amount of quiet huddling until P14. Note that after P9–10, 20% to 40% of the observation period (8 min) is spent in noncontact behavior. [From Leonard (1982). Copyright 1982 by the American Psychological Association. Reprinted by permission of the publisher.]

the strategy for behavioral thermoregulation (Leonard, 1982; see Section 3.1). At moderate-to-low ambient temperatures and when core body temperature is declining, neonatal hamster pups engage in active rooting behavior, where each pup in turn dives into the warmest part of the pile (Figs. 4A and 7). At higher ambient temperatures, when body temperature is maintained or rising, however, pups root less and may even separate (Figs. 4B and 7). Eventually, thermal stimuli lose control of huddling as pups develop homoiothermy. A passive, quiet form of huddling emerges in the second week postpartum (Leonard, 1982) (Figs. 4D and 7) that is similar to the "filial" huddling of rat pups that is influenced by conspecific odors (Alberts and Brunjes, 1978). Establishment of huddle sites by a litter of pups dispersed in a small open field seems to involve the same transition (Schoenfeld and Corwin, 1982; see Section 2.1.2). Following dispersal, pups younger than P12 form huddles with the first littermates encountered, usually two to three pups to a huddle, and immediately engage in active rooting, as hamster pups of this age typically do when tested at an ambient temperature of 22°C (Leonard, 1982). Little exploration precedes huddling in these young pups. By P12, all pups come to one site to establish a single, passive huddle after about 6 to 8 min of extensive exploration, during which they make frequent contact with each other but do not huddle (cf. Leonard, 1982). Although prehuddling exploration is influenced by the presence of novel odors (see Section 2.1.1), the role of odors in the unitary, quiet huddling of older hamster pups has yet to be tested directly. However, the age when such huddling replaces active thermal rooting is the age at which preferences for nest odors peak (see Section 2.1.1).

Hamster pups begin to show components of play fighting as early as P14–15 (Dieterlen, 1959; Rowell, 1961; Daly, 1976), although under conditions of group housing, play fighting may appear somewhat later (Goldman and Swanson, 1975). Dieterlen (1959) suggests that the onset of play fighting depends on eye opening. According to Rowell (1961), play fighting "differs from the real thing in being slightly slower, with gentler movements, in being unaccompanied by chirping and squealing, and in that each bout typically turns into grooming by both the opponents." The lack of squealing and running away at the end of the interaction were also noted by Goldman and Swanson (1975). Play fighting may develop as an extension of the passive huddling of older pups (see above), particularly the mutual grooming and occasional food snatching that pups engage in while huddling (Rowell, 1961). Indeed, once huddling stops and the pups leave the maternal nest, play fighting continues until puberty (Rowell, 1961; Goldman and Swanson, 1975).

4.3. Dissolution of the Family Group

The major event in the break-up of the hamster family is the dissolution of the maternal nest in the fourth and fifth weeks postpartum (see below). However, there are antecedents to this break-up that occur even in the second and third weeks. These include the first "forays" by the pups outside of the maternal nest as early as P12 and the first consumption of nonmaternal solid food by pups as early as P9–10 (Daly, 1976; see Sections 2, 3). By these actions, the pups establish for the first time their developing ability to survive apart from the dam's care, well before they actually achieve such independence for good. We caution that the exact timetable in the wild is unknown. Since there is experimental evidence that restricted housing of a dam and her litter retards in some respects the achievement of independence by the pups (see below, and Section 4.4), it is conceivable that the dissolution of the family unit in the wild may begin as early as the third week. Indeed, when hamster pups are isolated before P20, they display many behaviors that might allow them to survive independent of the dam (Dieterlen, 1959; see Section 4.4). Moreover, although physiological thermogenesis does not completely mature until the end of the third week, it is sufficiently functional by the end of the second week so that pups begin to tolerate some heat loss when they preferentially engage in behaviors that are not strictly thermoregulatory, such as exploration instead of thermotaxis or huddling (see Section 3.1).

In the fourth week, the maternal nest, the focal point for the family unit, begins to lose its structure (Rowell, 1961). The female stops maintaining the original nest and moves instead to a new nest site away from the pups. The pups usually follow her to the new site, however, and she will typically move once again before she and her pups become completely independent of one another. Without frequent repair by the dam, these nests are effectively dismantled by the pups from their frequent movements in and out and their handling of the nest material as they practice building their own nests. The dam also stops retrieving the pups at about this time (Rowell, 1961); this may be either a consequence of or a factor contributing to the demise of the maternal nest. Overall, the dam generally avoids contact with her pups in the fourth and fifth weeks.

For their part, the pups, already capable of feeding themselves, initiate few bouts of nursing after about P25 (see Section 3.2.1). Instead, they show increasing evidence of their independence in their nest building, digging, and hoarding. Although there is disagreement about the exact days when these latter behaviors are first seen, all are commonly observed sometime during the fourth week postpartum (cf. Daly, 1976; Rowell, 1961; Dieterlen, 1959; Etienne et al., 1982; Crandall, 1980). It is generally agreed that hamster pups do not usually begin to establish truly independent nest sites until around P30 (Dieterlen, 1959; Rowell, 1961). However, they do exhibit mature nest building by P20–22 (Dieterlen, 1959; Daly, 1976). Dieterlen calls these early nests "kleine Spielnester" (p. 61) or "little play-nests," since they are not really maintained by the pups. Pups will handle nest material even earlier; for example, they begin to chew at nest shavings by P14 (Dieterlen, 1959; Daly, 1976; Rowell, 1961). A similar sequence and timetable occurs in the development of hoarding. Between P21

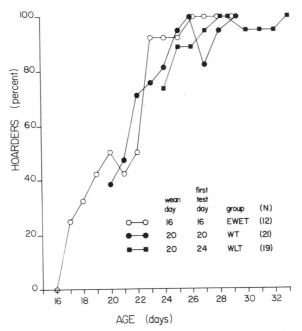

Figure 8. The ontogeny of hoarding behavior for three groups of hamster pups weaned and tested at different ages. Groups: EWET, early weaning, began testing early; WT, normal weaning, began testing at weaning; WLT, normal weaning, began testing late. [From Crandall (1980) courtesy of the author.]

and P25, the proportion of hamster pups capable of showing a coordinated sequence of hoarding behaviors increases from 40% to 95% (Crandall, 1980) (Fig. 8). The individual components of gathering, pouching, carrying, and unloading are seen in various combinations, without integration into a full hoarding sequence, as early as P14–17 (Crandall, 1980; Etienne *et al.*, 1982). Mature digging behavior appears at the same time that mature nest building is seen, around P20–24 (Dieterlen, 1959; Rowell, 1961; Daly, 1976). Unfortunately, the components of digging are not well documented and there has been little attempt to identify the earliest appearance of any fractionated components (cf. Dieterlen, 1959).

Coordination in the development of these three behaviors or "action patterns" is not too surprising when one considers that all three are used in establishing and maintaining a nest. Hamsters are burrowing animals, and in their natural habitat, their nests are made in burrows dug typically beneath cornfields (Dieterlen, 1959). To establish a nest in the wild, the hamster would have to (1) dig a network of burrows, (2) hoard nesting material, and (3) build a nest. The ability to hoard food in the burrow would be especially important to maintaining an independent nest site in a semiarid clime where food may become scarce (see Richards, 1966). Thus, a hamster pup would have difficulty leaving the maternal nest and moving to an independent site before all three of these behaviors had been established in its repertoire.

In the 20-odd days between weaning and puberty, hamster pups go through two transitional phases, each lasting about ten days. In the first, the pups establish independent nests, an effort that arises in part from a new agonistic relationship with the mother. However, the pups still affiliate—by play fighting—with one another. In the second phase, hamster pups gradually establish a largely agonistic relationship with all conspecifics as they develop their ability to mate successfully. Undoubtedly, play fighting serves in part as practice for adult agonistic encounters (Goldman and Swanson, 1975). Whether a transitional affiliative relationship with littermates also fosters success in the infrequent affiliative encounters of adulthood, such as mating and pup rearing, is unknown.

4.4. Effect of Early Social Isolation

Dieterlen (1959) used so-called Kaspar–Hauser (K–H) animals, which were reared without exposure to conspecifics and their odors, as a means to test whether behavioral development depends on early experience with social odors. He argued that such an experiment would be as important to the study of development in a macrosmatic animal as the use of acoustical isolation is to the study of song development in birds. However, although early sensory isolation or deprivation typically retards or otherwise interferes with development (e.g., Marler, 1981), Dieterlen's K–H hamsters were, if anything, somewhat precocious in the development of certain behavior patterns, such as nest building. Conclusions on the function of early sensory experience in the development of these behavior patterns are complicated by the fact that Dieterlen never succeeded in isolating pups from their mothers before P5–9 and thus all pups had about one week of sensory experience in the nest. Why, then, was there *any* effect of the procedure?

Perhaps Dieterlen's experiments constituted a test not of the effect of early sensory experience, but rather of the ability of hamster pups to survive when socially isolated at an early age, i.e., when weaned before weaning usually occurs. The fact that the K–H pups were precocious in showing nest building and hoarding in the second week postpartum instead of the fourth week was not, in our opinion, due to a failure of effective sensory isolation. It was due instead to a form of disinhibition (see Daly, 1976). The mature forms of nest building and hoarding are normally repressed by the presence of the dam and the maternal nest. However, the pups are capable of exhibiting these behaviors if they and the dam part company before the normal time for weaning. Indeed, the "normal" time in the wild may occur much earlier than in a laboratory setting.

Instances of repressed behavior are notable in other studies on developing hamsters. For example, Etienne et al. (1982) found a relatively low incidence of fully coordinated hoarding in pups housed with the dam through age P30. Once isolated at P30, however, these pups showed dramatic increases in hoarding. Crandall (1980) has recently found in our laboratory that as many as 40% of hamster pups weaned early are capable of coordinated hoarding by P21 and nearly all can hoard by P25 (Fig. 8). As another example, hamsters will show lateral flank gland marking as early as P21 when tested without the mother (Daly, 1976) and by P26 when tested where they can readily leave the maternal nest (Rowell, 1961), but they do not mark until about P40 when observed in a relatively small arena in which they reside with the dam and littermates (Goldman and Swanson, 1975).

We conclude then that hamster pups are prepared to survive if isolated from the dam in the third week postpartum. However, they can and perhaps normally do tolerate affiliation with the dam into the fifth week.

5. THE IMPACT OF A BRIEF GESTATION: IS THE NEWBORN GOLDEN HAMSTER PUP "IMMATURE" FOR ITS AGE?

The golden hamster has one of the shortest gestations of any placental mammal. However, there is little agreement as to the impact of this brief gestation on postnatal development. Some investigators, in finding certain features of development to be immature in the newborn hamster, have assumed or concluded that the postnatal period is comparable to the late gestational period of other species (Lakars and Herring, 1980; Richards, 1966). Others have focused on the similarity of developmental timetables between hamsters and other species (Campbell and Mabry, 1972) or even on the relatively faster pace of hamster development (Hall and Rosenblatt, 1979) and so either have not really considered whether the short gestation has an impact or have implicitly concluded that it has a limited impact on postnatal development compared with other species. In any event, it appears from these various reports that there is a fair degree of heterogeneity in maturational timetables for the postnatal golden hamster despite an extremely short gestation.

This issue has been treated more formally by Daly (1976). He studied species-typical behaviors in developing hamsters of three species that differ markedly in the length of gestation (golden, 16 days; Djungarian, 18 days; Chinese, 20 days) and looked for species differences as a function of both postconception and postnatal age. Postconception age differences were much more prevalent than postnatal age differences, leading to the general conclusion that development *speed* is a more significant variable among hamster species than stage of maturation at birth. In other words, golden hamsters develop more rapidly in a shorter period of gestation than Chinese hamsters to reach approximately the same level of maturation at birth as the Chinese. However, Daly also found that there were numerous exceptions to this general rule. For example, golden hamsters exhibited grooming and scratching behaviors two to three days after the other two species. They also opened their eyes two to three days later than the others. On the other hand, goldens showed independent defecation at an earlier age than the other two species.

Comparisons between developing golden hamsters and developing rats tell a similar story; that is, although a number of timetables are comparable, others diverge noticeably. As noted in previous sections, goldens and rats exhibit a number of landmarks at comparable times in postnatal development, e.g., in vestibular and olfactory development, in the control of early suckling behavior, in the sensory control of huddling behavior, in the timing of eye opening, and in the rise and fall of gross activity. Yet, golden hamsters are more vigorously thermotaxic at birth and eat solid food and leave the nest sooner than rats.

All of this leads us to conclude that the impact of a 16-day gestation on postnatal development of the golden hamster is rather limited or at least cannot be generalized. Although in some respects the newborn golden is decidedly immature and its subsequent development is somewhat retarded, in ways that are consistent with "early birth," its maturational state is nonetheless generally on a par with other hamster and rodent species at comparable postnatal ages.

6. SUMMARY AND CONCLUSIONS

The neonatal golden hamster presents us with an unusual mix of capabilities in its behavioral repertoire. Extreme exothermia coupled with strong thermotaxic ability in the first week of life is contrasted with the accelerated transition from suckling and nest sitting to the ingestion of solid food and forays from the nest in the second week of life. In this period of time, the hamster is seen as both less and more mature than the rat at comparable postnatal ages. Viewed another way, the hamster is rushed through fetal development in record time, born extremely dependent on its mother, normally an isolate, for warmth and nourishment, and then is rushed again through infancy to be prepared to establish independent nest sites well before a month is up. One wonders: Why the rush?

Perhaps the periods of fetal and neonatal development are timed in part to accommodate the female hamster, for which the period of maternity is in a sense an intrusion into an otherwise isolated existence. Combined with the relatively unproductive semiarid climate of the golden hamster's home, the social isolation that is characteristic of hamster species in general places great strains on the reproductive vigor of the golden hamster. To our knowledge, this combination of factors is unique to the golden hamster among rodent species. The overall acceleration of fetal and neonatal development may provide for the least drain of the dam's physiological resources, but it also may reduce the total period of maternity for an animal that is normally a social isolate and is otherwise highly aggressive toward pups and adult conspecifics alike.

Richards (1966) cites evidence that the shortened period of gestation may be a recent development in the evolutionary history of the golden hamster. Moreover, he suggests that the reliable incidence of cannibalism exhibited by both lactating and nonlactating females during the first few days of exposure to pups may indicate that the female hamster has not yet adjusted behaviorally to this evolutionary development. Nevertheless, as Siegel and Rosenblatt (1980) indicate, cannibalism usually involves no more than a small portion of the litter and reliably subsides after 48 hr of constant exposure to the neonate pups in the nest. This initial period of sensitization to the pups may be facilitated by their active thermal rooting, one of the few manifestations of "early" birth in the golden hamster. Having contributed to the optimal survival of the litter, such behavior by the pups may have reduced the pressure in the recent evolution of the golden hamster for significant adjustments in maternal behavior. In the rat, the opposite may have occurred. Rat maternal behavior is so well attuned to the needs of the litter (Leon *et al.,* 1978) that there is little reason for rat pups to display self-maintenance behaviors such as thermal rooting with the vigor shown by hamsters.

The patterns of convergence and divergence among mammalian species in timetables of development are important features of mammalian phylogeny even though it may be difficult at present to fathom their role in phylogenesis. Until there is evidence to the contrary, one must assume that convergent timetables derive from adjustments in development that occurred relatively early in phylogenesis and are common to many related species, whereas divergent timetables have their origin in developmental adjustments that have occurred relatively recently in phylogenesis and thus are most likely to be tied to the peculiar living conditions of present species (see Gottlieb, 1971). Although there is a growing body of data that can be used to make tentative statements about the significance of various developmental timetables in the hamster and other altricial mammals (e.g., Rosenblatt, 1976), interpretations are severely hampered by inadequate coverage of a wide range of behaviors and species and by differences in the procedures used for testing behavior in developing animals (see previous comments and those of Daly, 1976). For us, the ultimate goal is to achieve sufficient understanding of the adaptive significance of landmark events in behavioral development so that these landmarks can then be studied and understood in terms of comparable landmark events in neural development (Leonard, 1974a, 1975; see Oppenheim, 1981). Clearly, such an analysis of mechanisms of development can only be as sophisticated as the understanding of behavior is complete. We hope that future studies will be able to sort out the numerous sources of influence that impinge upon the behaviors of developing animals and to fill in many gaps in our present understanding of the phylogeny of development.

ACKNOWLEDGMENTS. We are grateful to Drs. Jim Corwin and Jim Crandall for their permission to present some of their unpublished work and for their comments and suggestions during the preparation of this manuscript. We are indebted to Drs. Lakars and Herring for sending us a camera-ready copy of their drawing of hamster mandibles. We also thank Ms. Lois Hager, who did a masterful job of battling the Foundation's new word processor on our behalf.

REFERENCES

Alberts, J. R., 1978a, Huddling by rat pups: Multisensory control of contact behavior, *J. Comp. Physiol. Psychol.* 92:220–230.
Alberts, J. R., 1978b, Huddling by rat pups: Group behavioral mechanisms of temperature regulation and energy conservation, *J. Comp. Physiol. Psychol.* 92:231–245.
Alberts, J. R., and Brunjes, P. C., 1978, Ontogeny of thermal and olfactory determinants of huddling in the rat, *J. Comp. Physiol. Psychol.* 92:897–906.
Alberts, J. R., and Leimbach, M. P., 1980, The first foray: Maternal influences in nest egression in the weanling rat, *Dev. Psychobiol.* 13:417–429.
Alberts, J. R., and May, B., 1980a, Development of nasal respiration and sniffing in the rat, *Physiol. Behav.* 24:957–963.
Alberts, J. R., and May, B., 1980b, Ontogeny of olfaction: Development of the rat's sensitivity to urine and amyl acetate, *Physiol. Behav.* 24:965–970.
Almli, C. R., and Fisher, R. S., 1977, Infant rats: Sensorimotor ontogeny and effects of substantia nigra destruction, *Brain Res. Bull.* 2:425–459.
Altman, J., and Sudarshan, K., 1975, Postnatal development of locomotion in the laboratory rat, *Anim. Behav.* 23:896–920.
Blass, E. M., Hall, W. G., and Teicher, M. H., 1979, The ontogeny of suckling and ingestive behaviors, *Prog. Psychobiol. Physiol. Psychol.* 8:243–299.
Bolles, R. C., and Woods, P. J., 1964, The ontogeny of behaviour in the albino rat, *Anim. Behav.* 12:427–441.
Campbell, B. A., and Mabry, P. D., 1972, Ontogeny of behavioral arousal: A comparative study, *J. Comp. Physiol. Psychol.* 81:371–379.
Campbell, B. A., and Raskin, L. A., 1978, Ontogeny of behavioral arousal: The role of environmental stimuli, *J. Comp. Physiol. Psychol.* 92:176–184.
Carter, C. S., and Marr, J. N., 1970, Olfactory imprinting and age variables in the guinea pig, *Anim. Behav.* 18:238–244.
Chalupa, L. M., 1981, Some observations on the functional organization of the golden hamster's visual system, *Behav. Brain Res.* 3:189–200.
Chalupa, L. M., Morrow, L., and Rhoades, R. W., 1978, Behavioral consequences of visual deprivation and restriction in the golden hamster, *Exp. Neurol.* 61:442–454.
Cornwell, C. A., 1975, Golden hamster pups adapt to complex rearing odors, *Behav. Biol.* 14:175–188.
Cornwell, C. A., 1976, Selective olfactory exposure alters social and plant odor preferences of immature hamsters, *Behav. Biol.* 17:131–137.
Cornwell-Jones, C. A., 1979, Olfactory sensitive periods in albino rats and golden hamsters, *J. Comp. Physiol. Psychol.* 93:668–676.
Cornwell-Jones, C. A., and Azar, L. M., 1982, Olfactory development in gerbil pups, *Dev. Psychobiol.* 15:131–137.
Cornwell-Jones, C. A., and Holder, C. L., 1979, Early olfactory learning is influenced by sex in hamsters but not rats, *Physiol. Behav.* 23:1035–1040.
Cornwell-Jones, C. A., and Kovanic, K., 1981, Testosterone reduces olfactory neophobia in male golden hamsters, *Physiol. Behav.* 26:973–977.
Cornwell-Jones, C. A., Stephens, S. E., and Dunston, G. A., 1982, Early odor preferences of rats are preserved by neonatal 6-hydroxydopamine, *Behav. Neural Biol.* 35:217–230.
Crandall, J. E., 1980, Functional and anatomical development of medial prefrontal cortex in the Syrian hamster, unpublished doctoral dissertation, University of Florida (*Diss. Abstr. Int.* 1981, 42:822B).

Crandall, J. E., and Leonard, C. M., 1979, Developmental changes in thermal and olfactory influences on golden hamster pups, *Behav. Neural Biol.* 26:354–363.

Daly, M., 1976, Behavioral development in three hamster species, *Dev. Psychobiol.* 9:315–323.

Devor, M., and Schneider, G. E., 1974, Attraction to home-cage odor in hamster pups: specificity and changes with age, *Behav. Biol.* 10:211–221.

Dieterlen, F., 1959, Das Verhalten des syrischen Goldhamsters (*Mesocricetus auratus* Waterhouse), *Z. Tierpsychol.* 16:47–103.

Etienne, A. S., Emmanuelli, E., and Zinder, M., 1982, Ontogeny of hoarding in the golden hamster: The development of motor patterns and their sequential coordination, *Dev. Psychobiol.* 15:33–45.

Fink, J. S., and Smith, G. P., 1980, Mesolimbicocortical dopamine terminal fields are necessary for normal locomotor and investigatory exploration in rats, *Brain Res.* 199:359–384.

Finlay, B. L., Marder, K., and Cordon, D., 1980, Acquisition of visuomotor behavior after neonatal tectal lesions in the hamster: The role of visual experience, *J. Comp. Physiol. Psychol.* 94:506–518.

Galef, B. G., Jr., and Clark, M. M., 1971, Parent-offspring interactions determine time and place of first ingestion of solid food by wild rat pups, *Psychon. Sci.,* 25:15–16.

Goldman, L., and Swanson, H. H., 1975, Developmental changes in pre-adult behavior in confined colonies of golden hamsters, *Dev. Psychobiol.* 8:137–150.

Gottlieb, G., 1971, Ontogenesis of sensory function in birds and mammals, in: *The Biopsychology of Development* (E. Tobach, L. R. Aronson, and E. Shaw, eds.), Academic Press, New York, pp. 67–128.

Grafe, M. R., and Leonard, C. M., 1982, Developmental changes in the topographical distribution of cells contributing to the lateral olfactory tract, *Dev. Brain Res.* 3:387–400.

Gregory, E. H., and Bishop, A., 1975, Development of olfactory-guided behavior in the golden hamster, *Physiol. Behav.* 15:373–376.

Gregory, E., and Pfaff, D. W., 1971, Development of olfactory-guided behavior in intact rats, *Physiol. Behav.* 6:573–576.

Hall, W. G., and Bryan, T. E., 1981, The ontogeny of feeding in rats: IV. Taste development as measured by intake and behavioral responses to oral infusions of sucrose and quinine, *J. Comp. Physiol. Psychol.* 95:240–251.

Hall, W. G., and Rosenblatt, J. S., 1979, Developmental changes in the suckling behavior of hamster pups: A comparison with rat pups, *Dev. Psychobiol.* 12:553–560.

Hall, W. G., Cramer, C. P., and Blass, E. M., 1977, The ontogeny of suckling in rats, *J. Comp. Physiol. Psychol.* 91:1141–1155.

Harder, W., 1949, Zur Morphologie und Physiologie des Blindarmes der Nagetiere, *Vert. Dtch. Zool. Ges.* 1949:95–109.

Heimer, L., and Wilson, R. D., 1975, The subcortical projections of the allocortex: Similarities in the neural connections of the hippocampus, the piriform cortex, and the neocortex, in: *Golgi Centennial Symposium: Perspectives in Neurobiology* (M. Santini, ed.), Raven Press, New York, pp. 177–193.

Henderson, N. D., 1981, A fit mouse is a hoppy mouse: Jumping behavior in 15-day-old *Mus musculus, Dev. Psychobiol.* 14:459–472.

Hissa, R., 1968, Postnatal development of thermoregulation in the Norwegian lemming and the golden hamster, *Ann. Zool. Fenn.* 5:354–383.

Johanson, I. B., and Hall, W. G., 1979, Appetitive learning in 1-day-old rat pups, *Science* 205:419–421.

Johanson, I. B., and Hall, W. G., 1980, The ontogeny of feeding in rats: III. Thermal determinants of early ingestive responding, *J. Comp. Physiol. Psychol.* 94:977–992.

Johnston, R. E., and Coplin, B., 1979, Development of responses to vaginal secretion and other substances in golden hamsters, *Behav. Neural Biol.* 25:473–489.

Kasamatsu, T., and Pettigrew, J. D., 1979, Preservation of binocularity after monocular deprivation in the striate cortex of kittens treated with 6-hydroxydopamine, *J. Comp. Neurol.* 185:139–162.

Kleitman, N., and Satinoff, E., 1982, Thermoregulatory behavior in rat pups from birth to weaning, *Physiol. Behav.* 29:537–541.

Lakars, T. C., and Herring, S. W., 1980, Ontogeny of oral function in hamsters (*Mesocricetus auratus*), *J. Morphol.* 165:237–254.

Leon, M., 1974, Maternal pheromone, *Physiol. Behav.* 13:441–453.

Leon, M., Croskerry, P. G., and Smith, G. K., 1978, Thermal control of mother-young contact in rats, *Physiol. Behav.* 21:793–811.

Leonard, C. M., 1974a, Degeneration argyrophilia as an index of neural maturation: Studies on the optic tract of the golden hamster, *J. Comp. Neurol.* 156:435–458.

Leonard, C. M., 1974b, Thermotaxis in golden hamster pups, *J. Comp. Physiol. Psychol.* 86:458–469.

Leonard, C. M., 1975, Developmental changes in olfactory bulb projections revealed by degeneration argyrophilia, *J. Comp. Neurol.* 162:467–486.

Leonard, C. M., 1978, Maturational loss of thermotaxis prevented by olfactory lesions in golden hamster pups (*Mesocricetus auratus*), *J. Comp. Physiol. Psychol.* 92:1084–1094.

Leonard, C. M., 1981, Some speculations concerning neurological mechanisms for early olfactory recognition, in: *The Development of Perception: Psychobiological Perspectives* (R. N. Astin, J. R. Alberts, and M. R. Petersen, eds.), Academic Press, New York, pp. 383–410.

Leonard, C. M., 1982, Shifting strategies for behavioral thermoregulation in developing golden hamsters, *J. Comp. Physiol. Psychol.* 96:234–243.

Macrides, F., 1975, Temporal relationships between hippocampal slow waves and exploratory sniffing in hamsters, *Behav. Biol.* 14:295–308.

Marler, P., 1981, Bird song: The acquisition of a learned motor skill, *Trends Neurosci.* 4:88–94.

Newman, R., and Winans, S. S., 1980, An experimental study of the ventral striatum of the golden hamster. II. Neuronal connections of the olfactory tubercle, *J. Comp. Neurol.* 191:193–212.

Okon, E. E., 1971, The temperature relations of vocalizations in infant golden hamsters and Wistar rats, *J. Zool.* 164:227–237.

Oppenheim, R. W., 1980, Metamorphosis and adaptation in the behavior of developing organisms, *Dev. Psychobiol.* 13:353–356.

Oppenheim, R. W., 1981, Ontogenetic adaptations and retrogressive processes in the development of the nervous system and behaviour: A neuroembryological perspective, in: *Maturation and Development: Biological and Psychological Perspectives* (K. J. Connolly and H. F. R. Prechtl, eds.), Lippincott, Philadelphia, pp. 73–109.

Pedersen, P. E., and Blass, E. M., 1982, Prenatal and postnatal determinants of the 1st suckling episode in albino rats, *Dev. Psychobiol.* 15:349–355.

Porter, R. H., and Etscorn, F., 1976, A sensitive period for the development of olfactory preference in *Acomys cahirinus, Physiol. Behav.* 17:127–130.

Pujol, R., Abonnenc, M., and Rebillard, J., 1975, Development and plasticity in the auditory system: Methodological approach and first results, in: *Aspects of Neural Plasticity* (F. Vital-Durand and M. Jeannerod, eds.), INSERM, Paris, pp. 45–54.

Richards, M. P. M., 1966, Maternal behavior in virgin female golden hamsters (*Mesocricetus auratus* Waterhouse): The role of the age of the pups, *Anim. Behav.* 14:303–309.

Rosenblatt, J. S., 1976, Stages in the early behavioural development of altricial young of selected species of non-primate mammals, in: *Growing Points in Ethology* (P. P. G. Bateson and R. A. Hinde, eds.), Cambridge University Press, Cambridge, pp. 345–383.

Routtenberg, A., Strop, M., and Jerdan, J., 1978, Response of the infant rat to light prior to eyelid opening: Mediation by the superior colliculus, *Dev. Psychobiol.* 11:469–478.

Rowell, T. E., 1961, The family group in golden hamsters: Its formation and break-up, *Behaviour* 17:81–94.

Rudy, J. W., and Cheatle, M. D., 1977, Odor-aversion learning in neonatal rats, *Science* 198:845–846.

Schneider, G. E., 1969, Two visual systems: Brain mechanisms for localization and discrimination are dissociated by tectal and cortical lesions, *Science* 163:895–902.

Schneider, G. E., 1970, Mechanisms of functional recovery following lesions of visual cortex or superior colliculus in neonate and adult hamsters, *Brain Behav. Evol.* 3:295–323.

Schoenfeld, T. A., and Corwin, J. V., 1982, Maturation of olfactory exploration in hamsters is correlated with late afferentation of the olfactory tubercle, paper presented at the meeting of the Association for Chemoreception Sciences, Sarasota, Florida.

Schoenfeld, T. A., Street, C. K., and Leonard, C. M., 1979, Maturation of Wallerian degeneration: An EM study in the developing olfactory tubercle, *Soc. Neurosci. Abstr.* 5:177.

Schwob, J. E., and Price, J. L., 1978, The cortical projection of the olfactory bulb: Development in fetal and neonatal rats correlated with quantitative variations in adult rats, *Brain Res.* 151:369–374.

Siegel, H. I., and Rosenblatt, J. S., 1980, Hormonal and behavioral aspects of maternal care in the hamster: A review, *Neurosci. Biobehav. Rev.* 4:17–26.

Small, R. K., and Leonard, C. M., 1983, Early recovery of function after olfactory tract section correlated with reinnervation of olfactory tubercle, *Dev. Brain Res.* 7:25–40.

Smotherman, W. P., 1982, Odor aversion learning by the rat fetus, *Physiol. Behav.* 29:769–771.

Swanson, L. W., Mogenson, G. J., and Wu, M., 1982, Evidence for a projection from the lateral preoptic area and substantia innominata to the mesencephalic locomotor region, *Soc. Neurosci. Abstr.* 8:173.

Teicher, M. H., and Blass, E. M., 1977, First suckling response of the new-born albino rat: The roles of olfaction and amniotic fluid, *Science* 198:635–636.

Tobach, E., 1977, Developmental aspects of chemoreception in the Wistar (DAB) rat: Tonic processes, *Ann. N.Y. Acad. Sci.* 290:226–269.

Welker, W. I., 1964, Analysis of sniffing of the albino rat, *Behaviour* 22:223–244.

Westrum, L. E., 1975, Electron microscopy of synaptic structures in olfactory cortex of early postnatal rats, *J. Neurocytol.* 4:713–732.

14

Biological Rhythms

LAWRENCE P. MORIN

The most widely studied category of hamster biological rhythms is that for which the expected period is 24 hr. Such rhythms are considered to be true "circadian rhythms" if the periodicity of the event deviates slightly from 24 hr in the absence of known time-giving cues (Zeitgebern). Without a Zeitgeber, rhythms such as that for locomotor activity will assume a frequency unconstrained by an external synchronizing agent. This "free run" (Fig. 1) is considered to demonstrate an endogenous, self-sustained oscillation with a period near 24 hr, thus, a circadian rhythm. Timing for such a rhythm is presumed to be derived from a circadian "clock." The existence of such clocks is now generally accepted, although a contrary view does exist (see Brown, 1976; Brown and Scow, 1978).

Numerous 24-hr rhythms have been described, but to the majority, the label "circadian" cannot yet be properly applied. The 24-hr rhythms can be divided roughly into three groups: "diurnal," having maximum amplitude during the daylight hours; "nocturnal," with maximum amplitude during the dark; and "crepuscular," with bimodal peak activity at dawn and dusk. These cannot be considered "circadian" rhythms until shown to be endogenous, self-sustaining oscillations. Certain animals can vary between all three rhythm types according to time of the year (Eriksson, 1973).

Numerous significant matters of rhythm definition and organization have been discussed by Menaker (1982). Of conceptual importance is the need for a distinction between rhythms that directly reflect on the rhythm-generating mechanisn and those that are hierarchically linked to such a clock mechanism. The latter have been designated "slave processes" (Menaker, 1982) and may or may not exert feedback control over the rhythm-generating mechanism. For example, cultured hamster adrenals demonstrate circadian rhythmicity in corticoid secretion (Andrews, 1971), yet the timing of this rhythm is normally regulated by a higher-order mechanism, probably in the hypothalamus (Moore and Eichler, 1972; Krieger, 1980). Whether adrenal corticoids

LAWRENCE P. MORIN • Department of Psychiatry, State University of New York, Stony Brook, New York 11794.

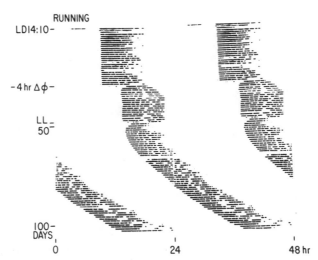

Figure 1. A clear, very precise record of hamster wheelrunning. Days are plotted vertically with the record created by pasting each 24-hr data strip below the preceding day's data. The original record has been photographically duplicated (hours 24 to 48) for visualization purposes. The initial photoperiod was 14 light : 10 dark. Activity was limited almost entirely to the dark phase. Note the precision of activity onsets compared with offsets. At day 30, the lights remained on an additional 4 hr forcing a phase delay (−4 hr Δ Ø) in the animal's locomotor rhythm as it re-entrained to the shifted light cycle. Re-entrainment occurred within two to three days. At day 48, the lights were left permanently on (LL; about 40 lux). An immediate phase delay occurred followed by three to four days of advancing transients. The rhythm then lengthened and eventually stabilized with a free-running period of 24.7 hr. The free-running period, tau, is calculated as the average time from the beginning of activity on day N to the beginning of activity on day N + 1. Alpha, the phase during which activity is most likely to occur, is measured from average activity onset to average offset. During the last 20 days, alpha was about 8.7 hr.

have any influence on the period of the higher-order rhythm remains to be established. Other slave processes may be driven by a higher-order mechanism and have no inherent rhythmicity. Recognition of such parallel and hierarchial organization of rhythmic processes is essential to their understanding (Moore-Ede *et al.*, 1976).

Noncircadian classes of biological rhythms tend to suffer descriptive and organizational difficulties similar to those of circadian rhythms. All rhythms are characterized by period, phase, and amplitude measures. Of these, the period of the rhythm has usually been studied in the greatest detail. Extremely long cycles are known to exist in the populations of certain species (e.g., 10-yr population cycle of the snowshoe hare; Davis and Meyer, 1973). The annual body weight and hibernation cycle is well documented for certain rodents (Mrsovsky, 1975). This rhythm will free run (Pengelley *et al.*, 1976) and has been called "circannual." A detailed assessment of annual rhythms is offered by Mrsovsky (1978). The particular form of annual reproductive rhythmicity for the hamster is described in detail by Reiter (Chapter 5, this volume).

"Infradian" rhythms are those with periods falling between the circadian and circannual ranges. An example is the menstrual cycle (about 28 days). The hamster estrous cycle is likewise an infradian rhythm (about 4 days) with its expression dependent on the interaction of a circadian clock with endocrine organ events (Alleva et al., 1971) as explained below.

High-frequency rhythms (with periods less than the circadian range) are called "ultradian." This class includes events measurable in widely differing time units: alpha or theta EEG waves (milliseconds), respiration (seconds), or meal timing (minutes).

It is obvious that the anatomical bases for all rhythm types are not the same. Theta rhythmicity in rats may be derived from two separate hippocampal cell populations (Krug et al., 1981). However, relatively little is known about neuroanatomical sites generating any rhythm. The exception is the role played by the suprachiasmatic nuclei in circadian rhythm generation (Moore and Eichler, 1972; Stephan and Zucker, 1972; Stetson and Watson-Whitmyre, 1976; Rusak, 1977a). The anatomical relationship of the suprachiasmatic nuclei to the visual system is now well established (Moore, 1983; Pickard and Silverman, 1981). Hence, for circadian rhythms, there is an anatomical rationale for the relationship between the measured rhythm and the time giving environmental light-dark (LD) cycle.

1. CIRCADIAN RHYTHMICITY AND REGULATION BY LIGHT

1.1. Entrainment and Phase Response Curves

The typical form of rhythm expression is not a free run. Rather, the internally generated rhythm synchronizes with the actions of an appropriate external oscillator or Zeitgeber. For hamsters, the only documented Zeitgeber is the light-dark photoperiod (Fig. 1). Similar to the presumptive biological circadian clock, the sidereal day of the earth (time to rotate 360° on its axis) has a period near, but not exactly, 24 hr. Its period varies slightly with the earth's position in orbit around the sun, but averages about 23 hr 56 min 3.4 sec (Malloney, 1978). If the biological clock is considered to be coupled to the photoperiod by physical means (in this case by stimulation of photoreceptors and nervous connections; see below) and has a natural frequency relatively similar to the photoperiod, then the natural external photocycle becomes a driving oscillation that induces the internal clock to resonate with the same frequency as the driver. If either the extent of coupling or the similarity of natural frequency is lessened, then the biological clock will tend to oscillate independent of the photoperiod.

Photoperiodic control of hamster circadian rhythm phase is well established, Bruce (1960) having demonstrated that light cycles that are 24-hr long can entrain locomotor rhythmicity. In a LD 12 : 12 cycle, hamsters begin running shortly after lights off (Bruce, 1960; Elliott, 1974, 1976). Hamsters are sufficiently nocturnal that under this photoperiod, about 99% of all wheel revolutions occur in the dark phase (Zucker and Stephan, 1973).

A critical question for rhythm researchers has concerned the mechanisms by which an internal clock can become synchronized with an external clock. Pittendrigh (1974) has proposed and elaborated (see Pittendrigh and Daan, 1976a–c; Daan and Pit-

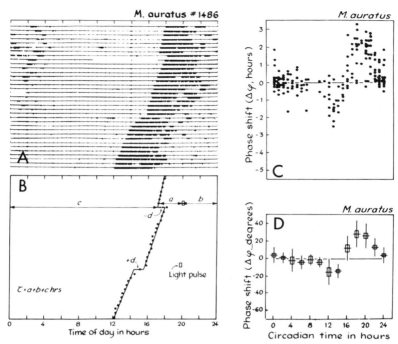

Figure 2. In (A), the free-running wheelrunning rhythm is shown for a hamster housed in constant darkness (DD). The running onsets are illustrated schematically in (B). At each rectangle in (B), the hamster received a 15-min light pulse. The circadian time of each pulse was calculated as the interval, a, from running onset to the pulse onset divided by the period, tau, plus 12 hr. The resultant changes in activity onset phases are indicated by the notations −d and +d, for a phase delay and a phase advance, respectively. (C) shows the phase response curve for a group of hamsters with several phase change points plotted for each animal. The magnitude of the phase shifts has been plotted with advance and delay shifts indicated in hours. In (D), the data have been transformed to degrees of shift and are displayed as mean changes according to the hour of the circadian day. The horizontal line through the rectangle indicates the mean; the vertical limits of the rectangle indicate the sample standard error and the vertical line shows the standard deviation of the sample. Circadian time 12 hr is the time of running onset. [Reprinted from Figs. 1 and 2 in Daan and Pittendrigh (1967a) with permission from Springer-Verlag, Inc.]

tendrigh, 1976a,b) a nonparametric model of entrainment by which light induces a phase advance or delay of the circadian clock. The relationship between the time of a light pulse and magnitude of phase advance or delay may be graphically represented as a phase response curve (PRC). Each animal has a PRC (Fig. 2) that consists of two major divisions: (1) The subjective nighttime during which light stimulation will induce a phase response and (2) the subjective daytime that occurs for hamsters during normal daylight hours. The latter phase of the circadian day is considered to be the period of several hours during which photostimulation will induce minimal phase advances or delays. The mean PRC generally shows no phase change between circadian time 23.5 and 8 to 12 hr (the "dead zone") (Fig. 2). Note in Fig. 2, however, that there is extremely wide variability in phase response at all times of day. Much of this has been attributed to measurement error, but individual differences in phase response to light undoubtedly exist (Daan and Pittendrigh, 1976a; Stetson and Gibson, 1977; Ellis *et al.*, 1982; Aschoff *et al.*, 1982). The individual differences appear so great, in fact, that this feature might account for as much of the variability in phase response as the underlying curve itself (Daan and Pittendrigh, 1976a), although a recent PRC report indicates that variability between individuals may be substantially less than previously found (Takahashi and Zatz, 1982; Takahashi *et al.*, 1984). At the very least, the average PRC cannot be expected to accurately predict the phase response of particular individuals.

Entrainment, according to the nonparametric model, is presumed to exist when the Zeitgeber affects the circadian clock, changing its phase (psi), such that change in psi = T - tau where T is the period of the Zeitgeber and tau is the period of the internal clock or pacemaker. When animals are exposed to light "pulses" of 15 min to 2 hr in length during the subjective night (Fig. 3) (DeCoursey, 1964; Elliott, 1974; Daan and Pittendrigh, 1976a), phase advances or delays will occur according to the circadian time at which an individual was stimulated. If a single pulse delivered every 24 hr (T = 24) to an animal with tau = 24.2 hr induces a change in psi = + 0.2 hr, then entrainment will result. A phase-delay-inducing pulse delivered at the appropriate point in the animal's PRC can precisely counteract a similar phase-advance-inducing pulse (Daan and Pittendrigh, 1976a). Thus, hamsters are able to entrain to two brief light pulses positioned 14 hr apart (a "skeleton" photoperiod) and rhythmically behave as if the photoperiod is LD 14 : 10 (Pittendrigh and Daan, 1976a,b,c; Dann and Pittendrigh, 1976a). Light pulses of 1 sec duration are sufficient to entrain adult male hamsters (Earnest and Turek, 1983b). The brightness of the entraining light has little effect on psi in hamsters (within a tested range of 0.01 to 1.0 lux) (Pohl, 1976). When carefully assessed, however, a phase shift response-light intensity relationship can be determined for specific wavelengths (Takahashi *et al.*, 1984). For example, when $10^{15} cm^{-2} sr^{-1}$ photons of 574 nm light are administered as a 15-min pulse at circadian time 18, an approximate 30-min phase advance can be induced in a DD-housed male hamster. In contrast, 515- or 476-nm light delivered at the same intensity parameters will induce an approximate 1.5-hr phase advance (Takahashi *et al.*, 1984).

The logic behind PRC involvement with entrainment can work two ways: if light pulses in otherwise dark conditions can effect phase changes, the dark pulses occurring in LL should also yield similar, but opposite sign, changes in phase. Recent efforts to generate a PRC using dark "pulses" have proved successful (Boulos and Rusak, 1982; Ellis *et al.*, 1982). Either 2-hr or 6-hr dark "pulses" were able to induce lasting phase shifts of up to 3.5 hr after 2-hr dark or 6.5 hr after 6-hr dark. The shape of the curve was generally similar to that created from light pulses in DD, but in the opposite direction. There was no obvious distinction between immediate (observed the day after the pulse) and steady-state changes in phase (compare DeCoursey, 1964) nor was there readily apparent asymmetry between advance and delay portions of the curve (Boulos and Rusak, 1982). The phase advance capability of a 3-hr dark pulse has also been shown (Ellis *et al.*, 1982), but these investigators failed to find significant phase delays.

The utilization of dark pulses in PRC experiments may be relevant to work showing that some hamsters can generate a circadian rhythm of self-selected darkness

(Warden and Sachs, 1974). The rhythm of self-selected darkness can be either in phase with the locomotor rhythm or out of phase, depending on the individual hamster. Still other individuals appear unable to generate a darkness self-selection rhythm and are aperiodic in lighting selection (Warden and Sachs, 1974; Warden, 1978). How the self-selected darkness may effect the timing of running or any other variable is unknown.

In Fig. 2D, the amplitude and area of the phase advance portion of the PRC is noticeably greater than the delay portion. This raises the question as to the mode of action of constant light (LL) on the circadian period. In a simple interpretation, LL should cause a net decrease in tau according to the daily summation of phase advances and delays across the PRC. In the hamster, if LL has any effect, it is to lengthen the circadian period (Daan and Pittendrigh, 1976b). This is in accordance with "Aschoff's Rule," which states that tau in LL is greater than tau of DD for nocturnal animals, but is the reverse for diurnal animals (Pittendrigh, 1960). Moreover, the rule states that tau of nocturnal animals is expected to increase parametrically as intensity increases. For *Mus musculus, Peromyscus leucopus,* and *P. maniculatus,* this relationship is clearly true (Aschoff, 1960; Daan and Pittendrigh, 1976b). Only two studies have systematically examined whether LL intensities differentially affect hamster circadian period length (Daan and Pittendrigh, 1976b; Aschoff et al., 1973). Neither study convincingly demonstrates that tau increases with brightness of LL. On the other hand, very dim (0.005 to 0.01 lux) or dark conditions seem to facilitate activity rhythms with periods shorter than found under brighter LL (1.0 or 200 lux) (Aschoff et al., 1973; Fitzgerald et al., 1978). The failure of the hamster circadian period to clearly respond to varying light intensities, as it does for other rodents, has been specifically attributed to the species differences in PRC shape (Daan and Pittendrigh, 1976b).

Swade (1969) and Daan and Pittendrigh (1976b) suggest that a "velocity response curve" (VRC) can account for the above parametric effects of light on nocturnal animals. According to the velocity response hypothesis, the effect of LL on period length is based on a presumed relationship between the PRC and duration of light stimulation: light delays (or advances) a given circadian phase yielding longer (or shorter) exposure of that phase to further light (see Daan and Pittendrigh, 1976b, for illustrations and derivation of the VRC). A prediction derived from this concept is that animals (such as hamsters) with relatively large advance and small delay portions of the PRC would not lengthen tau in LL to the same extent as would another rodent (e.g., *Peromyscus maniculatus*) with the opposite PRC characteristics. Comparison of data from the hamster with that from the maniculatus shows that the tau of *P. maniculatus* increases under LL conditions by at least twice the amount shown by hamsters (Daan and Pittendrigh, 1976b).

One potential difficulty with the velocity response approach is its reliance on the exact shape of the PRC. DeCoursey (1964) and Elliott (1974) have shown that the phase shifts obtained during the circadian cycle immediately following the light pulse generate a PRC that has approximately equal advance and delay portions. The Daan and Pittendrigh (1976a) data and a second PRC from DeCoursey (1964) showing larger advance than delay portions are derived from the steady-state phase shift based on a rhythm that often is not stable for several days after a light pulse. This suggests that at least two different phenomena are being assessed by the different PRC methods. Clearly, some form of realignment of the rhythm-generating system and restabilization occurs during the days after a phase shift. The operational mode of the nonparametric model for a laboratory-housed animal receiving daily lights-on and lights-off transitions would seem to be the immediate form of the PRC. Perhaps the steady-state PRC form is more appropriate for burrow-housed animals that may not perceive phase-advancing or - delaying light pulses on a daily basis.

The velocity response conceptualization may also have generalized importance with respect to the tonic administration of certain drugs. The identical logic that predicts species differences in tau response to LL would likewise predict differences in

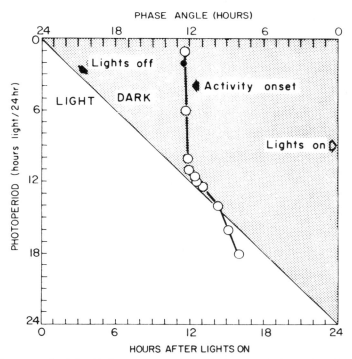

Figure 3. The presumed relationship between onset of the wheelrunning rhythm and the light-dark cycle (Elliott, 1976). Under normal laboratory conditions, there is an abrupt transition between darkness and light phases of the day (diagonal line). The hours of light in a given photoperiod are scaled on the ordinate; running onsets relative to the moment of lights on are indicated on the abscissa. The graph shows that running onsets are controlled only to some extent by the photoperiod. Generally, onset occurs after the lights go off, but under light durations > 14 hr, running will start in the light. At the other extreme, with long durations of dark, hamsters generally do not start wheelrunning until 12 to 13 hr before the lights go on. However, several exceptions to the latter rule have been documented (see text). (Reprinted with permission of the Federation Proceedings.)

tau response to certain drugs. Furthermore, individuals that overrespond or underrespond to LL or tonic drug administration should have PRCs that are correspondingly deviant from the species mean (Daan and Pittendrigh, 1976b).

Bruce (1960) showed that certain photoperiods would not necessarily entrain hamsters. For example, LD 6 : 18 permitted long free runs during the dark phase for some animals. Other animals entrained to LD 6 : 18 with activity beginning shortly before lights off. More recently, Elliott (1974, 1976) studied the relationship between activity onset and the time of lights on (phase angle difference; psi) for hamsters entrained to various photoperiods for up to 12 weeks. These results are depicted in Fig. 3. Other investigators have reported that tau for some animals appears to free run for prolonged periods within the 18-hr dark period, approaching a psi of 15 to 17 hr, not predicted from Elliott's data (Ellis and Turek, 1979; Bittman, 1978; Morin, 1982). These examples may reflect diminished coupling between the photoperiod and the circadian rhythm.

The Zeitgeber period (T) need only approximate 24 hr in order to entrain a circadian rhythm. Such non-24-hr T cycles using short light periods (e.g., LD 1 : 23.3) have been used to test the limits of clock entrainment by the Zeitgeber (Pittendrigh and Daan, 1976b). Entrainment will be lost when the Zeitgeber period deviates too much from the circadian period. In one experiment with male hamsters, the estimated lower limit of entrainment (50%) to a 1-hr light pulse was between T = 23.1 and 23.2 hr with the upper limit between T = 24.7 and 24.8 hr (Elliott, 1974). Using 14-hr light per day and slowly decreasing the duration of dark per day, T cycles as short as 21.5 hr were able to entrain female hamsters (Carmichael *et al.*, 1981). Therefore, the limits of entrainment appear to be a function of T-cycle length, the length of the light period, and probably the illumination intensity.

1.2. Aftereffects

Two forms of spontaneous variance in tau have been analyzed by Pittendrigh and Daan (1976a); short-term day-to-day frequency instability and long-term gradual change in frequency. Aschoff (1978) has provided an extensive review of some of these issues for a wide variety of animals. Age alone appears to contribute to an increase in hamster circadian rhythm frequency, although this result may be confounded by continuous access to wheels (Pittendrigh and Daan, 1974). More importantly, tau is extremely susceptible to constraints imposed by an individual's photic history. The alterations in the circadian system stemming from this history are called "aftereffects" (Pittendrigh, 1960).

Because of the susceptibility of tau to photic history, great care must be taken to insure stable and describable conditions prior to tau measurement. Furthermore, the probable impact of those conditions should be assessed before engaging in an experiment. As emphasized by Aschoff (1978) and Pittendrigh and Daan (1976a), disregard of aftereffects can greatly diminish the value of certain circadian rhythm experiments.

Four types of aftereffects on tau have been documented (Pittendrigh and Daan, 1976a; Aschoff, 1978): (1) those related to the ratio of light to dark; (2) those related to

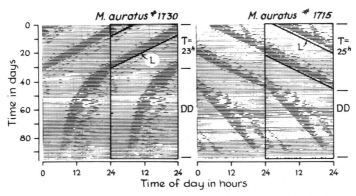

Figure 4. After effects of T cycles in two hamsters (Pittendrigh and Daan, 1976b). Before being released into constant darkness (DD), hamster #1730 was entrained to T = 23 hr and #1715 was entrained to T = 25 hr. The duration of the light phase was 15 min in each case. Note that in DD, the short T cycle has induced a relatively short free-running period, whereas the long T cycle has induced a relatively long free-running period. (Reprinted with permission of Springer-Verlag, Inc.)

the period (T) of the Zeitgeber; (3) those related to a phase shift of the Zeitgeber; and (4) those related to the presence of constant light. For hamsters in DD, tau is apparently unaffected by the prior LD ratio. With T = 24, tau (DD) is approximately the same whether the preceding photoperiod is 1 : 23 or 18 : 6. Similarly, a history of LL does not affect subsequent tau (Pittendrigh and Daan, 1976a). However, tau is markedly dependent on prior T (Fig. 4), although the effect has apparently been fully demonstrated in only a few hamsters (Pittendrigh and Daan, 1976a). Pittendrigh (1960) has reported that for four hamsters, T = 25 induced an aftereffect change in tau about 0.18 hr longer than following T = 23 hr. Elliott (1974) has clearly shown that following exposure of hamsters to any one of several long or short T cycles, the previously entrained locomotor rhythm will persist during 48 hr of DD with the period of the entraining cycle.

More extensively documented have been the aftereffects induced by Zeitgeber phase shifts. During the determination of a PRC, phase shifts necessarily occur, but the relationship between the shift size or direction and the aftereffects is ambiguous. Among hamsters exposed to 15-min light pulses, phase advances that were greater than 0.67 hr were followed by an average 0.04 hr decrease in tau. However, equally long phase delays also decreased tau by an average 0.01 hr, not significantly different from the change after a phase advance (Pittendrigh and Daan, 1976a). This is similar to what has been found with *Peromyscus maniculatus,* but contrasts with the results from *Mus musculus* and *P. leucopus* (Pittendrigh and Daan, 1976a). Longer exposure to light can induce larger phase changes and fairly large aftereffects on tau as shown in Figs. 5 and 6. Clearly, a one-time phase shift of the Zeitgeber has a strong effect on subsequent period (Morin, 1981a). Dark pulses (2- or 6-hr duration) are also able to induce aftereffects on tau in accordance with the direction of phase shift (Boulos and Rusak, 1982; Ellis *et al.,* 1982).

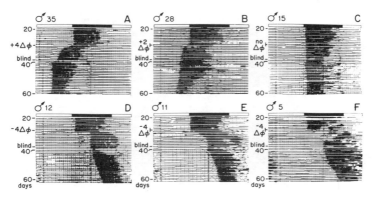

Figure 5. Examples of wheelrunning rhythm responses to shifts of the light-dark cycle (Morin, 1980). Each animal was initially entrained to LD 14 : 10. At day 28, each sustained a 4- or 2-hr phase advance of the LD cycle [+4 or +2 △ ∅ in (A) and (B) above], no phase change [no △ ∅ in (C)] or a 4-hr phase delay of the LD cycle [−4 △ ∅ in (D), (E), and (F)]. After ten days in the shifted LD cycle, each animal was blinded to achieve DD. A phase advance induced periods less than 24 hr; no phase change yielded periods slightly greater than 24 hr and phase delays yielded periods significantly greater than 24 hr. The group aftereffects are shown in Fig. 6. Note also the three distinct patterns for achieving re-entrainment of activity onset after the phase delay: in (D), re-entrainment was very rapid; in (E) it was relatively slow and in (F) the pattern was somewhat unclear. Each pattern was found with the same approximate frequency. (Reprinted with permission of Physiology and Behavior. Copyright 1980, Pergamon Press, Ltd.)

1.3. Phase Shifts and Transients

When a phase shift occurs, it is usually followed by one or more cycles of rapidly changing durations ("transients") that persists until steady state is achieved (Pittendrigh and Daan, 1976a). Typically, re-entrainment of activity onset after a phase advance requires more days (hence, more transients) than after a phase delay ("asymmetry-effect") (Aschoff *et al.*, 1975). Figure 5 shows that a one-time 4-hr phase advance was not entirely completed within ten cycles for any of nine animals. Re-entrainment of activity onset was accomplished within ten cycles by 9 of 9 and 9 of 10 animals given +

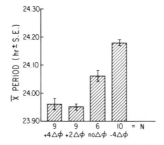

Figure 6. Mean period of the free-running rhythm of hamsters blinded for three weeks and subjected to a light cycle phase shift (△ ∅) ten days prior to blinding. For further description of the treatment groups see Morin (1980) and Fig. 5.

2-hr or + 4-hr phase shifts, repectively (Morin, 1981a). However, re-entrainment of offset of locomotor activity may be more rapid after a phase advance than a phase delay (Sisk and Stephan, 1981).

When both activity phase onset and termination are examined, the pattern of phase shifting is recognized as more complex than previously realized. Stephan *et al.* (1982) have confirmed that 6-hr phase delays are associated with rapid activity onset re-entrainment and slower re-entrainment of activity offset. In contrast, 6-hr phase advances produce speedy re-entrainment of activity termination, but slower re-entrainment by activity onset. Despite the effect of shift direction, the latency to re-entrainment by activity onset after a phase advance (about 11.5 days) still seems to be slower than the latency to re-entrainment by activity offset after a phase delay (about 6 days). This apparently is not true when considering activity phase offset after a phase advance and activity phase onset after a phase delay. Latency to re-entrainment in each case was about 1.5 days (Stephan *et al.*, 1982). The "asymmetry-effect" appears to be contingent on the variable assessed (activity phase onset or offset) and the phase shift direction.

The pattern of re-entrainment of wheelrunning onset after a phase delay varies from animal to animal (Morin, 1981a). Three basic modes of re-entrainment have been called "immediate," "gradual," and "overshift" according to the pattern of transient cycles (Stephan *et al.*, 1982). Immediate re-entrainment (Fig. 5D) is accomplished without transients. Gradual re-entrainment (Fig. 5E) is characterized by a number of transient cycles with running onsets progressively approaching the re-entrainment phase position. Overshift re-entrainment (Fig. 5F) exists when activity onset starts later than the final phase position of re-entrainment and the animal displays advancing transients to reach that position (Morin, 1981a; Stephan *et al.*, 1982).

1.4. Masking

The phenomenon called "masking" (Aschoff, 1960) can affect the interpretation of phase-shift data, particularly re-entrainment rates, but has not been well studied. Masking has been defined as "a 'direct' influence of the exogenous agent [. . .] on the observable function [. . .] without necessarily changing the underlying biological mechanism, i.e., the circadian oscillator" (Aschoff *et al.*, 1973). Masking can be either positive (facilitating the occurrence of an event) or negative (inhibiting the occurrence of an event). Aschoff and colleagues (1973) note that general tilt cage activity of hamsters increases as the light to dark transition approaches. On the other hand, wheelrunning recorded simultaneously with general activity starts after the lights go out. The wheelrunning onset that might otherwise occur in the light may have been prevented from occurring ("masked") by the presence of that light. The incidence of general activity, such as drinking (Sisk and Stephan, 1981), is much greater than wheelrunning during the light phase and may indicate that those behavioral rhythms are relatively resistant to masking.

There have been few direct tests of masking. Constant light does appear to reduce overall locomotor activity (Aschoff, 1960), suggesting a general negative effect of light on the rhythm amplitude. In one experiment, re-entrainment of wheelrunning following a 12-hr phase shift was examined. Four days after the shift, the animals were placed in DD and their running records examined for "unmasked" activity. Such unmasking was not found, suggesting that the observed, rapid phase shift was real and not an artifact of missing data created by negative masking (Sisk and Stephan, 1981). There is a definite need for more research into the issue of masking since its ubiquity is presumed, but not established.

1.5. Activity and Rest Phases

The phase of the circadian day during which locomotion occurs has been termed the "activity phase" or alpha. Correspondingly, the rest phase or rho refers to the daily period without locomotion. The terms have been widely used with different meanings (Aschoff, 1965; Davis and Menaker, 1980; Pittendrigh and Daan, 1976a), but the concensus is that alpha + rho = tau, the period of the circadian day. This equation is rather straight forward when hamsters do 99% of their running at night (Zucker and Stephan, 1973), but becomes more complicated when running becomes distributed through other portions of the day (e.g., after SCN lesions, rhythm splitting, or spontaneous rhythm desynchrony). A probabilistic definition of the activity phase may be more accurate in certain instances than is the activity–rest dichotomy (Morin, 1980). Clearly, hamsters are rarely, if ever, active through the entire "activity phase." Alpha more realistically refers to the circadian time during which activity is most likely to occur.

Using the dichotomous definition, alpha in DD has been shown to change substantially according to prior lighting conditions. With entrainment to long or short T cycles, alpha may "compress" to a shorter duration. Subsequent release into DD may result in "decompression" (Pittendrigh 1974; Pittendrigh and Daan, 1976a). However, using many more animals, Elliott (1974) failed to obtain alpha compression unless T was at the limits of entrainment. Duration changes in alpha or rho do not appear necessary for entrainment (Elliott, 1974).

Long-day photoperiods tend to force hamster locomotor activity into the short duration dark period. Under natural photoperiod conditions in the arctic or southern Germany, hamsters produce a maximal alpha of about 12 hr for sunlight duration of 10 hr or less. From 10 to 16 hr sunlight, alpha decreases inversely with the hours of sunlight; little further alpha decrease is found as the duration of sunlight increases to 24 hr (Daan and Aschoff, 1975). This appears to affect alpha in subsequent DD. Release into DD resulted in alpha being longer for eight of nine animals with a LD 1 : 23 history compared with an 18 : 6 history (Pittendrigh and Daan, 1976a). This aftereffect of photoperiod on alpha contrasts with the failure of photoperiod to influence aftereffects on tau.

The duration of the activity phase has been repeatedly measured, but is the subject of great variability (see below). The "characteristic" hamster running rhythm, entrained or not, has a precise onset and is bimodal with the first peak broader than the second. Alpha is measured from onset of the first peak to offset of the second. However, various conditions (especially hormonal state; see below) can effect the precision of activity onset. Offset of activity is usually extremely variable under the best of conditions (Fig. 9) (Morin and Cummings, 1981).

There seem to be at least two general types of running patterns. The first is the bimodal pattern described above (Pittendrigh and Daan, 1976c). In one experiment, this pattern occurred in 18 of 29 blind male hamsters and alpha measured 3.4 hr. The remaining 11 animals showed only one broad running phase that endured 5.7 hr (Morin and Cummings, 1981). Clearly, the animals with the unimodal running pattern had longer activity phases, but this is not necessarily translatable into greater distance run, although such a relationship has been found (Morin and Cummings, 1981). The significance of the two types of running patterns for PRC and tau generation is currently unknown. It is also important to note that different methods for measuring activity can give different types of running pattern results (Aschoff *et al.*, 1973).

1.6. Splitting

In addition to its parametric effects on tau, constant light can also induce a complex rhythmic phenomenon called "splitting" (Pittendrigh, 1960). Initial descriptions were that an apparently stable rhythm would suddenly break into two components, each with its own frequency. After a number of days, the two components would stably resynchronize with each other, but 180° out of phase and with a common period less than that prior to splitting (Pittendrigh, 1967, 1974). A presumptive evening oscillation ("E"; formerly called the "N" oscillation) would become temporarily desynchronized from a presumptive morning ("M") oscillation (Pittendrigh and Daan, 1976c), assume a shorter period and pictorially "cross" over the longer period "E" oscillation, eventually reaching a stable antiphase position. A two-oscillator model based on this conceptualization has been developed (Daan and Berde, 1978).

Splitting has been found to occur in light intensities of 40 to 200 lux, but not 1 lux. The average latency in LL to splitting is about 55 days (Pittendrigh and Daan, 1976c; Morin, 1980; Morin and Cummings, 1982; Earnest and Turek, 1982). A dark "pulse" provided in LL-housed animals may facilitate splitting (Ellis *et al.*, 1982). The stable antiphase position takes an average of 12 days to achieve (Morin and Cummings, 1982).

There is a significant correlation between prior tau and latency to splitting or tau during the split conditions (Morin, 1980). Where the phase relationship between running onset of the split components has been measured, the mean was about 180°; although for 3 of 25 animals, tau deviated from the mean by two or more standard deviations. The animals with split rhythms also had longer taus prior to splitting than animals with rhythms that did not split (Morin and Cummings, 1982). Fusion of the two components may occur after 150 to 200 days of LL (Pittendrigh and Daan, 1967c), but is infrequent, occurring in about 5% of animals (Earnest and Turek, 1982). In one set of animals, mean tau before splitting was 24.49 hr; during the split it was 23.98 hr. Animals with split rhythms that were maintained in LL for over nine months produced a median period of 23.78 hr. Certain animals had remarkably short periods, the range being 23.44 to 24.16 hr (N = 8) (Morin and Boulos, unpublished).

Exposure to constant darkness fuses the split rhythm components and eliminates the split condition induced by LL within one to four days. In DD, the fused wheelrunning period lengthens about 0.16 hr compared with about 0.07 hr for unsplit animals transferred from LL to DD (Earnest and Turek, 1982). Re-exposure to LL will usually result in rapid (1 to 4 days) resumption of the split. The effectiveness of LL exposure in reinstating splitting is apparently contingent on light being initiated 4 to 5 hr after activity onset. Constant light started at other circadian times was ineffective. Constant light re-exposure was also ineffective if splitting had not previously occurred (Earnest and Turek, 1982).

The split locomotor rhythm of males has been entrained to 2-hr dark pulses using T cycles of 24, 24.23, and 24.72 hr (Boulos and Morin, 1982). In some animals, however, one limb of the split remained free running. In addition, entrainment of the split rhythm could be accomplished by phasing the dark "pulse" with either one of the two split limbs. A wide variety of complex rhythm phenomena, such as phase jumps and modulation of the amount of running associated with each limb, are frequently found after such dark "pulses" (Boulos and Morin, 1982; Lees and Hallonquist, 1982). These results support the idea that the circadian rhythm system is composed of at least two populations of coupled oscillators each with its own bidirectional PRC.

Individual differences are emphasized by the splitting studies: no two individuals are alike except in the most general ways. Thus, although attempts to model hamster running rhythms using the "E" and "M" oscillatory system have been reasonably successful with respect to the generalities, they remain incomplete with respect to individuals.

Rhythm complexity among rodents was noted at length by Swade and Pittendrigh (1967) and has been modeled by Winfree (1967, 1971) using a large number of oscillators. Generally, however, the tendency has been to ignore the complex nature of circadian rhythmicity in favor of a simplified rhythm theory based on the presumptive "E" and "M" oscillators (Daan and Berde, 1978). Recently, several studies have resurrected the conceptualization that hamster circadian rhythms are indeed more complex than can be explained by a two-oscillator model (David and Menaker, 1980; Morin, 1980; Morin and Cummings, 1981, 1982; Boulos and Morin, 1982; Lees and Hallonquist, 1983).

As splitting occurs, there is frequently evidence of additional rhythmic locomotion. The suggestion has been made that splitting might be causally related to the appearance of "secondary rhythmic components" (Morin, 1980). In one study (Morin, 1980), (1) the expected pictorial "crossing" of the "E" and "M" components was never observed in any of 35 hamsters showing splitting and (2) a mild to strong additional rhythmic component was observed in 43% of these animals at the time of splitting. Display of this extra activity (the "B" component) (Fig. 7) is clearly not necessary for splitting, although its appearance may offer clues as to the organization of physiological processes that promote splitting. As a result of this ambiguity of the splitting derivation, recent nomenclature has sought to avoid linking visual running record characteristics with hypothesized functional or origins (Morin and Cummings, 1982; Lees and Hallonquist, 1983). Thus, as illustrated in Fig. 7, one limb of the split is designated

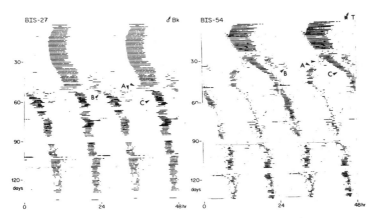

Figure 7. Splitting of the wheelrunning rhythm of two male hamsters housed in about 40 lux constant light (Morin and Cummings, 1982). The "A" limb of split activity abruptly advanced from the phase of former circadian activity. The "C" limb continued for several days with a period similar to that before the split. Additional activity not directly associated with either "A" or "C" has been labeled "B". See text for a discussion of the relationship between the current terminology and the designation of "E" and "M" oscillations. Splitting began about days 30 and 49 and stabilized about days 55 and 69 for animals BIS-54 and BIS-27, respectively. In each case, the two major components of the split rhythm were about 180° antiphase. Note that the "A" limb is fairly clear for BIS-54, whereas the relationship of "B" to the split is uncertain. For animal BIS-27, the visualizability of "A" and "B" are reversed. (Reprinted with permission of Physiology and Behavior. Copyright 1982, Pergamon Press, Ltd.)

the "C" limb because it tends to continue for several cycles with the same period as the prior, unsplit rhythm, The "A" limb is the one that abruptly advances, having a period temporarily shorter than the "C" limb. However, although generally useful, this mode of description is not always applicable (Morin and Cummings, 1982).

Several related types of activity records also suggest that the "E" and "M" conceptualization is incomplete: (1) splitting can occur with one component becoming visually obvious without any immediately (8 to 10 days) antecedent evidence of rhythmicity; (2) sometimes it is not possible to discern the individual identity of the two components (i.e., which is "E" and which is "M"); (3) certain records of splitting contain components that apparently never showed the pictorial "crossing" (Figs. 2 to 4 in Morin and Cummings, 1982); (4) it seems impossible to discern whether the "B" component of "extra" activity has any more or less control of the split than the "A" or "C" components (Morin and Cummings, 1982); and (5) each limb of a split record may, on rare occasions, split again (Boulos and Morin, 1982).

After the split has occurred, records frequently provide evidence of one or more rhythmic components traversing (having periods greater or less than 24 hr) the split behavioral system (Morin, 1980; Morin and Cummings, 1982). The extent of this multioscillatory activity is highly variable across animals. In certain individuals, however, a strong argument can be made that temporal coincidence between oscillatory events is necessary to account for the complex rhythmic patterns in wheelrunning. The temporal pattern (running onset time, duration, and time of running offset) during the "activity phase" often changes systematically across days under the apparent influence of several coincident oscillations. Overt rhythmic activity appears dictated by the joint function of two or more oscillations with activity presence and absence being equally rhythmic (Morin, 1980; Morin and Cummings, 1981, 1982; Davis and Menaker, 1981). This observation serves to emphasize the need for understanding the similar rhythmicity of both rest and activity: the patterning of the one is dependent on the pattern of the other.

An important practical ramification of these observations relates to the use of phase reference points in rhythm analysis. For example, in constant dark conditions, a circadian-range oscillation of intense locomotor activity can often be seen running through the activity phase of the primary circadian activity rhythm that has a different frequency. Also, the bulk of the activity frequently shifts from the early subjective night to the late subjective night phase (Morin and Cummings, 1981; Davis and Menaker, 1980). In both cases, the onset of alpha is highly variable and often not easily predictable. Therefore, calculations of phase relationships between endogenous and/or exogenous rhythms can be extremely difficult. A practical consequence of this variability is the extreme difficulty in generating a phase response curve from these animals. An apparently useful phase reference point under certain conditions (e.g., activity onset in LD) may be less useful than other phase points (e.g., midpoint of activity or offset of activity) under other conditions (Daan and Aschoff, 1975).

Under LL conditions, rather than split or remain stably rhythmic, activity patterns of certain animals devolve into a fractionated pattern designated "aperiodic" (Pittendrigh, 1974). An alternative perspective is that such animals remain periodic, but with an uncoupling of circadian oscillations. For example, one animal in LL had individual running bouts that free ran with slightly different circadian periods, gradually yielding a spectral analyzed "ultradian" pattern of running (Morin and Cummings, 1982). Because the ultradian rhythm pattern was apparently derived from a disintegrating circadian rhythm, it is probably that the ultradian pattern represents desynchronized groups of circadian-range oscillations. This raises the possibility that a series of activity bouts composing the normal circadian activity phase is actually the result of multiple, closely phased and coupled circadian range oscillations.

The regulation of locomotion within the activity phase has recently been studied in animals with split activity rhythms. It is possible to artificially enhance the amount of activity associated with one limb of split rhythm by giving a 1-hr daily dark pulse simultaneously with that limb. The increased activity may remain when the dark pulses cease or it may be transferred to the other limb (Boulos and Morin, 1982). This facility to associate running with time of a dark pulse could account for the discovery that splitting in LL can sometimes be induced by a single dark pulse (Ellis *et al.*, 1982). Such an explanation would rely on the existence of one or more "invisible" oscillations (i.e., without associated locomotion) that suddenly become visible through the action of the dark pulse. Thus, splitting could be understood, not as the divergence of two coupled oscillations, but as the new association of activity with an already existing freerunning or phaselocked (e.g., 180° antiphase) oscillation or set of oscillations.

2. CHEMICAL AND TEMPERATURE CONTROL OF CIRCADIAN RHYTHMICITY

Richter (1965) made extensive manipulations of rodent central nervous and endocrine systems in an effort to describe sites responsible for the clocklike control of the locomotor activity rhythm. The generalization that emerged from these studies was that the period of the rhythm was more or less immutable. Circadian locomotor rhythm generation remained despite loss of the adrenals, thyroids, ovaries, or testes. Recently,

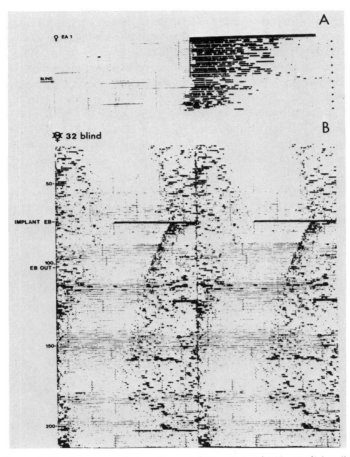

Figure 8. Changes in female hamster wheelrunning periodicity associated with estradiol availability (adapted from Morin *et al.*, 1977). In (A), the locomotor record is shown relative to the dark phase (dark bar at top of figure) of an LD 14 : 10 cycle. Beginning with day 1 of the record, onsets were alternately advanced for two consecutive days then relatively delayed ("scalloping"). The second day of advanced onset was always followed by a positive estrous smear indicating ovulation several hours earlier and estradiol secretion during the preceding 36 hr. Scalloping continued during the 20 days recorded after blinding. In (B), a blind, ovariectomized animal showed an abrupt decrease in tau after a subcutaneous implant of a Silastic capsule containing estradiol benzoate (EB). The shorter circadian period was maintained for many weeks after the hormone was removed. (Copyright 1977 by the American Association for the Advancement of Science.)

however, female hamsters have been shown to have earlier onsets of locomotor activity on estrous cycle days of high estradiol concentration (Fig. 8A). The phenomenon ("scalloping") occurs in LD 14 : 10 and DD (Morin et al., 1977a,b; Finkelstein et al., 1978). The fact that scalloping is demonstrable in DD shows that it is not caused by diminished sensity to light (Aschoff, 1978). Ovariectomy has been shown to induce a phase delay of activity onset (Widmaier and Campbell, 1980) and estradiol benzoate (EB), chronically administered in Silastic® capsules, induces a phase advance in ovariectomized animals entrained to LD 14 : 10 or 16 : 8 (Morin et al., 1977a,b; Widmaier and Campbell, 1980). The phase change was reversed with the removal of the capsule. In four separate groups from three experiments, EB shortened tau by an average 6.8 min when given to ovariectomized hamsters that were blind or housed in constant dim illumination (Fig. 8B) (Morin et al., 1977b; Takahashi and Menaker, 1980; Zucker et al., 1980a). Thus, acute changes in phase and period of wheelrunning appear related to the presence of estradiol. Interestingly, the short taus occurring with EB treatment persist after removal of the hormone (Morin et al., 1977b). Stimulation by light (Morin et al., 1977b) or progesterone, a well-established inhibitor of estradiol-facilitated behavior (Morin, 1977), may neutralize the effects of EB on tau and psi (Takahashi and Menaker, 1980). In the longer term, tau shortens and psi becomes increasingly positive in association with estrous cycle resumption during the annual hamster reproductive cycle (Morin, 1982).

In male hamsters, neither EB nor testosterone has observable effects on tau (Zucker et al., 1980a; Morin and Cummings, 1981, 1982). The sex difference in the response to EB is dependent on the neonatal hormone environment for its expression to the extent that the adult circadian period of neonatally androgenized genetic females will not shorten in response to EB (Zucker et al., 1980a). Male and female hamsters may also differ in their ability to entrain to long-period T cycles and in the shape of their respective PRC's. The PRC shapes predict the observation that females are least likely to entrain to T = 24.75 and that females initiate locomotor activity significantly in advance of males. The measured PRC's of females are also much more variable than those of males (Davis et al., 1983). The precise roles of neonatal or adult hormones in the display of these differences has not been determined. Although large phase changes occur slowly in males exposed to prolonged short photoperiods (LD 6 : 19 or 2 : 22) (Ellis and Turek, 1979; Bittmann, 1978), these have not yet been related to any hormone changes as they have for females (Morin, 1982).

It is useful to know that pinealectomy has neither an effect on the circadian period of male hamsters (Morin and Cummings, 1981) or on the PRC or duration of the activity phase (Aschoff et al., 1982). In two studies, however, there is the indication that total activity may decline after pinealectomy (Morin and Cummings, 1981; Aschoff et al., 1982). Pinealectomized females are reported to accomplish a 12-hr phase shift more rapidly than controls (Finkelstein et al., 1978), but small phase shifts induced by single light pulses in DD apparently were not faster than expected (Aschoff et al., 1982).

The rhythm splitting often observed following prolonged exposure to LL is also subject to some hormonal control. Castrated or castrated, progesterone-implanted females each have a 65% to 85% incidence of splitting or other multioscillatory free-running rhythm in LL. Only about 14% to 25% of EB-treated castrated females have similar records (Morin, 1980; Morin and Cummings, 1982). Incidence is about 60% for intact or castrated adult males and neither testosterone nor EB significantly change the male response to LL. Furthermore, sex differences are evident in the manner in which splitting occurs and in the clarity of the split rhythms (Morin and Cummings, 1982).

When implanted with EB, precision of running onset by blind, ovariectomized females becomes more precise and there appears to be more activity early in alpha (Morin et al., 1977b). Several laboratories have measured the amount or duration of wheelrunning under the influence of estradiol. The number of daily wheel revolutions by intact females has been shown to vary with the estrous cycle, the greatest number occurring during the night of cycle day 4 after two days of estrogen stimulation (Richards, 1966; Baranczuk and Greenwald, 1973). Another group of entrained intact females failed to statistically show this cyclic variation (Morin et al., 1977b), although all animals showed scalloping and certain individuals appeared to have clear estrous-cycle-related changes in amount of activity. In the same study, entrained, ovariectomized animals suffered an approximate 20% reduction in the number of wheel revolutions regardless of receiving EB, progesterone, or empty Silastic capsule treatment. However, EB treatment of blind ovariectomized animals was associated with an increase in revolutions. Blank capsules did not effect the number of revolutions, but progesterone decreased activity (Morin et al., 1977b).

Other researchers have indirectly estimated activity according to some index (e.g., "percentage of bins with activity") or calculated the number of activity minutes per day covered by event-recorder ink. The latter method has proved to correlate, under the conditions used, with the number of daily wheel revolutions by males (Morin and Cummings, 1981). The ability to accurately compare measures remains a significant problem.

Using the "percentage of bins" method, intact females showed estrous-cycle-related changes, with more activity occurring on days 3 and 4 (Finkelstein et al., 1978; Takahashi and Menaker, 1980). Ovariectomy diminished and estradiol restored the amount of activity (Widmaier and Campbell, 1980; Takahashi and Menaker, 1980). Chronic progesterone can overcome the facilitatory actions of estradiol on the amount of activity (Takahashi and Menaker, 1980). Large changes in the activity index may be induced by the surgical procedure alone in long photoperiod control animals (Widmaier and Campbell, 1980). Intact or ovariectomized plus estradiol groups in long photoperiod did not appear to differ in their amount or distribution of activity; removal of the endogenous or exogenous hormones greatly reduced the activity index. However, removal of an empty capsule from an already castrated animal also resulted in an approximate 33% loss of activity. It is important to note that animals that had undergone gonadal regression in LD 6 : 18 did not increase activity in response to estradiol. Nevertheless, these animals, which were running at only 50% of long photoperiod estradiol-treated levels, did diminish activity (about 43%) following hormone removal. Controls reduced activity by 17% (Widmaier and Campbell, 1980).

The activity phase in LD 16 : 8 is characterized by sharp onsets when estradiol is present. In LD 6 : 18, the activity of anestrous animals without endogenous estradiol is quite diffusely distributed with quite variable onsets. Exogenous estradiol renders the onset more precise and advances the bulk of activity toward the time of alpha onset (Widmaier and Campbell, 1980). One disturbing item in this study was the report that seven weeks following removal of Silastic capsules containing estradiol, the level of serum estradiol was highly correlated with the amount of running (Widmaier and Campbell, 1980). This may explain the after effects of estrogen in free-running animals (Morin et al., 1977b), although the patterns of activity in that study would suggest otherwise. The prolonged existence of estradiol in serum after its presumed removal certainly merits consideration in future experiments.

Among males exposed to short photoperiod or DD, activity onset becomes highly variable as do offset and other points between (Ellis and Turek, 1979; Morin and Cummings, 1981). A variety of measures have been used to assess the male activity pattern under conditions of androgen presence or absence. The measures include counts of wheel revolutions, minutes of activity per day, minutes per running bout, precision of activity phasing during each day, number of daily running bouts, and hourly distribution of daily activity (Ellis and Turek, 1979; Morin and Cummings, 1981). In a series of experiments that did not always measure wheel revolutions, all measures tended to be altered by presence or absence or androgen, but the precision of activity onset (Fig. 9) was found to be the most reliable index of hormone state (Morin and Cummings, 1981). Androgen presence before gonadal regression or after spontaneous recrudescence markedly increases alpha onset precision in LD 6 : 18 or DD-housed animals and is associated with an increased number of daily wheel revolutions (Ellis and Turek, 1979; Morin and Cummings, 1981). However, during the regression phase of the annual reproductive cycle, testosterone may be unable to facilitate locomotor activity (Ellis and Turek, 1981). This would be a further indication that the effectiveness of gonadal steroids as behavior facilitators becomes diminished during prolonged exposure to short photoperiod (compare Morin et al., 1977a; Morin and Zucker, 1978; Campbell et al., 1978; Widmaier and Campbell, 1979).

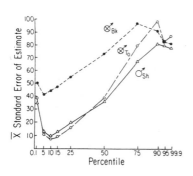

Figure 9. Mean index (standard error of estimate; SEE) of variability in activity rhythm precision according to hormone status of long-term blind male hamsters (Morin and Cummings, 1981). The time of completing a given percentage of the animal's daily activity was assessed across about ten circadian days and a least squares best fit line was calculated for each percentile of activity. The SEE provides an index of variability about the line that is independent of slope. The figure shows that for the tenth percentile (shortly after activity onset), castrated males without replacement testosterone (♂ BK) had an SEE of about 45 min, whereas intact (♂ Sh) or castrated males given testosterone (♂ To) had SEEs of about 10 min. (Reprinted with permission of Physiology and Behavior. Copyright 1981, Pergamon Press, Ltd.)

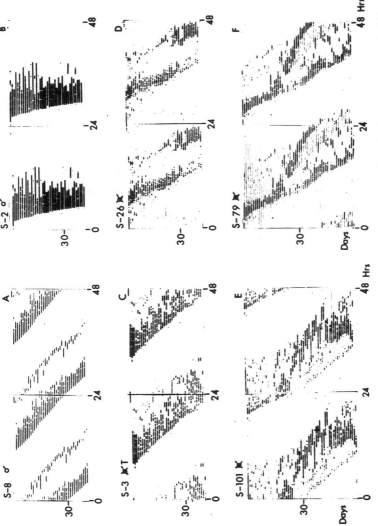

Figure 10. Wheelrunning records of long-term blinded, male hamsters that were intact (♂), castrated, testosterone-treated (⊗ T), or castrated (⊗) (Morin and Cummings, 1981). Animals with endogenous or exogenous androgen (A, B, and C) showed clear, precise running onsets unlike the animals without androgen (D, E, and F). Note the bimodal activity distribution in (A) and unimodal distribution in (B). In (D), (E) and (F), activity demonstrated forms of rhythmicity within alpha, the general activity phase. (Reprinted with permission of Physiology and Behavior. Copyright 1981, Pergamon Press, Ltd.)

Analysis of grouped data (Fig. 9) can lead to an unfortunate loss of the pictorial impact. Figure 10 shows a variety of running patterns of blind males, with and without androgen. An analysis that looks only at grouped data would obscure the clear differences in periodicity within alpha that are observed in many records (Morin, 1980; Morin and Cummings, 1981, 1982). Furthermore, it has been argued that these periodicities are, in part, responsible for the overall circadian period (Morin, 1980; Davis and Menaker, 1980). Among males, absence of androgen appears to facilitate the visualization of additional periodicities, but is not necessary (Morin and Cummings, 1981). In numerous animals, the phase of the densest locomotor activity becomes increasingly delayed from the expected alpha onset phase and appears to free run or assume a new, temporarily stable phase within alpha. Under some circumstances, the altered running pattern seems to be characterized by a phase delay of several hours. This may last 5 to 20 days then abruptly disappear (Davis and Menaker, 1980; Morin and Cummings, 1981). The concept of homogeneous or bimodal activity phase is not supported by these data.

Heavy water (D_2O; deuterium oxide) has the greatest effect on phase and period of any tested chemical. The summary offered by Daan and Pittendrigh (1976b) is that D_2O lengthens the period, but appears not to differentially affect phases of the cycle. Change in tau in DD among animals given 20% D_2O in drinking water is about 0.9 hr. Period change differences between mice and hamsters in response to D_2O seem related to tissue concentrations of D_2O. Among hamsters exposed to an LD cycle, the locomotor rhythm will entrain, or not, depending on the concentration of D_2O consumed. The phase angle difference of running onset becomes progressively delayed as D_2O concentration is increased (Richter, 1977; Eskes and Zucker, 1978), until entrainment fails when animals consume 10% to 20% D_2O. When entrainment fails, the rhythm may remain "relatively coordinated" with the light cycle. In this circumstance, the rhythm will systematically vary in phase with respect to the LD cycle, but the exact period will temporarily lengthen or shorten according to the influence of light on the PRC. Relative coordination (Pittendrigh and Daan, 1976b) between the free-running locomotor rhythm and the LD cycle often occurs with D_2O concentrations in the 20% range (Richter, 1977). Tau during 100% D_2O consumption is about 27.5 hr (Richter, 1977). Tests with 5% D_2O show that tau lengthens by an amount independent of the prior period measured in LL or DD. Therefore, although homeostasis of tau appears to exist when animals are transferred from DD to LL, it is not evident during deuteration (Fitzgerald et al., 1978).

Tau also lengthens after hypophysectomy in male or female hamsters (Zucker et al., 1980b). Because gonadal hormones apparently do not affect tau of male hamsters, hypophysectomy must exert its action on the circadian period via some other branch of the neuroendocrine system.

The thyroid gland has been implicated in normal rhythm regulation. Female hamsters that had sustained an approximately 90% thyroidectomy were given thiourea, a thyroid-function-blocking agent. This treatment combination induced a 0.2 hr increase in tau. Neither nonthyroidectomized animals given thiourea nor thyroidectomized animals without thiourea showed any period changes (Beasley and Nelson, 1982).

Lithium significantly lengthens tau in rats and disrupts their entrainment to long T cycles (Kripke and Wyborney, 1980; McEachron et al., 1981). A similar effect of lithium on tau has been observed in hamsters (I. Zucker, unpublished manuscript cited by McEachron et al., 1981).

Tricyclic antidepressant drugs are able to modify circadian period and phase. Clorgyline-treated males entrained to LD 14 : 10 began running onsets 1 hr later than controls and did not phase advance as fast as controls following a 4-hr advance of the light cycle (Craig et al., 1981). Clorgyline and imipramine induce an average 0.61 and 0.36 hr, respectively, lengthening of tau in DD housed males (Wirz-Justice and Campbell, 1982).

Body temperature has also been manipulated in hamsters in an effort to alter circadian phase and period. Dropping core temperature to 14 to 17 °C for about 3 hr in two animals caused the activity rhythm in DD to delay by about 0.67 hr (Rawson, 1960). These results are consistent with more extensive observations from Peromyscus. Notably, Rawson (1960) generated a phase-response curve for Peromyscus using cooling as the stimulus. Only phase delays were found and those delays that were greater than 10-min duration occurred exclusively during the period from 3 hr prior to activity onset until 6 hr after onset. Richter (1975) obtained large phase delays (1 to 3.75 hr) in hamsters that lost respiratory activity and heart rate when cooled to 0 to 2 °C. Tau was also slowed in most of these animals. All animals in this experiment were drinking a 10% v/v ethanol solution and were blind. Ethanol itself is known to lengthen the period of locomotor rhythms in hamsters (Zucker et al., 1976). Recently, however, Gibbs (1983) maintained blind, freerunning male hamsters at colonic temperatures ranging from 10°–20°C for 3–24 hr. The magnitude of the resultant phase delays varied from about 1 hr in response to colonic temperature held at 12°C for 3 hr to 13 hr in response to 24 hr with the same hypothermia. Q_{10} for the rate of the hamster clock appears to be between 1.08 and 1.34 and the hamster's clock is more temperature compensated than is the rat's (Gibbs, 1983).

3. NEURAL CONTROL

In 1972, two sets of investigators reported that destruction of the suprachiasmatic nuclei (SCN) abolishes circadian periodicity of rat locomotor activity, drinking, and adrenal corticosterone rhythms in the rat (Moore and Eichler, 1972; Stephan and Zucker, 1972). These studies rapidly followed the first clear demonstration of direct retinal projections to the SCN of the hypothalamus (Moore and Lenn, 1972). Eichler and Moore (1974) found a similar retinohypothalamic tract in hamsters. Most neuroanatomic studies of the SCN have involved rats, but work with hamsters is becoming more frequent (Pickard, 1980, 1982; Pickard and Silverman, 1981; Card and Moore, 1984).

A clear description of the hamster SCN has become available (Card and Moore, 1984). The hamster SCN sit bilaterally over the caudal third of the optic chiasm, each immediately lateral to the third ventricle. Dimensions of each nucleus are approximately 650 μm × 600 μm × 300 μm (length × height × width), but they are also fused ventromedially. The extent of each nucleus is generally well delimited, but distinct boundaries are not apparent caudally. From this less distinct SCN region, fibers extend dorsally and caudally (Card and Moore, 1984). Both the SCN and their exiting fiber systems stain heavily for vasopressin, somatostatin, vasoactive intestinal polypeptide, neuropeptide Y, GABA, and serotonin (Card and Moore, 1984). The relationship of these substances to hamster rhythmicity is only now being investigated (Albers *et al.*, 1984).

Summarizing the known connections with each hamster SCN, there appears to be input via a large number of fibers from the retina, lateral septum, preoptic periventricular nuclei, SCN, paraventricular thalamic nuclei, and ventral subiculum. Input is also received from the medial preoptic area, periventricular hypothalamus, ventromedial hypothalamus, ventral lateral geniculate nuclei, zona incerta/lateral terminal nuclei, and dorsal nucleus lateral lemnicus; other regions provide less input (Pickard, 1982). The ventral lateral geniculate in particular provides SCN afferents containing neuropeptide Y (Card and Moore, 1982; Moore *et al.*, 1984). Suprachiasmatic nuclei efferents in hamsters are less well described, but presumably are similar to those in rats (Stephan *et al.*, 1981; Berk and Finkelstein, 1981). Efferents include projections to the ventromedial hypothalamic, dorsomedial hypothalamic, premamillary and interpeduncular nuclei; to the basal preoptic anterior hypothalamic and supraoptic areas; and to the paraventricular hypothalamic nuclei, septal nuclei, habenula and midbrain central gray (Stephan *et al.*, 1981; Berk and Finkelstein, 1981; Moore, 1983).

Anatomical understanding has permitted lesion and knife-cut studies of the functional importance of the SCN and their connections. Ablation of the SCN disrupts hamster circadian rhythmicity in dim LL (Stetson and Watson-Whitmyre, 1976; Rusak, 1977a). Ultradian oscillations have frequently been observed subsequent to the lesion (Rusak, 1977a). No work has been published concerning rhythmicity in DD after SCN lesions.

Attempts to describe anatomical locations of other SCN connections that mediate rhythm expression have been performed almost entirely with rats and have been only partially successful. Suprachiasmatic nuclei isolation obliterates circadian behavioral rhythmicity (Stephan and Nunez, 1979). Neither Halasz-type semicircular cuts just rostral to the SCN nor parasaggital cuts have any effect on drinking or eating rhythms (Nunez and Stephan, 1977; Nishio *et al.*, 1979; Van Den Pol and Powley, 1979; Dark, 1980). In contrast, Halasz-type cuts (which also sever dorsal connections) made retrochiasmatically in rats eliminate behavioral and endocrine rhythms (Moore and Eichler, 1976; Nunez and Stephan, 1977; Van Den Pol and Powley, 1979; Dark, 1980). This type of cut may permit distinction between SCN rhythm generation and entrainment functions (Dark, 1980). Paraventricular hypothalamic nucleus lesions which block gonadal regression during short photoperiod do not alter entrainment or circadian wheelrunning rhythmicity by male hamsters (Pickard and Turek, 1983).

Rate of re-entrainment by rats is also slowed following unilateral blinding, but this effect is not seen in hamsters (Stephan *et al.*, 1982). This species difference has been attributed to anatomical differences in the distribution of retinohypothalamic fibers to the SCN (Moore, 1973; Pickard and Silverman, 1981).

There are no reports showing isolation of the SCN from direct retinal efferents.

However, direct electrical stimulation of hamster or rat SCN can affect running onset phase in a manner analogous to the effect of light (Rusak and Groos, 1982). Lesions of the ventromedial hypothalamus may abolish rat circadian rhythms synchronized to timed meals (Krieger, 1980; Mistelberger, 1982). Lesions of the lateral hypothalamus (LHth) or of the primary optic tracts (POT) that impinge on the LHth results in long circadian locomotor (hamster) and drinking (rat) rhythms (Rusak, 1977b; Rusak and Boulos, 1981; Rowland, 1976). Lateral hypothalamus lesions may also abolish daily rat wheelrunning and temperature rhythms (Moore and Eichler, 1976). Rate of re-entrainment by hamsters is also retarded (Rusak, 1977b) and nocturnality of rats increased (Rowland, 1976) by such lesions. Similar slow rates of re-entrainment occur after large lateral geniculate nucleus lesions in male hamsters (Zucker et al., 1976), but this has not been replicated in females (Carlisle, 1975).

The functional significance of connections between the two SCN has recently been examined in hamsters. Unilateral SCN lesions will eliminate an existing split in the locomotor rhythm. Large changes in tau of split or unsplit animals usually accompany the lesions, but the direction or magnitude of change seem relatively uncorrelated with the extent of unilateral SCN damage (Pickard and Turek, 1982, 1984). Small lesions seem unable to eliminate the split or markedly alter the period. However, unilateral SCN lesions prior to splitting may be unable to prevent subsequent splitting (Davis and Gorski, 1981) and midsagittal cuts separating the two SCN fail to eliminate stable phasing of the two split components (Sisk and Turek, 1982). Thus, the data available cannot yet resolve questions concerning the functional significance of the inter-SCN connection.

Acetylcholine has been implicated as a neurotransmitter capable of regulating circadian rhythm phase in rats and mice (Zatz, 1979; Zatz and Brownstein, 1979; Zatz and Herkenham, 1981). Similarly, intraventricular carbachol exerts phase control of the male hamster running rhythm, and the effect is circadian phase dependent (Earnest and Turek, 1983a). It is not yet certain whether these effects are through an action on the SCN or some other locus. Stimulation of the hamster suprachiasmatic region (but apparently not the lateral ventricle) by nanogram quantities of a putative peptide transmitter, neuropeptide Y, is able to generate a phase response curve without after-effects on period (Albers et al., 1984). This suggests that the LGN afferent system containing neuropeptide Y (Card and Moore, 1982; Moore et al., 1984) may ordinarily exert phase control over circadian rhythms.

The presumptive retinal photoreceptor for phase shifting of the circadian locomotor rhythm has maximal spectral sensitivity at 500 nm. Particularly unusual is the observation of an equal rhythm phase shift response to light pulse durations up to 45 sec as long as the quanta of light stimulation is held constant. The reciprocal relationship between intensity and duration for phase shifting apparently endures 15 to 450 times longer than for rods or cones, respectively (Takahashi et al., 1984).

3.1. Photoperiodism

Gaston and Menaker (1967) established that male hamsters require more than 12 hr of light per day in order to photoperiodically maintain gonadal activity. It is now well established that hamsters use a circadian clock to measure photoperiodic time: a short period of light (e.g., 6 hr) can be stimulatory or nonstimulatory to the reproductive system according to the phase of the animal's circadian day during which it occurs (Elliott *et al.*, 1972; Stetson *et al.*, 1975; Elliott, 1974, 1976). Similarly, intraventricular carbachol provided at specific phases of subjective night will induce gonadal regression (Earnest and Turek, 1983a). A circadian clock is also used to assess photic information necessary for breaking the refractoriness to further gonadal regression in short photoperiod after spontaneous recrudescence (Stetson *et al.*, 1976; Chapter 5, this volume). Lesions of the SCN remove the antigonadal effects of short photoperiod (Stetson and Watson-Whitmyre, 1976; Rusak and Morin, 1976).

The mechanism by which this photoperiodic time measurement is achieved remains obscure. Two rhythm-based models have been proposed to account for the phenomenon (Pittendrigh and Daan, 1976c). The external coincidence model (Bunning hypothesis, Bunning, 1936) postulates a dual role for light: it acts as an entraining agent and, when occurring during the photosensitive phase, will stimulate gonadal activity (Pittendrigh, 1974). This model requires the presence of light for maintenance of gonadal activity in normal adult hamsters. The internal coincidence model (Pittendrigh, 1974) similarly requires that light act as an entraining agent. However, no direct effect of "photostimulation" is presumed necessary. Instead, through its entrainment function, light may establish appropriate phase relationships (coincidences) among internal rhythms. The product of such coincidence will necessarily be maintenance of reproductive activity.

Unfortunately, the two hypotheses have proved rather difficult to separate experimentally. One study utilized a variety of phase shifts in an attempt to affect subsequent gonadal regression in DD. Animals were shifted then blinded five to ten days later. About 45% of the animals that experienced a 4-hr phase delay failed to achieve the expected testicular regression (i.e., equal to the range of unshifted controls) following blinding. Moreover, the effect was specific to the direction of the phase change because 4- or 2-hr phase advances did not interfere with gonadal regression after blinding. Thus, the phase delay induced an "aftereffect" on the gonadal response to DD in some animals (Morin, 1981a). This is contrary to the prediction of the external coicidence model that requires that without light, gonadal activity must cease in a normal fashion. The data are consistent with the internal coincidence model and a nonrhythm-based endocrine model called the "carry-over effect" (Follett *et al.*, 1967). The latter has been invoked to account for the retarded gonadal regression observed when hamsters have access to running wheels (Elliott, 1974; Eskes and Zucker, 1978; Borer *et al.*, 1981; Morin, 1982).

A provocative experiment related to the internal or external coincidence question has been performed by Warden (1978). Male hamsters taken from LD 14 : 10 were given access to running wheels and allowed to self-select their lighting for four to seven weeks. Among animals rhythmic in their lighting selection, all of those that self-selected darkness in phase with their running rhythm had testes less than 850 mg, whereas all with rhythmic dark selection out of phase with running had gonads greater than 850 mg. Most notable is the fact that none of the animals with regressed testes ever experienced total darkness during the full "photosensitive" phase. Recent work has also shown that under certain lighting conditions, hamsters will entrain locomotor activity as if in a long photoperiod, but the gonads may or may not regress (Earnest and Turek, 1983b). These three experiments (Warden, 1978; Morin, 1981a; Earnest and Turek, 1983b) raise serious questions about the "photostimulatory" role of light as proposed by the external coincidence model.

4. OTHER RHYTHMS

4.1. Reproductive Activity

The estrous cycle of rodents is reasonably well understood (see Chapter 3, this volume). Everett et al. (1949) demonstrated that a "critical period" regulated the daily timing of rat gonadotropin release and consequent ovulation. This timing appears to depend on some form of 24-hr clock. Alleva et al. (1971) utilized the remarkable four-day precision of the hamster estrous cycle to show that it is regulated by a circadian clock. In the experiment, the onset of sexual receptivity occurred at almost exactly 96-hr (4-day) intervals when the animals were entrained to an LD 14 : 10 cycle. When placed in LL, the animals showed free-running estrous cycles with lordosis occurring at an average interval significantly greater than 96 hr. More importantly, each animal had a simultaneously recorded circadian wheelrunning rhythm equal to the estrous cycle length divided by four (Alleva et al., 1971). Using extreme T cycles, the locomotor rhythm has been entrained to a 21.5-hr photoperiod. Simultaneously monitored estrous cycles were 85.7 hr long (Carmichael et al., 1981).

Although extremely short T cycles can reduce the length of the hamster estrous cycle, they appear to have no effect on the length of gestation. A T cycle consisting of 14 L : 9 D failed to significantly shorten the duration of pregnancy (Carmichael and Zucker, 1982).

Female hamsters in dim LL given heavy water (D_2O) greatly lengthened the circadian locomotor period (Fitzgerald and Zucker, 1976). Nevertheless, the estrous cycle remained a four times multiple of the activity period. Under a variety of entrainment or photoperiod conditions, the preovulatory gonadotropin surge was found to correlate with the timing of activity onset and not with the time of lights off, lights on, or middarkness (Stetson and Gibson, 1977; Stetson and Anderson, 1980; Moline et al., 1981). With progressively shorter T cycles, female hamsters tend to start running and estrus onset later with respect to lights off (Carmichael et al., 1981; Carmichael and Zucker, 1982). Together, the above experiments show that estrous cyclicity and the circadian locomotor rhythm are regulated by a common circadian clock. Further support for this proposition is also derived from observations of intact females that lose clear circadian running rhythmicity. Such animals become reproductively acyclic at the same time (Fitzgerald and Zucker, 1976; Stetson et al., 1977). Females sustaining SCN lesions also simultaneously lose circadian and estrous cycles (Stetson and Watson-Whitmyre, 1976).

An interesting observation of lordosis onset timing has been made in hamster females with split locomotor rhythms. Lordosis onset was found to be associated with either of the two major limbs of the split condition and, in one animal, first with one limb, then with the other (Swann and Turek, 1982). The results support a clear association between the circadian locomotor rhythm system and the clock mechanism timing the hamster estrous cycle. Because either limb of the split-running record can apparently be associated with the timing of estrous cycle events, the results also suggest a large amount of redundancy among the components of the system that does the timing. Two reports suggest, however, that lordosis onset need not be tied expressly to the onset phase of a locomotor rhythm (Carmichael et al., 1981; Swann and Turek, 1982).

Male hamsters and rats also demonstrate rhythms in sexual behavior. Larsson (1958) and Dewsbury (1968) reported that male rats are most likely to copulate at night. In fact, when a male and female are housed together in constant dim light with free-running activity rhythms, copulation only occurs when the male's subjective night is coincident with the female's behavioral estrus phase (Richter, 1970). Similarly, intact hamsters are more likely to copulate during the dark phase of the photoperiod (Morin and Zucker, 1978; Eskes, 1982). The better nocturnal copulatory behavior can be eliminated by SCN lesions (Eskes, 1982). With testosterone replacement given to castrated animals via Silastic capsules, the increased likelihood of copulation during the dark is maintained despite the fact that there is considerably less circulating androgen during the dark. The lower androgen level is presumably caused by higher nocturnal metabolism of the steroids. One interpretation of these data is that at certain times of day male hamsters become relatively refractory to the behavioral-activating effects of androgen (Morin and Zucker, 1978).

Similarly, when exposed to prolonged short photoperiod that induces testicular regression (see Chapter 5, this volume), males also become relatively refractory to sexual behavior facilitation by androgen (Morin et al., 1977a; Morin and Zucker, 1978; Campbell et al., 1978). Therefore, two methods are available for terminating reproduction during the months with short days: (1) photoperiod-dependent hypothalamic-pituitary-gonadal response that includes increased sensitivity to negative feedback by steroids on gonadotropin-regulating systems (Turek and Campbell, 1979), and consequent decreased gonadal activity and loss of spermatogenesis (Reiter, 1974, Chapter 5, this volume); and (2) greatly diminished copulation probability induced by a decreased sensitivity to the behavior facilitating effects of androgens.

Among females, short photoperiod induces loss of estrous cycles, but appears to leave the sexual behavior response to steroid hormones unaffected. Sequential estradiol benzoate (EB) and progesterone injections will facilitate sexual behavior in females made reproductively acyclic by either short photoperiod or melatonin treatment (Morin et al., 1976; Morin, 1982b). In the latter case, the EB dose was sufficiently low to have permitted detection of photoperiod-induced changes in behavioral responsivity to the hormone. Nevertheless, the synergistic action of combined EB and progesterone may have overwhelmed any low-level melatonin-induced inhibition of estradiol activity. For example, estradiol in a Silastic capsule is more effective in facilitating wheelrunning by long photoperiod housed females than by those housed under a short photoperiod (Widmaier and Campbell, 1980).

4.2. Drinking

Hamsters drink in a rhythmical fashion with about 80% of intake occurring at night (Zucker and Stephan, 1973; Rusak, 1977b). Access to wheels reduces nocturnal intake to about 60% (Zucker and Stephan, 1973). The drinking rhythm persists in constant lighting conditions and phase shifts after a shift of the LD cycle (Rusak, 1977b; Sisk and Stephan, 1981). It is abolished by SCN lesions (Rusak, 1977b).

Following a reversal of an LD 12 : 12 cycle, hamsters require 25 to 40 days to re-establish preshift levels of nocturnal water intake (Zucker and Stephan, 1973). This compares with six to nine days needed for wheelrunning re-entrainment. The drinking rhythm re-entrains following a 6-hr phase advance or delay in six to nine days if animals are without wheels (Sisk and Stephan, 1981). Access to wheels permits re-entrainment of the drinking rhythm following a 6-hr delay in about 4.3 days, but entrainment after a 6-hr phase advance takes about 12 days. Re-entrainment of wheelrunning after a phase delay (about 1.1 days) is faster than re-entrainment of drinking, but there is no difference between the behaviors in latency to re-entrainment after a phase advance (Sisk and Stephan, 1981). The periods of the two rhythms in DD are similar (Sisk and Stephan, 1981) and have similar phasing and periodicity when they simultaneously split (Shibuya et al., 1980).

4.3. Body Temperature

Rats have a well-documented rhythm in body temperature (Satinoff et al., 1982), but there are few investigations with hamsters. Under LD 12 : 12, a male housed with four cagemates maintained a day–night rhythm, with telemetered temperature about 1 °C higher at night. However, individual housing of the same animal markedly damped or eliminated the rhythm (Moller and Bojsen, 1974). An isolated individual was reported to have a daily rhythm in rectal temperature (Chaudhry et al., 1958). Rectal temperatures of individual animals of either sex, but unknown housing or lighting conditions, were obtained four times per day for three to five days, with and without forced exercise. In each case, rectal temperature was elevated at night. The rhythm amplitude was greater in males than females, but was attenuated by forced exercise for either sex (Folk et al., 1961). The Turkish hamster (Mesocricetus brandti) apparently has a 2° day/night difference in telemetered temperature. The nocturnal rise in temperature anticipates the dark phase of LD 16 : 8 with the maximum temperature occurring about 4.2 hr after lights off (Albers et al., 1982).

4.4. Feeding

The literature on hamster feeding is sparse. Initial reports concerning its rhythmicity indicated that males eat about 55% of their food during the 12-hr dark phase

(Zucker and Stephan, 1973). This has been confirmed (Silverman and Zucker, 1976; Toates, 1978), but whether the day/night difference represents a circadian rhythm is unclear. Experiments that have monitored meal patterns for one or more days in LD, DD, or LL reveal that during the early subjective night, individual meals are smaller and more frequent than at other times whether or not animals have wheel access (Morin, 1981b). This patterns shifts with a phase shift of the LD cycle, free runs in LL or DD, and splits when wheelrunning splits (Morin, 1981b), but disappears when circadian running rhythmicity is lost following an SCN lesion (Gustafson and Morin, unpublished). Thus, a certain characteristic of the eating pattern appears to be a circadian rhythm. It is conceivable that the circadian rhythm in the meal pattern is the result of time competition between mutually exclusive behaviors, such as general locomotion or drinking.

Hamsters eat about 15 meals spread over 24 hr (Borer et al., 1980; Rowland, 1982; Morin, 1981b). The individual meals do not entrain to the LD cycle, they are not affected by light, and they do not free run in synchrony with the wheelrunning rhythm. However, similar to voles (Lehmann, 1976; Daan and Slopsema, 1978), hamster meals persist with periodicity in the ultradian range. Apparently unlike voles, there are several significant ultradian periods simultaneously evident. Thus, hamster meals seem to be timed by internal clocks. This timing persists even during food deprivation and refeeding (Morin, 1981b). The feeding behavior is clearly distinguishable from the running, gnawing, and hoarding rhythms of hamsters (see below).

4.5. Aggression

Male hamsters show a rhythm in aggression with more offensive/defensive behaviors occurring during a phase 1 to 2 hr after lights off. The rhythm appears to be circadian since it completes a 12-hr phase shift within one to four weeks and persists in dim LL (Landau, 1975a). Adrenalectomized animals maintained on exogenous adrenal corticoid therapy do not show an aggression rhythm although the locomotor rhythm is sustained (Landau, 1975b). Thus, rhythmic control of hamster aggression may be through the known circadian rhythm in hamster adrenal corticoid secretion (Andrews, 1968).

4.6. Other Rhythms

Gnawing on a wooden dowel passively protruding into the hamster's cage is manifested as a circadian rhythm with phase and period properties similar to those of wheelrunning (Morin, 1978). Hamsters will also hoard 45 mg Noyes pellets in a day/night rhythm, with approximately ten times more being hoarded at night (Toates, 1978).

Heart rate measured before and after forced exercise also has an apparent day/night rhythm. Resting rate in partially restrained, unexercised animals was 412 beats/min at midday and 457 beats/min during the night (Folk et al., 1961).

In one of the very first demonstrations of a circadian rhythm in hamsters, Bunning (1958) showed that gut motility maintained an approximately 24-hr rhythm in vitro.

The rat shows enhanced paradoxical sleep when maintained under 25-min light : 5-min dark cycles (Lisk and Sawyer, 1966). This phenomenon has been replicated with hamsters (Tobler and Borbely, 1977), but with the interesting difference that contrary to experience with rats the increase in paradoxical sleep occurs during the light.

The hamster cheek pouch has been used to study rhythms in epithelial mitotic activity (Brown and Berry, 1968; Moller *et al.*, 1974; Moller, 1978). A light cycle of LD 12 : 12 was found to be an inadequate Zeitgeber for this rhythm (Moller *et al.*, 1974). Guicking (1970) reported synchronization of locomotor activity with white noise and this may also be an effective Zeitgeber with mitotic activity (Moller, 1978). When 60 dB white noise occurred in phase with light during an LD 12 : 12 cycle, epithelial mitoses were maximal during the midportion of the day (Moller, 1978).

Female hamsters are differentially affected by an insulin injection according to time of day. All animals died after being given 10 IU insulin 5 hr after lights off. The percent of animals surviving progressively increased across the day with 100% survival after injection 1 hr before lights off (Burns and Meier, 1981).

Rhythms of brain tryptophan and serotonin (5-HT) have been measured in hamsters (Morgan *et al.*, 1976; Philo *et al.*, 1977). Telencephalic 5-HT concentration varied systematically with time of day, less being found during the late night and early morning than at other times. In the brainstem, 5-HT peaked at midmorning and night. Tryptophan in the telencephalon and brainstem was maximal in late afternoon, but reached a second peak at lights on.

Prolactin given to female hamsters exerts a differential effect on the ovaries and uterine weight according to time of day of injection. Given at 0200 for eight days, prolactin enhances uterine weight to 156% of that found when the hormone is given at 0800. The critical event phasing the effectiveness of prolactin may be peak adrenal corticoid secretion (Joseph and Meier, 1974).

In vitro, the hamster adrenal shows a circadian rhythm of corticosterone synthesis for as long as ten days (Andrews, 1968, 1971). There is a phase-specific response to adrenocorticotropic hormone (ACTH), with the greatest increase in ACTH-stimulated corticoid synthesis occurring when levels are ordinarily low (Andrews, 1980). The rhythm of O_2 consumption by cultured adrenals is, however, not affected by ACTH (Andrews and Folk, 1964). Progesterone in the culture medium will also provoke a phase response similar to that of ACTH, but of vastly lower amplitude (Andrews, 1980). These studies are of particular interest because they demonstrate (1) the capacity of a substrate to be rhythmic in the absence of neural regulation, and (2) that certain rhythms can be phasically controlled, whereas others in the same organ are independent of that control.

5. SUMMARY AND FUTURE DIRECTIONS

The study of hamster biological rhythms has focused almost entirely on those with periods in the circadian range. The two general exceptions to this rule are the study of (1) the infradian estrous cycle that has been clearly shown to rely on the interaction of a circadian clock and endocrine events to time a four-day ovulatory rhythm and (2) the annual reproductive cycle that is also the product of a circadian timing system that measures day length and particular neuroendocrine responses to the measured day. There has been little research concerning hamster ultradian rhythms.

The hamster has been a species of choice for circadian rhythm research because its wheelrunning records have been characterized as having clear activity onsets under entrained or free-running conditions. Thus, onset timing has been extremely simple to measure. This generalization of rhythm precision appears to be true primarily for males with active gonads. Recent research shows that several hormones and drugs can also modify the period, phase, or amount of activity in the circadian locomotor cycle. These experiments have further shown that timing of running within the active phase is more complex than can be explained by a two oscillator model. However, the general characteristics of freerunning, split, and entrained locomotor rhythms are fairly well captured by a two-oscillator model and support Pittendrigh's "nonparametric model of entrainment."

Lesions of the SCN abolish several circadian rhythms, but only recently have there been anatomical studies of the hamster's SCN. The systematic functional analysis of connections with the SCN has not been done, nor have there been any published neurophysiological papers concerning circadian neural rhythmicity in and around the SCN. Such experimentation might well demonstrate different pathways modulating the rhythmicity of different behavioral or physiological systems. What is clearly lacking in the field of mammalian rhythms is a knowledge of rhythm physiology that can link the regularities of entrained or free-running rhythms (e.g., phase response curves) with the period and phase irregularities that can result from the various hormonal, photoperiod, or neural manipulations. Interest in the neuroanatomy and neurophysiology of mammalian rhythms is increasing and should soon produce considerable insight into the understanding of rhythmic phenomena.

ACKNOWLEDGMENTS. Preparation of this manuscript was supported by NICHD grant HD16231. E. Gustafson and M. L. Gavin provided helpful criticisms. The author is very grateful to D. Caselles for her extensive typing assistance.

REFERENCES

Albers, H. E., Carter, D. S., Darrow, J. M., and Goldman, B. D., 1982, Circadian rhythms of body temperature and wheelrunning in the Turkish hamster (Mesocricetus brandti), *Society for Neuroscience 12th Annual Meeting* Minneapolis, Minnesota, Abstr. #14. 11.

Albers, H. E., Ferris, C. F., Leeman, S. E., and Goldman, B. D., 1984, Avian pancreatic polypeptide phase shifts hamster circadian rhythms when microinjected into the suprachiasmatic region, *Science* 223:833–835.

Alleva, J. J., Waleski, M. V., and Alleva, F. R., 1971, A biological clock controlling the estrous cycle of the hamster, *Endocrinology* 88:1368–1379.

Andrews, R. V., 1968, Temporal secretory responses of cultured hamster adrenals, *Comp. Biochem. Physiol.* 26:179–193.

Andrews, R. V., 1971, Circadian rhythms in adrenal organ cultures, *Gegenbours Morph. Jahrb. Leipzig* 117:89–98.

Andrews, R. V., 1980, Phase response profile of hamster adrenal organ cultures treated with ACTH and exogenous steroid, *Comp. Biochem. Physiol.* 67A: 257–277.

Andrews, R. V., and Folk, G. E., 1964, Circadian metabolic patterns in cultured hamster adrenal glands, *Comp. Biochem. Physiol.* 11:393–409.

Aschoff, J., 1960, Exogenous and endogenous components in circadian rhythms, *Cold Spring Harbor Symp. Quant. Biol.* 25: 11–27.

Aschoff, J., 1965, Circadian vocabulary, in: *Circadian Clocks* (J. Aschoff, ed.), North Holland, Amsterdam, pp. x–xix.

Aschoff, J., 1978, Circadian rhythms: Influences of internal and external factors on the period measured in constant conditions, *Z. Tierpsychol.* 49: 225–249.

Aschoff, J., Figala, J., and Poppel, E., 1973, Circadian rhythms in locomotor activity in the golden hamster (Mesocricetus auratus) measured with two different techniques, *J. Comp. Physiol. Psychol.* 85: 20–28.

Aschoff, J., Hoffmann, K., Pohl, H., and Wever, R., 1975, Re-entrainment of circadian rhythms after phase-shifts of the Zeitgeber, *Chronobiologia* 2: 23–78.

Aschoff, J., Gerecke, U., Von Goetz, C., Groos, G. A., and Turek, F. W., 1982, Phase responses and characteristics of free-running activity rhythms in the golden hamster: Independence of the pineal gland, in: *Vertebrate Circadian Systems* (J. Aschoff, S. Daan, and G. Groos, eds.), Springer-Verlag, Berlin, pp. 129–140.

Baranczuk, R. and Greenwald, G. S., 1973, Peripheral levels of estrogen in the cyclic hamster, *Endocrinology* 92: 805–812.

Beasley, L. J., and Nelson, R. J., 1982, Thyroid gland influences the period of hamster circadian oscillations, *Experientia* 28: 870–871.

Berk, M. L., and Finkelstein, J. A., 1981, An autoradiographic determination of the efferent projections of the suprachiasmatic nucleus of the hypothalamus, *Brain Res.* 226: 1–13.

Bittman, E. L., 1978, Photoperiodic influences on testicular regression in the golden hamster: Termination of scotorefractoriness, *Biol. Reprod.* 17: 971–977.

Borer, K. T., Rowland, N., Mirow, A., Borer, R. C., Jr., and Kelch, R. P., 1980, Physiological and behavioral responses to starvation in the golden hamster, *Am. J. Physiol.* 236:E105–E112.

Borer, K. T., Campbell, C. S., Gordon, L., Jorgenson, K., and Tabor, J., 1981, Exercise reinstates estrous cycles in hamsters maintained in short photoperiod, Society for Neuroscience 11th Annual Meeting, Los Angeles, Abstr. #70.14.

Boulos, Z., and Morin, L. P., 1982, Entrainment of split circadian rhythms in hamsters. Society for Neuroscience 12th Annual Meeting, Minneapolis, Abst. #151.10.

Boulos, Z., and Rusak, B., 1982, Circadian phase response curves for dark pulses in the hamster, *J. Comp. Physiol.* 146:411–417.

Brown, F. A., 1976, Evidence for external timing of biological clocks, in: *An Introduction to Biological Rhythms* (J. D. Palmer, Ed.), Academic, New York, pp. 209–279.

Brown, F. A., and Scow, K. M., 1978, Magnetic induction of a circadian cycle in hamsters, *J. Interdiscip. Cycle Res.* 9:137–145.

Brown, J. M., and Berry, R. J., 1968, The relationship between diurnal variation of the number of cells in mitosis and of the number of cells synthesizing DNA in the epithelium of the hamster cheek pouch, *Cell Tissue Kinet.* 1:23–33.

Bruce, V. G., 1960, Environmental entrainment of circadian rhythms, *Cold Spring Harbor Symp. Quant. Biol.* 25: 29–47.

Bunning, E., 1936, Die endonome Tagesperiodik als Grundlage der photoperiodischen Reaktion, *Ber. Dtsch. Bot. Ges.* 54: 590–607.

Bunning, E., 1958, Das Weiterlaufen der "physiologischen Uhr" im Saugerdarm ohne zentrale Steuerung, *Naturwissenschaften* 45: 68.

Burns, J. T., and Meier, A. H., 1981, A circadian rhythm in insulin overdose in the golden hamster (Mesocricetus auratus), in: *Chronopharmacology and Chronotherapeutics* (C. A. Walker, K. F. A. Soliman, and C. M. Winget, eds.), Florida A & M University Foundation, Tallahassee, pp. 315–318.

Campbell, C. S., Finkelstein, J. S., and Turek, F. W., 1978, The interaction of photoperiod and testosterone on the development of copulatory behavior in castrated male hamsters, *Physiol. Behav.* 21: 409–415.

Card, J. P., and Moore, R. Y., 1982, Ventral lateral geniculate nucleus efferents to the rat suprachiasmatic nucleus exhibit avian pancreatic polypeptide-like immunoreactivity, *J. Comp. Neurol.* **206:** 390–396.

Card, J. P., and Moore, R. Y., 1984, The suprachiasmatic nuclei of the golden hamster: Immunohistochemical analysis of cell and fiber distribution, *Neuroscience,* in press.

Carlisle, G., 1975, Entrainment of circadian activity rhythms of female golden hamsters after lesions in the lateral geniculate nucleus, unpublished undergraduate honors thesis in psychology, Univ. California, Berkeley.

Carmichael, M. S., and Zucker, I., 1982, Entrainment to non-24-hr days and gestation length of golden hamsters, *J. Reprod. Fertil.* **66:** 691–693.

Carmichael, M. S., Nelson, R. J., and Zucker, I., 1981, Hamster activity and estrous cycles: Control by a single versus multiple circadian oscillator(s), *Proc. Natl. Acad. Sci.* **78:** 7830–7834.

Chaudhry, A. P., Halberg, F., Keeman, C. E., Harner, R. N., and Bittner, J. J., 1958, Daily rhythms in rectal temperature and in epithelial mitoses of hamsterpinna and pouch, *J. Appl. Physiol.* **12:**221–224.

Craig, C., Tamarkin, L., Garrick, N., and Wehr, T. A., 1981, Long-term and short-term effects of clorgyline (a monoamine oxidase type A inhibitor) on locomotor activity and pineal melatonin in hamster, Society for Neuroscience 11th Annual Meeting, Los Angeles, Abst. #229.14.

Daan, S., and Aschoff, J., 1975, Circadian rhythms of locomotor activity in captive birds and mammalian activity rhythms, *J. Theor. Biol.* **70:**592–597.

Daan, S., and Pittendrigh, C. S., 1976a, A functional analysis of circadian pacemakers in nocturnal rodents. II. The variability of phase response curves, *J. Comp. Physiol.* **106:**253–266.

Daan, S., and Pittendrigh, C. S., 1976b, A functional analysis of circadian pacemakers in nocturnal rodents. III. Heavy water and constant light: Homeostasis of frequency? *J. Comp. Physiol.* **106:**267–290.

Daan, S., and Slopsema, S., 1978, Short-term rhythms in foraging behavior of the common vole, Microtus arvalis, *J. Comp. Physiol.* **127:** 215–227.

Dark, J., 1980, Partial isolation of the suprachiasmatic nuclei: Effects on circadian rhythms of rat drinking behavior, *Physiol. Behav.* **25:** 863–873.

Davis, F. C., and Gorski, R. A., 1981, Functional symmetry of the suprachiasmatic nuclei, Society for Neuroscience 11th Annual Meeting, Los Angeles, Abstr. #18.13.

Davis, F. C., and Gorski, R. A., 1982, Perinatal entrainment of hamster circadian rhythms, Society for Neuroscience 12th Annual Meeting, Minneapolis, Abstr. #14.10.

Davis, F. C., and Menaker, M., 1980, Hamsters through time's window: Temporal structure of hamster locomotor rhythmicity, *Am. J. Physiol.* **239:** R149–R155.

Davis, F. C., Darrow, J. M., and Menaker, M., 1983, Sex differences in the circadian control of hamster wheelrunning activity, *Am. J. Physiol.* **244:** R93–R104.

Davis, G. J., and Meyer, R. K., 1973, FSH and LH in the Snowshoe hare during the increasing phase of the 10-year cycle, *Gen. Comp. Endocrinol.* **20:** 53–60.

DeCoursey, P. J., 1960, Phase control of activity in a rodent, *Symp. Quant. Biol.* **25:** 49–54.

DeCoursey, P. A., 1964, Function of a light response rhythm in hamsters, *J. Cell. Comp. Physiol.* **63:** 189–196.

Dewsbury, D. A., 1968, Copulatory behavior of rats—Variations within the dark phase of the diurnal cycle, *Commun. Behav. Biol.* **A1:** 373–377.

Earnest, D. J., and Turek, F. W., 1982, Splitting of the circadian rhythm of activity in hamsters: Effects of exposure to constant darkness and subsequent re-exposure to constant light, *J. Comp. Physiol.* **145:** 405–411.

Earnest, D. J., and Turek, F. W., 1983a, Role for acetylcholine in mediating effects of light on reproduction, *Science* **219:** 77–79.

Earnest, D. J., and Turek, F. W., 1983b, Effect of one-second light pulses on testicular function and locomotor activity in the golden hamster, *Biol. Reprod.* **28:** 557–565.

Eichler, V. B., and Moore, R. Y., 1974, The primary and accessory optic systems in the golden hamster, Mesocricetus arautus, *Acta Anat.* **89:**359–371.

Elliott, J. A., 1974, Photoperiodic regulation of testis function in the golden hamster: Relation to the circadian system, Ph.D. Dissertation, University of Texas.

Elliott, J. A., 1976, Circadian rhythms and photoperiodic time measurement in mammals, *Fed. Proc.* 35:2339–2346.

Elliott, J. A., Stetson, M. H., and Menaker, M., 1972, Regulation of testis function in golden hamsters: A circadian clock measures photoperiodic time, *Science* 178:771–773.

Ellis, G. B., and Turek, F. W., 1979, Changes in locomotor activity associated with the photoperiodic response of the testes in male golden hamsters, *J. Comp. Physiol.* 132:277–284.

Ellis, G. B., and Turek, F. W., 1981, Testosterone and the photoperiod interact to regulate daily locomotor activity in male golden hamsters, *Fed. Proc.* 40: 307.

Ellis, G. B., McKlveen, R. E., and Turek, F. W., 1982, Dark pulses affect the circadian rhythm of activity in hamsters kept in constant light, *Am. J. Physiol.* 242: R44–R50.

Eriksson, L. O., 1973, Spring inversion of the diel rhythm of locomotor activity in young sea-going trout (Selmo trutta trutta L.) and Atlantic salmon (Salmo salar L.), *Aquilo Ser. Zool.* 14: 68–79.

Eskes, G. A., 1982, Significance of daily cycles in sexual behavior of the male golden hamster, in: *Vertebrate Circadian Systems* (J. Aschoff, S. Daan, and G. Groos, eds.), Springer-Verlag, New York, pp. 347–353.

Eskes, G. A., and Zucker, I., 1978, Photoperiodic regulation of the hamster testis: Dependence on circadian rhythms, *Proc. Natl. Acad. Sci.* 75: 1034–1038.

Everett, J. W., Sawyer, C. H., and Markee, J. E., 1949, A neurogenic timing factor in control of the ovulatory discharge of luteinizing hormone in the cyclic rat, *Endocrinology* 44: 234–250.

Finkelstein, J. S., Baum, F. R., and Campbell, C. S., 1978, Entrainment of the female hamster to reversed photoperiod: Role of the pineal, *Physiol. Behav.* 21: 105–111.

Fitzgerald, K. M., and Zucker, I., 1976, Circadian organization of the estrous cycle of the golden hamster, *Proc. Natl. Acad. Sci.* 73: 2923–2927.

Fitzgerald, K., Zucker, I., and Rusak, B., 1978, An evaluation of homeostasis of circadian periodicity in the golden hamster, *J. Comp. Physiol.* 123: 265–269.

Folk, G. E., Schellinger, R. R., and Snyder, D., 1961, Day-night changes after exercise in body temperature and heart rates of hamsters, *Iowa Acad. Sci.* 68: 594–602.

Follett, B. K., Farner, D. S., and Morton, M. L., 1967, The effects of alternating long and short photoperiods on gonadal growth and pituitary gonadotropins in the white-crowned sparrow, Zonotrichia leucophrys gambelii, *Biol. Bull.* 133: 333–342.

Gaston, S., and Menaker, M., 1967, Photoperiodic control of hamster testis, *Science* 158:925–928.

Gibbs, F. P., 1983, Temperature dependence of the hamster circadian pacemaker, *Am. J. Physiol.* 244:R607–R610.

Guicking, A., 1970, Uber den Einfluss von Schall auf die tagesperiodische Aktivitat des Goldhamsters. I, *J. Interdiscip. Cycle Res.* 1:323–334.

Joseph, M. M., and Meier, A. H., 1974, Circadian component in the fattening and reproductive responses to prolactin in the hamster, *Proc. Soc. Exp. Biol. Med.* 146:1150–1155.

Krieger, O. T., 1980, Ventromedial hypothalamic lesions abolish food-shifted circadian adrenal and temperature rhythmicity, *Endocrinology* 106: 649–654.

Kripke, D. F., and Wyborney, V. G., 1980, Lithium slows rat circadian activity rhythms, *Life Sci.* 26:1319–1320.

Krug, M., Brodemann, R., and Ott, T., 1981, Identical responses of the two hippocampal theta generators to physiological and pharmacological activation, *Brain Res. Bull.* 6:5–11.

Landau, I. T., 1975a, Light-dark rhythms in aggressive behavior of the male golden hamster, *Physiol. Behav.* 14:767–774.

Landau, I. T., 1975b, Effects of adrenalectomy on rhythmic and non-rhythmic aggressive behavior in the male golden hamster, *Physiol. Behav.* 14:775–780.

Larsson, K., 1958, Age differences in the diurnal periodicity of male sexual behavior, *Gerontologia* 2:64–72.

Lee, J. G., Hallonquist, J. D. and Mrosovsky, N., 1983, Differential effects of dark pulses on the two components of split circadian activity rhythms in golden hamsters, *J. Comp. Physiol.* 153: 123–132.

Lehmann, V., 1976, Short-term and circadian rhythms in the behaviour of the vole, Microtus agnestis (L.), *Oecologia* 23:185–199.

Lisk, R. D., and Sawyer, C. H., 1966, Induction of paradoxical sleep by lights off stimulation, *Proc. Soc. Exp. Biol.* 123: 664–667.

Malloney, E. S., 1978, *Dutton's Navigation and Piloting,* 13th Edition, United States Naval Institute, Annapolis, Maryland, pp. 359–416.

McEachron, D. L., Kripke, D. F., and Wyborney, V. G., 1981, Lithium promotes entrainment of rats to long circadian light-dark cycles, *Psychiatr. Res.* 2:511–519.

Menaker, M., 1982, The search for principles of physiological organization in vertebrate circadian systems, in: *Vertebrate Circadian Systems* (J. Aschoff, S. Daan, and G. A. Groos, eds.), Springer-Verlag, New York, pp. 1–12.

Mistelberger, R., 1982, Entrainment to food and light schedules in VMH lesioned rats, Society for Neuroscience i2th Annual Meeting, Minneapolis, Abstr. #151.13.

Moline, M. L., Albers, H. E., Todd, R. B., and Moore-Ede, M. C., 1981, Light-dark entrainment of proestrous LH surges and circadian locomotor activity in female hamsters, *Horm. Behav.* 15:451–458.

Moller, U., 1978, Interaction of external agents with the circadian mitotic rhythm in the epithelium of the hamster cheek pouch, *J. Interdiscip. Cycle Res.* 9:105–114.

Moller, U., and Bojsen, J., 1974, The circadian temperature rhythm in Syrian hamsters as a function of the number of animals per cage, *J. Interdiscip. Cycle Res.* 5:61–69.

Moller, V., Larsen, J. K., and Faber, M., 1974, The influence of injected tritiated thymidine on the mitotic circadian rhythm in the epithelium of the hamster cheek pouch, *Cell Tissue Kinet.* 7:231–239.

Moore, R. Y., 1973, Retinohypothalamic projections in mammals: A comparative study, *Brain Res. Bull.* 49:403–409.

Moore, R. Y., 1983, Organization and function of a CNS circadian oscillator: The suprachiasmic hypothalamic nucleus, *Fed. Proc.,* 42:2783–2789.

Moore, R. Y., and Eichler, V. B., 1972, Loss of circadian adrenal corticosterone rhythm following suprachiasmatic lesions in the rat, *Brain Res.* 42:201–206.

Moore, R. Y., and Eichler, V. B., 1976, Central neural mechanisms in diurnal rhythm regulation and neuroendocrine responses to light, *Psychoneuroendocrinology* 1:265–279.

Moore, R. Y., Gustafson, E. L., and Card, J. P., 1984, Identical immunoreactivity of afferents to the rat suprachiasmatic nucleus with antisera against avian pancreatic polypeptide, molluscan cardioexcitatory peptide and neuropeptide Y, *Cell Tiss. Res.* 236:41–46.

Moore, R. Y., and Lenn, N. J., 1972, A retinohypothalamic projection in the rat, *J. Comp. Neurol.* 146:1–14.

Moore-Ede, M. D., Schmelzer, W. S., Kass, D. A., and Herd, J. A., 1976, Internal organization of the circadian timing system in multicellular animals, *Fed. Proc.* 35:2333–2338.

Morgan, W. W., Pfeil, K. A., Reiter, R. J., and Gonzales, E., 1976, Comparison of changes in tryptophan and serotonin in regions of the hamster and the rat brain over a twenty-four hour period, *Brain Res.* 117:77–84.

Morin, L. P., 1977, Progesterone: Inhibition of rodent sexual behavior, *Physiol. Behav.* 18:701–715.

Morin, L. P., 1978, Rhythmicity of hamster gnawing: Ease of measurement and similarity to running activity, *Physiol. Behav.* 21:317–320.

Morin, L. P., 1980, Effect of ovarian hormones on synchrony of hamster circadian rhythms, *Physiol. Behav.* 25:741–749.

Morin, L. P., 1981a, An effect of photoperiod history on reproductive function and a circadian rhythm of male hamsters, *Physiol. Behav.* 27:89–94.

Morin, L. P., 1981b, Ultradian rhythms in hamster and rat eating, *Soc. Neurosci. 11th Ann. Mtng.,* Los Angeles, Abst. #19.9.

Morin, L. P., 1982a, Phase and period of female hamster running rhythms during the annual reproductive cycle. Society for Neuroscience 12th Annual Meeting, Minneapolis, Abst. #151.9.

Morin, L. P., 1982b, Acute or longterm melatonin fails to block estradiol benzoate plus progesterone facilitation of hamster receptivity, *Conference Reprod. Behav.,* E. Lansing (Abstract).

Morin, L. P., and Cummings, L. A., 1981, Effect of surgical or photoperiodic castration, testosterone replacement or pinealectomy on male hamster running rhythmicity. *Physiol. Behav.* 26:825–838.

Morin, L. P., and Cummings, L. A., 1982, Splitting of wheelrunning rhythms by castrated or steroid treated male and female hamsters, *Physiol. Behav.* 29:665–675.

Morin, L. P., and Zucker, I., 1978, Photoperiodic regulation of copulatory behavior in the male hamster, *J. Endocrinol.* 77:249–258.

Morin, L. P., Fitzgerald, K. M., Rusak, B., and Zucker, I., 1977a, Circadian organization and neural mediation of hamster reproductive rhythms, *Psychoneuroendocrinology* 2:73–98.

Morin, L. P., Fitzgerald, K. M., and Zucker, I., 1977b, Estradiol shortens the period of hamster circadian rhythms, *Science* 196:305–307.

Mrsovsky, N., 1975, The amplitude and period of circannual cycles of body weight in golden-mantled ground squirrels with medial hypothalamic lesions, *Brain Res.* 99:97–116.

Mrsovsky, N., 1978, Circannual cycles in hibernators, in: *Strategies in Cold: Natural Torpidity and Thermogenesis* (L. C. H. Wang and J. W. Hudson, eds.), New York, Academic Press, pp. 21–66.

Nishio, T., Shiosaka, S., Nakagawa, H., Sakumoto, T., and Satoh, K., 1979, Circadian feeding rhythm after hypothalamic knife-cut isolating suprachiasamatic nucleus, *Physiol. Behav.* 23:763–769.

Nunez, A. A., and Stephan, F. K., 1977, The effects of hypothalamic knife cuts on drinking rhythms and the estrus cycle of the rat, *Behav. Biol.* 20:224–234.

Pengelley, E. T., Asmundson, S. J., Barnes, B., and Aloia, R. C., 1976, Relationship of light intensity and photoperiod to circannual rhythmicity in the hibernating ground squirrel, Citellus lateralis, *Comp. Biochem. Physiol.* 53A:273–277.

Philo, R., Rudeen, P. K., and Reiter, R. J., 1977, A comparison of the circadian rhythms and concentrations of serotonin and norepinephrine in the telencephalon of four rodent species, *Comp. Biochem. Physiol.* 57C:127–130.

Pickard, G. E., 1980, Morphological characteristics of retinal ganglion cells projecting to the suprachiasmatic nucleus: A horseradish perioxidase study, *Brain Res.* 183:458–465.

Pickard, G. E., 1982, The afferent connections of the suprachiasmatic nucleus of the golden hamster with emphasis on the retinohypothalamic projection, *J. Comp. Neurol.* 211:65–83.

Pickard, G. E., and Silverman, J.-A., 1981, Direct retinal projections to the hypothalamus piriform cortex, and accessory optic nuclei in the golden hamster as demonstrated by a sensitive anterograde horseradish perocidase technique, *J. Comp. Neurol.* 196:155–172.

Pickard, G. E., and Turek, F. W., 1982, Splitting of the circadian rhythm of activity is abolished by unilateral lesions of the suprachiasmatic nuclei, *Science* 215:1119–1121.

Pickard, G. E., and Turek, F. W., 1983, The suprachiasmatic nuclei: Two circadian clocks? *Brain Res.* 268:201–210.

Pittendrigh, C. S., 1960, Circadian rhythms and circadian organization of living systems, *Cold Spring Harbor Symp. Quant. Biol.* 25:159–182.

Pittendrigh, C. S., 1967, Circadian rhythms, space research and manned space flight, in: *Life Sciences and Space Research V*, pp. 122–134, Amsterdam, North Holland.

Pittendrigh, C. S., 1974, Circadian oscillations in cells and the circadian organization of multicellular systems, in: *The Neurosciences, Third Study Program* (F. O. Schmitt and F. G. Worden, eds.), MIT Press, Cambridge, pp. 437–358.

Pittendrigh, C. S., and Daan, S., 1974, Circadian oscillations in rodents: A systematic increase of their frequency with age, *Science* 186:548–550.

Pittendrigh, C. S., and Daan, S., 1976a, A functional analysis of circadian pacemakers in nocturnal rodents. I. The stability and lability of spontaneous frequency, *J. Comp. Physiol.* 106:223–252.

Pittendrigh, C. S., and Daan, S., 1976b, A functional analysis of circadian pacemakers in nocturnal rodents. IV. Entrainment: Pacemaker as clock, *J. Comp. Physiol.* 106:291–331.

Pittendrigh, C. S., and Daan, S., 1976c, A functional analysis of circadian pacemakers in nocturnal rodents. V. Pacemaker structure: A clock for all seasons, *J. Comp. Physiol.* 106:333–355.

Pohl, H., 1976, Proportional effects of light on entrained circadian rhythms of birds and mammals, *J. Comp. Physiol.* 112:103–108.

Rawson, K. S., 1960, Effects of tissue temperature on mammalian activity rhythms, *Cold Spring Harbor Symp. Quant. Biol.* 24:105–113.

Reiter, R. J., 1974, Circannual reproductive rhythms in mammals related to photoperiod and pineal function: A review, *Chronobiologica* 1:365–395.

Richards, M. P. M., 1966, Activity measured by running wheels and observations during the oestrous cycle, pregnancy and pseudopregnancy in the golden hamster, *Anim. Behav.* 14:450–458.

Richter, C. P., 1965, *Biological Clocks in Medicine and Psychiatry,* Thomas, Springfield, Illinois, p. 10.

Richter, C. P., 1970, Dependence of successful mating in rats on functioning of 24-hour clocks of the male and female, *Commun. Behav. Biol.* A5:1–5.

Richter, C. P., 1975, Deep hypothermia and its effect on the 24-hour clock of rats and hamsters, *Johns Hopkins Med. J.* 136:1–10.

Richter, C. P., 1977, Heavy water as a tool for study of the forces that control length of period of the 24-hour clock of the hamster, *Proc. Natl. Acad. Sci.* 74:1295–1299.

Riley, J. N., Card, J. P., and Moore, R. Y., 1981, A retinal projection to the lateral hypothalamus in the rat, *Cell Tissue Res.* 214:257–269.

Rowland, N., 1976, Endogenous circadian rhythms in rats recovered from lateral hypothalamic lesions, *Physiol. Behav.* 16:257–266.

Rusak, B., 1977a, The role of the suprachiasmatic nuclei in the generation of circadian rhythms in the golden hamster, *Mesocricetus auratus, J. Comp. Physiol.* 118:145–164.

Rusak, B., 1977b, Involvement of the primary optic tracts in mediation of light effects on hamster circadian rhythms, *J. Comp. Physiol.* 118:165–172.

Rusak, B., and Boulos, Z., 1981, Pathways for photic entrainment of mammalian circadian rhythms, *Photochem. Photobiol.* 34:267–273.

Rusak, B., and Groos, G., 1982, Suprachiasmatic stimulation phase shifts rodent circadian rhythms, *Science* 215:1407–1409.

Rusak, B., and Morin, L. P., 1976, Testicular responses to photoperiod are blocked by lesions of the suprachiasmatic nuclei in golden hamsters, *Biol. Reprod.* 15:366–374.

Satinoff, E., Liran, J., and Clapman, R., 1982, Aberrations of circadian body temperature rhythms in rats with medial preoptic lesions, *Am. J. Physiol.* 242:R35–R357.

Shibuya, C. A., Melnyk, R. B., and Mrosovsky, N., 1980, Simultaneous splitting of drinking and loco-motor activity rhythms in golden hamsters, *Naturwissenschaften* 67:45–56.

Silverman, H. J., and Zucker, I., 1976, Absence of post-fast food compensation in golden hamster (*Mesocricetus auratus*), *Physiol. Behav.* 17:271–285.

Sisk, C. L., and Stephan, F. K., 1981, Phase shifts of circadian rhythms of activity and drinking in the hamster, *Behav. Neural Biol.* 33:334–344.

Sisk, C. L., and Turek, F. W., 1982, Role of the inter-connection of the suprachiasmatic nuclei in the hamster circadian system, Society for Neuroscience 12th Annual Meeting, Minneapolis, Abstr. #14.9.

Stephan, F. K., and Nunez, A. A., 1979, Elimination of circadian rhythms in drinking, sleep and tem-perature by isolation of the suprachiasmatic nuclei, *Behav. Biol.* 20:1–16.

Stephan, F. K., and Zucker, I., 1972, Circadian rhythms in drinking behavior and locomotor activity of rats are eliminated by hypothalamic lesions, *Proc. Natl. Acad. Sci. U.S.A.* 69:1583–1586.

Stephan, F. K., Berkley, K. J., and Moss, R. L., 1981, Efferent connections of the rat suprachiasmatic nucleus, *Neuroscience* 6:2625–2641.

Stephan, F. K., Donaldson, J. A., and Gellert, J., 1982, Retinohypothalamic trait symmetry and phase shifts of circadian rhythms in rats and hamsters, *Physiol. Behav.* 29:1153–1159.

Stetson, M. H., and Anderson, P. J., 1980, Circadian pacemaker times gonadotropin release in free-running female hamsters, *Am. J. Physiol.* 238:R23–R27.

Stetson, M. H., and Gibson, J. T., 1977, The estrous cycle in golden hamsters: A circadian pacemaker times preovulatory gonadotropin release, *J. Exp. Zool.* 201:289–294.

Stetson, M. H., and Watson-Whitmyre, M., 1976, Nucleus suprachiasmaticus: The biological clock in the hamster? *Science* 191:197–199.

Stetson, M. H., Elliott, J. A., and Menaker, M., 1975, Photoperiodic regulation of hamster testis: Circadian sensitivity to the effects of light, *Biol. Reprod.* 13:329–339.

Stetson, M. H., Matt, K. S., and Watson-Whitmyre, M., 1976, Photoperiodism and reproduction in golden hamsters: Circadian organization and termination of photorefractoriness, *Biol. Reprod.* 14:531–537.

Stetson, M. H., Watson-Whitmyre, M., and Matt, K. S., 1977, Circadian organization in the regulation of reproduction: Timing of the 4-day estrous cycle of the hamster, *J. Interdiscip. Cycle Res.* 8:350–352.

Swade, R. H., 1969, Circadian rhythms in fluctuating light cycles: Toward a new model of entrainment, *J. Theor. Biol.* 24:227–239.

Swade, R. H., and Pittendrigh, C. S., 1967, Circadian locomotor rhythms of rodents in the arctic, *Am. Nat.* 101:431–466.

Swann, J., and Turek, F. W., 1982, Cycle of lordosis behavior in female hamsters whose circadian activity rhythm has split into two components, *Am. J. Physiol.* 243:R112–R118.

Takahashi, J. S., and Menaker, M., 1980, Interaction of estradiol and progesterone: Effects on circadian locomotor rhythm of female golden hamsters, *Am. J. Physiol.* 239:R497–R504.

Takahashi, J. S., and Zatz, M., 1982, Regulation of circadian rhythmicity, *Science* 217:1104–1111.

Takahashi, J. S., DeCoursey, P. J., Bauman, L., and Menaker, M., 1984, Spectral sensitivity of a novel photoreceptive system mediating entrainment of mammalian circadian rhythms, *Nature* 308:186–188.

Toates, F. M., 1978, A circadian rhythm of hoarding in the hamster, *Anim. Behav.* 26:631.

Tobler, I., and Borbely, A. A., 1977, Enhancement of paradoxical sleep by short light periods in the golden hamster, *Neurosci. Lett.* 6:275–277.

Turek, F. W., and Campbell, C. S., 1979, Photoperiodic regulation of neuroendocrine-gonadal activity, *Biol. Reprod.* 20:32–50.

Van Den Pol, A. N., and Powley, T., 1979, A fine-grained anatomical analysis of the role of the rat suprachiasmatic nucleus in circadian rhythms of feeding and drinking, *Brain Res.* 160:307–326.

Warden, A. W., 1978, Circadian rhythms of self-selected lighting in golden hamsters: Relation to gonadal condition, *Chronobiologia* 5:28–38.

Warden, A. W., and Sachs, B. D., 1974, Circadian rhythms of self-selected lighting in hamsters, *J. Comp. Physiol.* 91:127–134.

Widmaier, E. P., and Campbell, C. S., 1980, Interactions of estradiol and photoperiod on activity patterns in the female hamster, *Physiol. Behav.* 24:923–930.

Winfree, A. T., 1967, Biological rhythms and the behavior of populations of coupled oscillators, *J. Theor. Biol.* 16:15–42.

Winfree, A. T., 1971, Comment, in: *Biochronometry* (M. Menaker, ed.) National Academy of Science, Washington, D.C., pp. 150–151.

Wirz-Justice, A., and Campbell, I. C., 1982, Antidepressant drugs can slow or dissociate circadian rhythms, *Experientia* 38:1301–1309.

Zatz, M., 1979, Photoentrainment, pharmacology, and phase shifts of the circadian rhythm in the rat pineal, *Fed. Proc.* 38:2596–2601.

Zatz, M., and Brownstein, M. J., 1979, Intraventricular carbachol mimics the effects of light on the cricadian rhythm in the rat pineal gland, *Science* 203:358–360.

Zatz, M., and Herkenham, M. A., 1981, Intraventricular carbachol mimics the phase-shifting effect of light on the circadian rhythm of wheel-running activity, *Brain Res.* 212:234–238.

Zucker, I., and Stephan, F. K., 1973, Light-dark rhythms in hamster eating, drinking and locomotor behaviors, *Physiol. Behav.* 11:239–250.

Zucker, I., Rusak, B., and King, R. G., 1976, Neural bases for circadian rhythms in rodent behavior, in: *Advances in Psychobiology* Volume 3 (A. H. Riesen and R. F. Thompson, eds.), John Wiley and Sons, New York, pp. 35–74.

Zucker, I., Fitzgerald, K. M., and Morin, L. P., 1980a, Sex differentiation of the circadian system in the golden hamster, *Am. J. Physiol.* 238:R97–R101.

Zucker, I., Cramer, C. P., and Bittman, E. L., 1980b, Regulation by the pituitary gland of circadian rhythms in the hamster, *J. Endocrinol.* 85:17–25.

15

Regulation of Energy Balance in the Golden Hamster

KATARINA T. BORER

1. ANATOMY AND PHYSIOLOGY OF HAMSTER GASTROINTESTINAL TRACT

This selective review of structure and function of hamster gastrointestinal tract is presented on the assumption that peculiarities of this system hold the clue to peculiarities of hamster feeding and energy regulation. Where information on the golden hamster was not available, data from closely related species are substituted. The hamster gastrointestinal tract is outlined in Fig. 1. Not shown in Fig. 1 are the mouth region with cheek pouches and salivary glands. Hamsters have internal cheek pouches as do some squirrels, three other genera of true hamsters, and three genera of African Muridae and African gerbil. They use pouches to transport food. Pouches are highly vascular and can be used as a site for endocrine gland homografts (Handler and Shepro, 1968).

1.1. Stomach

Unlike the monogastric rat, the hamster has a compartmentalized stomach consisting of a forestomach and a glandular stomach. The forestomach bears structural and functional similarities to the rumen in herbivores. It resembles rumen ultrastructurally (Takahashi and Tamate, 1976), harbors ruminant-type microorganisms (Mangold, 1929) in greater concentrations than are found in the glandular stomach (Kunstyr, 1974), and is the site of production and absorption of volatile fatty acids (Hoover *et al.*, 1969).

Microorganism inocula from hamster forestomach and cecum can ferment cellulose

---

KATARINA T. BORER • Department of Kinesiology, University of Michigan, Ann Arbor, Michigan 48109.

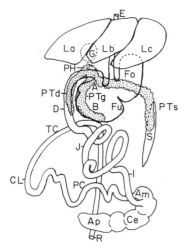

Figure 1. Hamster gastrointestinal tract. Abbreviations: A, antrum of glandular stomach; Am, ampulla of cecum; Ap, apex of cecum; B, body of glandular stomach; Ce, body of cecum; CL, colonic loop; D, duodenum; E, esophagus; Fo, forestomach (esophageal diverticulum); Fu, fundus of glandular stomach; G, gall bladder; I, ileum; J, jejunum; La, liver, right cranioventral part; Lb, liver, left cranioventral part; Lc, liver, dorsocaudal part; PC, proximal colon; PH, pancreas head; PTd, pancreas tail, duodenal lobe; PTg, pancreas tail, gastric lobe; PTs, pancreas tail, splenic lobe; R, rectum; S, spleen; TC, transverse colon. [Adapted after Snipes, 1979 and Jewell and Charipper (1951).]

wall from alfalfa (Ehle and Warner, 1978) as efficiently as inocula from bovine rumen (Banta *et al.*, 1975), but are less efficient in comparison to bovine rumen when straw is used as substrate. On ingestion, food first enters the forestomach and remains there during the first 10 to 60 min (Ehle and Warner, 1978). Although the forestomach may play an important role in digestion of plant foodstuffs, it appears to do so in conjunction with the cecum, since its removal at weaning does not affect digestion of synthetic or plant food or the animal's subsequent growth (Ehle and Warner, 1978). The forestomach shows greater fluctuations in its content between meals (41.5%) than the glandular stomach (17.8%) (Rowland, 1982).

In the glandular stomach, gastrin-secreting cells are more numerous in the antrum (gastrin concentration 179.4 ± 15 pg/mg) than in the fundus (gastrin concentration 54.1 ± 28.9 pg/mg) (Borer *et al.*, 1983b). Gastrin-secreting cells, as well as glandular cells, originate in the isthmus, below the gastric pit area (Fig. 2) (Hattori and Fujita, 1976) and migrate to the lower pyloric gland within five days (Fujimoto *et al.*, 1980).

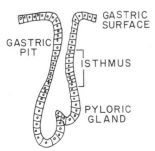

GASTRIC SURFACE

GASTRIC PIT

ISTHMUS

PYLORIC GLAND

Figure 2. Structure of the mucosa in hamster pyloric antrum. Isthmus is the generative cell zone. [Redrawn after Hattori and Fujita (1976).]

The lifespan of gastrin-secreting cells is 10 to 15 days (Fujimoto *et al.*, 1980), whereas mucin-containing glandular cells turn over in 14 days (Hattori and Fujita, 1976). Cells that line the surface epithelium of the glandular stomach originate in the upper part of isthmus and are sloughed off the epithelial surface within a week (Hattori and Fujita, 1976).

In the hamster, the fundus and body of the stomach appear to have greater concentration of somatostatinlike immunoreactivity (SRIF-LI, 2418.9 ± 405 pg/mg and 1986.9 ± 325 pg/mg, respectively) than the antrum (156.9 ± 45.9 pg/mg) (Borer *et al.*, 1983b). In contrast, in the rat, the antrum has equal or higher concentrations of SRIF-LI than the fundus (Arimura *et al.*, 1975; Kronheim *et al.*, 1976; McIntosh *et al.*, 1978).

1.2. Intestine

A number of studies report fine structure of hamster intestine (Balas *et al.*, 1980; Buschmann and Manke, 1981a,b; Dinda *et al.*, 1979; Fox *et al.*, 1978a,b,c). Morphological changes in the intestine are induced by withdrawal of trophic and differentiating influences of pancreatic juice (Balas *et al.*, 1980), by starvation, and by lipid meal (Buschmann and Manke, 1981a,b). Water and glucose transport (Fox *et al.*, 1978a,b) and disaccharidase (Dinda *et al.*, 1979) and peptidase activities (Dinda and Beck, 1982) are inhibited following the ingestion and absorption of ethanol.

1.3. Cecum

The cecum of the golden hamster resembles that of the dwarf hamster (Snipes, 1979a). It is as prominently developed in the hamster as in the herbivorous vole (Snipes, 1979b) that belongs to the same family as hamsters, Cricetidae, but to a separate subfamily, Microtinae. Cecum, in conjunction with the forestomach, harbors microflora that promotes fermentation of cellulose and allows hamsters to digest plant foodstuff.

1.4. Liver and Gallbladder

In the hamster, the liver is the principal site of lipid synthesis (Ho, 1979; Trayhurn, 1980; Tsai *et al.*, 1982), but some lipid synthesis goes on in white and brown adipose tissues as well. Lipid synthesis in hamsters is inhibited during acute exposure to cold (Denyes and Carter, 1961; Rowland, 1984). Hamster bile composition resembles that of man (Anderson *et al.*, 1972) more than the bile of rat or mouse (Wachtel *et al.*, 1969). It consists mostly of the glycine and taurine conjugates of cholic and chenodeoxycholic acids with lesser amounts of deoxycholate and lithocholate. There is no synthesis of muricholic acids from cholesterol or 26-hydroxy-cholesterol and α-hydroxycholesterol in hamster. After bile-duct ligation, the bulk of bile acid pool can be sulfated and excreted in urine within four days (Galeazzi and Javitt, 1977).

1.5. Pancreas

The pancreas is a thin sheet of ribbonlike tissue. Its head lies in the curvature of the duodenum and its tails lie along the spleen, stomach, and duodenum (Fig. 1) (Jewell and Charipper, 1951; Takahashi *et al.*, 1977). Pancreatic juice containing digestive enzymes is secreted into the duodenum at a rate of between 250 to 450 μg protein/hr per 100 g body weight (Helgeson *et al.*, 1980a). Secretin induces increase in the enzyme concentration and the rate of flow of pancreatic juice, whereas pancreozymin stimulates protein release alone (Helgeson *et al.*, 1980b). Islets of Langerhans are more numerous in the tails than in the head of pancreas (Jewell and Charipper, 1951), are embedded deeply inside the pancreatic tissue, and are relatively large (300 μm in diameter) compared with those of rat and man (Ogrowsky *et al.*, 1980).

1.6. Synopsis

Several features of hamster digestive tract point to this animal's ability to digest vegetable material. Thus, although the hamster is classified as an omnivore and although it is capable of predation (Jacobs, 1945; Polsky, 1974, 1976, 1977), some structural features of its gastrointestinal tract and some features of its feeding behavior and endocrine responses to feeding (see below) characterize it as a herbivore rather than an omnivore.

2. ENDOCRINE INFLUENCES ON SOMATIC GROWTH, BODY FAT, AND FEEDING IN THE GOLDEN HAMSTER

Interest in the role of hormones in energy regulation in the rat (Bray, 1974) has been greatly stimulated by the observation of parallel increases in body fat and serum insulin concentration that follow the destruction (Bernardis and Frohman, 1971; Frohman and Bernardis, 1968; Frohman *et al.*, 1969; Goldman *et al.*, 1974; Hales and Kennedy, 1964; Hustvedt and Løvø, 1972) or isolation (Tannenbaum *et al.*, 1974) of the medial basal hypothalamus (MBH), and by the reciprocal changes in serum concentration of sex hormones and the body fat level (Landau and Zucker, 1976; Wade, 1972, 1976). The failure to replicate consistently these two phenomena in the hamster (see Sections 4 and 2, respectively) and the propensity of hamsters to display weight increases unaccompanied by increased food consumption has discouraged research in this area in this species.

A putative role of individual hormones in energy balance of hamsters is summarized below. Greater weight was given to studies documenting correlations between weight changes and hormonal changes than to studies demonstrating weight changes in response to administration of pharmacological doses of hormones.

2.1. Insulin

Insulin does not appear to play the same role in the regulation of energy balance in the hamster as it does in the rat due to an apparent difference in the neural controls of insulin release and a difference in tissue sensitivity to anabolic actions of the hormone.

Destruction (Bernardis and Frohman, 1971; Frohman and Bernardis, 1968; Frohman et al., 1969; Goldman et al., 1974; Hales and Kennedy, 1964; Hustvedt and Løvø, 1972; Løvø and Hustvedt, 1978; Rohner et al., 1977) or isolation (Tannenbaum et al., 1974) of the MBH in the rat leads to hyperinsulinemia, accumulation of excess fat and weight whether (Louis-Sylvestre et al., 1980; Løvø and Hustvedt, 1978; Rohner et al., 1977) or not (Løvø and Hustvedt, 1978; Han and Frohman, 1970) such animals are allowed to consume increased amounts of food immediately before or after surgery. In contrast, lesions of the MBH in the hamster do not consistently induce increases in weight gain (Malsbury et al., 1977; Marks and Miller, 1972) and it is not known whether they are associated with hyperinsulinemia. On the other hand, a limbic forebrain circuit that involves fibers interconnecting the hypothalamus and hippocampus through rostromedial septum controls somatic growth and fat synthesis in the hamster but not in the rat (Docke, 1977; Murphy et al., 1972). Thus, a lesion of rostromedial septum (Borer et al., 1977), bilateral hippocampal transection (Borer et al., 1979a), or a septohypothalamic cut (Borer et al., 1979b) all lead to acceleration of skeletal and ponderal growth. The latter two procedures (but not the septal lesion) also lead to accumulation of excess fat. All three are associated with increases in serum concentrations of insulin (Table 1) and growth hormone (GH). Hyperinsulinemia persists even when animals are fasted before blood collection. Septal lesions in hypophysectomized hamsters induce no increases in weight gain or blood insulin concentration (Borer et al., 1977). Growth hormone is also a potent secretagogue of insulin in the hamster in vitro (Curry et al., 1975).

In all of these circumstances in the hamster, insulin oversecretion and increased weight gain were associated with increased secretion of GH and may depend on positive energy balance that accompanies rapid growth. In contrast, hyperinsulinemia and accumulation of excess fat are observed in hypophysectomized rats with lesions of MBH (Han and Frohman, 1970) demonstrating the independence of the neural controls of insulin release from the neural controls of growth. Although female rats with isolated MBH also oversecrete GH and display increased skeletal growth (Mitchell et al., 1972, 1973), these two phenomena appear to be independent of the hyperinsulinemic and obesifying effects of such knife cuts in the rat.

The following additional evidence suggests that insulin may facilitate somatic growth as well as fat synthesis in the hamster rather than fat synthesis alone. Increased serum insulin concentration (Table 1), along with increased GH concentrations accompany rapid somatic growth (Borer and Kelch, 1978) and repletion of body fat stores (Tsai et al., 1982) during the first few days following the termination of growth-inducing disk exercise. In addition, weight increases induced by administration of protamine zinc insulin (PZI) in hamsters were in two (Rowland, 1978; Borer and Stemmer, unpublished data; Table 1) out of three reports (Ritter and Balch, 1978) shown to be permanent rather than reversible as in the rats (Borer and Stemmer, 1974; Hoebel and Teitelbaum, 1966). This suggests that insulin injections, which are known to elicit GH release in man (Roth et al., 1963), may stimulate somatic growth as well as lipogenesis in the hamster, whereas they predominantly stimulate lipogenesis in the rat.

Table 1. Relationship between Serum Insulin Concentration, the Rate of Ponderal Growth, and Body Fat in the Hamster

Treatment	Serum insulin (ng/ml)[a]	Ponderal growth (g/day)	Body fat (%)	Reference
Sedentary female	2.3–3.0 (1.5)[b]	0.1–0.4	13–18	Borer and Kelch, 1978a Borer et al., 1977 Moffatt, 1983
Septal lesions in female	15.0 (7.1)	1.9	14.2	Borer et al., 1977
Hippocampal transections in female	16.4 (4.4)	1.8	17.9	Borer et al., 1979a
Septohypothalamic cuts in female	7.5 (3.6)	2.5	24.1	Borer et al., 1979b
Hypophysectomized female	1.0 (0.8)	0	21.1	Browne and Borer, 1978 Borer et al., 1977
Septal lesions in hypo-physectomized female	1.4 (1.0)	−0.1	20.2	Borer et al., 1977
Exercising female	3.1 (1.9)	1.2	7.7	Borer and Kelch, 1978 Moffatt, 1983
Retired exercising female	5.6 (1.5)	0.7	19.1	Borer and Kelch, 1978 Moffatt, 1983
Recovery from weight loss in female	2.1 (0.7)	2.7	—	Borer and Kelch, 1978
Catch-up growth in female	7.0 (2.3)	4.1	—	Borer and Kelch, 1978
Protamine-zinc insulin injections in male and female	830 ng/kg	2.0		Rowland, 1978

[a] Measured with double-antibody radioimmunoassay utilizing radioiodinated porcine insulin as tracer, purified rat insulin (Eli Lilly Laboratories, 22.6 μU/ng) as standard, and antiserum against porcine insulin developed in a guinea pig. Assay sensitivity was 1.3 ± 0.3 pg per assay tube.

[b] Values for animals killed in *ad libitum*-fed state are outside parentheses, those from animals killed after a 4-hr fast, within parentheses.

In vitro evidence indicates that the adipose tissue of herbivorous rabbit, guinea pig, and hamster is four to six times less responsive to the lipogenic influence of insulin than the adipose tissue of omnivorous rat (DiGirolamo and Rudman, 1966). Thus an apparent equivalence in PZ insulin-induced weight increases in the hamster and the rat (Table 2) may reflect different anabolic processes in the rat and the hamster. Weight increases in the hamster may represent accretion of body components other than the body fat alone. Until body composition changes are carefully documented in insulin-injected and brain-lesioned hamsters, it will not be possible to decide whether hamsters differ from rats primarily in the neural controls over insulin release, in tissue sensitivity to lipogenetic actions of this hormone, as Table 2 and DiGirolamo and Rudman (1966) data suggest, or in both.

Rapid weight gain that accompanies increased serum insulin and GH concentrations in the hamster is associated with normophagia or with transient and dispropor-

Table 2. Increases in the Rate of Weight Gain in Injected Relative to Uninjected Hamsters and Rats in Response to Protamine Zinc Insulin

Hamster			Rat		
Dose U/kg	Increase in weight gain (%)	Reference	Dose U/kg	Increase in weight gain (%)	Reference
5	0	Borer and Stemmer, 1974[a]	7	0.2	Borer and Stemmer, 1974
10	0.5	Borer and Stemmer, 1974[a]	14	0.4	Borer and Stemmer, 1974
20	0.5	Borer and Stemmer, 1974[a]	20	0.6	Lotter and Woods, 1977
25	1.0	Rowland, 1978	40	0.9	MacKay et al., 1940
80	1.7	Ritter and Balch, 1978	50	2.1	MacKay et al., 1940
			60	3.8	MacKay et al., 1940

[a]Borer, K. T., and Stemmer, P. Unpublished data 1974. Four groups of female hamsters (Con Olson, Madison, Wis.) on a diet of Purina laboratory chow and sunflower seeds and weighing 105 g at the start, received 5, 10, and 20 U/kg of protamine zinc insulin or 0.9% saline, SC, during 14 successive days. Two groups of Holtzman Sprague-Dawley rats on a diet of Purina lab chow and weighing 270 g at the start received PZI (7 U/kg per day during 10 days and 14 U/kg per day during subsequent 28 days), or 0.9% saline SC.

tionately small hyperphagia. Thus, following septal lesions (Borer *et al.*, 1977), there is an eightfold increase in ponderal growth during the first four postoperative weeks and only 1.2-fold increase in food intake during ten postoperative days; following hippocampal transections, the rate of weight gain increased tenfold, whereas food consumption increased 1.3-fold during the seven postoperative weeks (Borer *et al.*, 1979a); during the first week after termination of voluntary running (Moffatt, 1984) ponderal growth rate is 33 times greater and food consumption 1.2 times greater than in sedentary controls; during the catch-up growth that takes place during early recovery from exercise with food restriction (Borer and Kelch, 1978), ponderal growth rate is ten times greater and food consumption is 1.4 times greater than in sedentary controls. All of this evidence suggests that acceleration in the rate of weight gain that accompanies increased serum insulin and GH concentrations in hamsters results primarily from physiological processes and not from increases in energy intake.

Injections of insulin have traditionally been used as a diagnostic tool for glucoprivic control of feeding. Both the short-acting (Iletin) and long-acting (PZ) insulin can stimulate increased food consumption in the hamster. The effect is not dose dependent, since all of the pharmacological doses tested (10 to 80 U/kg, DiBattista, 1983, 1984; Ritter and Balch, 1978; 100 U/kg, Rowland, 1978) of Iletin stimulate intake of approximately 1 g of food only, which is equivalent to a single additional meal. Feeding response occurs 3–5 hr postinjection at the three lower doses (DiBattista, 1984; Ritter and Balch, 1978) and 6 hr postinjection at the higher dose (Rowland, 1978). Not only is feeding unresponsive to systematic increases in the dose of short-acting insulin, but the 1.4- to 1.5-fold increases in food consumption following chronic injections of long-acting insulin fail to match threefold to fourfold increases in the rate of weight gain (Ritter and Balch, 1978; Rowland, 1978). Thus, as after weight-gain-promoting lesions of the limbic forebrain, increases in weight gain that accompany increases in serum concentrations of insulin in hamsters result more from physiological actions of insulin rather than from substantial increases in energy consumption.

Insulin without a doubt plays an essential role in regulation of fat stores in the hamster as in other mammals. This is evident with feeding regimens which involve intermeal intervals (IMIs) that are longer than 2-hr ones preferred by ad-bitum feeding hamsters. If hamsters are given four 1-hr exposures to food separated by 5-hr IMIs, they are unable to maintain their energy balance and will keep losing weight on this regimen and eventually die of excessive weight loss (Borer et al., 1979c). At least part of the failure in weight maintenance is induced by deficient insulin responses to feeding in intermittently fed hamsters. A 24-hr exposure to 5-hr IMIs leads to an 18% reduction of plasma insulin concentration, and to disappearance of postprandial insulin response to meal ingestion (Borer et al., 1985). In addition, stimulation by insulin of lipogenesis may be deficient in intermittently fed hamsters, as hepatic fatty acid synthetase shows a 30% reduction in hamsters maintained on 5-hr IMIs for a day (Borer et al., 1985). Significant decline in serum insulin concentration occurs during 6 to 8 hr of acute starvation. Within an hour of refeeding, serum insulin returns to the elevated plateau (Borer et al., 1979c; Borer, unpublished data). Concentrations of insulin in ad libitum refed hamsters subjected to a 24-hr fast are 30% lower than the values from animals that were not subjected to a previous food deprivation (Borer, unpublished data). Starvation diabetes has been described in both the hamster (Feldman and Lebowitz, 1973; Feldman et al., 1979) and the rat (Grey et al., 1970; Permutt and Kipnis, 1975; Zawalich et al., 1979), but it remains to be clarified whether the hamster is more susceptible to it, and whether this phenomenon accounts for this species' inability to regulate its body fat stores at relatively mild regimens of food restriction.

Hamsters regain lost body fat without increasing daily energy consumption provided they are allowed to feed at their preferred two-hourly IMIs (Borer et al., 1979c). During the recovery from fat loss, the rate of weight gain is proportional to the magnitude of preceding weight loss, indicating that energy deficit induces increases in the rate of fat synthesis. Increased lipogenesis occurs in underweight hamsters under the conditions of decreased plasma insulin concentrations (Borer et al., 1985) suggesting that the rapid weight recovery observed in these animals must depend more on increased sensitivity of the liver and white adipose tissue to lipogenic actions of this hormone than on increased concentrations of serum insulin. A similar relationship has been described between body fat levels in the rat and the sensitivity of its adipocytes to lipogenic action of insulin (Olefsky, 1977a,b).

Other relevant information on the biology of hamster insulin release is that hamsters are resistant to diabetogenic action of alloxan (House and Tassoni, 1957; Nace et al., 1956) and they tolerate poorly intravascular administration of alloxan and streptozotocin (Sak and Beaser, 1962), but will become diabetic if they receive IP streptozotocin (50 mg/kg) on three successive days (Phares, 1980).

2.2. Growth Hormone

Growth hormone plays an important role in the control of weight gain in adult hamsters. Adult hamsters that display low rates of somatic growth, can be stimulated to significantly increase their skeletal and ponderal growth by exposure to voluntary running activity (Borer and Kelch, 1978) or by damage to rostromedial septum (Borer et al., 1977), by transection of hippocampus (Borer et al., 1979a), or by septohypothalamic cuts (Borer et al., 1979b). Oversecretion of GH appears to be necessary for acceleration of ponderal growth in exercising hamsters and in hamsters with the limbic forebrain lesions because increases in weight gain are present only when increased concentrations of GH are noted (Borer and Kelch, 1978; Borer et al., 1977, 1979a,b) and not when hypophysectomized hamsters are exposed to exercise (Browne and Borer, 1978) or given septal lesions (Borer et al., 1977). Since oversecretion of insulin accompanies oversecretion of GH in hamsters with damage to the limbic forebrain, both of these hormones may be necessary for acceleration of the rate of weight gain.

Additional support for this notion comes from experiments involving administration of exogenous GH and insulin (K. T. Borer, 1975, unpublished data). Female hamsters obtained from Con Olson, Madison, Wisconsin, weighing 90 to 100 g were maintained on a diet of sunflower seeds and Purina chow pellets. Groups of six animals were matched by weight and injected subcutaneously on alternate days with the following hormones:

1. Seven injections of rat GH (NIAMDD rat GH B2 0.9. IU/mg, 1 IU/hamster).
2. Sixteen injections of rat GH (1 IU/hamster).
3. Twelve injections of pig GH (Calbiochem) (1 IU/hamster).
4. Sixteen injections of bovine GH (NIH GH B17, 0.9 IU/mg, 1 IU/hamster).
5. Sixteen injections of protamine zinc insulin (2 IU/hamster).
6. Sixteen injections of both bovine GH (1 IU/hamster) and PZ (2 IU/hamster).

Four groups of controls, for experiments 1 through 6, received SC injections of saline. Skeletal growth was evaluated from length increments, determined by stretching hamsters over a tape measure (Borer and Kuhns, 1977) and from the difference between starting and final weight measurements. In addition, food consumption and the rate of ponderal growth were monitored during the time injections were given. The effects of hormone injections on body weight are shown in Fig. 3. Injection of GH induced transient acceleration in the rate of weight gain and lasting increments in linear growth. In accordance with the variable biological effectivness of this highly species-specific hormone, rat GH induced greatest acceleration in the rate of weight gain (1.2 ± 0.1 versus 0.7 ± 0.1 g/day, $P < 0.01$ in group 1; 1.5 ± 0.1 versus 0.5 ± 0.1 g/day, $P < 0.001$ in group 2), whereas pig GH (0.8 ± 0.1 versus 0.6 ± 0.1 g/day, $P < 0.05$) and beef GH (0.8 ± 0.1 versus 0.4 ± 0.1 g/day, $P < 0.02$) were less potent. Protamine zinc insulin at 2 IU/hamster every two days was about as potent as bovine GH or porcine GH injections (0.7 ± 02 g/day) and the effects of GH and PZI were not additive (0.9 ± 0.1 g/day). Rat GH induced significant skeletal growth (0.5 ± 0.1 versus 0.0 ± 0.2 cm, $P < 0.02$), whereas the body elongation induced by nonrodent

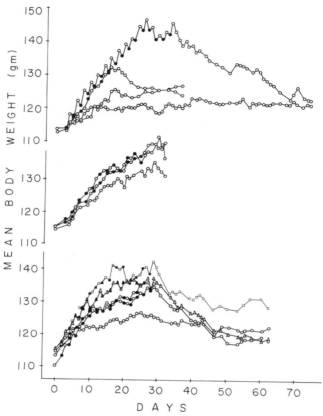

Figure 3. Weight changes in groups of six female hamsters maintained on a diet of Purina lab chow and sunflower seeds and subjected to subcutaneous injections, every two days, of 1 IU of growth hormone from several different species: rat (NIAMDD GH B2, solid circles, top) (Calbiochem, solid circles, center), beef (NIH B17, solid circles, bottom) and to injections of protamine zinc insulin (2 IU) alone (solid triangles, bottom) and in conjunction with injections of bovine GH (solid squares, bottom). Two independent groups of hamsters received injections of rat GH (top), porcine GH (center), and bovine GH (bottom). Control hamsters received subcutaneous injection of saline every two days.

GHs did not reach statistical significance. Increases in body weight induced by GH injections were transient and were dissipated shortly after the injections were discontinued, but combined injections of GH and PZI induced a significant and permanent increase in the body weight level (17.3 ± 2.3 versus 7.2 ± 3.0 g, P < 0.05). Weight increases induced by GH injections were not associated with increases in food consumption (1: 23.2 ± 0.8 versus 24.9 ± 0.8 Cal/100 g/day; 2: 24.2 ± 1.0 versus 26.5 ± 1.6 Cal/100 g/day). Without the body composition information to help clarify the nature of large transient increases in body weight, it is difficult to interpret the effects of GH injections on the control somatic growth. The combined growth-induction studies and injection data suggest that both high GH and high insulin titers may be necessary for rapid somatic growth in the hamster.

There is no evidence, from either the injection or growth-induction studies that weight gain associated with increased GH concentrations in the serum results from increased food consumption. Increases in food consumption seen during voluntary exercise and following lesions of limbic forebrain do not match in magnitude increases in the rate of weight gain (see Section 2.1). Much of the hyperphagia noted in exercising hamsters reflects the energy cost of running, because the hypophysectomized hamsters display it in proportion to the amount of voluntary running they do (Brown and Borer, 1978). And the limited hyperphagia that accompanies striking weight increases in hamsters with lesions of limbic forebrain could be due to increases in concentrations of circulating insulin rather than GH.

In the rat, neither exercise nor lesions of limbic forebrain induce oversecretion of GH and increased somatic growth. Surgical isolation of the MBH is associated with increased plasma GH concentrations (Mitchell *et al.*, 1973) and skeletal growth (Mitchell *et al.*, 1972, 1973) in the female, but not in the male rat (Rice *et al.*, 1976). Thus it is reasonable to conclude that the hamster differs from the rat in the organization of its neural controls of GH release. Growth hormone secretion responses to exercise and stress in the hamster follow the pattern exhibited by primates and not the pattern seen in rats (Borer *et al.*, 1982b). The sensitivity of hamster's GH secretory mechanism to physical activity may provide the way this rodent coordinates its growth to the onset of energy-abundant seasons of spring and summer. Seasonal differences in the rate of growth have been reported for hamsters maintained under the natural illumination prevailing in the northern latitudes (Granados, 1951). Granados suggested the difference in the seasonal growth rates between the two sexes (females grow faster than males in autumn, males grow faster than the females in winter, spring and summer) may explain discrepant reports of body size differences in this species (female larger: Bruce and Hindle, 1934; Gerall and Thiel, 1975; Hamilton and Hogan, 1944; Kowalewski, 1969; Swanson, 1967, 1970; male larger: Ben-Menahem, 1934; Bond, 1945).

We have observed seasonal differences in the rate of growth in the female hamster that could not be attributed to the length of the photoperiod (K. T. Borer, C. S. Campbell, J. Tabor, and J. K. Jorgenson, 1981, unpublished data; Borer *et al.*, 1981). Two groups of female hamsters from Engle Laboratory Animals matched by weight at the outset were assigned to 10L:14D or to 16L:8D photoperiods for 15 weeks, between October and February (winter hamsters). Another two groups of female hamsters from the same source and of similar starting age and weight were similarly treated between March and July (summer hamsters). Body lengths were measured at the start and at the end of the photoperiodic exposure. Estrous cycles were monitored during weeks 12 to 15 by examination of vaginal secretion (Orsini, 1961) to evaluate the effectiveness of photoperiodic manipulation. After 15 weeks, animals were killed by decapitation during the second hour after the onset of light for determinations of serum and pituitary concentrations of GH, prolactin (PRL), and for determination of body fatness. The results are summarized in Table 3. Short-day photoperiod induced anestrus

Table 3. Effects of Season and Photoperiod on Body Composition, Somatic Growth, and Anterior Pituitary Function in Female Hamsters

	Summer		Winter		
	Short-day (n = 8)	Long-day (n = 8)	Short-day (n = 7)	Long-day (n = 8)	Significance (P < 0.05)
Number of estrous cycles	5.9 ± 0.6[a]	6.9 ± 0.1	1.0 ± 0.7	6.4 ± 0.3	S,P,S × P[b]
Weight increment (g/15 weeks)	39.6 ± 0.7	31.9 ± 2.3	15.1 ± 3.7	10.6 ± 2.4	S,P
Length increment (cm/15 weeks)	1.45 ± 0.06	1.73 ± 0.12	1.21 ± 0.06	1.24 ± 0.08	S
Percent body fat	15.4 ± 0.5	11.3 ± 0.9	14.7 ± 0.9	11.2 ± 0.8	P
Serum concentrations					
of growth hormone	20.7 ± 4.2	62.1 ± 16.6	11.8 ± 2.1	12.3 ± 4.2	S
prolactin (ng/ml)	48.2 ± 17.0	36.4 ± 2.9	8.5 ± 3.9	20.9 ± 2.9	S,P,S × P
Pituitary growth hormone					
μg/AP	38.6 ± 6.3	42.5 ± 3.6	11.1 ± 2.7	15.3 ± 1.3	S
μg/mg AP	10.3 ± 1.1	9.4 ± 0.8	3.8 ± 0.8	3.8 ± 0.5	S
Pituitary prolactin					
μg/AP	22.1 ± 5.3	27.1 ± 2.9	8.1 ± 4.1	50.2 ± 7.1	P,S × P
μg/mg AP	5.6 ± 1.0	6.2 ± 0.6	2.6 ± 1.2	14.8 ± 1.9	P,S × P

[a] Mean ± S.E.M.
[b] S = effects of the season, P = effects of the photoperiod, S × P = interaction between season and photoperiod, two-way analysis of variance.

in winter (Borer et al., 1983b) but not in summer hamsters. Likewise, serum and pituitary PRL concentrations and pituitary PRL content were reduced as a result of short-day photoperiodic exposure in winter (Borer et al., 1983b), but the same treatment was ineffective in summer. In contrast, percentage of body fat was significantly higher in short-day than in long-day hamsters in either season. Growth rate was primarily influenced by the season and not by the photoperiod. Summer hamsters displayed greater ponderal and skeletal growth, higher concentrations of GH in serum, and the anterior pituitary and greater pituitary GH content than the winter animals. The mechanism of this seasonal effect on the rate of growth is unexplained. However, similar seasonal effects in hamster growth were reported by Hoffman (1983). Animals of either sex blinded at age 42 days throughout the year and checked for growth eight weeks later showed peak growth (in excess of growth in sighted animals) in late summer and early spring and least growth in late winter and early spring. In addition, animals of either sex blinded in May failed to undergo reproductive regression. In the natural environment, photoperiod may exert an indirect influence over hamster rate of growth. Stimulatory photoperiods facilitate running and physical activity in hamsters (Ellis and Turek 1979; Widmaier and Campbell, 1981a), and increased levels of running facilitate somatic and skeletal growth in this species (Borer and Kuhns, 1977).

2.3. Prolactin

Since PRL plays an important role in the control of growth and fattening in some seasonally breeding birds and mammals, it is of interest to establish the role of this hormone in growth and regulation of body fat in the seasonally breeding hamster. In the sheep, long photoperiod and increases in serum PRL concentration have been linked to stimulation of somatic growth (Schanbacher and Crouse, 1981). Prolactin also stimulates somatic growth in young rats (Sinha and Vanderlaan, 1982). However, in the hamster there is no evidence that photoperiod or changes in serum PRL concentration influence the rate of ponderal and linear growth (Table 3).

Prolactin stimulates premigratory fattening in the White-throated Sparrow when it is administered in a specific temporal relationship to corticosterone injections (Meier and Davis, 1967). Meier postulates that corticosterone entrains circadian rhythms of photosensitivity in mammals and birds (Meier, 1972; Meier and MacGregor, 1972). Corticosterone induces photosensitive periods during which exposure to light stimulates secretion of pituitary gonadotropins and PRL. In an experiment designed to test this hypothesis in the female hamster, Joseph and Meier (1974) secured some evidence that corticosterone may entrain circadian rhythms of PRL release. When PRL release occurs 12 hr after corticosterone, anestrus is induced. When peak PRL occurs 20 hr after (or 4 hr before) corticosterone surge, synthesis of fat in the liver and accumulation of fat in the adipose tissue are reduced. Joseph and Meier (1974) speculate that peak corticosterone release occurs at midday in female hamsters. In 12L : 12D female hamsters, the highest frequency and amplitude of PRL pulses occur during the first 6 hr following the onset of light (Borer et al., 1982a), although just the opposite pattern of serum PRL concentration was noted in 16L : 8D hamsters killed by decapitation (Widmaier and Campbell, 1981b).

Photoperiod has a significant influence on the body fat content of female hamsters (Table 3). Thus, animals maintained on a diet of Purina formulab 5008 in 12 to 16 hr of light per day are significantly leaner than animals maintained in 8 to 10 hr of light per day. Exposure to short-day photoperiods induces body weight gain after a delay of four to six weeks (Bartness and Wade, 1984; Campbell et al., 1983), is expressed in female but not the male when fed laboratory chow (Wade, 1983), and is magnified when animals of either sex are fed a fat-enriched diet (33% by weight) (Bartness and Wade, 1984; Wade, 1983). Since serum PRL, measured during the second hour after the onset of light, is significantly higher in the early part of the 12-hr day in long-day animals (Borer et al., 1982a), Meier's hypothesis could hold true. In the hamster, presence of high serum PRL several hours prior to midday corticosterone peak, a condition which may prevail in long photoperiods, could lead to reduced fat synthesis in the liver and adipose tissue. Additional evidence supporting Meier's hypothesis comes from another experiment in which we have manipulated the concentration of serum PRL (Borer et al., 1983b; Kandarian, 1983). Three groups of hamsters (n = 5) were matched by weight and maintained 15 weeks in nonstimulatory short-day photoperiod (8L : 16D). All animals became anestrous between the 11th and 12th weeks of short-day photoperiodic exposure. At the start of week 12, hamsters were either exposed to sham surgery or

Table 4. Changes in Endocrine and Somatic Variables[a]

	Control	Bromocriptine	Pituitary	p
Serum prolactin ng/ml	11.3 ± 1.4	7.8 ± 1.5	111.7 ± 12.7	0.05
Body fat %	13.9 ± 1.9	9.4 ± 0.7	6.9 ± 1.7	0.05
Food intake g/100g	6.2 ± 0.3	7.1 ± 0.5	6.7 ± 0.6	NS
Basal metabolic rate ml O_2/kg per hr	2.89 ± 0.54	3.25 ± 0.39	2.38 ± 0.2	NS
NE-stimulated metabolic rate	7.26 ± 0.62	7.45 ± 0.89	5.59 ± 0.8	NS
Fecal energy KCal/day	8.9 ± 1.3	9.8 ± 1.2	8.9 ± 1.1	NS

[a]Female hamsters maintained in short-day photoperiod (8L:16D) and given daily injections (SC) of peanut oil (control group n = 5), bromocriptine (250 μg, n = 5), and peanut oil following the implantation of four anterior pituitaries under the kidney capsule at the start of injections (pituitary group n = 5).

received four anterior pituitaries under the kidney capsule from female hamster donors (PIT group), a procedure that increased serum PRL concentrations tenfold (Table 4). Between weeks 12 and 16, all hamsters received daily SC injections of peanut oil or of 2-α bromocriptine (CB 154, Sandoz, Inc., 250 μg in 0.05 ml of peanut oil), a procedure that induced some reduction is serum PRL concentration (Table 4). On the 32nd day of injections, animals were killed by decapitation during the second hour after the onset of light to collect blood for determinations of serum PRL concentrations. Food intake, fecal energy context, and resting and norepinephrine-stimulated (0.8 mg/kg per IP) oxygen consumption were measured during days 22 through 31 of injection, respectively, and none were affected by manipulations of serum PRL concentration. Carcass fat was also measured. Figure 4 and Table 4 show that significant increments in serum PRL concentration (pituitary implant group) were associated with significant

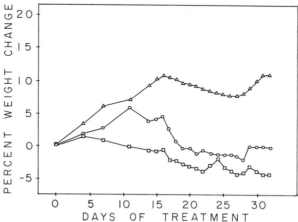

Figure 4. Weight changes in groups of five female hamsters maintained in nonstimulatory short-day photoperiod (8L:16D) and given daily SC injections of peanut oil (control group, open circles), 250 μg of bromocriptine (open triangles), or peanut oil following the implantation of four anterior pituitaries under the kidney capsule at the start of injections (open squares). (Data from Borer *et al.*, 1983b and Kandarian, 1983.)

reduction in body weight and in percentage of body fat, whereas low serum PRL levels were accompanied by higher body fat, and weight increments were higher in control and bromocriptine-treated animals.

2.4. Gonadal Hormones

Gonadal hormones exert stronger, and a generally inhibitory, influence over the rate of ponderal and skeletal growth in the male than in the female hamsters. The adult female hamster is usually heavier and longer than the adult male (Bruce and Hindle, 1934; Gerall and Thiel, 1975; Hamilton and Hogan, 1944; Kowalewski, 1969, Swanson, 1967, 1970). Thus, the removal of testes, before or after puberty, leads to faster linear and ponderal growth in postpubertal hamsters and to body weights and lengths in the range of values seen in adult intact female hamsters (Kowalewski, 1969; Manda and Matsumoto, 1973a; Swanson, 1967). The removal of ovaries usually has little if any effect on linear and ponderal growth (Borer, 1982; Fleming, 1978; Manda and Matsumoto, 1973b; Swanson, 1967), although infrequently a modest increase in the rate of ponderal growth is reported (0.6 versus 0.3% per day, Morin and Fleming, 1972; also Gerall and Thiel, 1975). These data suggest that steroid hormones produced by the male gonad exert an inhibitory influence over the rate of somatic growth, and that the female gonad has little impact on the body size or on the skeletal growth in the hamster. Inhibitory influence of the testis over the rate of hamster weight gain is very small in the short-day photoperiod and is greatly amplified in the long-day photoperiod (Granneman and Wade, 1982). This effect is accomplished by physiological processes and in absence of any change in food consumption (Granneman and Wade, 1982).

Experiments involving administration of pharmacological doses of gonadal steroids reveal a variety of effects on weight, fatness, and growth in hamsters. The results differ according to age and sex of hamsters and are difficult to interpret without a knowledge of the likelihood that the animals will encounter such hormone titers in the course of their lifespan.

A possibility that exogenous steroids may have an effect on hamster growth was explored by Manda and Matsumoto (1973a–f). They have noted that naturally occurring vegetable foodstuffs of hamsters, such as alfalfa, contain estrogenic substances (1973d); that the concentration of these substances is higher during the first-bloom stage than during later growing stages (1973f; and that 0.01 μg of diethylstilbestrol or a dose of alfalfa extract of equivalent potency administered in food stimulates significant linear and ponderal growth, an increase in body fat content (1973a,e), and increased food consumption and feed efficiency (1973c) in young (days 30 to 90) intact male or ovariectomized female hamsters. The same treatment has no effect on linear and ponderal growth and reduces the percentage of body fat in young intact female hamsters or in adult hamsters of both sexes regardless of their gonadal status (1973b).

Administration of pharmacological doses of gonadal steroids to sexually mature gonadectomized hamsters reveals an inhibitory effect of both testosterone and estradiol on body weight (Kowalewski, 1969) and on percentage of body fat (Borer, 1982). At equivalent doses, testosterone appears to be more effective in lowering the body weight of hamsters than estradiol (Kowalewski, 1969). Male hamster appears to be more sensitive to the weight-reducing effect of testosterone than the female hamster (Kowalewski, 1967; Zucker et al., 1972).

Hamsters appear remarkably insensitive to the weight-suppressing effects of estradiol. At doses of about 35 μg/kg, estradiol induces a weight loss of 0.4% per day in the hamster (Morin and Fleming, 1978). Comparable weight losses (about 0.7% per day) are obtained in the rat with estradiol dose of about 1.5 μg/kg. The 20-fold difference in the sensitivity to weight-suppressing effects of estradiol between the hamster and the rat most likely reflects the difference in the affinity of brain, anterior pituitary, and uterine tissues for estradiol in the two species (Feder et al., 1974).

Male gonad interferes with the increased fat synthesis that takes place in hamsters maintained on a high fat diet (Wade, 1982, 1983). Such dietary obesity is more strongly expressed in the female than in the male hamster and in short-day than in the long-day photoperiod (Wade, 1983). Since exposure to short-day photoperiod leads to reduced concentrations of gonadal hormones in male (Berndtson and Desjardins, 1978) and female (Widmaier and Campbell, 1981b), hamsters' dietary obesity seems to be curbed by the antilipogenic actions of gonadal hormones in hamsters, and male sex steroids appear more potent than the female sex steroids in this process. Replenishment of body fat stores following food restriction is likewise suppressed in male hamsters in whom testicular function is maintained by stimulatory photoperiod (Granneman and Wade, 1982).

Gonadal hormones do not exert a strong influence on food consumption in the golden hamster. Removal of gonads has no effect on food consumption in the hamsters (Zucker et al., 1972). On the other hand, if food intakes from several estrous cycles are averaged to reduce data variability, food consumption is greater on the day of vaginal discharge following the night of nocturnal estrus and on the following day than during the two preceeding days of the cycle (Miceli and Fleming, 1983; Morin and Fleming, 1978). Serum estradiol reaches its highest circulating concentrations (about 150 pg/ml) at midday on the day preceding the estrous vaginal discharge and starts to rise (90 pg/ml) the day before. Lowest values for serum estradiol (10 to 25 pg/ml) are recorded on the day of vaginal discharge and the day after it (Saidapur and Greenwald, 1978; Widmaier and Campbell, 1981b). Thus the modest variation in quantity of food eaten during hamster estrous cycle reflects a slight modulatory role of endogenous estradiol titers. Gonadal hormones appear to suppress selection of fat (and to a lesser extent, protein) in the female hamster (Miceli and Fleming, 1983).

Gonadal hormones also do not exert a strong influence over physical activity of golden hamsters. Measures of voluntary activity in rotating drums (Bond, 1945) and of various natural forms of locomotor activity (Richards, 1966) reveal a modulatory influence of changing sex hormone titers during the estrous cycle. However, removal of the gonads and replacement with pharmacological doses of sex steroids has little if any effect on the levels of voluntary disk running in hamsters of either sex (Borer, 1982) in striking contrast to the almost complete dependence of voluntary activity in the rat on the activating influence of sex hormones (Wade, 1976). Administration of pharmacological doses of progesterone leads to hyperphagia and weight gain in gonadectomized adult hamsters of either sex (Zucker eet al., 1972). Similar effect is obtained in prepubertal (36 to 47 day old) male but not female hamsters. Prepubertal female hamsters remain unresponsive to the 5-mg dose of progesterone in the absence of pituitary as well. Progesterone also promoted weight gain, food consumption and survival of adrenalectomized hamsters.

Neither hyperphagia nor increased weight gain could be related to increased concentrations of circulating progesterone in the hamster. Progesterone concentration is increased during pregnancy (between 20 and 30 ng/ml) and lactation (between 5 and 20 ng/ml) (Baranczuk and Greenwald, 1974). Food intake remains normal throughout pregnancy and weight gain increases during the last five days of fetal development (Fleming, 1978; Zucker et al., 1972). During lactation, hamsters lose weight and display hyperphagia that is proportional to the size of the litter and hence reflects variable energy drain of the consumed milk (Fleming, 1978). During the last four days of pregnancy and throughout lactation, hamsters display increased incidence of food hoarding (Fleming, 1978) and increased selection of fat and protein (which favors better weight maintenance during lactation) (Fleming and Miceli, 1983). In addition, high levels of voluntary running on an activity disk are suddenly curtailed during the last two days of pregnancy, and they gradually return to prepregnancy level during the 21 days of lactation (Borer, 1982). The mechanisms underlying alterations in feeding, weight gain, hoarding, and levels of physical activity during pregnancy and lactation in the hamsters remain unexplained.

2.5. Thyroid and Adrenal Hormones

Information on the role of thyroid and adrenal hormones in energy regulation in the hamster is scarce. Injection of 30 μg thyroxine (T_4)/100 g body weight over a two-week period has no apparent effect on hamster weight or body fat. Administration of triiodothyronine (T_3) (110 μg/100 ml drinking water) over a two-week period induces a 15% to 20% weight loss and unsaturation of body fat (Kodama and Pace, 1963). Relative increase in the proportion of oleic acid (18 : 1) and decrease in palmitic acid (16 : 1) in T_3-treated hamsters resembles the pattern seen in cold-exposed (Fawcett and Lyman, 1954; Kodama and Pace, 1963, 1964) or food-restricted hamsters and suggests an increased utilization of saturated fatty acids under all of these conditions of increased energy need. Administration of adrenocorticotropic hormone (ACTH) (3 mg/100 g), cortisol (3 mg/100 g), or epinephrine (30 μg/100) for two weeks had little effect on body weight and body fat of hamsters (Kodama and Pace, 1963). Adrenal demedullation resulted in slightly enhanced weight gain over an extended period of time (7% in four months) (Borer, 1982).

2.6. Synopsis

Endocrine influences on somatic growth, body fat, and feeding in the hamster differ from other rodents in several respects. In the hamster, increases in serum insulin are usually associated with high growth hormone/titers and accompany accelerated somatic growth. Hormone-induced acceleration in ponderal growth largely reflects physiological mechanisms favoring energy storage and little if any change in energy intake in this species. Prolactin is lipolytic under stimulatory photoperiod. Triiodothyronine is also lipolytic and promotes unsaturation of hamster body fat.

Testosterone is catabolic, but gonadal hormones in general exert minor influence over feeding, food selection, growth, energy regulation, or physical activity in this species.

3. PHYSIOLOGICAL AND BEHAVIORAL RESPONSES OF HAMSTERS TO SIGNS OF IMPENDING AND ACTUAL DISTURBANCES IN ENERGY BALANCE

The amount of daylight, environmental temperature, opportunity for physical activity, food abundance, and caloric density of food all undergo annual fluctuations in hamster's natural habitat. Since they either signal impending or represent actual changes in the amount of food energy available to this rodent, they have evolved into potent modulators of energy balance in the hamster. The hamster energy regulatory mechanism is exceptionally sensitive to variations in these variables. The hamster also displays greater propensity for metabolic rather than ingestive responses to fluctuations in these variables, an idiosyncratic trait that distinguishes it from other rodents.

3.1. Photoperiod

Hamster meals occur about every 2 hr throughout the day (Borer et al., 1979c), hence there is no strong circadian photoperiodic influence over feeding (Zucker and Stephan, 1973). In the hamster, the length of the photoperiod mainly affects mechanisms of body fat storage, rather than growth (Table 3) or food consumption. In contrast, long-day photoperiod appears to have direct stimulatory effect on the growth of sheep (Schanbacher and Crouse, 1981) and cattle (Peters and Tucker, 1978; Peters et al., 1981). In the short-day photoperiod, lipid synthesis in white and brown depots is greatly enhanced in the presence of little if any change in food consumption (Wade, 1982). It is more strongly expressed in the female than in the male hamster (Wade, 1983) and is greatly amplified on diets enriched with fat (Wade, 1983; Wade and Bartness, 1984a).

Photoperiodic enhancement of fat deposition in the hamster seems to be mediated in part by reduced energy expenditure. Thus, resting energy expenditure is about 11% lower in short-day hamsters fed fat-enriched diet, but their rate of fat deposition is increased by about 200% (Wade, 1982) suggesting the operation of physiological mechanisms in photoperiodic stimulation of fat synthesis in this species. Cold exposure is more effective than voluntary wheelrunning in preventing photoperiodic fattening in hamsters provided with fat-enriched diets (Bartness et al., 1984).

Thus, hamsters seem to have environmentally responsive mechanisms for fat storage and utilization. A decrease in daylength promotes fat storage especially in the presence of fat-rich diets (Wade, 1982, 1983) and in parallel with declining titers of testosterone (Wade, 1983). Photoperiod-induced fattening is associated with increases in the size of brown adipose tissue and in its capacity for NE-induced (and cold-induced) thermogenesis (Wade, 1982). Cold exposure selectively taps the enlarged adipose stores and the accumulated body fat protects the hamster from cold-induced losses of lean body mass (Bartness *et al.*, 1984). Likewise, reduction in estradiol titers that characterizes different phases of female hamster reproductive biology and accompanies seasonal reduction in daylight, promotes selection of dietary fat (Fleming and Miceli, 1983; Miceli and Fleming, 1983) and fat synthesis (Wade, 1983; Bartness and Wade, 1984). Such environmentally directed and hormonally controlled changes in fat synthesis anticipate the circumstances of increased energy expenditure during lactation and exposure to cold.

A hormonal and neural basis of photoperiodic fattening in golden hamsters is incompletely understood. It involves a direct photoperiodic influence over PRL (Widmaier and Campbell, 1981b) and thyroid hormone release (Vriend *et al.*, 1979) as well as the indirect influence of photoperiod over fat deposition (Bartness and Wade, 1984) and over PRL (Widmaier and Campbell, 1981b) and thyroid hormone release that requires the mediation of pineal hormones (Hoffman *et al.*, 1982; Steger *et al.*, 1982; Vaughan *et al.*, 1982; Vriend, 1983; Vriend and Reiter, 1977; Vriend *et al.*, 1977, 1979). The role of PRL and thyroid hormones in stimulating fat mobilization and inducing fat loss was discussed in the preceding section. The hamster thus appears to be equipped with complementary mechanisms favoring fat deposition, one involving direct photoperiodic influence over the neuroendocrine control of PRL and thyroid hormone release and the other involving photoperiodic stimulation of pineal hormone release and mediated by the effects of pineal melatonin on gonadal hormone release.

A remarkable lack of congruence in the photoperiodic effects on body weight (fat) is found in the closely related Siberian hamsters in whom decreasing daylight promotes weight and fat losses (Wade and Bartness, 1984b).

3.2. Ambient Temperature

Hamsters are capable of striking physiological and modest behavioral responses to variable ambient temperature that serve to defend body temperature. After prolonged exposure to extreme ambient temperature, hamsters extend the range and intensity of their physiological adjustments.

In the cold, hamsters either defend normal body temperature (38 to 38.5 °C) or hibernate and maintain reduced body temperature (5 to 6 °C). The thermoregulatory response to cold has metabolic, cardiovascular, respiratory, endocrine, and behavioral components. Nonacclimated hamsters show proportional increases in metabolic rate at temperatures between the thermoneutral zone (25 °C) and -15 °C and cold-adapted hamsters between 25 °C and -30 °C (Fig. 5). When the exposure to cold exceeds the hamster's capacity to maintain stable body temperature, it will increase its heat production sixfold, double its heart rate, and triple its respiratory rate as its core temperature declines between 38 and 32 °C (Adolph and Lawrow, 1951). Hamsters will reduce their heat loss through selective vasoconstriction which produces a temperature gradient of

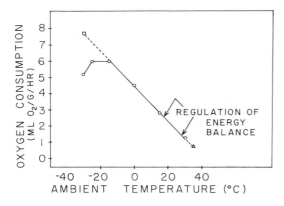

Figure 5. Metabolic responses of hamsters to changes in ambient temperature. Normal body fat levels are maintained between the thermoneutral zone (25 °C) and 15 to 20 °C. At lower and higher ambient temperatures hamsters lose body fat. Data are from animals adapted to 22 °C (open circles), 6 °C (open square), or 35 °C (open triangle). [Redrawn from Cassuto (1958); Kodama and Pace (1964); and Pohl (1965).]

several degrees between the warmer body core and cooler skin surface and between the warmer chest and head region and cooler posterior parts of the body (Pohl, 1965). Hamsters also reduce heat loss by piloerection and by assuming rounded posture that reduces their surface area (Adolph and Lawrow, 1951). Heat production is also accomplished through nonshivering thermogenesis (Jansky, 1973) and at temperatures below 15 °C in noncold-adapted animals through shivering as well. In cold-adapted animals, shivering starts at temperatures below 0 °C (Pohl, 1965). Nonshivering thermogenesis results from sympathetic (Young *et al.,* 1982) and endocrine stimulation of heat production in the brown adipose tissue (Smith and Horwitz, 1969), muscle, and liver (Jansky, 1973). Thyroid hormones and catecholamines increase nonshivering thermogenesis in hamsters (Cassuto and Amit, 1968; Cassuto *et al.,* 1970; Nedergaard and Lindberg, 1979; Rabi and Cassuto, 1976a,b) presumably by increasing activity of mitochondrial oxidative enzymes in liver and muscle (Chaffee *et al.,* 1961) and by reduced coupling of heat production to oxidative phosphorylation in the brown adipose tissue (Nicholls, 1979).

Thermoregulatory response to cold in hamsters involves increased mobilization of metabolic fuels through glycogenolysis, gluconeogenesis (Rabi *et al.,* 1977), and lipolysis (Denyes and Baumber, 1965; Denyes and Carter, 1961; Minor *et al.,* 1973). Hamsters maintain stable body fat levels at ambient temperatures between 30 and 15 °C (Fig. 5) (Kodama and Pace, 1964). When ambient temperature falls below 15 °C threshold, the rate of lipid mobilization exceeds the rate of lipid synthesis in proportion to the fall in ambient temperature (Fig. 6). At any given temperature, stable levels of reduced body fatness are achieved within a week and maintained thereafter (Kodama and Pace, 1964). In addition, at temperatures below 15 °C hamster body fat becomes unsaturated due to an increase in the proportion of oleic acid (18 : 1) and decrease in the proportion of palmitic acid (16 : 0) (Fawcett and Lyman, 1954; Kodama and Pace, 1963, 1964). The change in composition of fatty acids is thought to reflect increased utilization of saturated fatty acids in cold-exposed hamsters. Although semistarvation and administration of thyroid hormones induces a similar change in the saturation of hamster body fat, fat loss accounts for relatively greater proportion of weight loss (72%)

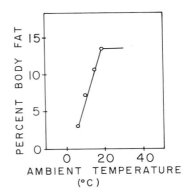

Figure 6. Effect of ambient temperature on hamster body fat.
[Redrawn from Kodama and Pace (1964).]

in cold-exposed than in starved (28%) or T$_3$-treated hamsters (34%) (Kodama and Pace, 1964).

Cold-induced lipolysis is mediated by increased sympathetic activation of white adipose tissue (Jones and Musacchian, 1976) or by its stimulation by increasing titers of ACTH (Rudman and Shank, 1966), thyroid hormones (Bauman *et al.,* 1968; Cassuto *et al.,* 1970; Rabi *et al.,* 1977), and catecholamines (Guidicelli *et al.,* 1977; Hissa *et al.,* 1974; Rabi *et al.,* 1977). Hamster white adipose tissue is as sensitive to the lipolytic action of ACTH as is the rat adipose tissue (0.01 to 0.1 μg/ml) (Rudman and Shank, 1966), but is ten times less responsive (0.01 μg/ml) than the rat adipose tissue (0.001 μg/ml to 0.001 g/ml) to the antilipolytic actions of insulin (Rudman and Shank, 1966). In contrast to the rat, and as in man, hamster white adipose tissue has both α- and β-adrenergic receptors (Aktories *et al.,* 1980, 1981; Jakobs and Aktories, 1981; Pecquery *et al.,* 1979; Pecquery and Guiducelli, 1980) that participate in the activation of adenylate cyclase and lipolysis.

Thermoregulatory response to cold in the hamster involves a significant increase in food and water consumption (Fig. 7) (Adolph and Lawrow, 1951; Bartness *et al.,* 1984; Bauman *et al.,* 1968; Lyman, 1954; Minor *et al.,* 1973).

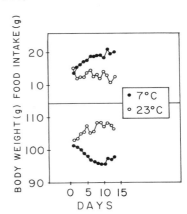

Figure 7. Changes in food consumption and body weight in hamsters maintained at 23 °C (open circles) or 7 °C (solid circles). [Redrawn from Minor *et al.* (1973).]

Energy consumption, however, does not match the energy expenditure in the cold-exposed hamster. Weight losses consistently accompany cold exposure and range between 0.2% and 0.5% per day at temperatures between 5 and 10 °C (Lyman, 1948; Minor et al., 1973; Panuska and Wade, 1958). Thus, the thermoregulating hamster appears to replace only about one third as much energy as it loses in efforts to preserve constant body temperature.

If given the opportunity to run in the activity wheels, cold-exposed hamsters do not display a distinct change in behavior from what is seen at 20 to 23 °C: there are reports of increases, no change, and slight decreases in the levels of running activity (Adolph and Lawrow, 1951; Lyman, 1954; Rowland, 1984). In such animals, fat synthesis in the brown adipose tissue is greatly enhanced, highlighting its thermoregulatory role, whereas lipogenesis in the liver and white adipose tissue is reduced reflecting the prevailing influence of fuel mobilization (Rowland, 1984).

The hibernating response to cold temperature has not received much systematic attention. A significant number of hamsters are reported to die rather than enter hibernation when exposed to low ambient temperatures (Adolph and Lawrow, 1951; Lyman, 1954). Presence of an activity wheel and presence of powdered diet that precludes hoarding are both said to delay induction of hibernation (Lyman, 1954). Hamster hibernation is characterized by episodes of inactivity and reduced body temperature interrupted by awakening during which hamster eats and drinks. Weight declines throughout the hibernatory period (Lyman, 1948, 1954; Panuska and Wade, 1958).

Prolonged exposure to cold leads to metabolic and endocrine adaptations that increase hamster's ability to withstand cold. Adaptations to chronic cold involve increases in the weight of brown adipose tissue and liver (Chaffee et al., 1962; Rabi and Cassuto, 1976a), increased mitochondrial respiration in brown adipose tissue and liver (Chaffee et al., 1961), greater thermogenic response to norepinephrine (Vybiral and Jansky, 1974, but also see conflicting evidence from Nedergaard, 1982), increased rate of lipolysis (Rabi et al., 1977; Sicart et al., 1978) due to increased sympathetic tone and higher titers of thyroid hormones and catecholamines, and increased unsaturation of fatty acids (Minor et al., 1973; Sicart et al., 1978). Cold-adapted hamsters are less able to withstand high ambient temperatures than warm-adapted hamsters (Cassuto and Chaffee, 1966).

In a hot environment, hamsters increase fluid consumption and reduce food consumption (Chayoth and Cassuto, 1971). Prolonged exposure to 35 °C leads to weight loss and reduction in liver weight (Cassuto and Chaffee, 1966). Brown adipose tissue shows reductions in mitochondrial protein and fat-free mass (Rabi and Cassuto, 1976a). Heat-acclimated hamsters display reduced basal heat production (Cassuto and Amit, 1968; Cassuto, 1968; Chayoth and Cassuto, 1971b), reduced mitochondrial respiration in the BAT (Rabi and Cassuto, 1976a), reduced rate of lipolysis (Rabi et al., 1977), increased glycogen synthesis in liver due to decreased activity of glucose 6-phosphatase (Chayoth and Cassuto, 1971a,b), and decreased magnitude of response to epinephrine and triiodothyronine (Cassuto and Amit, 1968). Several of these metabolic changes can be reversed by administration of triiodothyronine. Thus T_3 administration increases liver glycogenolysis (Cassuto *et al.*, 1974), increases activity of glucose-6-phosphatase (Cassuto *et al.*, 1970), increases the mass of BAT (Cassuto *et al.*, 1970), and increases mitochondrial respiration in BAT and liver (Rabi *et al.*, 1977; Cassuto and Amit, 1968). Heat-adapted hamsters are less able to withstand cold than animals maintained at 20 to 23 °C or cold-adapted hamsters. Thus adaptation to heat in the hamster appears to involve reduced activation of sympathetic nervous system (Jones and Musacchia, 1976), reduced endocrine thermogenic responses, as well as reduced food consumption.

3.3. Fast and Glucoprivation

The prevailing view is that animals initiate a meal in response to a central or peripheral signal of reduced nutrient availability and that a depletion of body energy reserves facilitates the frequency and size of subsequent meals. Hamster responses to fasting challenge both of the implicit premises of this view: that meal initiation is reactive rather than anticipatory and that the compensation for the energy deficit has to be behavioral, i.e., an increase in the rate of food consumption.

3.3.1. Hamster Responses to Energy Deficit

Hamsters subjected to a variety of regimens of food deprivation do not alter the amount or pattern of feeding from that observed in nondeprived state. They do, however, increase food hoarding. Thus, when their access to food is restricted to 2 hr per day, hamsters consume between 10% and 15% of their normal daily food intake (Kutscher, 1969), or the amount they would have consumed in 2 hr had their access to food been continuous. A hamster will normally consume about 25 Cal/kg in one meal during a 2-hr period whether it feeds on Purina chow pellet or sweet condensed milk (diluted 3 : 1 with water). Hamsters given 2-hr per day access to liquid diet tend to eat a slightly smaller meal (16 Cal/kg) than the *ad libitum*-fed hamsters (Rowland, 1982).

When they are alternately fed and deprived of food every 12 hr, hamsters will consume a quantity of food that is appropriate for a 12-hr period (Borer *et al.*, 1979c; Silverman and Zucker, 1976) and their behavior will not be modified by continued weight loss over period of days. The same type of response is observed whether the animals are fed during the 12 hr of light or during the 12 hr of dark (Silverman and Zucker, 1976).

When they are alternately fed and deprived of food every 24 hr (and occasionally fed for 24 hr after a 36-hr fast), hamsters consume a quantity of food that is appropriate for a 24-hr period and continue to lose weight over a period of days (Borer *et al.*, 1979c; Silverman and Zucker, 1967; Šimek, 1968, 1969, 1980; Šimek and Petrasék, 1974). When such intermittent feeding is extended for many weeks, the food intake is normal or subnormal during the first five weeks, but increases by about 35% during the subsequent 10 to 15 weeks (Šimek, 1968, 1969, 1980; Šimek and Petrasék, 1974). The intermittent schedule that most effectively elicits hyperphagia consists of alternating two-week periods of more severe (food available during two days per week) and less severe (food available during three days per week) regimens of restriction (Šimek,

1969). Such intermittent feeding schedule also induces within 15 to 20 weeks hypertrophy of forestomach (110%), stomach, and small intestine (60% to 85%; Šimek, 1969).

If one extends the length of hamster's normal 2-hr intermeal interval (IMI) to 5, 7, 10, or 11 hr, hamster's food consumption is proportional to the number of hours of access to food and is inversely proportional to the duration of IMIs (Borer et al., 1979c). Hamsters will consume 1 to 1.3 g of food during a 1-hr exposure to food. Thus, since 5-hr, 7-hr, and 11-hr IMIs will allow 4-, 3-, and 2-hourly exposures to food, respectively, hamsters consume 4.9, 3.8, and 2.3 g/day under such feeding schedules (Borer et al., 1979c). A second contiguous hour of exposure to food does not increase the consumption of food beyond intake incurred during the first hour, demonstrating the operation of an obligatory 2-hr fast before the hamster will initiate another meal. The hamster adheres to a rigid pattern of about 12 daily meals separated by 2-hr IMIs. This feeding pattern is unaffected by either acute fast or profound depletion of body energy reserves (Borer et al., 1979c).

The principal cause of weight loss in hamsters maintained on relatively modest deprivation schedules such as 5-hr IMIs is their failure to increase their food consumption by adding meals or increasing their size. As a consequence, hamsters subjected to a 5-hr feeding schedule, ingest a little less than one half of their ad libitum energy intake (Borer et al., 1979c). A secondary reason for their failure to maintain body energy balance, is a pattern of deficient insulin and lipogenic enzyme responses to feeding. Thus, extension of IMI from the preferred 2-hr ones to 5-hr, reduces plasma insulin concentration by 18%, blocks the postprandial peak of insulin release, reduces the activity of hepatic lipogenic enzyme fatty acid synthetase (FAS) by 30%, but raises the activity of hepatic malic enzyme by 60% (Borer et al., 1985). In the face of reduced energy intake, and compromised insulin release and stimulation of lipogenesis, hamster's preferred defense mechanism, reduction in energy expenditure, is quantitatively inadequate (see below).

These findings raise two questions. Do hamsters show normal endocrine and metabolic changes associated with food deprivation? Since they do not increase the size and frequency of meals, by what means do hamsters keep in positive energy balance and recover from weight deficits?

During acute food deprivation, hamsters display a pattern of endocrine and metabolic changes that are typical of mammals in general and of rodents in particular (Table 5). A 50% decline in liver glycogen content and serum insulin concentration, a decrease in blood glucose concentration, and a significant rise in serum free fatty acids and ketones all occur within 6 to 8 hr of acute starvation (Rowland, 1982, 1983).

Hamsters regulate their energy balance and recover from weight deficit by a physiological rather than ingestive mechanism. The prerequisite for maintenance of energy balance and for recovery from weight loss is that hamsters be allowed continuous access to food and their preferred 2-hr IMIs. If this condition is met, hamsters will consume normal predeprivation quantities of food and gain weight at rates that are proportional to the magnitude of their weight deficit (Fig. 8). Thus for each 10% of weight deficit, hamster weight gain increases by 0.9 g.

Hamsters can modify their rate of weight recovery in accordance with the magni-

Table 5. Endocrine and Metabolic Changes in Hamsters Subjected to Acute Energy Deficit

Variable	Ad libitum value	Nature of change	Magnitude of change %	Time to 50% change	Reference
Serum insulin (ng/ml)	2–3	↓	23	5 hr	Borer *et al.*, 1979
Glucose mg/100 ml	100–120	↓	24	4 hr	Borer *et al.*, 1979
FFA nM/ml	17	↑	550	6 hr	Borer *et al.*, 1979
Ketones nM/ml	37	↑	154	4.5 hr	Borer *et al.*, 1979
Liver glycogen g/100 g	2.2	↓	100	6 hr	Rowland, 1982
Stomach weight (g)	2.6	↓	50	3 hr	Rowland, 1982

tude of their energy deficit by reducing their rate of energy expenditure (Borer *et al.*, 1985). A reduction in resting metabolic rate of about 0.06 ml O_2/g·h for each percent of weight loss, allows the hamster that is 10% underweight to reduce its energy expenditure by 25%, and the hamster that is 20% underweight, by about 50%. The respective energy savings of about 9 and 17.5 Kcal should allow the 10% and 20% underweight hamsters fat resynthesis at the rates of 1 and 2.3 g/day, which is of the order of magnitude actually observed (Fig. 8). A 75% reduction in locomotion also aids underweight hamsters in recovering from energy deficit without hyperphagia, as does an increase in the activity rate of hepatic FAS. The magnitude of increase in the FAS activity is modest (less than 100% increase after 20% weight loss) compared to increases in lipogenic enzyme activities reported for intermittently-fed rats (De Bont *et al.*, 1975). Thus, increases in the rate of fat synthesis represent, at best, only an auxilliary mechanism for recovery from energy deficit in the hamster. Brown adipose tissue does not contribute to the reduction in energy expenditure in underweight hamsters, but

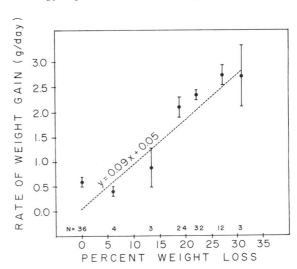

Figure 8. Relationship between magnitude of weight loss and rate of compensatory weight gain during first seven days after return of ad libitum food. [Reproduced from Borer *et al.* (1979c) with permission from the American Physiological Society.]

actually adds to it through increases in heat production which are proportional to the magnitude of energy deficit (Borer et al., 1985). Brown adipose tissue thus appears to serve as an auxilliary body warmer in a fuel deficient hamster, and not as a component of energy regulatory mechanism.

Hamsters maintained on intermittent access to food over long periods of time, show evidence of metabolic adaptations to such conditions. In such animals, liver glycogen (Šimek, 1980) and liver lipid content (Šimek, 1975) is greater than in ad-libitum fed animals suggesting an increase in the efficiency of fuel storage. Other manifestations of both increased fat synthesis and fat utilization in chronically intermittently starved hamsters are an increased respiratory quotient and heat production on meal ingestion (Šimek, 1975), increased oxygen consumption by liver and kidney homogenates (Šimek, 1974), and increased activity of liver succinic dehydrogenase (Šimek, 1968).

An exception to the postfast feeding response of hamsters just described involves hamster feeding behavior with food rich in lipid. Although hamsters maintain a normal daily energy intake when they are offered a high-fat in addition to a low-fat diet choice, they will show an 80% preference for a high-fat dietary choice and in effect modify their diet composition from a high-carbohydrate (55%) low-fat (4% to 6%) one to a high-fat (50%) low-carbohydrate (19%) one (Borer, al. 1974; Borer and Kooi, 1975). With such a dietary choice, hamsters display a 50% increase in food consumption while they recover from a weight deficit (Borer and Kooi, 1975; Borer et al., 1979c). The basis for this dietary influence on postfast feeding response in hamsters is not known.

Experiments with various regimens of protracted food restriction demonstrate that the hamster has a limited ability to recover from weight losses and catch up to weight levels of nonaeprived controls (Borer and Kelch, 1978; Borer and Kooi, 1975; Silverman and Zucker, 1976). In part this stems from the hamster's failure to make up for arrest of the slow continuous somatic growth throughout the period of deprivation (Borer and Kelch, 1978). In part this failure appears to be related to excessively long IMIs. Thus, least successful weight recovery was noted in hamsters subjected to 24-hr IMIs and most successful weight recovery occurred in hamsters given daily restricted food ration (Borer et al., 1979c). We hypothesize that excessively long IMIs may have a deleterious effect on the hamster's capacity for weight recovery by inducing starvation anorexia (Hamilton, 1969), starvation diabetes (Feldman and Lebowitz, 1973; Feldman et al., 1979), or by decimating the bacteria and protozoa in the forestomach and cecum as is known to occur during prolonged starvation in ruminants (Meiske et al., 1958).

In designing studies involving food deprivation in hamsters, one should be mindful of the limited size of their body fat depot and of the fact that starvation leads to catabolism of body compartments other than fat in hamsters (Kodama and Pace, 1964). Figure 9 illustrates the relationship between the magnitude of weight loss and the percentage of body fat in female hamsters maintained on Purina Formulab chow. When this group of 17 females lost 20% of their starting weight of 122 g, there was an almost complete depletion of their body fat. Body weight losses in excess of 22% are frequently associated with signs of severely compromised energy production as manifested by hypothermia, unconsciousness, and whole-body tremors (Borer et al., 1979c). Thus, by

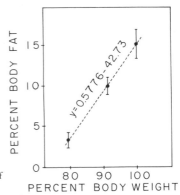

Figure 9. The relationship between body weight and percentage of body fat. [Data from Borer *et al.* (1981).]

virtue of its reliance on physiological mechanisms of weight recovery and on food hoarding, its modest store of lipid, and its limited ability to withstand long IMIs, hamsters should not be subjected to extreme food deprivations. They probably do not encounter such deprivations in their natural environment, and do not appear behaviorally or physiologically equipped to deal with them.

3.3.2. Hamster Responses to Glucoprivation

Administration of insulin, 2-deoxy-D-glucose (2-DG) and 5-thioglucose (5-TG) has been used as a dianostic test for the glucoprivic nature of feeding. Insulin is assumed to produce glucoprivation by facilitating the removal of glucose and other nutrients from the blood and by favoring their sequestration into the body energy stores or lean body mass, whereas 2-DG and 5-TG competitively inhibit the glycolytic pathway. In the rat and several other mammals, administration of both agents facilitates feeding (Epstein *et al.*, 1975), but 2-DG is ineffective in gerbils (Rowland, 1978) and cats (Jalowiec *et al.*, 1973).

In the hamster, administration of 2-DG at doses between 50 and 750 mg/kg has no effect on food consumption (DiBattista, 1982; Rither and Balch, 1978; Rowland, 1983; Sclafani and Eisenstadt, 1980; Silverman, 1978), but produces significant hyperglycemia (200% to 400% increase in blood sugar concentration) (DiBattista, 1982; Ritter and Balch, 1978) and in the concentration of free fatty acids (DiBattista, 1982). At doses of 1000 and 1500 mg/kg (Ritter and Balch, 1978; Rowland, 1978; Sclafani and Eisenstadt, 1980), 2-DG leads to a 36% reduction in food intake, lethargy, and other signs of debilitation along with hyperglycemia and a rise in serum ketones (Rowland, 1983). Likewise, administration of 5-TG in the dose range of 100 to 500 mg/kg has no effect on hamster food consumption within 5 hr following drug administration, but is accompanied with significant increases in blood sugar and free fatty acid concentrations (DiBattista, 1982).

Administration of short-acting insulin Iletin at doses between 20 and 100 U/kg leads to significant hypoglycemia (60% reduction blood sugar level) and to a 0.7 to 1.1

g increase in food consumption during a 5-hr postinjection period that is not dose dependent (DeBattista, 1982; Ritter and Balch, 1978; Rowland, 1978, 1983). Administration of long-acting PZ insulin at 25 U/kg induces significant hypoglycemia (blood glucose level of 48 mg/100 ml), a 63% increase in meal frequency, most of it during the dark, and a 30% increase in daily food consumption (Rowland, 1978). At a dose of 80 U/kg of PZ insulin, hamsters increase their daily food consumption by about 45% (Ritter and Balch, 1978). Combination of food deprivation and insulin (10 to 50 U/kg, DiBattista, 1983; 100 U/kg, Rowland, 1983) or 2-DG (1 g/kg, Rowland, 1983) is again ineffective in increasing food consumption in hamsters despite the robust physiological responses of fuel mobilization and fuel storage, respectively, in response to these manipulations.

These findings prompt two conclusions. First, hamsters detect glucoprivation induced by 2-DG and 5-TG and respond to it by sympathoadrenal activation of gluconeogensis and glycogenolysis, but not by a change in feeding behavior. And second, the feeding response to insulin administration is stimulated by some other than a glucoprivic action of this hormone since there is a lack of congruence between the feeding responses to 2-DG, 5-TG, and insulin, absence of a dose-dependent response to insulin, and a relatively long delay (4 to 6 hr) between the administration of insulin and the appearance of hyperphagia.

3.4. Physical Activity

Hamsters' propensity for unusually high levels of voluntary running on spinning or rotating activity devices compared with other mammals (Borer, 1980) can be used to assess the role of muscular energy expenditure in feeding and energy regulation in this species. A mature female hamster (120 g) will run between 11 and 16 km on an activity disk or in a rotating wheel each night, whereas a mature female rat (300 g) will run between 0.2 and 4.6 km (Borer, 1980). Such high levels of running activity are associated with a 54% increase in daily energy consumption during the first week of exercise (57.5% increase during the active period and a 36.5% increase during rest (Moffatt, 1984). Yet, there is no evidence that the energy cost of exercise provides a direct stimulus for increased food consumption. During the first week of exercise, active hamsters consume normal (Borer, 1974) or slightly lower (−8%) (Moffatt, 1984) quantities of food and undergo a 24% depletion of body fat stores. From the second week on, voluntary running is associated with a 30% to 40% increase in daily food consumption (Borer, 1974; Borer and Kelch, 1978; Moffatt, 1984; Tsai et al., 1981) even though body fat content continues to decline throughout four weeks of running to a 55% deficit (Fig. 10). It is not likely that body fat depletion provides a stimulus for increased food consumption either, since it is entirely without an effect in food restricted hamsters. Instead it is most probable that an alteration in endocrine and metabolic function established by several days of voluntary running provides the necessary stimulus for increased food consumption. This alteration does not depend on exercise-induced acceleration of somatic growth (Borer and Kuhns, 1977; Borer and Kelch, 1978), because exercise induces hyperphagia in hypophysectomized hamsters in

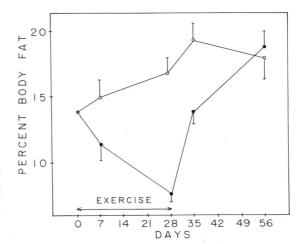

Figure 10. Body fat changes in female hamsters exposed to voluntary disk running (arrow, solid circles) or sedentary existence (open circles). Each circle represents data from seven animals. [Data from Moffatt (1984).]

proportion to the level of physical activity (Fig. 11) (Browne and Borer, 1978). Food consumption in exercising hamsters is characterized by increased meal frequency, decreased meal size, significant exercise-associated depletion of liver glycogen, and enhanced diurnal lipogenesis in liver and white and brown adipose tissues (Rowland, 1984). Thus it is most probable that exercise shortens the IMIs by causing a more rapid dissipation of stimuli that contribute to the usual 2-hr obligatory postprandial fast in hamsters.

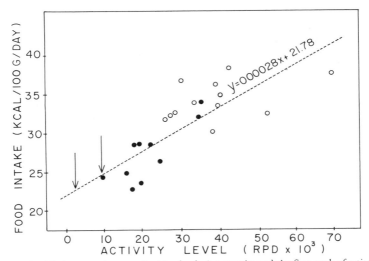

Figure 11. Relationship between the mean food intake during exercise and the first week of retirement and mean activity levels in hypophysectomized hamsters (solid circles) and hamsters with intact pituitaries (open circles). [Reproduced from Browne and Borer (1978) with permission from Pergamon Press.]

3.5. Variable Caloric Density and Texture of Food

Hamsters demonstrate a clear-cut ability to maintain constant caloric intake when presented with food stuffs of different caloric density. When their dietary choice consists of Purina pellets and sunflower seeds, sedentary hamsters display a strong preference (80%) for seeds, whereas exercising hamsters consume equal proportions of the two foodstuffs. Sedentary hamsters consume about 30 Cal/100 g per day and exercising hamsters about 50 Cal/100 g per day whether they are feeding on pellets alone or on a choice of pellets and seeds (Borer, 1974). Daily caloric intake remains the same whether hamsters consume Purina chow containing 3.4 to 3.6 Cal/g or sweet condensed milk (diluted 3 : 1) containing 1.2 Cal/ml (Rowland, 1982). Hamsters maintain constant caloric intake and double food consumption when their liquid diet is diluted 50% with water (Silverman and Zucker, 1976) or when they are presented with palatable 32% sucrose solution (Wade, 1982). Compensation for caloric dilution of liquid diet is accomplished within 24 hr. Compensation is considerably slower (4 days) and less complete (64% increase in food consumption) when the powdered diet is 50% diluted with kaolin (Silverman and Zucker, 1976).

Besides this ability for behavioral compensation for caloric dilution, hamsters also have an interesting physiological response to high-fat diets. When they have access to a high-fat diet, hamsters display an increase in the rate of weight gain and fat synthesis without an attendant increase in daily caloric intake. Thus hamsters with a choice of sunflower seeds and pellets gain weight between three and five times faster than hamsters maintained on a diet of pellets only, yet over a period of several weeks such animals consume fewer than the 30 Cal/100 g per day eaten by hamsters maintained on chow (Borer, 1974).

Wade (1982, 1983) has investigated the basis of this dietary obesity in hamsters fed a diet containing 15% protein, 36% carbohydrate, and 36% fat. Development of obesity takes place in hamsters eating normal quantities of food and appears to be mediated inpart by a 9% to 16% reduction in resting oxygen consumption and by other, as yet undetermined process. Although high-fat diet stimulates development of BAT mass and its thermogenic capacity, actual heat production by the BAT of fat-diet fed hamsters may be reduced as suggested by a 58% reduction in the lipoprotein lipase activity in this tissue (Wade, 1982). The role of photoperiod and photoperiod-induced changes in prolactin, gonadal, and thyroid hormones in dietary obesity in hamsters was discussed in a previous section.

Obesity also develops, without acceleration of somatic growth, when hamsters are fed a diet with a 39% rather than about 20% protein content (Borer et al., 1979d). Although the role of texture in hamster feeding and weight regulation has not been rigorously examined, results from separate studies where powdered or pelleted diets were used suggest that hamsters gain weight significantly faster (0.8 g/day, Borer et al., 1979d) on a powdered than on a pelleted diet (0.1 to 0.3 g/day) (Borer, 1974). Furthermore, admixture of water to powdered diet increases its acceptance by hamsters. Thus, female hamsters weighing between 105 and 120 g will show large and sustained weight gains on a diet of wet mash (K. T. Borer, 1974, unpublished data). The increase in body weight is between 9 and 10 g on day 1, most probably reflecting gut filling. However, when wet mash was available during an 8- and 15-day period, the rate of weight gain increased to 0.8 to 0.9 g/day from 0.2 g/day on pelleted diet. On removal of wet mash, there was a 4-g weight loss on day 1, rapid weight gain for another ten days (0.85 g/day), followed by a rapid weight loss (0.6 g/day) for the next 48 days to a prewet mash body weight. Thirteen-day exposure to activity disks blocked rapid weight gains stimulated by wet mash (weight change −0.4g/day).

3.6. Synopsis

The basic design of hamster mechanisms of feeding and weight regulation is anticipatory of momentary and seasonal energy shortages. The hamster's propensity to hoard food and its rigid and regular two-hourly rhythm of feeding seem to complement this rodent's inability to maintain body weight and accumulate body fat if its IMIs are extended beyond 2 hr. The hamster's basic response to energy shortage is to reduce energy expenditure and increase the efficiency of energy storage rather than to increase food consumption. Where increases in food consumption occur, as in cold exposure and exercise, they are not adequate to counteract the extent of lipid mobilization and body fat loss that these stimuli induce and appear to act as an emergency back-up system rather than as an adequate compensatory mechanism. Light has no significant circadian effect on feeding, but has important seasonal modulatory influence over body fat. Reduction in daylength promotes fat synthesis and enhances development of obesity, and this is amplified when the animals have access to high-fat diets. The latter two characteristics thus may ensure adequate filling of body fat stores and anticipates seasonal energy shortages. No set of simple unifying principles underlying feeding in hamsters can be discerned. Neither glucoprivation, reduction body fat content, or increased energy expenditure will promptly and consistently increase daily food consumption, meal frequency, or meal size in hamsters. Hamster feeding gives the appearance of a centrally programmed ultradian rhythm of standard-sized meals, the spacing of which may be imposed in part by a strong peripheral negative feedback. We suspect that mechanical, chemical, or hormonal signals associated with meal processing constitute this putative peripheral feedback, and that these signals are dissipated more rapidly in those circumstances (cold, exercise, insulin injections) where increased meal frequency and food consumption are found. Mechanisms controlling feeding behavior appear to be integrated with mechanisms controlling energy storage, and it is the latter rather than the former that respond to energy deficits in a compensatory fashion.

4. BRAIN MECHANISMS IN HAMSTER ENERGY REGULATION

The information on the brain mechanisms underlying hamster feeding and weight regulation is limited. It consists pricipally of data from lesioning and knife-cut experiments, most of it descriptive rather than analytical and with poor anatomical and neurochemical documentation of the nature of neural damage inflicted.

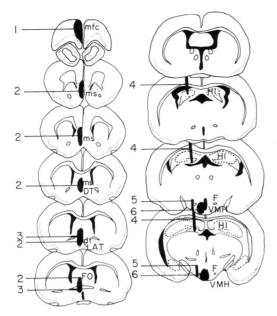

Figure 12. Neurosurgical procedures (solid outlines identified by numbers) that induce increased weight gain, reduce levels of spontaneous running, and reduce food hoarding and nest building. Abbreviations: CP, cerebral peduncles; dt, nucleus of diagonal tract; DT, diagonal band of Broca; F, columns of fornix; Fo, fornix; Hi, hippocampus; LAT, lamina terminalis; LH, lateral hypothalamus; mfc, medial frontal cortex; ML, medial lemniscus; ms, medial septal nucleus; VMH, ventromedial hypothalamic nucleus; ZI, zona incerta. [Data from: 1 = Shipley and Kolb (1977); 2 = Borer *et al.* (1977); Janzen and Bunnell (1976); Shipley and Kolb (1977); Sodetz and Bunnell (1970); 3 = Borer *et al.* (1979b); 4 = Borer *et al.* (1979a); 5 = Malsbury *et al.* (1978, 1979); Sorrentino and Reiter, (1971); 6 = Borer (1974, unpublished data); Malsbury *et al.* (1977), Marks and Miller (1972).]

4.1. *Weight Gain, Feeding, Hoarding, Nest Building, and Spontaneous Activity*

Damage in a number of diverse regions of hamster brain seems to induce a cluster of deficits that usually appear together. This implies that lesions damage a neural mechanism that serves this cluster of behaviors and physiological processes or that lesions damage independent but spatially overlapping neural systems that separately control individual functions. Thus, increased weight gain, reduced levels of spontaneous running, reduced food hoarding and nestbuilding generally appear after : damage (Fig. 12) to medial frontal cortex (Shipley and Kolb, 1977), diagonal band of Broca (Borer *et al.,* 1977; Janzen and Bunnell, 1976), rostromedial septum (Borer *et al.,* 1977; Janzen and Bunnell, 1976; Shipley and Kolb, 1977; Sodetz and Bunnell, 1970), semicircular anterior cuts with Halasz knife at the level of suprachiasmatic nucleus (Malsbury *et al.,* 1978), complete hypothalamic island containing ventromedial hypothalamus (VMH) in particular and medial basal hypothalamic nuclei (MBH) in general (Malsbury *et al.,* 1978; Sorrentino and Reiter, 1971), parasagittal cuts just lateral to the columns of the fornix at the level of anterior hypothalamus and VMH (Malsbury *et al.,* 1979), lesions of medial anterior hypothalamus and preoptic area (Malsbury *et al.,* 1977), lesions of MBH and VMH (Borer, 1974, unpublished data; Malsbury *et al.,* 1977; Marks and Miller, 1972), septohypothalamic cuts below the anterior commissure (Borer *et al.,* 1979b), hippocampal transection (Borer *et al.,* 1979a), hippocampal lesions (Shipley and Kolb, 1977), and lesions of posterolateral hypothalamus including the ascending dorsal and ventral noradrenergic bundles (Canguilhem *et al.,* 1977; Malan, 1969). The last lesion was done on the European hamster (*Cricetus cricetus*), which shows discrete seasonal cycles of weight gain and loss and may not have identical brain controls of energy stores as the golden hamster.

Four aspects of the above cluster of deficits deserve comment. First, in most instances where this issue was examined, lesions induced striking and prolonged weight gains accompanied by normal food consumption or with disproportionately small hyperphagia. Thus, following septal lesions (Borer *et al.*, 1977) there is an eightfold increase in ponderal growth during the first four postoperative weeks and only a 1.2-fold increase in food intake during ten postoperative days; following hippocampal transections, the rate of weight gain increased tenfold, whereas food consumption increased 1.3-fold during the seven postoperative weeks (Borer *et al.*, 1979a); and following septohypothalamic cuts, weight gain was five times faster and food intake was 1.1 to 1.2 times greater during the first postoperative month (Borer *et al.*, 1979b).

Second, weight increases represent a variable mixture of fat accumulation and somatic growth. Thus weight gain observed following septal lesions appears to represent acceleration of somatic growth only (Borer *et al.*, 1977, 1983), whereas hippocampal transections (Borer *et al.*, 1979a), septohypothalamic cuts (Borer *et al.*, 1979b), and VMH lesions (Borer, 1974, unpublished data) induce acceleration of somatic growth as well as development of obesity.

Third, weight increments observed after some of these procedures often depend on the presence of high levels of fat in the diet. Thus increased weight gains were observed following VMH lesions in male hamsters when they had access to preferred sunflower seeds and not when they were maintained on pellet diet (Marks and Miller, 1972), and other studies (Malsbury *et al.*, 1977) invariably used sunflower seeds in conjunction with pellets.

Finally, the hypoactivity that accompanies lesions of rostromedial septum (Borer *et al.*, 1983a), VMH (Borer, 1974, unpublished data), hippocampal transections (Borer *et al.*, 1979a), and septohypothalamic cuts (Borer *et al.*, 1979b) is related to acceleration in the rate of weight gain rather than to neurosurgically induced motor impairment or to shortages in metabolic fuels available for exercise. Thus when compelled by electric shock, septal-lesioned hamsters will run as fast and as long as neurologically intact hamsters. They have comparable concentrations of glycogen in liver and muscle, same body fat content, and similar ability to oxidize metabolic substrates in muscle homogenates (Borer *et al.*, 1983a). What distinguishes them from neurologically intact hamsters is reduced motivation to run or reduced ability of physical activity to mobilize metabolic fuels in a way that will sustain rapid and prolonged bouts of running characteristic of normal hamsters. We found no evidence of a reduction in basal sympathetic outflow as assessed by basal metabolic rate. Rather, we found evidence of reduced peripheral sensitivity to catecholamines as assessed by oxygen consumption following an injection of noradrenaline (Borer *et al.*, 1983a). Our conclusion was that the limbic forebrain circuit that encompasses fibers interconnecting hypothalamus and hippocampus is responsible for linking spontaneous running and weight regulation in a nonhomeostatic fashion. Hamsters that gain weight rapidly become hypoactive (Borer, 1980, 1982; Borer *et al.*, 1983a), whereas underweight hamsters increase their physical activity in proportion to the magnitude of their weight loss (Borer, 1982). Septal lesions abolish this negative correlation between the body energy content and the magnitude of physical energy expenditure. It would thus appear that lesions that produce the described cluster of deficits damage the mechanism in the golden hamster that is anticipatory of the seasonal energy shortages and is responsive to seasonal changes in daylength. This mechanism promotes increases in running in response to increases in daylight (Ellis and Turek, 1979; Widmaier and Campbell, 1981b) and development of obesity in response to decreases in daylight and high-fat content of the diet. By virtue of their association, this mechanism also appears to mediate hoarding and nest building, although the relationship of these two behaviors to photoperiod has not been extensively examined in the hamsters.

4.2. Weight Loss, Anorexia, and Somnolence

Destruction of three regions of hamster brain leads to rapid weight losses. There has been little effort to clarify whether such weight losses result from a disturbance in behavioral or physiological mechanisms of weight regulation. Rapid weight loss and death within ten days was noted in hamsters sustaining lesions of orbitofrontal cortex (Bunnell *et al.*, 1970; Shipley and Kolb, 1977). Likewise, lesions of basolateral amygdala (Matalka, 1967) or larger amygdalar lesions that include damage to stria terminalis (Shipley and Kolb, 1977) lead to aphagia and weight losses. Finally, lesions of subthalamic area and zona incerta lead to a period of somnolence, hypothermia, and catatonia. If animals are nursed through this period by hand-feeding wet mash and by supplying them with sunflower seeds, they will resume spontaneous food consumption.

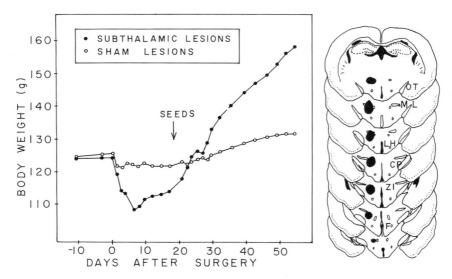

Figure 13. Body weight changes in hamsters of both sexes with electrolytic lesions of subthalamic area (open circles) (n = 6) and in sham-operated hamsters (solid circles) (n = 8, left). Arrow denotes the time sunflower seeds were added to Purina chow. Right: unilateral reconstruction of brain damage in hamsters which received subthalamic lesions bilaterally. Legend is the same as in Fig. 12. [Unpublished data, K. Borer (1974).]

Such animals display enhanced rate of weight gain (and normal levels of physical activity when activity disks are supplied) in spite of normal daily food consumption (Borer, 1974, unpublished data) (Fig. 13). There is insufficient evidence to conclude whether subthalamic lesions damage fiber pathways mediating antagonistic weight regulatory functions, or whether such lesions damage brain areas where integration of facilitatory and inhibitory weight regulatory mechanisms takes place.

4.3. Neurochemistry and Pharmacology of Feeding and Weight Regulation in Hamsters

There has been almost no work on the neurochemistry of weight regulatory mechanisms in the hamster. In the case of one segment of the circuit that suppresses somatic growth and development of obesity and that facilitates voluntary running (rostromedial septal lesions, hippocampal transections, and septohypothalamic cuts), a very similar pattern of neurochemical deficits is obtained after all three neurosurgical procedures. In each instance, there is a 30% to 60% depletion of serotonin and nor-epinephrine in the hippocampus, a 9% to 30% depletion of serotonin and nor-epinephrine in cerebral cortex, and no change in dopamine in either of the four brain regions sampled (cortex, corpus striatum, diencephalon, and hippocampus) or in nor-epinephrine and serotonin in diencephalon and corpus striatum (Borer et al., 1979e). In addition, hippocampal transection is associated with a significant depletion of acetyl-choline and somatostatin from hippocampus (Borer et al., 1982c, 1983b). Additional research will be necessary to determine whether depletion of individual neurotransmit-ters is fortuitous or contributory to the cluster of deficits seen after such limbic lesions.

High levels of voluntary activity in the hamsters appear to be mediated in part by release of endogenous opiates (Potter et al., 1983) because administration of naltrexone reduces the levels of voluntary disk activity. Furthermore, facilitation of growth hor-mone release by exercise depends on the secretion of endogenous opiates (Nicoski and Borer, 1983).

A number of pharmacological agents attenuate (piribedil 5 and 10 mg/kg, fenfluramine 5 mg/kg, sulpiride 20 mg/kg, and clonidine 0.05 and 0.1 mg/kg) or completely suppress (chlorpromazine 8 and 16 mg/kg, fenfluramine 10 mg/kg, and apomorphine 2 and 4 mg/kg) hoarding, whereas meprobamate (100 mg/kg) stimulates food ingestion by hamsters (Poignant and Rismondo, 1975).

Diurnal fluctuations in diencephalic norepinephrine and serotonin (increases in daytime) and in telencephalic DOPAC and 5HIAA (increases during nighttime) were reported by Rowland (1984).

4.4. Synopsis

Lesions of extensive areas of forebrain lead to a cluster of deficits that suggest some commonality of function: acceleration in the rate of weight gain especially on a diet rich in lipids, hoarding and nest-building deficits, reduced levels of voluntary running, and occasional obesity and hyperphagia. As this pattern of deficits appears when one damages structures at substantial distances from each other (medial frontal cortex, rostral septum, hippocampus, medial basal hypothalamus), it is evident that an extensive hypothalamic-limbic circuit underlies these energy regulatory functions. Noradrenergic, serotonergic, cholinergic, and somatostatinergic innervation of hippocampus and cerebral cortex is probably an integral component of this mechanism. It is also probable that the natural function of this circuit is to anticipate seasonal shortages of energy and to facilitate searching for food (increased voluntary running) or fattening and weight gain depending on the nature of photoperiodic signals. Lesions of orbitofrontal cortex and corticomedial amygdala lead to weight losses, whereas damage to subthalamus and zona incerta lead to somnolence and weight loss followed by accelerated rate of weight gain. Much additional work is needed to clarify the anatomy, neurochemistry, and pharmacology of the hamster weight-regulatory mechanism.

ACKNOWLEDGMENTS. Personal work described in this review could not have been done without the financial support of the Weight Watchers Foundation, National Institute of Mental Health, and the National Science Foundation, or the interactions with colleagues Robert Kelch, Neil Rowland, Lawrence Kuhns, Alan Tsai, and Richard Katz, and students Alice Kooi, Paul Stemmer, Mary White, Lynne Dolson, Sarah Holder Browne, John Hallfrisch, Nancy Peters, Janet Peugh, Arwin Mirow, Susan Kandarian, Jeanne Nichols, and Caren Potter. I also thank Kathleen O'Connor, Jayne Blemly, Mara Markovs, and Jenny Stockton for skilled technical assistance and Beverly Ballard for typing the manuscript.

REFERENCES

Adolph, E. F., and Lawrow, J. W., 1951, Acclimatization to cold air, hypothermia and heat production in the golden hamster, *Am. J. Physiol.* **166**:62–74.

Aktories, K., Schultz, G., and Jakobs, K. H., 1980, Regulation of adenylate cyclase in hamster adipocytes, *Naunyn-Schmied. Arch. Pharmacol.* **312**:167–173.

Aktories, K., Schultz, G., and Jakobs, K. H., 1981, The hamster adipocyte adenylate cyclase system. II. Regulation of enzyme stimulation, and inhibition by monovalent cations, *Biochim. Biophys. Acta* **676**:59–67.

Anderson, K. E., Kok, E., and Javitt, N., 1972, Bile acid synthesis in man: Metabolism of 7α-hydroxycholesterol-^{14}C and 26-hydroxycholesterol-H^3, *J. Clin. Invest.* **51**:112–117.

Arimura, A., Sato, H., Dupont, A., Nishi, N., and Schally, A. V., 1975, Somatostatin: Abundance of immunoreactive hormone in rat stomach and pancreas, *Science* **189**:1007–1009.

Balas, D., Senegas-Balas, F., Bertrand, C., Frexinos, J., and Ribet, A., 1980, Effects of pancreatic duct ligation on the hamster intestinal mucosa. Histological findings, *Digestion* **20**:157–167.

Banta, C. A., Warner, R. G., and Robertson, J. B., 1975, Protein nutrition of the golden hamster, *J. Nutr.* **105**:38–45.

Baranczuk, R., and Greenwald, G. S., 1974, Plasma levels of oestrogen and progesterone in pregnant and lactating hamsters, *J. Endocrinol.* **63**:125–135.

Bartness, T. J., and Wade, G., 1984, Photoperiodic control of body weight and energy metabolism in Syrian hamsters (*Mesocricetus auratus*): Role of pineal gland, melatonin, gonads and diet, *Endocrinology* **114**:492–498.

Bartness, T. J., Ruby, N. F., and Wade, G. N., 1984, Dietary obesity in exercising or cold-exposed Syrian hamsters, *Physiol. Behav.* **32**:85–90.

Bauman, T. R., Anderson, R. R., and Turner, C. W., 1968, Thyroid hormone secretion rates and food consumption of the hamster (*Mesocricetus auratus*) at 25.5°C and 4.5°C, *Gen. Comp. Endocrinol.* **10**:92–98.

Ben-Menahem, H., 1934, Notes Sur l'elevage du hamster de Syrie, *Arch. Inst. Pasteur Alger.* 12:403.

Bernardis, L. L., and Frohman, L. A., 1971, Effects of hypothalamic lesions at different loci on development of hyperinsulineuria and obesity in the weanling rat, *J. Comp. Neurol.* 141:107–116.

Berndtson, W. E., and Desjardins, C., 1974, Circulating LH and FSH levels and testicular function in hamsters during light deprivation and subsequent photoperiodic stimulation, *Endocrinology* 95:195–205.

Bond, C. R., 1945, The golden hamster (*Cricetus auratus*): Care, breeding and growth, *Physiol. Zool.* 18:52–89.

Borer, K. T., 1974, Absence of weight regulation in exercising hamsters, *Physiol. Behav.* 12:589–597.

Borer, K. T., 1980, Characteristics of growth-inducing exercise, *Physiol. Behav.* 24:713–720.

Borer, K. T., 1982, The nonhomeostatic motivation to run in the golden hamster, in: *Changing Concepts of the Nervous System* (A. R. Morrison and P. L. Strick, eds.), Academic Press, New York, pp. 539–567.

Borer, K. T., and Kelch, R. P., 1978, Increased serum growth hormone and somatic growth in exercising adult hamsters, *Am. J. Physiol.* 234:E611–E616.

Borer, K. T., and Kooi, A. A., 1975, Regulatory defense of the exercise-induced weight elevation in hamsters, *Behav. Biol.* 13:301–310.

Borer, K. T., and Kuhns, L. R., 1977, Radiographic evidence for acceleration of skeletal growth in adult hamsters by exercise, *Growth* 41:1–13.

Borer, K. T., Kelch, R. P., White, M. P., Dolson, L., and Kuhns, L. R., 1977, The role of septal area in the neuroendocrine control of growth in golden hamsters, *Neuroendocrinology* 23:133–150.

Borer, K. T., Kelch, R. P., Peugh, J., and Huseman, C., 1979a, Increased serum growth hormone and somatic growth in adult hamsters with hippocampal transections, *Neuroendocrinology* 29:22–23.

Borer, K. T., Peters, N. L., Kelch, R. P., Tsai, A. C., and Holder, S., 1979b, Contribution of growth, fatness and activity to weight disturbance following septohypothalamic cuts in adult hamsters, *J. Comp. Physiol. Psychol.* 93:907–918.

Borer, K. T., Rowland, N., Mirow, A., Borer, R. C., Jr., and Kelch, R. P., 1979c, Physiological and behavioral responses to starvation in the golden hamster, *Am. J. Physiol.* 236:E105–E112.

Borer, K. T., Hallfrisch, J., Tsai, A. C., Hallfrisch, C., and Kuhns, L. R., 1979d, The effects of exercise and dietary protein on somatic growth, body composition, and serum cholesterol in adult hamsters, *J. Nutr.* 109:222–228.

Borer, K. T., Trulson, M. E., and Kuhns, L. R., 1979e, The role of limbic system in the control of hamster growth, *Brain Res. Bull.* 4:239–247.

Borer, K. T., Campbell, C. S., Gordon, K., Jorgenson, K., and Tabor, J., 1981, Exercise reinstates estrous cycles in hamsters maintained in short photoperiods, Program, Society for Neuroscience 11th Annual Meeting, Los Angeles, California, October 18–23.

Borer, K. T., Kelch, R. P., and Corley, K., 1982a, Hamster prolactin: Physiological changes in blood and pituitary concentrations as measured by a homologous radioimmunoassay, *Neuroendocrinology* 35:13–21.

Borer, K. T., Kelch, R. P., and Hayashida, T., 1982b, Hamster growth hormone: Species specificity and physiological changes in blood and pituitary concentrations as measured by a homologous radioimmunoassay, *Neuroendocrinology* 35:349–358.

Borer, K. T., Segal, S., Vinik, A. I., and Shapiro, B., 1982c, Growth-inducing hippocampal transections affect limbic and gastrointential somatostatin and limbic choline acetyltransferase concentrations in hamsters, Program, Society for Neuroscience 12th annual meeting, Minneapolis, Minneapolis, October 31–November 5.

Borer, K. T., Allen, E. A., and Moffatt, R. V., 1982d, Not by food alone (regulation of energy balance in the hamster). Symposium, "Recent Advances in the Neuroscience of Ingestive Behavior." Program, Fifty-third annual meeting of the Eastern Psychological Association, Baltimore, Maryland, April 14–17.

Borer, K. T., Potter, C. D., and Fileccia, N., 1983a, Basis for hypoactivity that accompanies rapid weight gain in hamsters, *Physiol. Behav.* 30:389–397.

Borer, K. T., Campbell, C. S., Tabor, J., Jorgenson, K., Kandarian, S., and Gordon, L., 1983b, Exercise reverses photoperiodic anestrus in golden hamsters. *Biol. Reprod.* 29:38–47.

Borer, K. T., Shapiro, B., and Vinik, A. I., 1983b, A role for somatostatin in the control of hamster growth, *Brain Res. Bull.* 11:663–669.

Borer, K. T., Allen, E. R., Smalley, R. E., Lundell, L., and Stockton, J., 1985, Recovery from energy deficit in the golden hamsters, *Am. J. Physiol.* (in press).

Bray, G. A., 1974, Endocrine factors in the control of food intake, *Fed. Proc.* 33:1140–1145.

Browne, S. A. H., and Borer, K. T., 1978, The basis for exercise-induced hyperphagia in adult hamsters, *Physiol. Behav.* 20:553–557.

Bruce, H. M., and Hindle, E., 1934, The golden hamster *Cricetus (Mesocricetus) auratus* Waterhouse. Notes on its breeding and growth, *Proc. Zool. Soc. Lond.* 2:361–366.

Bunnell, B. N., Sodetz, F. J., and Shalloway, D. I., 1970, Amygdaloid lesions and social behavior in the golden hamster, *Physiol. Behav.* 5:153–161.

Buschmann, R. J., and Manke, D. J., 1981a, Morphometric analysis of the membranes and organelles of small intestinal enterocytes. I. Fasted Hamster, *J. Ultrastruct. Research* 76:1–14.

Buschmann, R. J., and Manke, D. J., 1981b, Morphometric analysis of the membranes and organelles of small intestinal enterocytes. II. Lipid-fed hamster, *J. Ultrastruct. Res.* 76:15–26.

Campbell, C. S., Tabor, J., and Davis, J. D., 1983, Small effect on brown adipose tissue and major effect of photoperiod on body weight in hamsters (*Mesocricetus auratus*), *Physiol. Behav.* 30:349–352.

Canguilhem, B., Schmitt, P., Mack, G., and Kempf, E., 1977, Comportement alimentaire, rhythmes circannuels ponderal et d'hibernation chez le hamster d'Europe porteur de lesions des faisceaux noradrenergiques ascendants, *Physiol. Behav.* 18:1067–1074.

Cassuto, Y., 1968, Metabolic adaptations to chronic heat exposure in the golden hamster, *Am. J. Physiol.* 214:1147–1151.

Cassuto, Y., and Amit, Y., 1968, Thyroxine and norepinephrine effects on the metabolic rates of heat-acclimated hamsters, *Endocrinology* 82:17–20.

Cassuto, Y., and Chaffee, R. J., 1966, Effects of prolonged heat exposure on the cellular metabolism of the hamster, *Am. J. Physiol.* 210:423–426.

Cassuto, Y., Chayoth, R., and Rabi, T., 1970, Thyroid hormone in heat-acclimated hamsters, *Am. J. Physiol.* 218:1287–1290.

Cassuto, Y., Chayoth, R., and Zor, V., 1974, Carbohydrate metabolism of heat-acclimated hamsters. IV. Hormonal control, *Am. J. Physiol.* 227:851–853.

Chaffee, R. R. J., Hoch, F. L., and Lyman, C. P., 1961, Mitochondrial oxidative enzymes and phosphorylations in cold exposure and hibernation, *Am. J. Physiol.* 201:29–32.

Chaffee, R. R. J., Atzet, J. E., and Kelly, K. H., 1962, Effect of multiple vs. single caging on body and organ growth during cold acclimatization, *Am. Zool.* 2:511–512.

Chayoth, R., and Cassuto, Y., 1971a, Carbohydrate metabolism in heat-acclimated hamsters. I. Control of glycogenesis in the liver, *Am. J. Physiol.* 220:1067–1070.

Chayoth, R., and Cassuto, Y., 1971b, Carbohydrate metabolism of heat-acclimated hamsters. II. Regulatory mechanisms of the intact animal, *Am. J. Physiol.* 220:1071–1073.

Curry, D. L., Bennett, L. L., and Li, A. H., 1975, Dynamics of insulin release of perfused hamster (*Mesocricetus auratus*) pancreases: Effects of hypophysectomy, bovine and human growth hormone and prolactin, *J. Endocrinol.* 68:245–251.

De Bont, A. J., Romsos, D. R., Tsai, A. C., Waterman, R. A., and Leveille, G. A., 1975, Influence of alterations in meal frequency on lipogenesis and body fat content in the rat, *Proc. Soc. Exp. Biol. Med.* 169:849–854.

Denyes, A., and Baumber, J., 1965, Comparison of serum total lipid during cold exposure in hibernating and nonhibernating mammals, *Nature (London)* 205:1063–1064.

Denyes, A., and Carter, J. D., 1961, Utilization of acetate 1-[^{14}C] by hepatic tissues from cold-exposed and hibernating hamsters, *Am. J. Physiol.* 200:1043–1046.

DiBattista, D., 1982, Effects of 5-thioglucose on feeding and glycemia in the hamster, *Physiol. Behav.* 29:803–806.

DiBattista, D., 1983, Food deprivation and insulin-induced feeding in the hamster, *Physiol. Behav.* 30:683–687.

DiBattista, D., 1984, Characteristics of insulin-induced hyperphagia in the golden hamster, *Physiol. Behav.* 32:381–387.

Di Girolamo, M., and Rudman, D., 1966, Species differences in glucose metabolism and glucose responsiveness of adipose tissue, *Am. J. Physiol.* 210:721–727.

Dinda, P. K., and Beck, I. T., 1982, Effect of ethanol on peptidases of hamster jejunal brush-border membrane, *Am. J. Physiol.* 242:G442–G447.

Dinda, P. K., Hurst, R. O., and Tibeck, I., 1979, Effect of ethanol on disaccharidases of hamster jejunal brush border membrane, *Am. J. Physiol.* 237:E68–E76.

Docke, F., 1977, A possible mechanism of the puberty-delaying effect of hippocampal lesions in female rats, *Endokrinologie* 69:258–261.

Ehle, F. E., and Warner, R. G., 1978, Nutritional implications of the hamster forestomach, *J. Nutr.* 108:1047–1053.

Ellis, G. B., and Turek, F. W., 1979, Changes in locomotor activity associated with the photoperiodic response of the testes in male golden hamsters, *J. Comp. Physiol.* 132:277–284.

Epstein, A. N., Nicolaidis, S., and Miselis, R., 1975, The glucoprivic control of food intake and glucostatic theory of feeding behavior, in: *Neural Integration of Physiological Mechanisms and Behavior* (G. J. Mogenson and F. R. Calaresu, eds.), University of Toronto Press, pp. 148–168.

Fawcett, D. W., and Lyman, C. P., 1954, The effect of low environmental temperature on the composition of depot fat in relation to hibernation, *J. Physiol.* 126:235–247.

Feder, H. H., Siegel, H., and Wade, G. N., 1974, Uptake of [6, 7-^{3}H] estradiol-17β in ovariectomized rats, guinea pigs, and hamsters: Correlation with species difference in behavioral responsiveness to estradiol, *Brain Res.* 71:93–103.

Feldman, J. M., and Lebowitz, H. E., 1973, Role of pancreatic monoamines in the impaired insulin secretion of the fasting state, *Endocrinology* 92:1469–1474.

Feldman, J. M., Henderson, J. H., and Blalock, J. A., 1979, Effect of pharmacological agents and fasting on pancreatic islet norepinephrine in the golden hamster, *Diabetologia* 17:169–174.

Fleming, A., 1978, Food intake and body weight regulation during the reproductive cycle of the golden hamster (*Mesocrieltus auratus*), *Behav. Biol.* 24:291–306.

Fleming, A. S., and Miceli, M. O., 1983, Effects of diet on feeding and body weight regulation and during pregnancy and lactation in the golden hamster (*Mesocricetus auratus*), *Behav. Neurosci.* 97:246–254.

Fox, J. E., McElligott, T. F., and Beck, I. T., 1978a, The correlation of ethanol-induced depression of glucose and water transport with morphological changes in the hamster jejunum in vivo, *Can. J. Physiol. Pharmacol.* 56:123–131.

Fox, J. E., McElligott, T. F., and Beck, I. T., 1978b, Effect of ethanol on the morphology of hamster jejunum, *Am. J. Digest. Dis.* 23:201–209.

Fox, J. E., Bourdages, R., and Beck, I. T., 1978c, Effect of ethanol on glucose and water absorption in hamster jejunum *in vivo*. Methodological problems: Anesthesia, nonabsorbable markers, and osmotic effect, *Am. J. Digest Dis.* 23:193–200.

Frohman, L. A., and Bernardis, L. L., 1968, Growth hormone and insulin levels in weanling rats with rostromedial hypothalamic lesions, *Endocrinology* 82:1125–1132.

Frohman, L. A., Bernardis, L. L., Schnatz, J. D. C., and Burek, L., 1969, Plasma insulin and triglyceride levels after hypothalamic lesions in weanling rats. *Am. J. Physiol.* 276:1496–1501.

Fujimoto, S., Hattori, J., Kimoto, K., Yamashita, S., Fujita, S., and Kawai, K., 1980, Tritrated thymidine autoradiographic study on origin and renewal of gastrin cells in antral area of hamsters, *Gastroenterology* 79:785–791.

Galeazzi, R., and Javitt, N. B., 1977, Bile acid secretion: The alternate pathway in the hamster, *J. Clin. Invest.* 60:693–701.

Gerall, A. A., and Thiel, A. R., 1975, Effects of perinatal gonadal secretions on parameters of receptivity and weight gain in hamsters, *J. Comp. Physiol. Psychol.* 89:580–589.

Giudicelli, Y., Agli, B., Brulle, D., and Nordman, R., 1977, Influence of α-adrenergic blocking agents on cyclic AMP, cyclic GMP and lipolysis in hamster white fat cells, *FEBS Lett.* 83:225–230.

Goldman, J. K., Bernardis, L. L., and Frohman, L. A., 1974, Food intake in hypothalamic obesity, *Am. J. Physiol.* 227:88–91.

Granados, H., 1951, Nutritional studies on growth and reproduction of the golden hamster (*Mesocricet auratus auratus*), *Acta Physiol. Scand.* 24 (suppl. 87):1–138.

Granneman, J. G., and Wade, G. N., 1982, Effects of photoperiod and castration on post-fast food intake and body weight gain in golden hamsters, *Physiol. Behav.* 28:847–850.

Grey, N. J., Goldring, S., and Kipnis, D. M., 1970, The effect of fasting, diet, and actinomycin D on insulin secretion in the rat, *J. Clin. Invest.* 48:881–887.

Hales, C. N., and Kennedy, G. C., 1964, Plasma glucose, non-esterified fatty acids and insulin concentration in hypothalamic-hyperphagic rats, *Biochem. J.* 90:620–624.

Hamilton, C. L., 1969, Problems of refeeding after starvation in the rat, *Ann. N. Y. Acad. Sci* 157:1004–1017.

Hamilton, J. W., and Hogan, A. G., 1944, Nutritional requirements of the Syrian hamster, *J. Nutr.* 27:213.

Han, P. W., and Frohman, L. A., 1970, Hyperinsulinemia in tube-fed hypophysectomized rats bearing hypothalamic lesions, *Am. J. Physiol.* 217:1632.

Handler, A. N., and Shepro, D., 1968, Cheek pouching technology: Uses and applications, in: *The Golden Hamster: Its Biology and Use in Medical Research*, (R. A. Hoffman, P. T. Robinson, and H. Magalhaes, eds.), The Iowa State University Press, Ames, Iowa, pp. 195–201.

Hattori, T., and Fujita, S., 1976, Tritiated thymidine autoradiographic study of cell migration and renewal in the pyloric mucosa of golden hamsters, *Cell Tissue Res.* 175:49–57.

Heldmaier, G., Steinlecher, S., Rafael, J., and Vsiansky, P., 1981, Photoperiodic control and effects of melatonin on nonshivering thermogenesis and brown adipose tissue, *Science* 212:917–919.

Helgeson, A. S., Pour, P., Lawson, T., and Grandjean, C. J., 1980a, Exocrine pancreatic secretion in the Syrian golden hamster *Mesocricetus auratus*. I. Basic values, *Comp. Biochem. Physiol.* 66A:473–477.

Helgeson, A. S., Pour, P., Lawson, T., and Grandjean, C. J., 1980b, Exocrine pancreatic secretion in the Syrian golden hamster *Mesocricetus auratus*. II. Effect of secretin and pancreozymin, *Comp. Biochem. Physiol.* 66A:479–483.

Hissa, R., Palokangas, R., and Vihko, V., 1974, Effects of propranolol, noradrenaline and insulin on fat mobilization in the golden hamster, *Comp. Gen. Pharmacol.* 5:207–212.

Ho, K. J., 1979, Circadian rhythm of cholesterol biosynthesis. Dietary regulation in the liver and small intestine of hamsters, *Int. J. Chronobiol.* 6:39–50.

Hoebel, B. G., and Teitelbaum, P., 1966, Weight regulation in normal and hypothalamic hyperphagic rats, *J. Comp. Physiol. Psychol.* 61:189–193.

Hoffman, R. A., 1983, Seasonal growth and development and the influence of the eyes and pineal gland on body weight of golden hamsters (*M. Auratus*), *Growth* 47:109–121.

Hoffman, R. A., Hester, R. J., and Towns, C., 1965, Effect of light and temperature on the endocrine system of the golden hamster (*Mesocricetus auratus* Waterhouse), *Comp. Biochem. Physiol.* 15:525–533.

Hoffman, R. A., Davidson, K., and Steinberg, K., 1982, Influence of photoperiod and temperature on weight gain, food consumption, fat pads and thyroxine in male golden hamster, *Growth* 46:150–162.

Hoover, W. H., Mannings, C. L., and Sheerin, H. W., 1969, Observations on digestion in the golden hamster, *J. Anim. Sci.* 28:349–352.

House, E. L., and Tassoni, J. P., 1957, Duration of alloxan diabetes in the hamster, *Endocrinology* 61:309–311.

Hustvedt, B. E., and Løvø, K., 1972, Correlation between hyperinsulinemia and hyperphagia in rats with rostromedial hypothalamic lesions, *Acta Physiol. Scand.* 84:29–33.

Jacobs, D. L., 1945, Food habits of the golden hamster, *J. Mammal.* 26:199.

Jakobs, K. H., and Aktories, K., 1981, The hamster adipocyte adenylate cyclase system. I. Regulation of enzyme stimulation and inhibition by manganese and magnesium ions, *Biochem. Biophys. Acta* 676:51–58.

Jalowiec, J. E., Panksepp, J., Shabelshelowitz, H., Zolovick, A. J., Stern, W. C., and Morgane, P., 1973, Suppression of feeding in cats following 2-deoxy-D glucose, *Physiol. Behav.* 10:805–807.

Jansky, L., 1973, Non-shivering thermogenesis and its thermoregulatory significance, *Biol. Rev.* 48:85–132.

Janzen, W. B., and Bunnell, B. N., 1976, Septal lesions and the recovery of function in the juvenile hamster, *Physiol. Behav.* 16:445–452.

Jewell, H. A., and Charipper, H. A., 1951, The morphology of the pancreas of the golden hamster, *Cricetus auratus*, with special reference to the histology and cytology of the islets of Langerhans, *Anat. Rec.* 111:401–415.

Jones, S. B., and Musacchia, X. J., 1976, Norepinephrine turnover in heart and spleen of 7-, 22- and 34°C-acclimated hamsters, *Am. J. Physiol.* 230:564–568.

Joseph, M. M., and Meier, A. H., 1974, Circadian component in the fattening and reproductive responses to prolactin in the hamster, *Proc. Soc. Exp. Biol. Med.* 146:1150–1155.

Kandarian, S., 1983, Exercise-induced reproductive and physiological changes in anestrus hamsters: The role of prolactin, unpublished masters thesis, University of Michigan.

Kodama, A. M., and Pace, N., 1963, Cold-dependent changes in tissue fat composition, *Fed. Proc.* 22:761–765.

Kodama, A. M., and Pace, N., 1964, Effect of environmental temperature on hamster body fat composition, *J. Appl. Physiol.* 19:863–867.

Kowalewski, K., 1969, Effect of pre-pubertal gonadectomy and treatment with sex hormones on body growth, weight of organs and skin collagen in hamsters, *Acta Endocrinol.* 61:48–56.

Kronheim, S., Berelowitz, M., and Pimstone, B. L., 1976, A radioimmunoassay for growth hormone release-inhibitory hormone: Method and quantitative tissue distribution, *Clin. Endocrinol.* 5:619–630.

Kunstyr, I., 1974, Some quantitative and qualitative aspects of the stomach microflora of the conventional rat and hamster, *Zbl. Vet. Med. A.* 21:553–561.

Kutscher, C. L., 1969, Species differences in the interaction of feeding and drinking, *Ann. N. Y. Acad. Sci.* 157:539–552.

Landau, T., and Zucker, I., 1976, Estrogenic regulation of body weight in the female rat, *Horm. Behav.* 7:29–39.

Lotter, E. C., and Woods, S. C., 1977, Injections of insulin and changes of body weight, *Physiol. Behav.* 18:293–297.

Louis-Sylvestre, J., LaRue-Achagiotis, C., and Le Magnen, J., 1980, Oral induction of the insulin hyperresponsiveness in rats with rostromedial hypothalamic lesions, *Horm. Metab. Res.* 12:671–676.

Løvø, A., and Hustvedt, B. E., 1978, Early effects of feeding upon hormonal and metabolic alterations in adult rats with rostromedial lesions, *Horm. Metab. Res.* 10:304–309.

Lyman, C. P., 1948, The oxygen consumption and temperature regulation of hibernating hamsters, *J. Exp. Zool.* 109:55–78.

Lyman, C. P., 1954, Activity, Food consumption and hoarding in hibernators, *J. Mammal.* 35: 545–552.

MacKay, E. M., Calloway, J. W., and Barnes, R. H., 1940, Hyperalimentation in normal animals produced by protamine insulin, *J. Nutr.* 20:59–66.

Malan, A., 1969, Controle hypothalamique de la thermoregulation et de ;'hibernation chez le hamster d'Europe, *Cricetus cricetus, Arch. Sci. Physiol.* 23:47–87.

Malsbury, C. W., Kow, L.-M., and Pfaff, D. W., 1977, Effects of medial hypothalamic lesions on the lordosis response and other behaviors in female hamster, *Physiol. Behav.* 19:223–237.

Malsbury, C. W., Strull, D., and Daood, J., 1978, Half-cylinder cuts anterolateral to the ventromedial nucleus reduce sexual receptivity in female golden hamsters, *Physiol. Behav.* 21:79–87.

Malsbury, C. W., Marques, D. M., and Daood, J. T., 1979, Sagittal knife cuts in the far-lateral hypothalamus reduce sexual receptivity in female hamsters, *Brain Res. Bull.* 4:833–842.

Manda, T., and Matsumoto, T., 1973a, Effect of estrogenic substances in herbage on the organ weight and body composition of intact female and orchiectomized hamsters, *Jpn. J. Zootech. Sci.* 44:11–18.

Manda, T., and Matsumoto, T., 1973b, Effect of estrogenic substances in herbage on the growth of the golden hamster, *Jpn. J. Zootech. Sci.* 44:1–10.

Manda, T., and Matsumoto, T., 1973c, Changes of organ weight and body composition of hamsters after oral administration of estrogenic substances in herbage, *Jpn. J. Zootech. Sci.* 44:367–374.

Manda, T., and Matsumoto, T., 1973d, Studies on estrogenic substances in herbage. V. Comparative test of growth promoting effect of various herbage extract on the golden hamster, *J. Jpn. Soc. Grassland Sci.* 19:389–393.

Manda, T., and Matsumoto, T., 1973e, Effect of estrogenic substances in herbage on organ weight and body composition of intact male and spayed hamsters, *Jpn. J. Zootech. Sci.* 44:97–104.

Manda, T., and Matsumoto, T., 1973f, Studies on estrogenic substances in herbage. VI. Seasonal variation of growth promoting effect of alfalfa extract on the golden hamster, *J. Jpn. Soc. Grassland Sci.* 19: 394–398.

Mangold, E., 1929, *Handbuch der Ernahrung und des Staffwechsel der Landwirtschaftlichen Nutztiere*, Bd. II., Julius Springer, Berlin, p. 321.

Marks, H. E., and Miller, C. R., 1972, Development of hypothalamic obesity in the male golden hamster (*Mesocricetus auratus*) as a function of food preference, *Psychon. Sci.* 27:263–265.

Matalka, E. S., 1967, The hoarding behavior and food intake of the hamster following hypothalamic and limbic forebrain lesions, Ph.D. dissertation, University of Florida.

McIntosh, C., Arnold, R., Bothe, E., Becker, H., Köbberling, J., and Creutzfeldt, W., 1978, Gastrointestinal somatostatin: Extraction and radioimmunoassay in different species, *Gut* 19:655–663.

Meier, A. H., 1972, Temporal synergism of prolactin and adrenal steroids in the regulation of fat storage, *Gen. Comp. Endocrinol.* (Suppl) 3:499–508.

Meier, A. H., and Davis, K. B., 1967, Diurnal variations of the fattening response to prolactin in the white-throated sparrow, *Zonotrichia albicollis*, *Gen. Comp. Endocrinol.* 8:110–114.

Meier, A. H., and MacGregor III, R., 1972, Temporal organization in avian reproduction, *Am. Zool.* 12:257–271.

Meiske, J. C., Salsburg, R. L., Hoefer, J. A., and Luecke, R. W., 1958, Effect of starvation and refeeding on some activities of rumen microorganisms in vitro, *J. Anim. Sci.* 17:774–781.

Miceli, M. O., and Fleming, A. G., 1983, Variation of fat intake with estrous cycle, ovariectomy and estradiol replacement in hamsters (*Mesocricetus auratus*) eating a fractionated diet, *Physiol. Behav.* 30:415–420.

Minor, J. G., Folk, G. E., and Dryer, B. L., 1973, Changes in triglyceride composition of white and brown adipose tissues during developing cold acclimation of the golden hamster, *Mesocricetus auratus*, *Comp. Biochem. Physiol.* 46B:375–385.

Mitchell, J. A., Smyrl, R., Hutchins, M., Schindler, W. J., and Critchlow, V., 1972, Plasma growth hormone levels in rats with increased naso-anal length due to hypothalamic surgery, *Neuroendocrinology* 10:31–45.

Mitchell, J. A., Hutchins, M., Schindler, W. J., and Critchlow, V., 1973, Increase in plasma growth hormone concentration and nasoanal length in rats following isolation of medial basal hypothalamus, *Neuroendocrinology* 12:161–173.

Moffatt, R. J., 1984, The effect of chronic exercise and retirement from chronic exercise and the regulation of body energy balance by the adult female golden hamster, Ph.D. dissertation, The University of Michigan.

Morin, L. P., and Fleming, A. S., 1978, Variation of food intake and body weight with estrous cycle, ovariectomy, and estradiol benzoate treatment in hamsters (*Mesocricetus auratus*), *J. Comp. Physiol. Psychol.* 92:1–6.

Murphy, H. M., Wideman, C. H., and Brown, T. S., 1972, Liver glycogen levels in rats with limbic lesions, *Physiol. Behav.* 8:1171–1174.

Nace, P. F., House, E. L., and Tassoni, O. P., 1956, Alloxan diabetes in the hamster. Dosage and blood curves, *Endocrinology* 58:305–308.

Nedergaard, J., 1982, Catecholamine sensitivity in brown fat cells from cold-acclimated hamsters and rats, *Am. J. Physiol.* 242:C250–C257.

Nedergaard, J., and Lindberg, O., 1979, Norepinephrine-stimulated fatty-acid release and oxygen consumption in isolated hamster brown-fat cells, *Eur. J. Biochem.* 95:139–145.

Nicholls, D. G., 1976, Hamster brown adipose tissue mitochondria. Purine nucleotide control of the ion conductance of the inner membrane, the nature of the nucleotide binding site, *Eur. J. Biochem.* 62:223–228.

Nicoski, D. R., and Borer, K. T., 1983, Growth-induced changes in the pulsatile pattern of growth hormone secretion in the hamster: The involvement of endogenous opiates, program, Society for Neuroscience 13th Annual Meeting, Boston, Massachusetts, November 6–11.

Ogrowsky, D., Fawcett, J., Althoff, J., Wilson, R. B., and Pour, P., 1980, Structure of the pancreas in Syrian hamsters. Scanning electron-microscopic observations, *Acta Anat.* 107:121–128.

Olefsky, J. M., 1977a, Insensitivity of large rat adipocytes to the antilipolytic effects of insulin, *J. Lipid Res.* 18:459–464.

Olefsky, J. M., 1977b, Mechanisms of decreased insulin responsiveness of large adipocytes, *Endocrinology* 100:1169–1177.

Orsini, M. W., 1961, The external vaginal phenomena characterizing the stages of the estrous cycle, pregnancy, pseudopregnancy, lactation, and the anestrous hamster, *Mesocricetus auratus* Waterhouse, *Proc. Anim. Care Panel* 11:193–206.

Panuska, J. A., and Wade, N. J., 1958, Hibernation in *Mesocricetus auratus, J. Mammal.* 39:298–299.

Pecquery, R., and Giudicelli, Y., 1980, Heterogeneity and subcellular localization of hamster adipocyte α-adrenergic receptors, *FEBS Lett.* 116:85–90.

Pecquery, R., Malagrida, L., and Guidicelli, Y., 1979, Direct biochemical evidence for the existence of α-adrenergic receptors in hamster white adipocyte membranes, *FEBS Lett.* 98:241–246.

Permutt, M. A., and Kipnis, D. M., 1975, Insulin biosynthesis and secretion, *Fed. Proc.* 34:1549–1555.

Peters, R. R., and Tucker, H. A., 1978, Prolactin and growth hormone responses to photoperiod in heifers, *Endocrinology* 103:229–234.

Peters, R. R., Chapin, L. J., Leining, K. B., and Tucker, H., 1978, Supplemental lighting stimulates growth and lactation in cattle, *Science* 199:911–912.

Phares, C. K., 1980, Streptozotocin-induced diabetes in Syrian hamsters: A new model of diabetes mellitus, *Experientia* 36:681–682.

Pohl, H., 1965, Temperature regulation and cold acclimation in the golden hamster, *Am. J. Physiol.* 20:405–410.

Poignant, J.-C., and Rismonad, N., 1975, Influence de l'administration de composes neurotropes sur le temps de ramassage d' un materiel alimentaire chez le Hamster et etude du comportement associe', *Psychopharmacologia* 43:47–52.

Polsky, R. H., 1974, Effects of novel environment on predatory behavior of golden hamsters, *Percept. Mot. Skills* 39:55–58.

Polsky, R. H., 1976, Conspecific defeat, isolation/grouping, and predatory behavior in golden hamsters, *Psychol. Rep.* 38:571–577.

Polsky, R. H., 1977, The ontogeny of predatory behavior in the golden hamster (*Mesocricetus a. auratus*). I. The influence of age and experience, *Behaviour* 61:26–57.

Potter, C. K., Borer, K. T., and Katz, R. V., 1983, Blockade of endogenous opiates reduces voluntary running but not self-stimulation in hamsters, *Pharmacol. Physiol. Behav.* 18:217–223.

Rabi, T., and Cassuto, Y., 1976a, Metabolic adaptations in brown adipose tissue of the hamster in extreme ambient temperatures, *Am. J. Physiol.* 231:153–160.

Rabi, T., and Cassuto, Y., 1976b, Metabolic activity of brown adipose tissue in T_3-treated hamsters, *Am. J. Physiol.* 231:161–163.

Rabi, T., Cassuto, Y., and Gutman, A., 1977, Lipolysis in brown adipose tissue of cold- and heat-acclimated hamsters, *J. Appl. Physiol.* 43:1007–1011.

Rice, R. W., Kroning, J., and Critchlow, V., 1976, The differences in the effects of surgical isolation of the medial basal hypothalamas on linear growth and plasma growth hormone levels in the rat, *Endocrinology* 98:982–990.

Richards, M., 1966, Activity measured by running wheels and observation during the estrous cycle, pregnancy and pseudo-pregnancy in the golden hamster, *Anim. Behav.* 14:450–458.

Ritter, R. C., and Balch, O. K., 1978, Feeding in response to insulin but not to 2-deoxy-D-glucose in the hamster, *Am. J. Physiol.* 234:E20–E24.

Rohner, T., Dufour, A. P., Karakash, C., LeMarchand, Y., Ruf, K. B., and Jeanrenaud, B., 1977, Immediate effect of lesion of the ventromedial hypothalamic area upon glucose-induced secretion in anesthetized rats, *Diabetologia* 13:239–242.

Roth, J., Glick, S. M., Yalow, R. S., and Berson, S. A., 1963, Hypoglycemia: A potent stimulus to secretion of growth hormone, *Science* 1940:987–988.

Rowland, N., 1978, Effects of insulin and 2-deoxy-D-glucose on feeding in hamsters and gerbils, *Physiol. Behav.* 21:291–294.

Rowland, N., 1982, Failure by deprived hamsters to increase food intake: Some behavioral and physiological determinants, *J. Comp. Physiol. Psychol.* 96:591–603.

Rowland, N., 1983, Physiological and behavioral responses to glucoprivation in the golden hamster, *Physiol. Behav.* 30:743–747.

Rowland, N., 1984, Metabolic fuel homeostasis in golden hamsters. I. Nycthemeral and exercise variables, *Am. J. Physiol.* 247: R57–R62.

Rowland, N., 1984, Nycthermeral variation in brain monoamines of Syrian hamsters: Relation to activity and energy homeostasis, *Brain Res.* 290:353–356.

Rudman, D., and Shank, P. W., 1966, Comparison of the responsiveness of perirenal adipose tissue of the rat, hamster, guinea pig and rabbit to the antilipolytic action of insulin, *Endocrinology* 79:565–571.

Saidapur, S. K., and Greenwald, G. S., 1978, Peripheral blood and ovarian levels of sex steroids in the cyclic hamster, *Biol. Reprod.* 18:401–408.

Sak, M. A., and Beaser, S. B., 1962, Alloxan diabetes mellitus in the golden hamster, *Mesocricetus auratus, Lab. Invest.* 11:255–260.

Schanbacher, B. D., and Crouse, J. D., 1981, Photoperiodic regulation of growth. A photosensitive phase during light-dark cycle, *Am. J. Physiol.* 241:E1–E5.

Sclafani, A., and Eisenstadt, D., 1980, 2-deocy-D-glucose fails to induce feeding in hamsters fed a preferred diet, *Physiol. Behav.* 24:641–643.

Shipley, J. E., and Kolb, B., 1977, Neural correlates of species typical behavior in the Syrian golden hamster, *J. Comp. Physiol. Psychol.* 91:1056–1073.

Sicart, R., Sable-Amplis, R., and Apid, R., 1978, Changes in lipid metabolism induced by starvation and cold exposure in the golden hamster (*Mesocricetus auratus*) *Comp. Biochem. Physiol.* 59A:335–338.

Silverman, H. J., 1978, Failure of 2-deoxy-D-glucose to increase feeding in the golden hamster, *Physiol. Behav.* 21:859–864.

Silverman, H. J., and Zucker, I., 1976, Absence of post-fast food compensation in the golden hamster (*Mesocricetus auratus*), *Physiol. Behav.* 17:271–285.

Šimek, V., 1968, Influence of intermittent fasting on morphological changes of the digestive system and on the activity of some enzymes in the golden hamster (*Mesocricetus auratus*), *Acta Soc. Zool. Bohemoslov.* 32:89–95.

Šimek, V., 1969, Influence of short-term and long-term intermittent fasting on morphological changes of the digestive system of the golden hamster (*Mesocricetus auratus*), *Acta Soc. Zool. Bohemoslov.* 33:151–161.

Šimek, V., 1974, Energy metabolism of golden hamsters adapted to intermittent fasting: Influence of season and sex, *Physiol. Bohemoslov.* 23:437–446.

Šimek, V., 1975, Effect of sex and season on the production and deposition of lipid and glycid reserves in the intermittently starving golden hamster, *Physiol. Bohemoslov.* 24:183–190.

Šimek, V., 1980, Effect of intermittent fasting followed by cold on growth, formation of reserves and energy metabolism in the golden hamster, *Physiol. Bohemoslov.* 29:167–172.

Šimek, V., and Petrasék, R., 1974, The effect of two time-different feeding regimens on food intake, growth rate and lipid metabolism in golden hamster (Rodentia), *Acta Soc. Zool. Bohemoslov.* 38:152–159. 38:152–159.

Sinha, N., and Vanderlaan, W. P., 1982, Effect on growth of prolactin deficiency induced in infant mice, *Endocrinology* 110:1871–1878.

Smith, R. E., and Horwitz, B. A., 1969, Brown fat and thermogenesis, *Physiol. Rev.* 49:330–425.

Snipes, R. L., 1979a, Anatomy of the cecum of the dwarf hamster (*Phodopus sungorus*), *Anat. Embryol.* 157:329–346.

Snipes, R. L., 1979b, Anatomy of the cecum of the vole *Microtus agrestis, Anat. Embryol.* 157:181–203.

Sodetz, F. J., and Bunnell, B. N., 1970, Septal ablation and the social behavior of the golden hamster, *Physiol. Behav.* 5:79–88.

Sorrentino, S., Jr., and Reiter, R. J., 1971, Lack of pineal-induced gonadal regression in dark-exposed and blinded hamsters after surgical isolation of the medial basal hypothalamus, *Gen. Comp. Endo. Crinol.* 17:227–231.

Steger, R. W., Bartke, A., and Goldman, B. D., 1982, Alterations in neuroendocrine function during photoperiod-induced testicular atrophy and recrudescence in the golden hamster, *Biol. Reprod.* 26:437–444.

Swanson, H. H., 1967, Effects of pre- and post-pubertal gonadectomy on sex differences in growth, adrenal and pituitary weights of hamsters, *J. Endocrinol.* **39:**555–564.

Swanson, H. H., 1970, Effects of castration at birth in hamsters of both sexes on luteinization of ovarian implants, estrous cycles and sexual behavior, *J. Reprod. Fertil.* **21:**183–186.

Takahashi, S., and Tamate, H., 1976, Light and electron microscopic observation of the forestomach mucosa in the golden hamster, *Tohoku J. Agric. Res.* **27:**26–39.

Takahashi, M., Pour, P., Althoff, J., and Donnelly, T., 1977, The pancreas of the Syrian hamster (*Mesocricetus auratus*). I. Anatomical study, *Lab. Anim. Sci.* **27:**336–342.

Tannenbaum, G. A., Paxinos, G. I., and Bindra, T., 1974, Metabolic and endocrine aspects of the ventromedial hypothalamic syndrome in the rat, *J. Comp. Physiol. Psychol.* **86:**404–413.

Trayhurn, P., 1980, Fatty acid synthesis in brown adipose tissue in relation to whole body synthesis in the cold acclimated golden hamster (*Mesocricetus auratus*), *Biochem. Biophys. Acta* **620:** 10–17.

Tsai, A. C., Bach, J., and Borer, K. T., 1981, Somatic, endocrine, and serum lipid changes during detraining in adult hamsters, *Am. J. Clin. Nutr.* **34:**373–376.

Tsai, A. C., Rosenberg, R., and Borer, K. T., 1982, Metabolic alterations induced by voluntary exercise and discontinuation of exercise in hamsters. *Am. J. Clin. Nutr.* **35:**943–949.

Vaughan, M. K., Powanda, M. C., Richardson, B. H., King, T. S., Johnson, L. Y., and Reiter, R. J., 1982, Chronic exposure to short photoperiod inhibits free thyroxine index and plasma levels of TSH, T_4, triiodothyronine (T_3) and cholesterol in female Syrian hamsters, *Comp. Biochem. Physiol.* **71A:**615–618.

Vriend, J., 1983, Evidence for pineal gland modulation of the neuroendocrine thyroid axis, *Neuroendocrinology* **36:**68–78.

Vriend, J., and Reiter, R. J., 1977, Free thyroxine index in normal, melatonin-treated and blind hamsters, *Horm. Metab. Res.* **9:**231–234.

Vriend, J., Sackman, J. W., and Reiter, R. J., 1977, Effects of blinding, pinealectomy and superior cervical ganglionectomy on free thyroxine index of male golden hamsters, *Acta Endocrinol.* **86:**758–762.

Vriend, J., Reiter, R. J., and Anderson, G. R., 1979, Effects of the pineal and melatonin on thryoid activity of male golden hamsters, *Gen. Comp. Endocrinol.* **38:**189–195.

Vybiral, S., and Jansky, L., 1974, Non-shivering thermogenesis in the golden hamster, *Physiol. Bohemoslov.* **23:**235–243.

Wachtel, N., Emerman, S., and Javitt, N. B., 1969, Metabolism of cholest-5-one-3β, 26-diol in the rat and hamster, *J. Biol. Chem.* **243:**5207–5212.

Wade, G. N., 1972, Gonadal hormones and behavioral regulation of body weight, *Physiol. Behav.* **8:**523–534.

Wade, G. N., 1976, Sex hormones, regulatory behavior and body weight, in: *Advances in the Study of Behavior*, Volume 6 (J. S. Rosenblatt, R. A. Hinde, E. Shaw, and G. C. Beer, eds.), Academic Press, New York, pp. 201–229.

Wade, G. N., 1982, Obesity without overeating in golden hamsters, *Physiol. Behav.* **29:**701–707.

Wade, G. N., 1983, Dietary obesity in golden hamsters: Reversibility and effects of sex and photoperiod, *Physiol. Behav.* **30:**131–137.

Wade, G. N., and Bartness, T. J., 1984a, Dietary obesity in hamsters: Effects of age, fat source and species, *Nutr. Behav.* (in press).

Wade, G. N., and Bartness, T. J., 1984b, Effects of photoperiod and gonadectomy on food intake, body weight, and body composition in Siberian hamsters, *Am. J. Physiol.* **246:**R26–R30.

Widmaier, E. P., and Campbell, C. S., 1981a, Interaction of estradiol and photoperiod on activity patterns in the female hamster, *Physiol. Behav.* **24:**923–930.

Widmaier, E. P., and Campbell, C. S., 1981b, The interaction of estradiol and day length in modifying serum prolactin secretion in female hamsters, *Endocrinology* **108:**371–376.

Wong, R., 1984, Hoarding versus the immediate consumption of food among hamsters and gerbils, *Behav. Processes* **9:**3–11.

Young, J. B., Saville, E., Rothwell, N. J., Stock, M. J., and Landsberg, L., 1982, Effect of diet and cold exposure on norepinephrine turnover in brown adipose tissue of the rat, *J. Clin. Invest.* **69:**1061–1071.

Zawalich, W. S., Dye, E. S., Pagliara, A. S., Rognstad, F., and Matschinsky, F. M., 1979, Starvation diabetes in the rat: Onset, recovery, and specificity of reduced responsiveness of pancreatic cells, *Endocrinology* 104:1344–1351.

Zucker, I., and Stephan, F. K., 1973, Light-dark rhythms in hamster eating, drinking and locomotor behaviors, *Physiol. Behav.* 11:239–250.

Zucker, I., Wade, G. N., and Ziegler, R., 1972, Sexual and hormonal influences on eating, taste preferences and body weight of hamsters, *Physiol. Behav.* 8:101–111.

16

Visual and Somatosensory Processes

BARBARA L. FINLAY and CLAIRE A. BERIAN

The hamster is an animal not specialized for vision that curiously has captured the attention of numerous specialists in vision. Although the initial choice of the hamster for Schneider's (1969) germinal paper was guided principally by the experimental convenience of the hamster's insatiable appetite for sunflower seeds and the ease of the neurosurgical approach to the midbrain, unforeseen advantages have emerged in studying hamsters. The analysis of hamster vision has forced a clearer understanding of the different reasons why comparative analyses of visual systems are useful. A researcher might choose to investigate hamster vision because the goal is to understand human vision. Since the hamster visual system is less elaborate than ours, one might hope to see the fundamental organization of mammalian vision somehow laid bare in the hamster. This is decidedly the context in which most work in rodent (principally rat) vision has been done to date. Alternatively, a researcher might be interested in the general design of sensory systems and in how the visual system is evolutionarily modified to fit the requirements of particular niches. By chance, it has turned out that the hamster is markedly less trainable than the rat, former principal representative of the "simple" mammalian visual system. The rat can usually be induced to perform simple versions of primate puzzles, whereas visuomotor tasks asked of the hamster must reflect its natural behavior more directly, and have included such things as recognition of seeds, crickets and other hamsters, and the ability to find holes and avoid barriers and threats. The resulting compilation of the natural visual capacities of a granivorous, predated-upon mammal primarily active at twilight makes a new sort of comparison possible to those of other well-studied vertebrates in different visual niches, such as frogs, monkeys, and cats. Basic design features of the vertebrate visual system versus niche specific adaptations are contrasted by these two approaches.

The following chapter will review the uses of vision and also somesthesis in the hamster and will summarize the anatomy and physiology of its visual system. Dissocia-

BARBARA L. FINLAY and CLAIRE A. BERIAN • Department of Psychology, Cornell University, Ithaca, New York 14853.

tions of visual function following damage to the brain will be analyzed. Commonalities and divergences in the organization of the hamster visual system with respect to visual system organization in other highly studied vertebrates will be noted.

1. USES OF VISION IN THE HAMSTER

The hamster regulates its activities by the amount of light in its environment. Its period of normal maximal activity begins at twilight and continues into early evening; in the laboratory this means that a rise in activity coincides with the "lights out" period. Independent of circadian rhythms, the hamster responds to high-light intensities by ceasing movement and exploration (Finlay et al., 1980a). In the immediate period pre- and post-eye opening, hamster pups are strongly photophobic, a condition that may serve to confine them to the burrow during daylight hours. The hamster shows increasing spatial resolution under increasing luminance up to 5×10^{-2} cd/m2, the limit of the human scotopic range, and thereafter is not benefited (or impaired) by increasing luminance (Emerson, 1980). This pattern is characteristic of an all-rod retina. In sum, the hamster is most active in dim light, where it shows its maximum acuity. This maximal acuity is fairly low. In a Y-maze grating discrimination task for food reward, hamsters show an acuity of 0.7 cycles/degree at maximum contrast (Emerson, 1980). (In comparison, human foveal acuity at maximum contrast is about 60 cycles/degree.)

Hamsters are slow to learn visual discrimination tasks; the hamsters reported in Emerson's study required 600 to 1050 trials to pick up a basic light/dark discrimination. We have observed comparable acquisition times in our own laboratory (Mort et al., 1980). Other investigators, using more extensive pretraining procedures, report somewhat more rapid acquisition (Chalupa et al., 1978). By contrast, the hooded rat can learn a more complicated pattern discrimination in less than a third of this number of trials (Wiesenfeld and Branchek, 1976). However, once criterion performance is reached, the optimal acuity for the rat is about the same as that of the hamster. It is unclear whether the hamster's slow learning reflects a general difficulty in learning arbitrary associations or whether visual cues are not salient to the hamster. The relative demands on these capacities in the rat's and hamster's respective niches should be investigated. Of course, it may also be the case that some other unidentified feature of this task depresses performance.

Given these impressive constraints, when the use of other sensory modalities is prevented hamsters show good visual guidance of their behavior under a variety of conditions. They visually recognize and approach food such as sunflower seeds and use vision to track and catch insects (Schneider, 1969; Finlay et al., 1980b). They appreciate the layout of food sources and gather seeds in an efficient route (Keselica and Rosinski, 1976). Their social behavior shows some evidence of visual involvement; the Syrian hamster has a black chest mark that can be displayed in aggressive encounters, and in such encounters shows rearing and piloerection in those same circumstances, apparent size-increasing actions that could serve as a visual signal (Johnston, personal

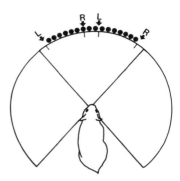

Figure 1. Dorsal view of the hamster's visual field showing the zone of binocular overlap. Dots indicate the range and overlap of maximal ganglion cell density in each eye.

communication). They scramble away from impending collision (or ingestion) in appropriate directions (Ayers, personal communication; Merker, 1980). They can detect cliffs, find holes, and avoid barriers while running (Keselica and Rosinski, 1976; Finlay *et al.*, 1980b). However, despite their ability to perceive depth visually, hamsters appear to be haptically dominated (as do other rodents). A lack of optical information for support is not aversive as long as there is tactual information for support. During locomotor activity, rodents use their forepaws and vibrissae to determine the availability of physical support when visual specification is unavailable (Schiffman, 1970, 1971).

Hamster vision is almost panoramic. Each eye views approximately a hemisphere, and the eyes are laterally placed such that there is a dorsal and rostral binocular zone of about 80° (as measured from photographs of the corneal vertex) and a corresponding blind zone caudally and ventrally (Fig. 1). This is identical to the rat binocular field (Hughes, 1977). Given the fact that there is no marked retinal specialization for central vision (Fig. 2) (see subsequent discussion) and that vision is panoramic, there is less necessity for eye movements than in animals with marked specializations of frontal eyes. Correspondingly, large eye movements are not seen, although there is ocular motility that could be employed both for saccades and for stabilization of the visual field during body movement (Mitchiner *et al.*, 1976). In rats and mice, the amplitude of spontaneous saccades is about 15 to 20 degrees (Bruckner, 1951), and it is likely that for the hamster it is similar. On electrical stimulation of the superior colliculus, saccades of up to 25 degrees, probably in excess of normal amplitude, have been observed in hamsters (Stein, 1981). Most of the hamster's visual acquisition of targets appears to be done by whole head and body movements rather than by eye movements. For example, upon entrance to an open arena, hamsters show a pattern of head movements that gives them information about the visual periphery (Mort *et al.*, 1980; Finlay *et al.*, 1980a).

1.1. Visual Development and the Role of Visual Experience

Visual experience is not necessary for the initial demonstration of most of the hamster's visuomotor capacities. Within two days of exposure to light (postnatal day 15), hamsters are able to recognize and turn toward sunflower seeds presented to them

(Finlay *et al.*, 1980a). Within three days, they avoid cliff edges. Older dark-reared hamsters can perform these tasks at once upon eye opening, and can learn a brightness discrimination in the normal amount of time (Chalupa *et al.*, 1978). Subsequent learning of a pattern discrimination in dark-reared animals is very slightly impaired, if at all. In contrast to almost every other mammal studied, the hamster has only slightly altered grating acuity after dark rearing. Why the hamster should be so relatively insensitive to dark rearing is not yet known, although investigators have hypothesized that this insensitivity has to do with the animal's lesser reliance on the visual cortex on which retinal specializations for high visual acuity are often dependent (Chalupa *et al.*, 1978). It is also possible that some peculiarity of its visual ecology requires lesser dependence on early visual stimulation.

Monocular deprivation produces more of an effect on acuity than does binocular deprivation, but again the effect is small compared with that in other animals that have been investigated, particularly the cat and the monkey. Since most of the effects of monocular deprivation are mediated by competition in the binocular segment, and since the hamster has a relatively small binocular field, this effect is not surprising (Emerson *et al.*, 1982).

Raising the animal in a visually aberrant environment—strobe rearing—produces much more dramatic effects than does dark rearing, both on ability to learn various visual discriminations, and on visual physiology (Chalupa and Rhoades, 1978).

2. ANATOMY AND PHYSIOLOGY OF THE HAMSTER VISUAL SYSTEM

2.1. The Retina

2.1.1. Distribution of Retinal Ganglion Cells

Two quantitative studies of ganglion cell distribution in the hamster retina have been performed with markedly different estimates of the number and spatial distribution of cells. Tiao and Blakemore (1976c) report an approximate total of 114,000 retinal ganglion cells, with all cell types maximally concentrated in a "crude area centralis" somewhat temporal to the optic disk. The maximal center/periphery difference was about 6/1. Our own laboratory has found approximately 80,000 retinal ganglion cells, no evidence of an area centralis and a differential spatial distribution of cells across the retinal surface by cell size (Sengelaub *et al.*, 1983). The maximal center/periphery difference was about 2 : 1. The principal difference in the two studies was the method of data collection; Tiao and Blakemore (1976c) used a somewhat informal method for choice of site for cell counting, apparently creating a sampling bias that elevated their central cell counts over the actual value.

The distribution we report can be seen in Fig. 2. There is a roughly horizontal band of elevated cell density in the inferior retina extending from the temporal to the nasal margins of the retina, peaking temporally. The lowest densities are found in extreme superior and inferior retina. This type of organization is perhaps more similar

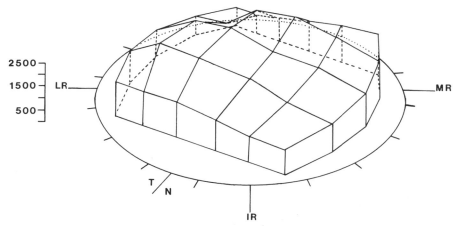

Figure 2. Density of neurons across the retinal surface in the retinal ganglion cell layer of the hamster. N, nasal; T, temporal; IR, inferior rectus insertion; MR, medial vector insertion; LR, lateral rectus insertion. Scale indicates cells/mm².

to the "visual streak" seen in rabbits (Oyster *et al.*, 1981) than to an area centralis, although the streak-to-periphery difference is less marked than that seen in rabbits. All size classes of neurons but the very largest have this distribution; the largest neurons are preferentially found in temporal retina, coincident with the zone of binocular overlap, and hence of ipsilateral projections. This is the same distribution found in other rodents (Cowey and Perry, 1979; Drager and Olsen, 1980).

On the basis of their ganglion cell density estimate, Tiao and Blakemore (1976c) predicted a limit to the spatial frequency resolution of the hamster of 1.8 cycles per degree. Our estimate is much lower (1.0 cycles per degree) and is closer to the behavioral limit (0.7 cycles per degree) found by Emerson (1980).

There is some evidence that the population of retinal ganglion cells is not morphologically and physiologically uniform. On the basis of somal diameter and axonal conduction velocity, the ganglion cells of the hamster retina can be classed into two and possibly three categories (Tiao and Blakemore, 1976c; Rhoades and Chalupa, 1978a, 1979). According to the criterion of somal diameter, there is a large population of cells averaging 8μm and another small population averaging 14μm, and a possible third group greater than 17 μm (Tiao and Blakemore, 1976c); according to the conduction velocity criterion, there is a large population of retinal ganglion cell axons with an average conduction velocity of about 6m/sec and another smaller group peaking at about 12m/sec, and again a possible high-velocity group. There is no direct proof in the hamster that these differential soma sizes and conduction velocities discriminate the same populations of retinal ganglion cells. For example, for those giant cells found preferentially distributed to the temporal retina of the hamster (Tiao and Blakemore, 1976c; Sengelaub *et al.*, 1983), the equation of soma size with conduction velocity is not required; soma size could also reflect the bilateral spread of the terminal arbor.

In other mammals, notably the monkey and the cat, retinal ganglion cells have been divided into various subclasses (x, y, and w; alpha, beta, and gamma) by multiple criteria, including morphology, receptive-field organization, spatial distribution, central targets, and neurochemistry (see Rodieck, 1979, for review). It is not yet known if the subclasses observed in hamsters reflect the same morphological and functional divisions.

In the cat (Fukada, 1977; Fukada and Stone, 1974) and monkey (Schiller and Malpeli, 1977; Bunt *et al.,* 1975) there is selective projection of the ganglion cell subclasses to the superior colliculus and geniculostriate system; in particular, both the largest, fastest-conducting and the smallest, slowest-conducting ganglion cells are thought to project to the superior colliculus. In the hamster, by contrast, it appears that all conduction velocity classes can be antidromically activated from the superior colliculus and all classes of somal sizes can be labeled by horseradish peroxidase (HRP) injections in the colliculus (Chalupa and Thompson, 1980).

Since the original tripartite classification of ganglion cells in cat retina, there has been a tendency to divide retinal ganglion cells of other animals studied into three classes, even in the absence of substantial evidence satisfying several criteria. It may be the case that in attempting to discern phylogenetic homologies at the cellular level, many investigators have too readily intrepreted their results as showing dichotomies and trichotomies that were not mutually exclusive nor exhaustive in the first place.

2.2. Superior Colliculus

2.2.1. General Function and Anatomy

The vertebrate optic tectum or superior colliculus is involved in the functions of orienting the animal to things of interest (particularly those off the body midline) and in some cases of organizing investigation of the environment (Ingle and Sprague, 1975; Finlay and Sengelaub, 1981). In most vertebrates, vision is represented in the colliculus, with other sensory modalities represented depending on which provide spatial information. These can include somesthesis and audition (Knudsen, 1982), the lateral line in teleosts (Bastian, 1981), and the infrared reception of the pit vipers and boids (Terashima and Goris, 1975; Haseltine *et al.,* 1977). In the hamster superior colliculus, the general pattern of sensory organization seen in the colliculus of most mammals is found. The superficial layers receive exclusively visual input. The deeper layers have fewer exclusively visual cells and receive auditory and somatic inputs that in some cases converge to yield bimodal or trimodal cells (Tiao and Blakemore, 1976a; Chalupa and Rhoades, 1977; Finlay *et al.,* 1978; Stein and Dixon, 1979). This basic laminar segregation is also found, for example, in monkey, mouse, and cat superior colliculus (Cynader and Berman, 1972; Drager and Hubel, 1976; Stein *et al.,* 1976). The species-specific adaptations of this pattern in hamsters include relatively less volume in the total structure devoted to vision and an extensive somatosensory representation nearly half of which is devoted to representation of the vibrissae.

These sensory maps are in general spatiotopic register such that defined spatial

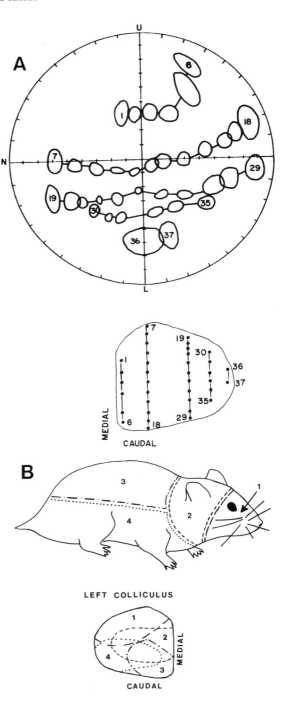

Figure 3. (A) Visual topography within the hamster superior colliculus. Each receptive field in the visual field map corresponds to a point in the superior colliculus represented by a dot. Lines connect rostral to caudal points in the colliculus, as well as corresponding nasal to temporal visual receptive fields. The visual field map is marked off in 10° increments and is centered on the optic disk, with the geometrical center of the eye indicated by a star. The horizontal and vertical meridians indicate the centers of the attachments of the external eye muscles. (B) Somatosensory topography within the hamster superior colliculus. Four somatic regions are represented: face (———); ear, neck and head (----------); upper body (—·—·—) and lower body (··········).

areas are found in columns perpendicular to the collicular surface (Fig. 3). The entire contralateral retina and at least the contralateral body surface and face are represented in each colliculus. In the rostral colliculus, the nasal visual field and the face and whiskers are represented; caudally, the temporal visual field and remaining body are found. The upper visual field and body are represented medially, and lower visual field, lower body, and feet are found laterally. The visual receptive fields are small (averaging about 10 degrees in the superficial layers), whereas the somatosensory receptive fields are larger, minimally subtending 30 to 40 degrees of visual field. Since the hamster makes relatively small eye movements, the problems of coding a dual sensory representation of space in which one receptor surface can move relative to the other are, for the main, avoided. However, it has been suggested that the lesser precision in the somatosensory topography relative to the visual topography assures congruence between visual and somatosensory information during normal movement of the eyes and body (Drager and Hubel, 1975; Finlay et al., 1978). The registration between visual and somatosensory information could only be disrupted if the animal assumed a highly atypical posture.

The superficial gray layer and the stratum opticum of the superior colliculus receive direct visual input from the ipsilateral and contralateral retina and from the primary visual cortex (Jhaveri and Schneider, 1974; Lent, 1982; Rhoades and Chalupa, 1979). As stated previously, all three size classes of cells found in the hamster retina innervate the hamster colliculus (Chalupa and Thompson, 1980).

2.2.2. Magnification of the Visual Field in the Superior Colliculus

In the hamster, at least 70 degrees of visual field nasal to the optic disk and at least 90 degrees temporal to the optic disk are represented in the superior colliculus (Finlay et al., 1978). A greater representation for central visual field than for peripheral visual field is seen: along the horizontal meridian of the visual field, for the central 30 degrees of visual field, each degree of visual angle occupies 0.025 mm of rostral to caudal distance in the colliculus, while in nasal and temporal visual field, one degree occupies 0.020 and 0.015, respectively. This changing magnification is reflected in the receptive field sizes of collicular neurons representing different parts of the visual field. Receptive field diameter is typically less than 10 degrees for cells representing the central 35 degrees of visual field, but in the periphery, receptive field diameters approach 20 degrees. This 2 : 1 ratio for receptive field sizes corresponds well to the 2 : 1 packing density gradient of ganglion cells, and since it has been shown that all size classes of retinal ganglion cells project to the colliculus, it seems likely that the colliculus in hamster receives an unbiased representation of both retinal topography and functional classes of ganglion cells. This is quite different from the situation in both cats and monkeys, where the peripheral visual field appears "overrepresented" in the colliculus with respect to total retinal ganglion cell distribution and where the particular functional class of "x cells" that carry fine-grained spatial and chromatic information are excluded.

The bulk of the retinotectal projection in the hamster is crossed. Projections from the ipsilateral retina are limited to clumps concentrated in parts of the rostral stratum opticum of the colliculus, with a minor degree of termination more dorsal, in the superficial gray, and deeper, in the intermediate gray (Frost and Schneider, 1976; Frost *et al.*, 1979). The temporal retina, source of the ipsilateral projection, projects bilaterally such that the same area of anterior visual field is represented in both colliculi. This projection pattern is characteristic of all rodents (Siminoff *et al.*, 1966; Drager and Hubel, 1976; Woolsey *et al.*, 1971) and, in fact, of all mammals but primates (Masland *et al.*, 1971; Feldon *et al.*, 1970; Ebbeson, 1970).

2.2.3. *Physiology: Visually Responsive Cells*

Receptive fields of single neurons in the hamster superior colliculus vary from 2 to 3 degrees in diameter in the superficial gray layer to almost the entire contralateral hemifield in some of the cells in the deeper laminae. In the superficial gray layer, the optimal stimuli for most neurons are smaller than the activating visual receptive field areas; stimuli subtending 10% or less of the receptive field diameter are often best (Stein and Dixon, 1979).

It should be emphasized that the configurational requirements for stimuli that activate collicular neurons are not strict: typically a wide variety of stimulus shapes, sizes, contrasts, and velocities will activate collicular cells. Colliculus cells are not selective for stimulus orientation. In this context, however, the majority of cells in the superficial layers of the superior colliculus prefer dark stimuli on light backgrounds. Slowly moving (less that 20 to 30 degrees per second) stimuli elicit the maximal response from most units, with the optimal velocity around 5 to 7 degrees per second. A smaller population of cells responsive to very fast movement has also been found (Tiao and Blakemore, 1976a). Stein and Dixon (1979) note that the general preference of colliculus cells for slowly moving stimuli is found both in animals that make few saccadic eye movements and in animals that make frequent eye movements (cats and monkeys). Therefore this property can be dissociated from oculomotor requirements. A psychophysics of "noticeability" of events in the visual periphery according to stimulus velocity has not been done for these animals, but would be of clear interest given known colliculus function, and the particular filtering properties of these neurons.

Some of the neurons in the hamster superior colliculus are sensitive to the direction of stimulus movement. Estimates of the proportion of these neurons vary from 10% (Tiao and Blakemore, 1976a) to 60% (Stein and Dixon, 1979), the variation produced by the particular criterion used for directionality. The direction preference exhibited by almost all collicular neurons that show a preference is upward and nasal (Rhoades and Chalupa, 1976; Chalupa and Rhoades, 1979). Stein and Dixon found that directional preferences of cells in the hamster colliculus can be negated by use of nonoptimal stimulus sizes or velocities. The direction selectivity seen is dependent on the integrity of the visual cortex (Rhoades and Chalupa, 1978c; Chalupa and Rhoades, 1979), substantially the only facet of visual receptive field organization in the colliculus dependent on the visual cortex. In addition, these cells tend to be innervated by slowly conducting retinal axons (less than 5 m/sec) (Rhoades and Chalupa, 1979). In summary, there is a substantial fraction of collicular cells that exhibits direction selectivity;

this selectivity is activated by a subset of the retinal input and depends on the visual cortex.

Substantial species differences exist in the presence of direction selectivity in the colliculus, the distribution of preferred directions, and the contribution of the visual cortex to this selectivity (Stein, 1981). Much effort has been expended to determine if species-typical requirements in vision give a clue to the function of these direction preferences. For example, the monkey superior colliculus has few directionally selective cells and the distribution of preferred directions is random (Cynader and Berman, 1972). The cat's superior colliculus shows direction selectivity for centifugal horizontal movement (Sterling and Wickelgren, 1969). In the mouse, approximately one quarter of the colliculus cells exhibit direction selectivity, usually for upward movement with a small nasal vector, as in the hamster (Drager and Hubel, 1975). Direction preference in the rabbit colliculus is for movement in either direction across the horizontal axis of the receptive field (Masland et al., 1971). No particularly satisfactory account of these species differences has been made. Stein (1981) has suggested that directional selectivity in the colliculus may mediate scanning patterns, but a cross-species analysis of patterns of head and eye movements in regard to this parameter has not been made.

The number of binocularly driven cells in the hamster colliculus has been reported to be few or none (Stein and Dixon, 1979; Finlay et al., 1978; Rhoades, 1979), although there has been some controversy about their total number (Tiao and Blakemore, 1976a). Binocular cells are located in rostral colliculus, in the deep part of the superficial gray area and below, and are dominated by the contralateral eye. Responsivity to stimulation of the ipsilateral eye increases shortly after contralateral eye removal, suggesting that normally the visual input from the ipsilateral eye is suppressed by the dominant contralateral eye (Chalupa and Henderson, 1980; Rhoades, 1980; Chalupa, 1981).

2.2.4. The Deep Layers of the Superior Colliculus: Multimodal Organization

In the intermediate and deep layers of the superior colliculus are found somatosensory, auditory, visual, and motor response-related neurons. The majority of somatosensory neurons are activated by light touch on the contralateral body. These receptive fields tend to be large and increase in size (in terms of absolute body surface) from the rostral to the caudal colliculus, where they can encompass a significant extent of both the contralateral and ipsilateral body surface (Chalupa and Rhoades, 1977).

Most of the somatosensory cells respond to light brushes of the vibrissae or body hair to light pressure, to air flow over the body surface, and to other gentle cutaneous stimuli and are rapidly adapting. These mechanoreceptive characteristics are quite similar to those found in the cat's somatosensory colliculus (Stein et al., 1976). Although most cells respond in an excitatory manner to somatic stimuli, some cells have high spontaneous activity that is inhibited by light somatic stimulation (Chalupa and Rhoades, 1977; Tiao and Blakemore, 1976a). None of the somatosensory cells have been found to respond to the direction of stimulus movement or any other complex receptive field property (Finlay et al., 1978).

In the vibrissal area of the somatosensory representation, a few cells can be found that respond to the stimulation of only one or two upper vibrissae, or to all of the lower vibrissae, or the rostral group or the caudal group. When receptive field can be localized to a small number of vibrissae, these vibrissae cross the visual fields of the cells mapped in the overlying superior colliculus (Chalupa and Rhoades, 1977).

In the mouse, the vibrissa subtend a larger part of the visual field than the hamster; they are about 10% longer in relation to body length than those of the hamster. Although there is no behavioral evidence to suggest that the mouse uses its vibrissae more or better than the hamster, it is interesting to note how closely rather subtle differences in gross morphology might be tracked in the spatiotopic organization of the colliculus. The detail of whisker representation in hamster colliculus is not as great as that seen in mouse colliculus, where the somatosensory representation of the whiskers occupies more relative extent in the colliculus and where the majority of somatosensory receptive fields are localized to less than six vibrissae, often to one or two (Drager and Hubel, 1975, 1976). Drager and Hubel have noted that the amount of area in the mouse colliculus representing individual vibrissae is consistently directly related to the area of the visual field subtended by the whiskers. Whiskers occupying a large sector of the visual field due to their length or position relative to the eye occupy a correspondingly large segment of the colliculus whisker representation. Drager and Hubel conclude that the somatosensory representation in the mouse colliculus is dictated by the visual projection in the overlying superficial gray rather than by peripheral tactile innervation density. However, in the owl, the visual and auditory fields "view" different angular extents and become incongruent in the caudal tectum; in boids and pit vipers, the infrared and visual maps are also in part incongruent for the same reason. It cannot, therefore, be a general principal that multimodal maps should be angularly congruent, but the animals studied so far are so diverse in preferred sensory modality, niche, and taxonomic relatedness that the reason for this variability in tectal organization is unclear.

Nociceptive neurons are also found in the intermediate and deep collicular layers. Some of these cells respond specifically to noxious mechanical (pinch, pin prick) and thermal stimuli, whereas others respond to innocuous somatic stimulation (Stein and Dixon, 1978). Nociceptive neurons have large receptive fields and respond to noxious stimuli with a sustained train of discharges. These responses can often endure for as much as two minutes after removal of the stimulus. The receptive fields are quite large and often bilateral with contralateral stimulation exerting a more pronounced effect than ipsilateral stimulation. In some cells, visual or acoustic input and nociceptive input converge. When convergence is noted, the visual receptive fields are discrete relative to the nociceptive fields that are often poorly delimited. Administering the narcotic etorphine profoundly suppresses responses to noxious stimuli while leaving responses to innocuous stimuli normal. The narcotic antagonist naloxone returns suppressed nociceptive responsivity to normal (Stein and Dixon, 1978). These nociceptive neurons have been found distributed throughout the intermediate and deep colliculus, the subjacent tegmentum, the pretectal area, and the periaqueductal gray. Stein and Dixon suggest that the collicular nociceptive neurons may play a role in alerting the animal to the general location of a painful stimulus, but that the large receptive field sizes of these neurons makes precise localization an improbable function for these cells.

The superior colliculus has also been implicated in the modulation of the lordosis posture in the hamster. Cells strongly responsive to tactile stimulation that trigger lordosis can be found in the intermediate and deep colliculus as well as the interpeduncular nucleus, adjacent tegmentum, and ventrolateral central gray. In the sexually receptive hamster, lumbosacral tactile stimulation elicits lordosis and tactile stimulation around the face interrupts lordosis. Some single collicular cells have been found that respond in opposing ways to face versus lumbosacral stimulation, and superior colliculus lesions abolish the opposing effects of facial and lumbosacral stimulation (Rose, 1982).

2.2.5. Somatosensory Input Pathways

Somatosensory information can come to the colliculus by a variety of routes, including the primary and secondary somatosensory cortices, lamina IV cells from all levels of the spinal cord, the contralateral dorsal column nuclei, nucleus of the spinal trigeminal tract, lateral cervical nucleus and internal basilar nucleus, and the contralateral superior colliculus. These various pathways have been extensively described by Rhoades (1981a,b) and will be summarized briefly here.

Both indirect or polysynaptic pathways and direct pathways to the colliculus exist. The indirect pathways include a variety of cortical and spinoreticular routes. The direct pathways arise from the contralateral dorsal column nuclei, internal basilar nucleus and lateral cervical nucleus, and the classical spinotectal tract originating in lamina IV of the spinal cord. Rhoades (1981a,b) has found that the response properties of spinotectal and corticotectal neurons are quite similar to those of the neurons in the intermediate and gray layers of the colliculus to which they project. The contribution of the cortical input to the response properties of colliculus cells appears to be minimal or subtle, since ablation of the ipsilateral primary and secondary cortex leaves the incidence and response properties of somatosensory cells in the colliculus apparently normal. Dorsal spinal transection, which disrupts most of the polysynaptic pathways described above, also has no effect on the responses of collicular cells. Rhoades suggests that the direct spinotectal tract, which travels in the ventrolateral spinal cord, must convey to the colliculus sensory information similar to that represented in the polysynaptic pathways. Finally, the receptive field of both corticotectal and spinotectal somatosensory cells are consistently smaller than those of collicular cells representing the same locus on the body, suggesting substantial convergence.

2.2.6. Output Pathways of the Superior Colliculus

The principal ascending target of the superior colliculus is the rostrolateral subdivision of the lateral posterior nucleus of the thalamus. This nucleus subsequently projects to circumstriate posterior neocortex. There is also an extensive projection to the lateral aspect of the ventral lateral geniculate body (Niimi et al., 1963) and a smaller projection to the pretectum (Crain and Hall, 1981). These projections originate in the superficial laminae.

The descending projections of the superior colliculus in the hamster have not yet been extensively described. A relatively small population of cells in the lateral part of the intermediate and deep layers of the hamster gives rise to a contralaterally destined tectospinal tract. In the hamster, as in the cat, the monkey, the tree shrew, and the rat, the tectospinal tract innervates only the upper cervical spinal cord, including neck musculature (Graham, 1977; Harting, 1977; Harting et al., 1973; Rhoades and De-llaCroce, 1980b; Waldron and Gwyn, 1969). In the hamster, most of the cells of origin of the tectospinal pathway respond only to somatosensory stimulation, particularly to light touch, and the receptive fields of these cells represent almost the entire body surface (Rhoades and DellaCroce, 1980b). In the cat, more than half of the cells giving rise to the tectospinal tract are responsive to visual stimuli (Abrahams and Rose, 1975). However, since it is well established that the colliculus is involved in the visual mediation of orientation, Rhoades and DellaCroce conclude that this tectospinal tract may be involved in the specific function of directing movements toward tactile stimuli and other pathways may subserve visuomotor functions. In other animals, the deep collicular layers have been shown to give rise to projections of a variety of juxta-oculomotor nuclei. It has been suggested that an indirect tectoreticulospinal tract may be the major functional link between the colliculus and the spinal cord (Anderson et al., 1971; Harting, 1977; Waldron and Gwyn, 1969). Presumably, tectoreticular, tecto-pontine, and tectobulbar connections exist in the hamster although no literature on such tracts is available. In sum, the colliculus in the hamster as in other mammals has a major ascending projection to the neocortex by way of the posterior thalamus and a probable variety of direct and indirect descending projections.

2.2.7. Stimulation and Lesion Experiments

Electrical stimulation of the superior colliculus in the hamster leads to an orienta-tion movement involving the whole body, including the pinnae, vibrissae, neck, eyes, and limbs, as it does in other animals. Species-typical alterations of this pattern of course exist: frogs snap, cats turn and orient their ears appropriately, monkeys move their eyes. In hamsters, if the stimulation is of long duration, or the stimulating electrode extends into the deeper collicular laminae, the stimulation often appears aversive, which is consistent with the observation of nociceptive cells in the deep colliculus (Stein, 1981; Palmer, 1980).

The effects of ablation of the superior colliculus in the hamster have been exten-sively studied (Schneider, 1969; Mort et al., 1980; Finlay et al., 1980; Merker, 1980b; Keselica and Rosinski, 1976) and confirm in some detail the results obtained in other rodents (Goodale and Murison, 1975; Ingle and Sprague, 1975). In sum, the colliculus appears to be involved in the mediation of attending and orienting to objects or events, particularly those in the posterior visual field. It is not involved in the gross control of ambulation, in avoidance of barriers and obstacles, or in object recognition.

In Schneider's (1969) now classic experiment contrasting the visual functions of the superior colliculus and visual cortex, two separable behaviors were found to be disrupted following collicular lesions. Hamsters failed to turn their heads to get sunflower seeds presented to them and failed to make a choice of patterns in a Y maze, although they showed they could recognize both the seeds and the Y-maze patterns if a spatial decision was not required. These two disorders were classed together to suggest that the hamsters were failing to make decisions regarding "place" while pattern vision was still intact. Subsequent research has shown that numerous aspects of spatial orientation are spared after colliculus lesions, and that the defect is both more multiple and more specific than this initial characterization (Keselica and Rosinski, 1976; Mort et al., 1980; Finlay et al., 1980b; Merker, 1980). The difficulty in orienting the head and body to things, whether they are sunflower seeds, crickets, exit holes, or various laboratory "distractors," is markedly more pronounced in the temporal periphery of the visual field than directly ahead of the animal. Hamsters can recognize and approach all of the above with normal or even "supernormal" accuracy and speed if they lie in the 80 degrees directly in front of the animal (Mort et al., 1980; Finlay et al., 1980b). After collicular lesions, a piece of active investigatory behavior is lost; on entrance to an open area, a normal hamster stops and looks left and right before advancing, whereas a colliculus-ablated hamster proceeds directly forward. Catching crickets, which usually requires successive reorientations, is difficult for colliculus-ablated hamsters (Finlay et al., 1980b) as is gathering collections of seeds (Keselica and Rosinski, 1976). In contrast, however, these animals do not bump into things with any unusual frequency; they recognize visual cliffs and guide their locomotion with regard to distant targets with normal accuracy, so a considerable fraction of spatial orientation capacity is spared. Hamsters with collicular lesions either as adults or juveniles show a hyperactivity unrelated to vision (Finlay et al., 1980a).

In summary, there is now a large mass of data on the anatomy and physiology of the hamster superior colliculus, as well as analysis of the behaviors dependent on it. A striking gap in this literature is the absence of information on collicular physiology while the hamster is behaving: practically nothing is known about the motor response of properties of their collicular cells and how eye, head, and whole body turns are coded. This gap is certainly due to technical difficulty alone, but the advances made in single neuron recording from small mobile animals may produce a complete picture in time.

2.3. Thalamocortical Organization of Vision and Somesthesis

2.3.1. Anatomy and Topography of Cortical Visual Areas

Three representations of the visual field have been identified in hamster posterior neocortex (Fig. 4). The primary visual cortex, cytoarchitectonic area 17, V1 (striate cortex), is bordered medially by area 18 or 18b (Vm) and laterally by area 18a (V2). Areas 29d and c, cingulate cortex medial to 18b, have also been reported to receive some visual input, and 29d is cytoarchitectonically quite similar to 18b (Vogt and Peters, 1981; Finlay and Slattery, 1984).

The dorsal lateral geniculate nucleus projects in a point by point manner to area

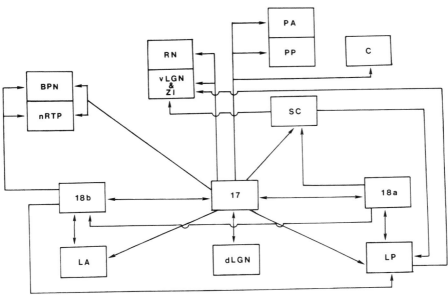

Figure 4. Schematic of visual cortical projections. Reciprocal projections are indicated by double arrows. Abbreviations: BPN, basilar pontine nucleus; C, caudate; dLGN, dorsal lateral geniculate nucleus; LA, lateral anterior nucleus; LP, lateral posterior nucleus; nRTP, nucleus reticularis tegmenti pontis; PA, anterior pretectal nucleus; PP, posterior pretectal nucleus; RN, reticular nucleus; vLGN, ventral lateral geniculate nucleus; ZI, zona incerta; 17, primary visual cortex (V1); 18a, lateral visual cortex (V2); 18b, medial visual cortex (Vm).

17, and as far as can be determined, projects there exclusively (Dursteler *et al.,* 1979; Garey and Powell, 1967). This is the same pattern that has been reported in other rodents and in primates, but not in cats (Drager, 1974; Rosenquist *et al.,* 1974; Van Essen, 1979). The visual field in the lateral geniculate is represented such that the superior-inferior axis of the visual field is represented rostrocaudally in the geniculate, and the nasotemporal axis mediolaterally. The magnification of the visual field in the geniculate appears roughly comparable to that observed in the colliculus (Schneider and Jhaveri, 1974), and it appears that a substantial portion if not all of the ipsilateral visual field is represented in geniculate and in cortex (Tiao and Blakemore, 1976b) (Fig. 5). This is unlike the situation in both cats and primates where the contralateral visual hemifield, not the contralateral retina, is represented in the lateral geniculate and striate cortex. The ipsilateral projection to the lateral geniculate from the retina is a band on the medial surface of the rostral part of the geniculate and is segregated from the contralateral projection; there are no other known functional or morphological laminae.

The visual field is represented topographically in area 17, with lower visual field rostral, upper visual field caudal, nasal visual field lateral, and temporal visual field medial. Approximately 80 degrees contralateral to the vertical midline and about 10

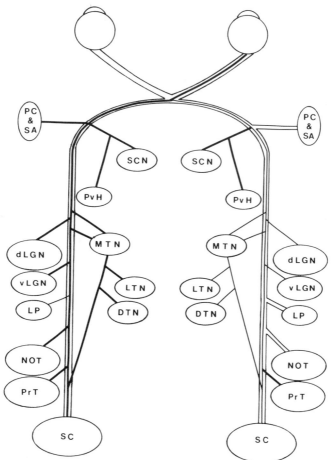

Figure 5. Schematic of retinofugal pathways, including retinocortical, retinohypothalamic, accessory optic, retinothalamic, and retinotectal projections. Thick lines represent dense projections, whereas thinner lines represent relatively sparse projections. Open lines indicate the lack of an identified retinal projection. Abbreviations: dlGN, dorsal lateral geniculate nucleus; DTN, dorsal terminal nucleus; LP, lateral posterior nucleus; LTN, lateral terminal nucleus; MTN, medial terminal nucleus; NOT, nucleus of the optic tract; PC, piriform cortex; PrT, pretectum; PvH, periventricular hypothalamus; SA, septal area; SC superior colliculus; SCN, suprachiasmatic nucleus; vLGn, ventral lateral geniculate nucleus. [Figure modified from the diagram of Pickard and Silverman (1981).]

degrees of ipsilateral field are represented. Thus, about 10 degrees on either side of the vertical midline are doubly represented in the lateral sector of area 17 in each hemisphere. (Tiao and Blakemore, 1976b). The same is seen in the mouse (Drager and Hubel, 1975). In the one study of hamster visual cortex physiology, the representation of visual field lower than 30 degrees from the horizontal meridian and greater than 60 degrees above the horizontal meridian could not be found, but this observation is suspect due to the difficulty of recording units under the sagittal and transverse sinuses along the medial and posterior declivity of the cortex (Tiao and Blakemore, 1976b).

Area 18a, lateral to 17, shows a second representation of the visual field reflected about the vertical midline represented on the lateral margin of striate cortex. The main thalamic projection to this cortical area is lateralis posterior (LP), which in turn receives its projections principally from the superior colliculus, the striate cortex, and a small projection directly from the retina. The projection from LP to 18a is topographically organized, although not to the precision seen in geniculostriate connectivity. Area 17 projects heavily to 18a, and 18a projects back to 17 and also to 18b (Fig. 4).

Area 18b's principal thalamic projection is from the lateral anterior nucleus (LA) (Fig. 4). This nucleus has not been well studied, but it is known that it receives a major projection from the striate cortex (Lent, 1982), and it has been shown to have pretectal afferents in the rat (Robertson et al., 1980) as well as input from the cingulate cortex (Beckstead, 1979).

Because of the partial representation of the ipsilateral hemifield, visual callosal projections are somewhat more extensive in hamsters than they are in cats and primates, where only 1 to 2 degrees of visual field adjoining the vertical meridian are doubly represented and thus represented in the corpus callosum (Rhoades and DellaCroce, 1980a).

2.3.2. Visual Cortex Physiology

In the striate cortex, there is an apparent trend toward disproportionate representation of the central visual field. Particularly, receptive field diameters of cortical cells representing peripheral visual field tend to be larger (up to about 30 degrees) than those representing central visual field, which are as small as 5 degrees (Tiao and Blakemore, 1976b). However, due to the difficulty of access to those areas of cortex representing peripheral visual field, the data base in support of this contention is not large.

Cells in striate cortex exhibit a very low rate of spontaneous activity and tend not to respond to any degree of change in whole-field illumination. Tiao and Blakemore have divided the types of receptive fields of visual cortical neurons into two classes: radially symmetric and asymmetrical. The majority of cells have radially symmetric response fields, and respond independent of stimulus orientation and direction of movement. A lesser number of cells have asymmetrical response fields, which tend to be smaller than the symmetric fields and exhibit bias for a particular stimulus orientation or vector of movement.

Among the radially symmetric fields are fields with "on" centers, fields with "off" centers, and fields with homogenous on-off responses. Several subclasses of cells with asymmetrical fields have also been described. Type 1 orientation selective fields respond best to flashed or moving bars or edges in a specific orientation, and sometimes are biased for one direction of movement. Preferred velocities vary from about 10 degrees/second to about 50 degrees/second. Type 2 orientation selective cells respond optimally to stimuli smaller than the width of the entire receptive field, are selective for the orientation of moving bars, and prefer velocities of about 20 to 30 degrees/second.

Type 3 orientation selective units are a very small number of cells that appear similar to lower order hypercomplex cells in cat striate cortex in that the response fields have inhibitory end zones.

Orientation sensitive cells in the hamster are more likely to prefer horizontal and vertical orientations; this has been observed in the rabbit and mouse as well (Chow *et al.*, 1971; Drager, 1975). Tiao and Blakemore note that types 1 and 2 cells are not as distinguished from each other in terms of velocity selectivity and orientation selectivity as are the simple and complex cells of cat striate cortex.

Ocular dominance ranking reveals that most binocular cells are dominated by the contralateral eye, although a few are dominated by the ipsilateral eye. Preliminary investigation (Tiao and Blakemore, 1976b) suggests that the hamster area 17 clusters cells according to ocular dominance, but it is not yet clear if the organization is really homologous to the pronounced ocular dominance organization of monkey striate cortex. Similarly, although the hamster cortex has some clustering of units with symmetrical and asymmetrical fields, the columnar organization of visual cortical cells according to orientation specificity is not as pronounced as that seen in cats and monkeys (Hubel and Wiesel, 1962, 1968).

2.3.3. Lesions of the Visual Cortex

If the visual cortex is removed, hamsters show some loss of visual capacity, a subtle loss compared with that seen in primates. They have difficulty doing horizontal-vertical pattern discriminations (Schneider, 1969, 1970) and they run into walls more than normal hamsters (Finlay *et al.*, 1980b). Clearly this is an as yet inadequate description of functions dependent on neocortical pathways.

2.3.4. Somatosensory Cortex

As with the mouse and the rat, the hamster has five rows of large mystacial vibrissae. These vibrissae project in an organized manner to the facial area of the contralateral primary somatosensory cortex. Within this area the vibrissae terminate in five rows of multicellular cytoarchitectonic units termed "barrels" in cortical lamina IV. This region is known as the posteromedial barrel subfield. It is interesting to note that the gerbil has seven rows of barrels corresponding to seven rows of mystacial vibrissae. The organization of cortical neurons thus reflects the arrangement of peripheral sense organs (Lee and Woolsey, 1975; Woolsey *et al.*, 1975). In the mouse each barrel is related in a 1 : 1 manner to discrete vibrissae and the number of peripheral fibers innervating the vibrissae is proportional to the number of neurons in the barrels. Thus the peripheral innervation density scales the magnification observed in cortical representation of the sensory surface. It is not yet known whether the same relationship exists in the hamster.

2.3.5. Other Retinal Connections

Pretectal nuclei appear to receive projections exclusively from the contralateral retina (Fig. 5). Retinal fibers terminate in the contralateral medial and lateral nuclei of the optic tract and the contralateral pretectal nucleus (Eichler and Moore, 1974). These respective nuclei correspond in turn to the posterior pretectal nucleus, nucleus of the optic tract, and olivary pretectal nucleus designated for other rodents (Scalia, 1972).

The accessory optic tract in hamsters also appears to comprise primarily fibers from the contralateral retina. The posterior accessory optic tract may be the homologue of the superior fasciculus of the rat accessory optic tract (Eichler and Moore, 1974). The fiber bundle travels with the primary optic tract originating in contralateral retina through the brachium of the superior colliculus. However, after exiting from the brachium, accessory optic fibers terminate in the medial terminal nucleus (MTN), lateral terminal nucleus (LTN), and dorsal terminal nucleus (DTN) contralateral to the retina of origin. There is also a small projection to the ipsilateral LTN and a minimal ipsilateral projection to the DTN (Eichler and Moore, 1974; Pickard and Silverman, 1981). Eichler and Moore note that in the hamster there appears to be no anterior accessory optic tract homologous to the inferior fasciculus of the accessory optic tract in the rat. This pathway in rats terminates in the MTN.

It has been established that the hamster has a direct retinohypothalamic connection, although the nucleus of termination is somewhat controversial. Printz and Hall (1974) conclude that the retinohypothalamic projections terminate in the ventromedial nucleus of the hypothalamus, with no terminals located in the suprachiasmatic nucleus. Eichler and Moore (1979) and Pickard and Silverman (1981) have shown retinal projections to the suprachiasmatic nucleus, but report none to the ventromedial nucleus. (The retinal projection is densest to the contralateral SCN, but there is a sparse ipsilateral projection.) Pickard and Silverman also noted terminations on the ventromedial border of the zona incerta and in small quantity in the rostral median eminence. It has been suggested that the retinohypothalamic connection may be important in the regulation of diurnal body rhythms and in the photic mediation of pineal-gonadal activity in the hamster (Eichler and Moore, 1974; Rusak, 1977).

3. SUMMARY

In the introduction, two different ways of viewing research on hamster vision were described. The first employed the hamster visual system as an example of a simple mammalian visual system, one more similar perhaps to the stem mammalian ancestor than other highly studied animals. The second approach viewed the hamster visual system as a concatenation of niche-specific functional adaptations. These descriptions together should give a complete account of the development and organization of the hamster visual system; the relative importance of each account is a matter for debate.

3.1. *The Hamster Visual System as a "Simple System"*

Given the wiring diagrams presented in Figs. 4 and 5, it should be admitted that extreme simplicity is not the first description one would offer of hamster visual anatomy. However, the basic distribution of retinal terminals is characteristic of almost every vertebrate visual system studied (Ebbeson, 1970). In thalamocortical anatomy, however, at a gross level, the hamster has fewer thalamic subdivisions and fewer separable neocortical divisions than felines and primates (Van Essen, 1979).

Overall, the predominant impression left by hamster visual anatomy is a lack of differentiation at every level of analysis. In the retina, local specializations in neuron density are minimal. In function or morphology, the retinal ganglion cells have not supported an extensive categorization, and show little differential central distribution, in marked contrast to that seen in more complex mammalian visual systems. Similarly, the best-described projection areas, particularly the superior colliculus and the ventral and dorsal lateral geniculates, receive an unbiased representation of the retinal surface and of retinal types. Within the dorsal lateral geniculate, although there is ocular segregation, no other lamination or obvious cellular differentiation is visible. The possible functional breakdowns of types of visual cortical cells seem impoverished compared with the differentiation seen in the cat, the monkey, or even the mouse.

This state of affairs is quite consistent with parcellation theories of the development of complex systems from simple systems (Ebbeson, 1980). Two differences in complex visual systems can be seen in comparison to this simple system: (1) cellular differentiation occurs within systems that are originally homogenous, and (2) these differentiated components *restrict* their range of projections compared with the original stem cells, both within and between structures. It should be emphasized that this need not necessarily be the case: the hamster visual system, with a more limited set of possible capabilities than a complex one, could reasonably be expected to have fewer and more specialized retinal projection patterns.

As a sidelight, the relative independence of the hamster visual system from the effect of visual deprivation and restriction might be used as evidence that simple systems require, in a fundamental way, less environmental facilitation of their construction, whereas complex systems are more plastic and more capable of "learning." It may be more parsimonious to view this phenomenon as the simple summation of numerous specification steps in the development of a complex visual system, any of which could require an environmental component, than to propose that simple and complex systems differ in kind in their environmental dependence.

3.2. Niche-specific Adaptations

Niche-specific adaptations in the hamster are more apparent in its visuomotor behavior directly and in the interaction of its sensory systems than within the anatomy or physiology of the visual and somatosensory systems separately. In behavior, such phenomena as stereotypic scanning patterns on entering an open arena, posturing in aggressive encounters, obvious ease in learning or otherwise acquiring insect predation and grain-gathering strategies (in stark contrast to the intense difficulty manifest in learning arbitrary visual associations) are examples of behaviors that in no way can be categorized as generalized or unspecified or characteristic of all animals with relatively simple visual systems. The spatiotopic relationship of the visual field and whiskers in the midbrain is an interesting example of a species-typical variation on a generalized vertebrate organizational theme.

How then can this apparently undifferentiated visual system be reconciled with a well-differentiated, niche-specific visuomotor repertoire? As a class, rodents have highly varied visuomotor repertoires (Finlay and Sengelaub, 1981), yet it seems quite likely that their visual systems would be indistinguishable except in simple size on the anatomical and physiological parameters described here. It seems clear that the question of how these visuomotor capabilities are produced is poorly addressed at the level of general anatomical connectivity and basic single neuron response properties. The circuitry, or computational program required to implement particular pieces of complex behavior must be addressed next. The substantial literature on hamster vision has made the need for this next step clear.

REFERENCES

Abrahams, V. C., and Rose, P. K., 1977, Projection of extra-ocular, neck muscle, and retinal afferents to superior colliculus in the cat: their connections to cells of origin of tectospinal tract, *J. Neurophysiol.* 38:10–18.

Anderson, M. E., Yoshida, M., and Wilson, V. J., 1971, Influences of cat superior colliculus on cat neck motoneurons, *J. Neurophysiol.* 34:898–907.

Bastiani, J., 1981, Visual and electrosensory responses in the optic tectum of a weakly electric fish, *Soc. Neurosci. Abstr.* 7:845.

Beckstead, R. M., 1979, An autoradiographic examination of corticocortical and subcortical projections of the mediodorsal-projection (prefrontal) cortex in the rat, *J. Comp. Neurol.* 184:43–62.

Bruckner, R., 1951, Spaltlampenmikroscopie und Ophthalmoskopie am Auge von Ratte und Maus, *Doc. Ophthalmol.* 5–6:452–554.

Bunt, A. H., Hendrickson, A. E., Lund, J. S., Lund, R. D., and Fuchs, A. F., 1975, Monkey retinal ganglion cells: morphometric analysis and tracing of axonal projections with a consideration of the peroxidase technique, *J. Comp. Neurol.* 164:265–286.

Chalupa, L. M., 1981, Some observations on the functional organization of the golden hamster's visual system, *Behav. Brain Res.* 3:198–200.

Chalupa, L. M., and Henderson, Z., 1980, Monocular enucleation in adult hamsters induces functional changes in the remaining ipsilateral retinotectal projection, *Brain Res.* 192:249–254.

Chalupa, L. M., and Rhoades, R. W., 1977, Responses of visual, somatosensory, and auditory neurones in the golden hamster's superior colliculus, *J. Physiol. (London)* 270:595–626.

Chalupa, L. M., and Rhoades, R. W., 1978, Directional selectivity in hamster's superior colliculus is modified by strobe-rearing but not by dark-rearing, *Science* 199:998–1001.

Chalupa, L. M., and Rhaodes, R. W., 1979, An autoradiographic study of the retinotectal projection in the golden hamster, *J. Comp. Neurol.* 186:561–570.

Chalupa, L. M., and Thompson, I., 1980, Retinal ganglion cell projections to the superior colliculus of the hamster demonstrated by the horseradish peroxidase technique, *Neurosci. Lett.* 19:13–19.

Chalupa, L. M., Morrow, A., and Rhoades, R. W., 1978, Behavioral consequences of visual deprivation and restriction in the golden hamster, *Exp. Neurol.* 61:442–454.

Chow, K. L., Masland, R. H., and Stewart, D. L., 1971, Receptive field characteristics of striate cortical neurons in the rabbit, *Brain Res.* 33:337–352.

Cowey, A., and Perry, V. H., 1979, The projection of the temporal retina in rats, studied by retrograde transport of horseradish peroxidase, *Exp. Brain Res.* 35:457–464.

Crain, B. J., and Hall, C. W., 1980, The organization of the lateral posterior nucleus of the golden hamster, *J. Comp. Neurol.* 193:351–370.

Crain, B. J., and Hall, W. C., 1981, The normal organization of the lateral posterior nucleus in the golden hamster and its reorganization after neonatal superior colliculus lesions, *Behav. Brain Res.* 3:223–228.

Cynader, M., and Berman, N., 1972, Receptive-field organization of monkey superior colliculus, *J. Neurophysiol.* **35**:187–201.

Drager, U. C., 1974, Autoradiography of tritiated proline and fucose transported transneuronally from the eye to the visual cortex in pigmented and albino mice, *Brain Res.* **82**:284–292.

Drager, U. C., 1975, Receptive fields of single cells and topography in mouse visual cortex, *J. Comp. Neurol.* **160**:269–290.

Drager, U. C., and Hubel, D., 1975, Responses to visual stimulation and relationship between visual auditory, and somatosensory inputs in mouse superior colliculus, *J. Neurophysiol.* **38**:690–713.

Drager, U. C., and Hubel, D. H., 1976, Topography of visual and somatosensory projections to mouse superior colliculus, *J. Neurophysiol.* **39**:91–101.

Drager, U. C., and Olsen, J. F., 1980, Origins of crossed and uncrossed retinal projections in pigmented and albino mice, *J. Comp. Neurol.* **191**:383–412.

Dursteler, M. R., Blakemore, C., and Garey, L. J., 1979, Projections to the visual cortex in the golden hamster, *J. Comp. Neurol.* **183**:185–204.

Ebbeson, S. O. E., 1970, On the organization of central vision pathways in vertebrates, *Brain Behav. Evol.* **3**:178–194.

Ebbeson, S. O. E., 1980, The parcellation theory and its relation to interspecific variability in brain organization, evolutionary and ontogenic development, and neuronal plasticity, *Cell Tissue Res.* **213**:179.

Eichler, V. B., and Moore, R. Y., 1974, The primary and accessory optic systems in the golden hamster, *Mesocricetus auratus, Acta Anat.* **89**:359–371.

Emerson, V. F., 1980, Grating acuity of the golden hamster: Effects of stimulus orientation and luminance, *Exp. Brain Res.* **38**:43–52.

Emerson, V. F., Chalupa, L. M., Thompson, I. D., and Talbot, R. J., 1982, Behavioral, physiological and anatomical consequences of monocular deprivation in the golden hamster (*Mesocricetus auratus*), *Exp. Brain Res.* **45**:168–178.

Feldon, S., Feldon, P., and Kruger, L., 1970, Topography of the retinal projection upon the superior colliculus of the cat, *Vision Res.* **10**:135–143.

Finlay, B. L., and Sengelaub, D. R., 1981, Toward a neuroethology of mammalian vision: Ecology and anatomy of rodent visuomotor behavior, *Behav. Brain Res.* **3**:133–149.

Finlay, B. L., and Slattery, M., 1984, Local differences in the amount of early cell death in neocortex predict adult local specializations, *Science* **219**:1349–1351.

Finlay, B. L., Schneps, S. E., Wilson, K. G., and Schneider, G. E., 1978, Topography of visual and somatosensory projections to the superior colliculus of the golden hamster, *Brain Res.* **142**:223–235.

Finlay, B. L., Marder, K., and Cordon, D., 1980a, Acquisition of visuomotor behavior after neonatal tectal lesions in hamster: The role of visual experience, *J. Comp. Physiol. Psychol.* **94**:506–518.

Finlay, B. L., Sengelaub, D. R., Berg, A. T., and Cairns, S. J., 1980b, A neuroethological approach to hamster vision, *Behav. Brain Res.* **1**:479–496.

Frost, D. O., and Schneider, G. E., 1976, Normal and abnormal uncrossed retinal projections in Syrian hamsters as demonstrated by Fink-Heimer and autoradiographic techniques, *Neurosci. Abstr.* **2**:812.

Frost, D. O., So, K.-F., and Schneider, G. E., 1979, Postnatal development of retinal projections in Syrian hamsters: A study using autoradiographic and anterograde degeneration techniques, *Neuroscience* **4**:1649–1677.

Fukada, Y., 1977, A three group classification of rat retinal ganglion cells: histological and physiological studies, *Brain Res.* **119**:327–344.

Fukada, Y., and Stone, J., 1974, Retinal distribution and central projections of y-, x-, and w-cells of the cat's retina, *J. Neurophysiol.* **37**:749–772.

Garey, L. J., and Powell, T. P. S., 1967, The projection of the lateral geniculate nucleus upon the cortex in the cat, *Proc. R. Soc. (London) Ser. B* **169**:107–126.

Goodale, M. A., and Murison, R. C. C., 1975, The effects of lesions of the superior colliculus on locomotor orientation and the orienting reflex in the rat, *Brain Res.* **88**:243–261.

Graham, J., 1977, An autoradiographic study of the efferent connections of the superior colliculus in the cat, *J. Comp. Neurol.* **173**:629–654.

Harting, J. K., 1977, Descending pathways from the superior colliculus: An autoradiographic analysis in the rhesus monkey (Macaca mulatta), J. Comp. Neurol. 173:583–612.

Harting, J. K., Hall, W. C., Diamond, I. T., and Martin, G. F., 1973, Anterograde degeneration study of the superior colliculus in Tupaia glis: Evidence for a subdivision between superficial and deep layers, J. Comp. Neurol. 148:361–386.

Haseltine, E., Kaas, L., and Hartline, P. H., 1977, Infrared and visual organization of the tectum of boid snakes, Soc. Neurosci. Abstr. 3:90.

Hubel, D. H., and Wiesel, T. N., 1962, Receptive fields, binocular interaction and functional architecture in the cat's visual cortex, J. Physiol. 160:167–287.

Hubel, D. H., and Wiesel, T. N., 1968, Receptive fields and functional architecture of monkey striate cortex, J. Physiol. 195:215–243.

Hughes, H. C., 1977, Anatomical and neurobehavioral investigations concerning the thalamocortical organization of the rat's visual system, J. Comp. Neurol. 175:311–336.

Ingle, D., and Sprague, J., 1975, Sensorimotor function of the midbrain tectum, Neurosci. Res. Prog. Bull. 13:167–287.

Jhaveri, S. R., and Schneider, G. E., 1974, Retinal projections in Syrian hamsters: Normal topography and alterations after partial tectum lesions at birth, Anat. Rec. 178:383.

Keselica, J. J., and Rosinski, R. R., 1976, Spatial perception in colliculectomized and normal golden hamsters, Physiol. Psychol. 4:511–514.

Knudsen, E. I., 1982, Auditory and visual maps of space in the optic tectum of the owl, J. Neurosci. 2:1177–1194.

Lee, K. J., and Woolsey, T. A., 1975, A proportional relationship between peripheral innervation density and cortical neuron number in the somatosensory system of the mouse, Brain Res. 99:349–353.

Lent, R., 1982, The organization of subcortical projections of the hamster's visual cortex, J. Comp. Neurol. 206:227–242.

Masland, R. H., Chow, K. L., and Stewart, D. L., 1971, Receptive-field characteristics of superior colliculus neurons in the rabbit, J. Neurophysiol. 34:148–156.

Merker, B. H., 1980, The sentinel hypothesis; A role for the mammalian superior colliculus, Ph.D. Dissertation, Massachusetts Institute of Technology.

Mitchener, J. C., Pinto, L. H., and Vanable, J. W., Jr., 1976, Visually evoked eye movements in the mouse (Mus musculus), Vision Res. 16:1169–1171.

Mort, E., Cairns, S., Hersch, H., and Finlay, B., 1980, The role of the superior colliculus in visually guided locomotion and visual orienting in the hamster, Physiol. Psychol. 8:20–28.

Niimi, K., Kanaseki, T., and Takimoto, T., 1963, The comparative anatomy of the ventral nucleus of the lateral geniculate body in mammals, J. Comp. Neurol. 121:313–323.

Oyster, C. W., Takahashi, E. S., and Hurst, D. C., 1981, Density, soma size and regional distribution of rabbit retinal ganglion cells, J. Neurosci. 12:1331–1346.

Palmer, D. S., 1980, Orienting elicited by superior colliculus stimulation in the hamster, Master's Thesis, Cornell University, New York.

Pickard, G. E., and Silverman, A. J., 1981, Direct retinal projections to the hypothalamus, piriform cortex, and accessory optic nuclei in the golden hamster as demonstrated by a sensitive anterograde horseradish peroxidase technique, J. Comp. Neurol. 196:155–172.

Printz, R. H., and Hall, J. L., 1974, Evidence for a retinohypothalamic pathway in the golden hamster, Anat. Rec. 179:57–66.

Rhoades, R. W., 1980, Response suppression induced by afferent stimulation in the superficial and deep layers of the hamster's superior colliculus, Exp. Brain Res. 40:185–195.

Rhoades, R. W., 1981a, Cortical and spinal somatosensory input to the superior colliculus in the golden hamster: An anatomical and electrophysiological study, J. Comp. Neurol. 195:415–432.

Rhoades, R. W., 1981b, Organization of somatosensory input to the deep collicular laminae in hamster, Behav. Brain Res. 3:201–222.

Rhoades, R. W., and Chalupa, L. M., 1976, Directional selectivity in the superior colliculus of the golden hamster, Brain Res. 118:334–338.

Rhoades, R. W., and Chalupa, L. M., 1978a, Conduction velocity distribution of the retinocollicular pathway in the golden hamster, Brain Res. 159:396–401.

Rhoades, R. W., and Chalupa, L. M., 1978b, Functional and anatomical consequences of neonatal visual cortical damage in the superior colliculus of the golden hamster, *J. Neurophysiol.* 41:1466–1494.

Rhoades, R. W., and Chalupa, L. M., 1979, Conduction velocity distribution of retinal input to the hamster's superior colliculus and a correlation with receptive field characteristics, *J. Comp. Neurol.* 184:243–264.

Rhoades, R. W., and DellaCroce, D. D., 1980a, Visual callosal connections in the golden hamster, *Brain Res.* 190:248–254.

Rhoades, R. W., and DellaCroce, D. D., 1980b, The cells of origin of the tectospinal tract in the golden hamster: An anatomical and electrophysiological investigation, *Exp. Neurol.* 67:163–180.

Robertson, R. T., Kaitz, S. S., and Robards, M. J., 1980, A subcortical pathway links sensory and limbic systems of the forebrain, *Neurosci. Lett.* 17:161–165.

Rodieck, R. W., 1979, Visual pathways, *Annu. Rev. Neurosci.* 2:193–225.

Rose, J. D., 1982, Midbrain distribution of neurons with strong, sustained responses to lordosis trigger stimuli in the female golden hamster, *Brain Res.* 240:364–367.

Rosenquist, A. C., Edwards, S. B., and Palmer, L. A., 1974, An autoradiographic study of the projections of the dorsal lateral geniculate nucleus and the posterior nucleus in the cat, *Brain Res.* 80:71–93.

Rusak, B., 1977, The role of the suprachiasmatic nuclei in the generation of circadian rhythms in the golden hamster, *Mesocricetus auratus, J. Comp. Physiol.* 118:145–164.

Scalia, F., 1972, The termination of retinal axons in the pretectal region of mammals, *J. Comp. Neurol.* 145:223–258.

Schiffman, H. R., 1970, Evidence for sensory dominance: Reactions to apparent depth in rabbits, cats and rodents, *J. Comp. Physiol. Psychol.* 71:38–41.

Schiffman, H. R., 1971, Depth perception as a function of age and photic condition of rearing, *J. Comp. Physiol. Psychol.* 76:491–495.

Schiller, P. H., and Malpeli, J. G., 1977, Properties and tectal projections of monkey retinal ganglion cells, *J. Neurophysiol.* 40:428–495.

Schneider, G. E., 1969, Two visual systems: Brain mechanisms for localization and discrimination are dissociated by tectal and cortical lesions, *Science* 163:895–902.

Schneider, G. E., 1970, Mechanisms of functional recovery following lesions of the visual cortex or superior colliculus in neonate and adult hamsters, *Brain Behav. Evol.* 3:285–323.

Schneider, G. E., and Jhaveri, S. R., 1974, Neuroanatomical correlates of spared or altered function after brain lesions in the newborn hamster, in: *Plasticity and Recovery of Function in the Central Nervous Systems* (D. G. Stein, J. J. Rosen, and N. Butters, eds.), Academic Press, New York, pp. 65–109.

Sengelaub, D. R., Windrem, M. S., and Finlay, B. L., 1983, Increased cell number in the adult hamster retinal ganglion cell layer after early removal of one eye, *Exp. Brain Res.* 52:269–276.

Siminoff, R., Schwassman, H. O., and Kruger, L., 1966, An electrophysiological study of the visual projection to the superior colliculus of the rat, *J. Comp. Neurol.* 127:435–444.

Stein, B. E., 1981, Organization of the rodent superior colliculus:some comparisons with other mammals, *Behav. Brain Res.* 3:175–188.

Stein, B. E., and Dixon, J. P., 1978, Superior colliculus cells respond to noxious stimuli, *Brain Res.* 158:65–73.

Stein, B. E., and Dixon, J. F., 1979, Properties of superior colliculus neurons in the golden hamster, *J. Comp. Neurol.* 183:269–284.

Stein, B. E., Magalhaes-Castro, B., and Kruger, L., 1976, Relationship between visual and tactile representations in cat superior colliculus, *J. Neurophysiol.* 39:401–419.

Sterling, P., and Wickelgren, B. G., 1969, Visual receptive fields in the superior colliculus of the cat, *J. Neurophysiol.* 32:1–15.

Terashima, S.-I., and Goris, R. C., 1975, Tectal organization of pit viper infrared reception, *Brain Res.* 83:490–494.

Tiao, Y.-C., and Blakemore, C., 1976a, Functional organization in the superior colliculus of the golden hamster, *J. Comp. Neurol.* 168:483–504.

Tio, Y.-C., and Blakemore, C. 1976b, Functional organization in the visual cortex of the golden hamster, *J. Comp. Neurol.* 168:459–482.

Tiao, Y.-C., and Blakemore, C., 1976, Regional specialization in the golden hamster's retina, *J. Comp. Neurol.* 168:439–458.

Van Essen, D. C., 1979, Visual areas of the mammalian cerebral cortex, *Annu. Rev. Neurosci.* 2:227–263.

Vogt, B. A., and Peters, A., 1981, Formal distribution of neurons in rat cingulate cortex: Areas 32, 24 and 29, *J. Comp. Neurol.* 195:603–625.

Waldron, H. A., and Gwyn, D. G., 1969, Descending nerve tracts in the spinal cord of the rat. I. Fibers from the midbrain, *J. Comp. Neurol.* 137:143–154.

Wiesenfeld, Z., and Branchek, T., 1976, Refractive state and visual acuity in the hooded rat, *Vision Res.* 16:823–827.

Wiesenfeld, Z., and Kornel, E. E., 1975, Receptive fields of single cells in the visual cortex of the hooded rat, *Brain Res.* 94:401–412.

Woolsey, C. N., Carlton, T. G., Kaas, J. H., and Earls, F. J., 1971, Projection of the visual field on superior colliculus of ground squirrel (*Citellus tridecemlineatus*), *Vision Res.* 11:115–127.

Woolsey, T. A., Welker, C., and Schwartz, R. H., 1975, Comparative anatomical studies of the sml face cortex with special reference to the occurrence of "barrels" in layer IV, *J. Comp. Neurol.* 164:79–94.

Appendix

Characteristics of Mesocricetus auratus

Approximate body weights:

Age (days)	Sex	Weight (g)
1	M	3
1	F	3
21	M	40
21	F	40
42	M	86
42	F	93
84	M	104
84	F	115
108	M	141
108	F	158

Average life span: 2 to 3 yr

Respiration rate: 74/min

Pulse rate: 450/min

Rectal temperature: 36 to 38°C

Standard laboratory photoperiod: 14 hours light—10 hours dark

Type of ovulation: Spontaneous

Estrous cycle length: 4 days

Length of receptive period (without mating): 12 to 29 hr

Time of ovulation (hours after onset of receptivity): 10 to 12 hr

Length of pseudopregnancy: 8 to 10 days

Length of pregnancy: 16 days

Postpartum estrus: None

Number of pups/litter: 8 to 12 (4 to 15)

Length of parturition: 1.5 to 2.5 hr

Sex ratio at birth: 106 M : 100 F

Pups begin eating solid food: 9 to 10 days   Eyes open: 12 to 14 days

Laboratory weaning age: 19 to 21 days
Sexual maturation:

Male   Earliest onset of copulatory behavior: 30 days
            Presence of sperm in penile smear: 42 days
            Maintenance of full reproductive activity: beyond 30 months

Female   Vaginal opening: 8 to 14 days   First ovulation: 26 to 30 days
            Earliest onset of copulatory behavior: 28 days
            Average onset of copulatory behavior: 34 days
            Recommended breeding age: 9 to 10 weeks
            Onset of reproductive deficiencies (cycle regularity, number of ova, maintenance of
                pregnancy): 40 weeks

Index